ENCYCLOPEDIA OF PHYSICS

CHIEF EDITOR

S. FLÜGGE

VOLUME VIa/4

MECHANICS OF SOLIDS IV

EDITOR

C. TRUESDELL

WITH 53 FIGURES

SPRINGER-VERLAG
BERLIN · HEIDELBERG · NEW YORK
1974

HANDBUCH DER PHYSIK

HERAUSGEGEBEN VON

S. FLÜGGE

BAND VIa/4

FESTKÖRPERMECHANIK IV

BANDHERAUSGEBER

C. TRUESDELL

MIT 53 FIGUREN

SPRINGER-VERLAG
BERLIN · HEIDELBERG · NEW YORK
1974

Professor Dr. SIEGFRIED FLÜGGE
Physikalisches Institut der Universität Freiburg i. Br.

Professor Dr. CLIFFORD AMBROSE TRUESDELL, III
The Johns Hopkins University, Baltimore, Maryland, USA

ISBN 3-540-06097-9 Springer-Verlag Berlin Heidelberg New York
ISBN 0-387-06097-9 Springer-Verlag New York Heidelberg Berlin

Contents.

Wave Propagation in Nonlinear Viscoelastic Solids.

By

JACE W. NUNZIATO, EDWARD K. WALSH, KARL W. SCHULER, and LYNN M. BARKER.

With 50 Figures.

I. Introduction.

1. Background. Scope of this article. An important aspect in the prediction of the behavior of solid viscoelastic materials concerns the ability to characterize their response to large dynamic loads. Essential to such a characterization is a constitutive theory which accounts for the effects associated with both finite deformations and long-range memory. In 1957–1964, COLEMAN and NOLL developed the theory of simple materials with fading memory which accounts for these effects as well as thermodynamic influences. Subsequent investigations by COLEMAN, GURTIN, HERRERA, CHEN, and others concerning wave propagation served to display the dynamic response of these materials and showed how the response was affected by material nonlinearities and internal dissipation. In particular, they utilized the theory of propagating singular surfaces to study the propagation of one-dimensional shock and acceleration waves and derived conditions governing their growth and decay. These conditions imply the existence of steady waves, and proofs of the existence of steady wave solutions in such dissipative media have been established by PIPKIN and GREENBERG.

The determination of the dynamic response of nonlinear viscoelastic solids involves, in addition to theoretical studies, experimental investigations which closely reproduce the same kinematical configuration and dynamical situation. Experimental developments in the field of shock wave physics have made it possible to produce large-amplitude plane waves by using a plate-impact loading configuration. In addition, instrumentation techniques based on laser interferometry have been developed by BARKER and HOLLENBACH which allow the fine structure of the wave profile to be observed with high resolution and accuracy. Due to the inertial confinement in such experiments, extremely large one-dimensional strains are readily produced which generally result in nonlinear behavior of the material sample. Utilizing these experimental techniques, NUNZIATO, SCHULER, and WALSH have investigated one-dimensional wave propagation in a particular viscoelastic solid and shown the correlation between experimental observations and the predictions of the general nonlinear theory of simple materials with fading memory.

In this article we give a unified presentation of theoretical and experimental results in nonlinear viscoelastic wave propagation. There is no attempt to cover the subject of viscoelastic wave propagation in its entirety; rather we concentrate primarily on those aspects of the subject which relate to motions of one-dimensional strain.[1] This article is related in several respects to the article, "Growth and Decay of Waves in Solids," by P. J. CHEN, which appears in the preceding part of this volume.[2] In that article CHEN utilizes the theory of propagating singular surfaces to examine the growth and decay of waves in various types of materials, including materials with memory, and shows that definitive results with regard to the qualitative behavior of waves can be obtained without having to appeal to any explicit constitutive assumption. Here our emphasis is somewhat different in that we wish to demonstrate more specific, quantitative results. In particular, we shall discuss in detail the experimental methods used to generate and observe one-dimensional waves and show that by observing entire steady shock wave profiles in a particular viscoelastic solid one can evaluate the material response functions necessary to characterize the solid completely in terms of a specific constitutive equation. With such a constitutive model, we can make quantitative predictions with regard to the growth and decay of waves and we show the correlation between these predictions and experimental observations for a variety of one-dimensional situations. We further show how many of these results can be extended to include thermodynamic influences.

2. Plan of this article. Notation. We begin in Chap. II with a complete description of the general experimental methods used in the generation and observation of one-dimensional waves. Then, as an introduction to the subsequent sections, the kinematics, balance laws, and the constitutive formulation for one-dimensional motions in materials with memory are given in Chap. III, and some results on the propagation of small amplitude acoustic waves are presented in Chap. IV. The propagation of steady shock waves including their existence in dissipative materials, their experimental observation, and the evaluation of steady wave solutions is considered in Chap. V. On the basis of the constitutive model and the response functions deduced here, we discuss in Chap. VI the growth and decay behavior of shock waves and in Chap. VII the growth and decay of acceleration waves. The discussion through Chap. VII deals only with mechanical effects; in Chap. VIII we extend these studies to show the thermodynamic influences on the propagation of shock and acceleration waves.

In this article we attempt to give a reasonably self-contained presentation of wave propagation in nonlinear viscoelastic solids. However, in those sections where we require results from the theory of singular surfaces, we merely present the main results without giving proofs. In these cases, the interested reader can refer to the comprehensive article by CHEN[2] or to the original papers for the analytical details.

In general, our notation conforms with that employed in "The Non-Linear Field Theories of Mechanics" by TRUESDELL and NOLL[3] and "Growth and Decay of Waves in Solids" by CHEN.[2] There is, however, one notable exception; that is, we shall follow the customary procedure in shock wave physics and take *stress and strain positive in compression*. This should not cause the reader any difficulty.

Acknowledgements. We acknowledge M. E. GURTIN's contributions to the theories of waves in dissipative media and express our gratitude to him for his numerousdiscussions and suggestions regarding many aspects of the work reported here. We thank C. TRUESDELL for

[1] A considerable amount of research has been done on the propagation of waves in viscoelastic rods, i.e., in the one-dimensional stress configuration. For a review of much of this work the interested reader may refer to one of the review articles by KOLSKY, e.g., KOLSKY [1958, *2*], [1960, *6*], [1962, *6*], [1964, *13*], [1965, *16*]. Also relevant is the article by BELL [1973, *2*] in this volume.

[2] CHEN [1973, *4*].

[3] TRUESDELL and NOLL [1965, *19*].

providing the opportunity to write this article. To B. M. BUTCHER, L. W. DAVISON, W. HERRMANN, D. E. MUNSON, and G. A. SAMARA, all of Sandia Laboratories, we express our gratitude for their continued encouragement and support. We are also indebted to P. J. CHEN and H. J. SUTHERLAND for helpful discussions and for their comments on a previous draft of this manuscript. In addition, we acknowledge R. E. HOLLENBACH's important contributions to the development of the experimental techniques reported in this article. Finally, we would like to thank Mrs. A. HOSTETTER for her excellent cooperation in the typing and preparation of the manuscript.

This work was supported by the U. S. Atomic Energy Commission. Also E. K. WALSH acknowledges the support of the University of Florida.

II. Experimental methods in one-dimensional wave propagation.

3. Introductory remarks. One important aspect of the determination of the dynamic response of solid materials involves the experimental investigation of wave propagation for a known kinematical configuration. In this article we are concerned primarily with the propagation of plane, one-dimensional waves and, in particular, their structure and velocity of propagation. Thus, in order to obtain experimental results which can be correlated with the results of the appropriate theoretical studies, it is necessary to consider three main problems with regard to the performance of the experiments. First, it is necessary to generate a wave in the material sample which closely reproduces the kinematical condition of one-dimensionality and has the desired structure. Second, changes in the structure of the wave profile as it propagates through the sample must be determined with sufficient resolution. Finally, an accurate measurement of the average velocity of propagation of the wave is required. As we shall see in subsequent chapters, applying current techniques to each of these phases of an experiment can yield excellent experimental results.

In this chapter we address these three problems and describe in some detail the methods of generating and observing one-dimensional, finite-amplitude waves. Particular emphasis will be placed on those techniques which are most applicable to the study of wave propagation in nonlinear viscoelastic solids.

4. Planar impact loading configuration. The generation of a one-dimensional plane wave in a material sample requires the application of a load uniformly and simultaneously over a planar surface of the sample. Various methods for applying this load have been considered. These include the detonation of an explosive placed against the sample, the impingement of a gasdynamic shock wave on the sample, the sudden absorption of electromagnetic energy in a region adjacent to the surface of the specimen, and the impacting of a flat flyer plate into a stationary target plate of the specimen material. The four techniques are, to a large extent, complementary, and each has its particular advantages. The explosives are unparalleled for loading large samples to high stresses.[1] Gasdynamic shock loading, on the other hand, can produce very low amplitude plane waves which cannot be conveniently or reproducibly generated by other experimental methods.[2] Since energy deposition methods involve "thermal loading," such techniques offer the unique opportunity to study the dynamic thermomechanical response of materials.[3] Of the four techniques, however, the plate impact method of loading probably offers the greatest flexibility. This is due to the fact that the loading applied

[1] Cf. RICE, MCQUEEN, and WALSH [1958, *4*], AL'TSHULER [1965, *1*].

[2] Cf. COMMERFORD and WHITTIER [1970, *8*].

[3] No general review of these techniques appears to exist. Results of representative studies employing lasers and electron beams have been presented by PERCIVAL [1971, *5*] and by PERRY [1970, *17*] and GAUSTER [1971, *4*], respectively.

to the sample can be carefully controlled by altering flyer plate parameters. In particular, the magnitude and duration of the load applied to the target can be carefully controlled by changing the material, velocity, or thickness of the flyer plate. The duration is also influenced by the diameters of the flyer and target plates since it is these dimensions which determine the length of time before lateral edge effects reach the center portion of the plates and alter the conditions of one-dimensional strain.[4] For a flyer plate diameter of the order of 100 mm, wave profiles spanning 5 to 20 μsec can be achieved.

Various methods, including explosive plane wave generators,[5] magnetically or electrically driven flying foils,[6] and exploding foils[7] have been employed to accelerate flyer plates. However, these methods subject the flyer plate to stresses comparable to those of interest in the experiment; thus, the flyer plate may not be stress-free when it impacts the target plate. In addition, the violent flyer accelerations associated with these methods can easily distort the flyer; thus the simultaneity and uniformity of the stress pulse imparted to the target plate are difficult to control.

A method for accelerating flyer plates which successfully overcomes these difficulties involves the use of a gun to accelerate a projectile with the flyer plate as a noseplate. All of the experimental data reported in this article were obtained using the compressed gas gun shown in Fig. 1.[8] It consists in three basic assemblies: the breech, the barrel, and the target chamber. The breech assembly, which is shown schematically in Fig. 2, contains a reservoir for the pressurized accelerating gas (usually air or helium at pressures up to 0.14 kbar) and a quick-acting slide valve (piston) to release the gas suddenly behind the projectile. In operation, the slide valve (piston) is first seated in the breech end of the barrel by applying air pressure to the rear end of the piston. The breech chamber is then loaded with air or helium to the desired pressure. To fire the gun, the seating pressure is vented and air pressure is applied to the firing chamber, causing a rapid retraction of the piston and release of the breech pressure behind the piston.

The smooth bore barrel is machined to close tolerances in order to guide the close-fitting projectile adequately during acceleration. The barrel of the gun shown in Fig. 1 is 4 m long with a 100 mm bore. The target chamber provides an isolated mounting for the target which permits its prealignment and assures that it does not become misaligned due to the recoil of the gun.

The projectile design represents a compromise between the need for long, rigid projectiles, which minimize the tilt and distortion of the noseplate, and the requirements of light weight and expendability (since for all but the lowest impact velocities the projectile is destroyed with each shot). For these reasons, projectiles are usually fabricated from one or two long, concentric, thin-wall tubes of alumin-

[4] It should be pointed out that a state of uniaxial strain along the centerline can be assured only if the target and flyer plates consist of isotropic materials. In the case of anisotropic materials, the state of stress and strain will of course depend upon the orientation of the symmetry planes with respect to the plane of impact.

[5] Cf. DUVALL [1961, *3*].

[6] Cf. JACOBSON [1967, *4*].

[7] Cf. GUENTHER, WUNSCH, and SOAPES [1962, *4*], and KELLER and PENNING [1962, *5*].

[8] Cf. BARKER and HOLLENBACH [1964, *1*], [1965, *3*]. Three basic types of guns have been used for wave propagation studies; compressed gas guns, powder guns, and two-stage, light gas guns. The compressed gas guns are typically used to achieve velocities up to 1.5 mm/μsec, the powder guns from about 0.5–3.0 mm/μsec, and the two-stage, light gas guns up to 7.0 mm/μsec. In this regard, also see HUGHES, OTTO, BLANKENSHIP, GOURLEY, and GOURLEY [1957, *3*], THUNBORG, INGRAM, and GRAHAM [1964, *16*], JONES, ISBELL, and MAIDEN [1966, *9*], ISBELL, SHIPMAN, and JONES [1968, *11*], KARNES [1968, *12*], and FOWLES, DUVALL, ASAY, BELLAMY, FEISTMANN, GRADY, MICHAELS, and MITCHELL [1970, *12*].

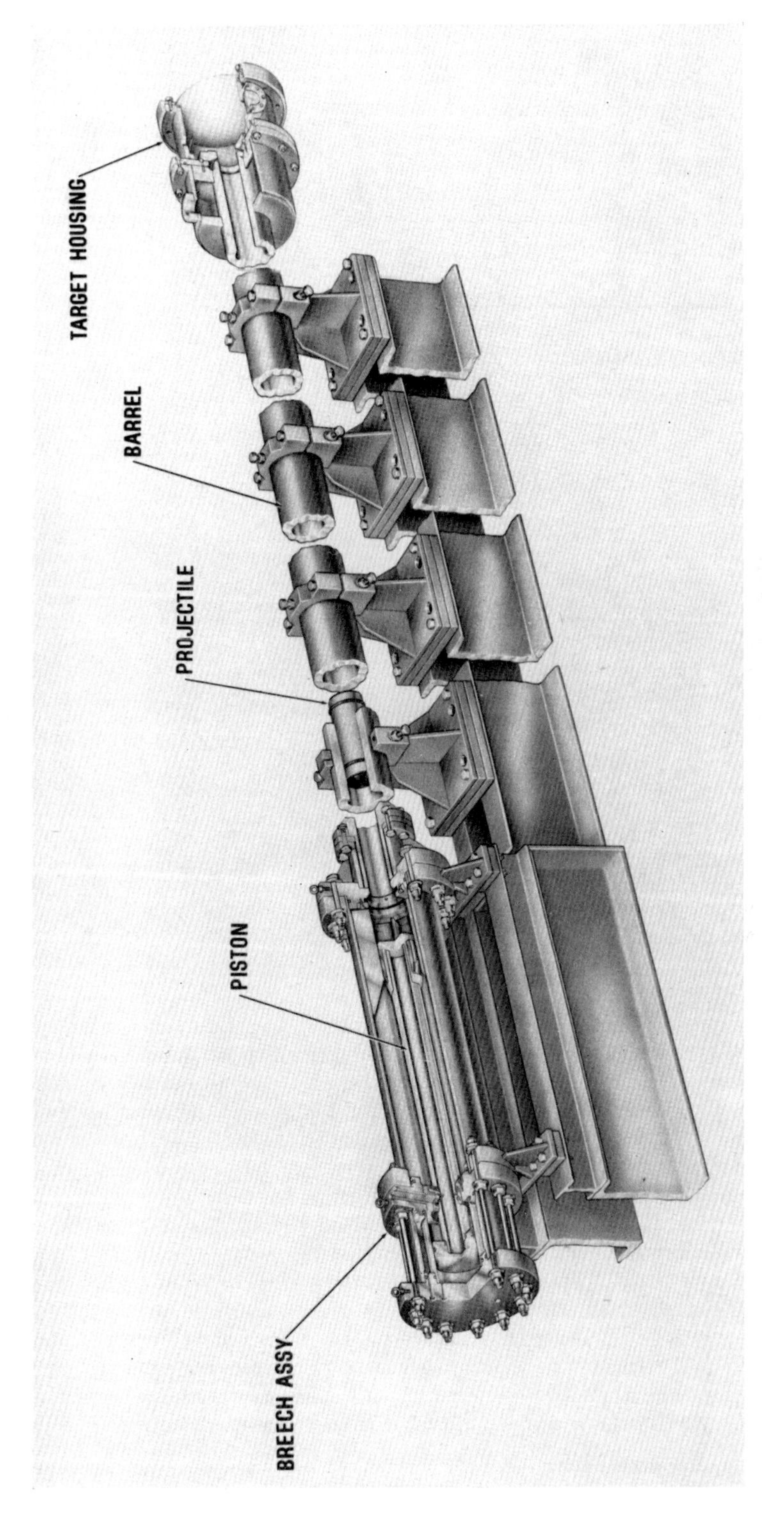

Fig. 1. A 100 mm compressed gas gun for plate impact experiments.

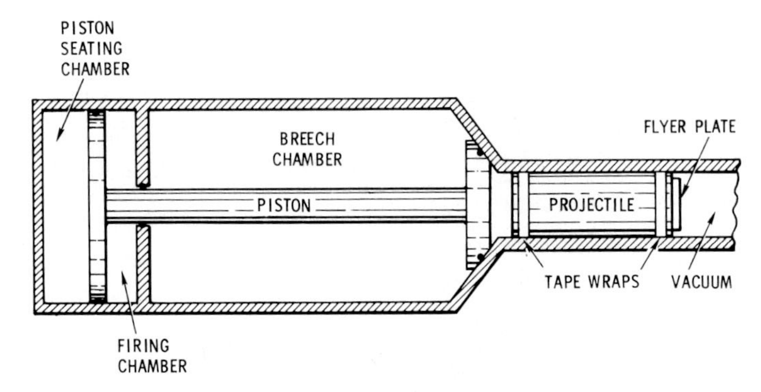

Fig. 2. Breech assembly of the 100 mm gas gun. The projectile is loaded through the muzzle of the gun.

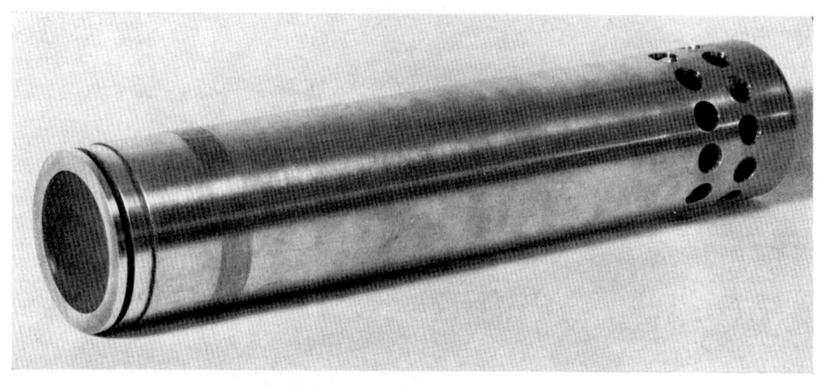

Fig. 3. A projectile for the 100 mm gas gun. The projectile length in this case is approximately 45 cm. The polymethyl methacrylate flyer plate at the left end is contained in an aluminum ring, the purpose of which is to short out the velocity and flush pins just prior to impact.

um or phenolic capped at each end by aluminum plates. A typical projectile is shown in Fig. 3.

Several precautions are necessary to reduce any build-up of air pressure between the target and projectile prior to impact, which would destroy the desired step function profile of the input wave. An expansion chamber is provided at the muzzle end of the gun, and both the expansion chamber and barrel are evacuated to less than 10^{-2} Torr prior to launching the projectile. Each end of the projectile is wrapped with tape to assure a close fit between the projectile and the barrel, thus minimizing any blow-by of the accelerating gas. In addition, openings are provided in the wall of the outer tube of the projectile (cf. Fig. 3); this ensures that any gases which may by-pass the rear seal must pressurize the interior of the projectile before significant pressure loading is applied to the front seal.

Since the projectile velocity determines the magnitude of the stress pulse applied to the target, it must be accurately measured. For projectile velocities up to 2 mm/µsec, electrically charged pins shorted by the moving projectile have been used to determine projectile velocities.[9] The pins are arranged either radially inward from the barrel or axially outward from the impact surface of the target

[9] For higher projectile velocities, flash X-rays are usually used; cf. ISBELL, SHIPMAN, and JONES [1968, *11*].

such that they contact the periphery of the projectile noseplate as it approaches the target. The distance between two successive pins is accurately premeasured and the time taken by the projectile to move from pin to pin is determined by using the shorting of the pins to start and stop a nanosecond counter. Projectile velocities determined from this type of data have accuracies of a few tenths of a percent.

A high degree of uniformity and simultaneity of loading is achieved by maintaining strict tolerances on the flatness of the impacting surfaces and by aligning the target plate parallel with the projectile noseplate. The flatness tolerance necessary to maintain a particular degree of planarity of the input wave is proportional to the impact velocity. For example, with an impact velocity of 0.06 mm/μsec and the impacting surfaces held flat to ±0.3 micron, the input wave is plane to within ±0.005 μsec. The flatness of the impact area is measured under a monochromatic light by observing the interference fringes produced between the polished specimen surface and an optical flat. A lapping machine is usually used in the preparation of flat surfaces.

Since the wave velocity in the target is typically one or two orders of magnitude greater than the projectile velocity, the degree of nonparallelism (tilt) existing between the plates at impact time is magnified by one or two orders of magnitude as the input wave is formed. This tilt is minimized by accurately positioning the target mounting ring to hold the target parallel with the projectile noseplate. With the projectile placed at the muzzle end of the barrel in the same position it will occupy at impact, the target mounting ring is aligned using an optical alignment technique. The alignment is accomplished by mounting an optically flat piece of glass on the mounting ring in place of the target. By use of monochromatic illumination, interference fringes are observed between the "impact" surface of the glass and the projectile noseplate. The target mounting ring is then adjusted to center the fringe pattern on the projectile noseplate, which is the condition for parallelism between the glass "target" and the center of the noseplate. The glass is then carefully replaced by the actual target, and the alignment is complete.

Although the same alignment procedure is used for all impact velocities, the tilt control achieved in practice is less precise for the faster velocities. This is due to the fact that lighter projectiles are required which are shorter and less rigid than those used at slower velocities. Fortunately, however, the tilt of the wave in the target varies inversely with the impact velocity. Thus, less impact tilt control is required at higher velocities. The impact tilts in the experiments reported in this article have generally been less than 0.002 radians per mm/μsec of impact velocity.

The tilt is measured by recording the contact of four electrical probes by the projectile plate at impact time.[10] The probes are located at $\pi/2$ radian intervals around the periphery of the impact area on the target and are set flush with the impact surface by the following electrical method. A small, optically flat, metal plate is placed over the pin in intimate contact with the impact surface of the target. The electrical potential difference to be used between the "flush pin" and the projectile during the shot is then placed between the pin and the plate, and the flush pin is adjusted so that it barely makes electrical contact with the plate. Pins set in this manner are flush with the impact surface of the target to an accuracy of ±0.5 microns. The simplicity of this procedure permits setting the pins just prior to mounting the target plate on the gun, thereby reducing the uncertainties caused by temperature variations or slight creep of the epoxy in which the pins are mounted. The fact that all four pins close within a fraction of a microsecond

[10] Cf. Barker and Hollenbach [1964, *1*].

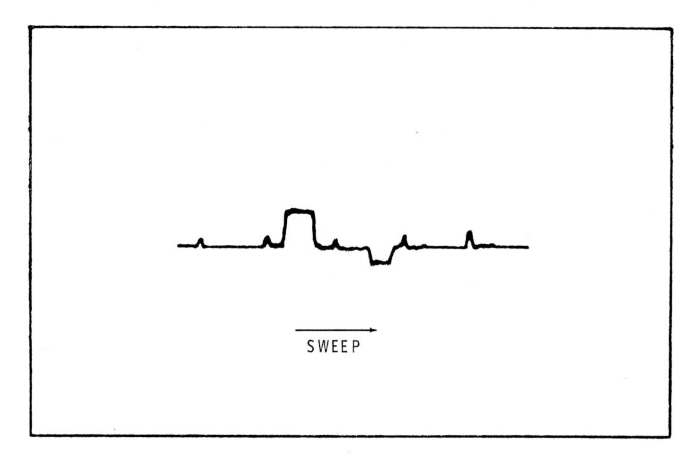

Fig. 4. Oscillogram record of the closures of four flush pins. (From BARKER and HOLLENBACH [1964, *1*].)

allows the use of fast oscilloscope sweep speeds (usually 0.1 μsec/cm) to record the closures; the fast speeds permit precise determinations of the flush pin closure times, and therefore of the tilt. Fig. 4 shows an oscilloscope record of the closure of four flush pins. For identification purposes the four pulses differ in amplitude and/or polarity. At fast sweep speeds, their relatively long voltage decay time constants (1.0 μsec) give the appearance of flat topped pulses. The time marks are 0.20 μsec apart so that in Fig. 4 the four pins have closed in 0.33 μsec.

5. Measurement of the transmitted wave profiles. There have been two generally successful approaches to the measurement of transmitted wave profiles generated in solid materials by planar impact. One has been to measure the free-surface velocity history of the target, and the other has been to use a stress transducer to measure the stress history at the interface between the impacted specimen and the transducer. Both the free surface and the stress transducer approaches usually suffer to a degree from the effect of the material boundary located at the measuring station. An ideal gauge measures the wave profile (either stress or particle velocity as a function of time) *within* the specimen where it is unaffected by boundaries. Although the boundary effect can be largely corrected for in non-ideal gauges, it nevertheless introduces a small ambiguity into the experimental results.

The two most common stress transducers used in wave propagation experiments are the quartz gauge, which makes use of the piezoelectric effect in crystal quartz,[11] and the manganin wire gauge, which makes use of the large piezoresistance coefficient in manganin.[12] Of the two methods, the quartz gauge is the more accurate, with a typical resolution of about ±3% in stress for stresses up to 20 kbar. A study

[11] The use of x-cut crystalline quartz as a gauge to measure stress-time profiles in solids was first suggested by NEILSON, BENEDICK, BROOKS, GRAHAM, and ANDERSON [1962, *7*]. A more complete treatment of the quartz gauge is given by GRAHAM, NEILSON, and BENEDICK [1965, *12*] and INGRAM and GRAHAM [1970, *13*].

[12] Manganin alloys have been used to make pressure measurements in hydrostatic experiments for many years as a result of the development and wide application of the manganin coil by BRIDGMAN (cf. BRIDGMAN [1964, *3*]). The use of manganin as a piezoresistance stress gauge for wave propagation experiments is due to BERNSTEIN and KEOUGH [1964, *2*] and FULLER and PRICE [1964, *10*]. In this regard, see the later studies of KEOUGH and WONG [1970, *15*] and LEE [1973, *9*].

by GRAHAM[13] reports very careful measurements of the nonlinear piezoelectric response and the second-, third-, and fourth-order longitudinal elastic constants of crystalline quartz. With these data even better stress resolution up to 40 kbar is possible. The time resolution of the quartz gauge is intrinsically sub-nanosecond. However, the tilt and non-planarity of the propagating wave causes an area averaging effect in the quartz gauge which usually limits the time resolution to several nanoseconds.

The manganin wire can in principle be used as an ideal in-material gauge to measure the stress profile inside the specimen. This is done by sandwiching the manganin wire between two pieces of the specimen material while keeping the layer containing the manganin wire thin enough to produce a negligible perturbation of the transmitted wave. In practice, however, this latter condition is difficult to achieve. Furthermore, the gauge output is not a single-valued function of the stress. Thus, manganin gauge results are usually only accurate to within ± 5 to 10% in stress and ± 25 to 50 nsec in time. The lower resolution of the manganin gauge relative to the quartz gauge offsets its advantage of in-material capability.

Several techniques have been used to measure specimen free-surface velocity histories. Most notable among these have been charged electrical probes to measure the time of arrival of the specimen surface at predetermined positions,[14] capacitance gauges in which the specimen surface is used as one of the plates of a capacitor such that the motion of the surface produces a change in capacitance,[15] and various optical configurations in conjunction with streak cameras.[16] If the target specimen is an insulator and if a transverse magnetic field is applied over the volume of the specimen, an electrical conductor embedded in the specimen becomes an in-material velocity gauge by virtue of the voltage generated as the conductor moves with the target material through the imposed magnetic field. This electromagnetic gauge has been used extensively in the investigation of detonation waves by Soviet researchers, but its time resolution is apparently rather poor, being of the order of 50 nsec.[17]

The most accurate of the velocity profile measuring techniques have been the laser interferometer systems developed by BARKER and HOLLENBACH.[18] With these techniques it is possible to achieve typical accuracies of $\pm 3\%$ in velocity with time resolutions of ± 2 to 10 nsec depending on the particular interferometer configuration and the speed of the fringe recording system. Furthermore, in transparent materials such as polymethyl methacrylate, the interferometer can often be used as an ideal in-material gauge. In-material laser interferometry was used to measure the transmitted wave profiles in all of the experiments reported in this article and, therefore, the techniques involved are described in considerable detail

[13] GRAHAM [1972, *11*].

[14] Cf. MINSHALL [1955, *1*].

[15] Cf. HUGHES, GOURLEY, and GOURLEY [1961, *6*], RICE [1961, *7*], and TAYLOR and RICE [1963, *3*].

[16] There are a number of specific optical streak-camera techniques which have been developed to measure free-surface velocity. For example, FOWLES [1961, *4*], and PEVRE, PUJOL, and THOUVENIN [1965, *18*] sensed the change in angular orientation of a reflecting specimen surface which resulted from the reflection of a plane wave inclined at an angle to the surface; DAVIS and CRAIG [1961, *2*] observed the motion of the surface-reflected image of a wire; EDEN and WRIGHT [1965, *9*] observed the progressive destruction of total internal reflection of a prism inclined at a small angle on the specimen surface; and AHRENS and RUDERMAN [1966, *2*] measured the change in angle of a foil immersed in a transparent liquid sample and inclined at an angle to the plane of the propagating wave.

[17] Cf. DREMIN and ADADUROV [1964, *8*] and JACOBS and EDWARDS [1970, *14*].

[18] BARKER and HOLLENBACH [1965, *3*], BARKER [1968, *2*], BARKER and HOLLENBACH [1972, *4*].

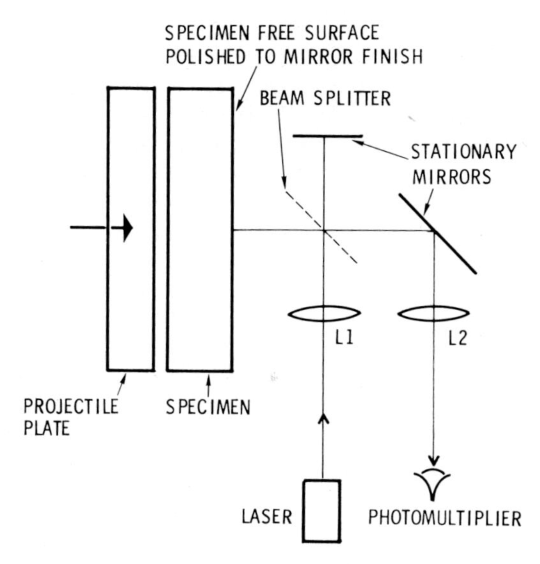

Fig. 5. The displacement interferometer. Lens L1 brings the laser beam to focus at the specimen free surface, and L2 recollimates the beam. The beam splitter, lenses, and stationary mirrors are mounted onto the specimen for stability of the interferometer. (From BARKER [1972, *3*].)

below. Because of their simplicity the free surface concepts are presented first; then the in-material configurations are described.

α) *The displacement interferometer.* In Fig. 5 we show a schematic diagram of an interferometer instrumentation system[19] in which the target specimen surface is polished to a mirror finish and used as one of the two mirrors of a Michelson displacement interferometer. The interferometer uses a beam splitter to divide the laser beam into two equal parts. The part which is reflected from the stationary mirror (cf. Fig. 5) is the reference beam, and the part reflected from the surface of the specimen is the signal beam. As the polished surface of the sample moves under the influence of the arriving stress wave, the frequency of the signal beam is changed slightly by the Doppler effect. When the reference beam and the signal beam are remixed at the beam splitter, the resultant beam is amplitude modulated at the difference frequency of the two beams. The difference frequency is simply the Doppler change in the frequency of the signal beam, since there is no shift in the reference beam frequency. The photomultiplier tube therefore senses light fringes at the Doppler shift frequency. A typical oscillogram record of the light fringes sensed by the photomultiplier is shown in Fig. 6.

Assuming that the signal beam is incident normal to the specimen surface, the Doppler shift in frequency is given exactly by

$$\nu_{\mathrm{d}}(t) = \frac{2 v_{\mathrm{s}}(t)}{\Lambda \left[1 - \frac{v_{\mathrm{s}}(t)}{c}\right]} \tag{5.1}$$

where $v_{\mathrm{s}}(t)$ is the reflecting surface velocity, Λ is the original wavelength of the laser light, and c is the velocity of light. In practice, the term v_{s}/c is less than 10^{-5} and is therefore negligible compared to unity. Thus, integrating (5.1) with respect

[19] Cf. BARKER and HOLLENBACH [1965, *3*].

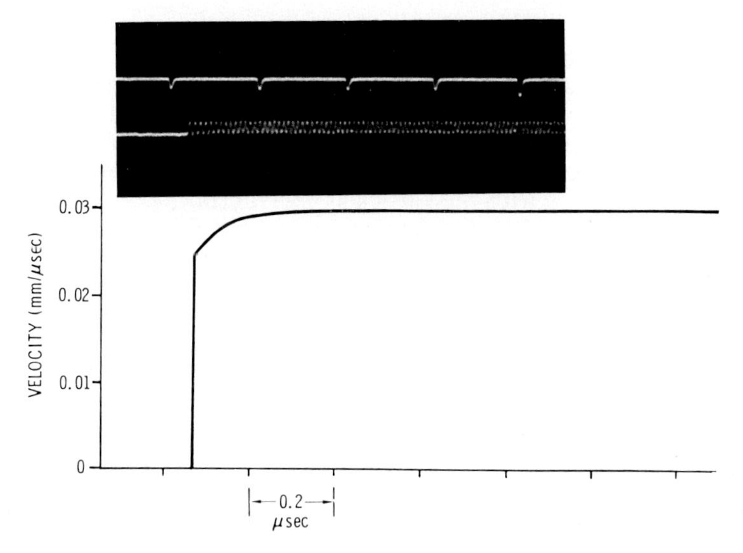

Fig. 6. Displacement interferometer data. The upper trace of the oscillogram is a timing comb having marks at 0.2 μsec intervals with every fifth mark accented. The lower trace is the light from the displacement interferometer. Below is the velocity profile obtained from the oscillogram record.

to the time yields the relation between the surface displacement $u_s(t)$ and the fringe count $f(t)$:

$$u_s(t) = \tfrac{1}{2} \Lambda f(t) . \tag{5.2}$$

The photomultiplier output amplitude $A(t)$ is a sinusoidal function of the fringe count:

$$A(t) = \tfrac{1}{2} A_{\max} \{1 + \sin[2\pi f(t) + v]\} \tag{5.3}$$

where $A_{\max}$ is the maximum amplitude and v is a phase angle. Eq. (5.3) is used in determining noninteger values of the fringe count. With this analysis, the light output of the beam can be converted to a velocity-time history at the surface of the specimen. The profile in Fig. 6 corresponding to the oscillogram record was obtained from a plate impact experiment on the polymer, polymethyl methacrylate.

The assumption that the signal beam is incident normal to the specimen surface is strictly true only for the central ray of the beam. The periphery of the beam has a small angle of incidence because of the lens which brings the beam to focus at the specimen free surface. Eq. (5.1) becomes quite complicated when it is modified to take the angle of incidence into account, but for the nonrelativistic case, in which terms involving v_s/c are neglected compared to unity, it becomes

$$\nu_d(t) = \frac{2 v_s(t) \cos\gamma}{\Lambda} \tag{5.4}$$

where γ is the angle of incidence of the light onto the surface. Although γ is always very small, the focusing lens makes the Doppler shift at the periphery of the laser beam slightly less than that at the center. In use, this leads to a gradual misalignment of the interferometer as the surface moves. An original alignment of either an all-dark or all-bright fringe pattern gradually changes to a bull's eye pattern because of the slight difference in fringe frequency with radial position in the beam. As a result, the interferometer has a depth-of-field limitation which is controlled by the f-number of the focusing lens through the maximum angle of incidence. The larger the effective f-number[20] of the lens, the greater will be the depth of field. On the other hand, the smaller the effective f-number of the focusing lens, the less sensitive the interferometer is to misalignment due to any slight tilt of the arriving wave. Thus, the depth-of-field and tilt-misalignment considerations present conflicting requirements on the effective f-number of the focusing lens, and the optimum f-number can vary with the particular experiment.

The extreme sensitivity of one complete fringe shift for each $\Lambda/2$ displacement of the free surface, plus the fact that one can easily estimate the fringe count to ± 0.1 fringe, gives the interferometer a distance resolution which is one to two orders of magnitude better than other free surface instrumentation techniques. However, the fine resolution of the displacement interferometer leads to a major problem, which is that of the fringe frequency. Eq. (5.1) shows that for the He—Ne laser wavelength of 632.8 nm the fringe frequency is 3.16 GHz per mm/μsec of free surface velocity. Many wave propagation experiments produce free surface velocities of one or more mm/μsec. Thus, there are severe difficulties in recording the resulting GHz fringe frequencies on an individual fringe basis. The limitations imposed by the frequency responses of photodetectors, broadband amplifiers, and oscilloscopes, together with the short recording times which result from the extremely fast oscilloscope sweep speeds necessary to resolve GHz fringe frequencies, place a practical limit of about 0.2 mm/μsec on the upper velocity capability of the displacement interferometer. Nevertheless, the displacement interferometer has proved to be a very valuable tool for wave propagation experiments in which the maximum velocities of interest are relatively small.

β) *The velocity interferometer.* One obvious way to reduce the fringe frequency problem of the displacement interferometer would be to use a laser which produces light at a longer wavelength (cf. Eq. (5.1)). Switching to the CO_2 laser with its 10.5 μm wavelength, for example, would reduce the fringe frequency by a factor of about 17. However, several problems arise when the wavelength is lengthened appreciably. For example, the resolution of the technique would suffer in proportion to the wavelength, and nanosecond-response detectors of 10.5 μm radiation require cryogenic temperatures and have low outputs.

Another way to reduce the fringe frequency is to beat the signal beam against a reference beam whose frequency is closer to that of the signal beam. Of course, the frequency of the reference beam must be precisely known at all times in order to use such a technique. The successful approach along these lines used the following reasoning. If there are no large jumps in velocity in the wave profile to be measured, the change in the Doppler frequency shift is small over a sufficiently small increment in time. Therefore, for the reference beam frequency one can use the identical frequency which comprised the signal beam at a known small time τ earlier. Since the reference beam frequency will always closely follow the signal beam frequency, the fringe frequency will be nonzero only during periods of changing Doppler frequency shift, i.e., during periods of surface acceleration. The necessity for recording and counting large numbers of fringes during periods of high, nearly constant surface velocities is thus eliminated.

The known reference beam frequency which closely follows the signal beam in frequency is obtained by reflecting the entire laser beam from the specimen free surface, and then splitting off half of the light and delaying it by a time τ before recombining it with the signal beam to form the fringes (cf. Fig. 7). This delay is

[20] The effective f-number is equal to the focal length of the lens divided by the diameter of the useful portion of the lens, which is the laser-beam diameter if the lens diameter is larger than that of the laser beam.

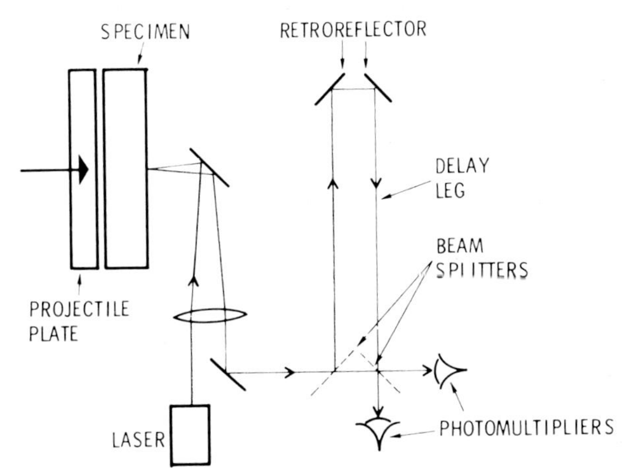

Fig. 7. The velocity interferometer. (From BARKER [1972, *3*].)

accomplished by sending the split-off beam around a delay leg of length $l = c\tau$.[21] Thus, before the free surface motion begins, a stationary fringe pattern is observed. During the first delay time interval τ after the start of the free surface motion the interferometer produces exactly the same fringe record as the displacement interferometer since the signal beam frequency is beating against the original laser beam frequency. Thereafter, however, the number of fringes to be recorded is greatly decreased. Total fringe counts of 5 to 20 are common, as opposed to hundreds for the displacement interferometer.

The sensitivity of this interferometer to misalignment due to tilt and free surface motion is discussed in some detail by BARKER,[22] and derivations relating the free surface velocity $v_s(t)$ to the fringe count $\mathscr{f}(t)$ have been given by BARKER and HOLLENBACH[23] and by CLIFTON.[24] Perhaps the most generally useful relation is

$$v_s\left(t - \frac{\tau}{2}\right) = \frac{\Lambda}{2\tau}\mathscr{f}(t), \tag{5.5}$$

which we derive below in the subsection on in-material interferometry.

Eqs. (5.1) and (5.2), which describe the displacement interferometer, show the fringe count $\mathscr{f}(t)$ to be proportional to the surface displacement. Eq. (5.5), however, shows that this interferometer configuration gives a fringe count which is proportional to the free-surface velocity. This interferometer has therefore been called a *velocity interferometer*. Both the displacement and the velocity interferometers provide continuous records of displacement and velocity, respectively, since the fringe count can be evaluated at any desired time during which data were recorded. As with the displacement interferometer, the value of $\mathscr{f}(t)$ can generally be determined to at least ± 0.1 fringe or better at any given time by using (5.3).

The velocity interferometer has a number of advantages over the displacement interferometer. First, it gives a direct reading of the velocity through the fringe count, so that it is no longer necessary to differentiate the displacement-time record to obtain the velocity. Second, as discussed previously, there are difficulties in varying the displacement interferometer proportionality constant $\Lambda/2$, whereas

[21] The delay-leg path lengths have most often been in the range of 1 to 10 m.
[22] BARKER [1968, *2*].
[23] BARKER [1968, *2*], BARKER and HOLLENBACH [1970, *2*], and BARKER [1971, *2*].
[24] CLIFTON [1970, *6*].

the velocity interferometer proportionality constant $\Lambda/2\tau$ is easily and continuously variable by varying the delay time τ. Third, instead of being limited to surface velocities less than 0.2 mm/μsec, there is, in principle, no limitation on the maximum velocity capability of the velocity interferometer. There is instead a limitation on the change in velocity during the time τ in order to keep the fringe frequency within the capabilities of the recording system:

$$|v_s(t) - v_s(t-\tau)| \leqq 0.2 \text{ mm}/\mu\text{sec}. \tag{5.6}$$

This restriction is far less severe than simply $v_s(t) \leqq 0.2$ mm/μsec. In cases where the acceleration is high, the delay time τ can be made correspondingly small in order to keep $|v_s(t) - v_s(t-\tau)|$ small. There are, of course, waves in which the acceleration appears to be infinite and for which the jump in velocity is far greater than 0.2 mm/μsec. In this case, the burst of fringes during the time τ after wave arrival is unresolved. Even here, however, it is usually possible to determine the number of fringes in the unresolved region by arranging, through the selection of τ, for the number of unresolved fringes to be small. Then, since the number of fringes between identical fringe positions on either side of the unresolved region must be an integer, it is nearly always possible from the boundary conditions of the experiment to say with confidence just what the integer must be. In a later figure (Fig. 10) we will show such a velocity interferometer oscillogram record of a wave in polymethyl methacrylate. This data will be discussed in more detail in the subsection on in-material interferometry.

Whenever very low velocities are to be measured, very long delay legs become necessary in order to achieve several fringes with the velocity interferometer. Therefore, the displacement interferometer still holds advantages for the velocity range below about 0.05 mm/μsec. Although velocities up to 0.2 mm/μsec have been measured with displacement interferometers, the advantages usually lie with velocity interferometry above about 0.1 mm/μsec.

γ) *Diffuse surface interferometry.* The velocity interferometer provided the first breakthrough in precision measurements of surface velocities in high-stress impact experiments. The main limitation of the velocity interferometer has been its requirement that the specimen surface being monitored retain a good mirror finish during the motion of interest. Many materials cannot be polished to a mirror finish, and others suffer a loss of their mirror finish as a result of the arrival of a strong wave. While this is not a problem with such materials as polymethyl methacrylate, it has made interferometry difficult or impossible to apply to some other wave propagation experiments.[25] This has led to the development by BARKER and HOLLENBACH[26] of an improved velocity interferometer system which is capable of monitoring any specimen surface, ranging from spectrally reflecting to diffusely reflecting. Its resolution is comparable to or better than that of the velocity interferometer instrumentation used in the experiments reported in this article. This improved velocity interferometer system is useful in studies of porous materials, geological materials, shock-induced phase changes, and the growth of the detonation wave in a high explosive. It has also been used in a large depth-of-field mode to measure the velocity history of a projectile accelerating through a long gun barrel.

δ) *In-material interferometry.* In order to use interferometry as an in-material gauge to measure wave profiles inside a material sample (as opposed to free surface

[25] This difficulty has sometimes been circumvented by using a buffer plate between the specimen and the mirror surface. See, for example, BARKER [1972, *3*].

[26] BARKER and HOLLENBACH [1972, *4*].

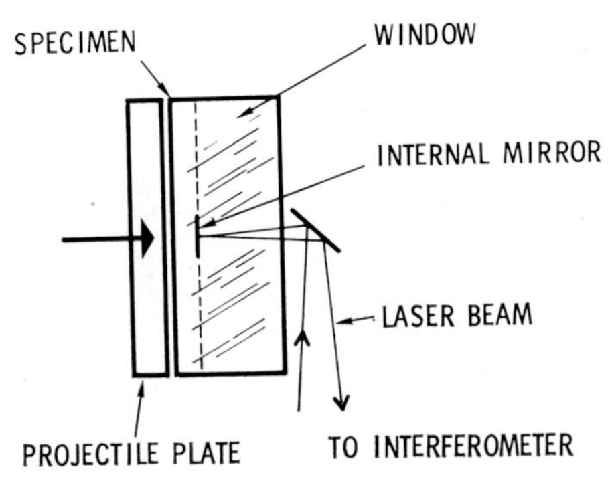

Fig. 8. Transparent specimen and window for obtaining in-material measurement of the particle velocity.

velocity measurements), it is necessary, of course, that the specimen be transparent to the laser beam. One must then embed a mirror inside the specimen material in such a way that the presence of the mirror results in a negligible perturbation in the wave profile. The sample with the embedded mirror is made by vapor depositing an aluminum mirror onto a "window" plate of the specimen material to a thickness of about 100 nm.[27] The mirrorized surface of the window is then bonded to the specimen plate by a very thin layer of epoxy cement (see Fig. 8). A photograph of a typical polymethyl methacrylate target with the embedded mirror is shown in Fig. 9. Such targets have epoxy bond layers less than 0.02 mm thick. Since the density and dynamic mechanical behavior of epoxy is close to that of polymethyl methacrylate, the epoxy has a negligible effect on the measured wave profile. The acoustic impedance[28] of the aluminum mirror is significantly different from that of polymethyl methacrylate. However, because of its thinness, the mirror reaches equilibrium with its surroundings in a fraction of a nanosecond. Thus, measurement of the velocity of the mirror is essentially a direct measurement of the particle velocity in the material sample.

Fig. 10 shows two in-material velocity interferometer oscillograms of a wave in a polymethyl methacrylate target and the particle velocity-time profiles deduced from the records. The impact velocity in the experiment of Fig. 10 was an order of magnitude greater than that of the displacement interferometer experiment of Fig. 5. During the 7-nsec delay time (τ) after the start of the motion (Fig. 10a), the wave produced several fringes which were too rapid to be tracked by the photomultiplier, but the number of unresolved fringes was determined by the method outlined previously. An unloading wave was also produced in the experiment of Fig. 10 when the impact-generated loading wave in the flyer reflected off the rear free surface of the polymethyl methacrylate flyer plate. The unloading wave then propagated through the flyer and into the target. The particle velocity history in the target during the passage of the unloading wave is shown in Fig. 10b under the corresponding oscillogram record.

One of the most important aspects of in-material interferometry concerns the variation of the index of refraction of the window material with density. As the wave propagates into the window, an expanding region in the window experiences a density change, and hence an index-of-refraction change. This can produce a shift in the laser light frequency over and above that caused by the motion of the

[27] The mirror diameter is usually about 12 mm.

[28] The acoustic impedance of a material is the product of the density and the longitudinal sound velocity.

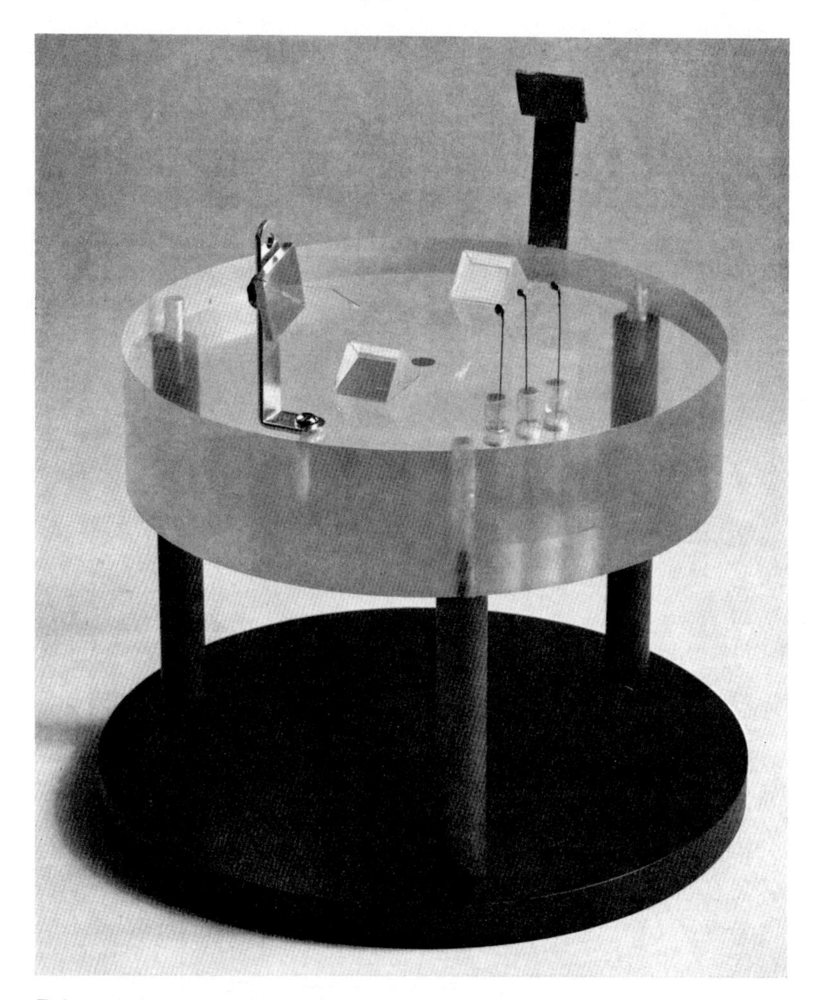

Fig. 9. Polymethyl methacrylate target with an internal mirror. The target diameter is 150 mm and the impact surface is on the bottom. The internal mirror is the small disk in the center; it lies in the plane of the bond layer which is visible at the near side of the target. The two prisms on the top provide entry and exit for the impact detecting beam (cf. Sect. 6).

embedded mirror. Additional experiments are required to evaluate the effect of this induced change in refractive index of the window material.

The approach used in measuring the index-of-refraction effect is illustrated in Fig. 11 where a Michelson interferometer is used to observe the motion of the impact surface of a transparent target specimen during symmetric impact.[29] In this case, the velocity v of the impact interface is just one-half of the projectile's impact velocity. The measured interferometer fringe frequency ν_m corresponding to v was sensed by the photomultiplier and recorded on an oscillograph. A number of similar experiments determined a v-vs-ν_m curve for the window material. It is assumed that for a given window material the fringe frequency ν_m is a single-valued function of v until the first wave reaches the free surface of the target window,

[29] The term "symmetric impact" implies that the projectile and target plates are made of the same material.

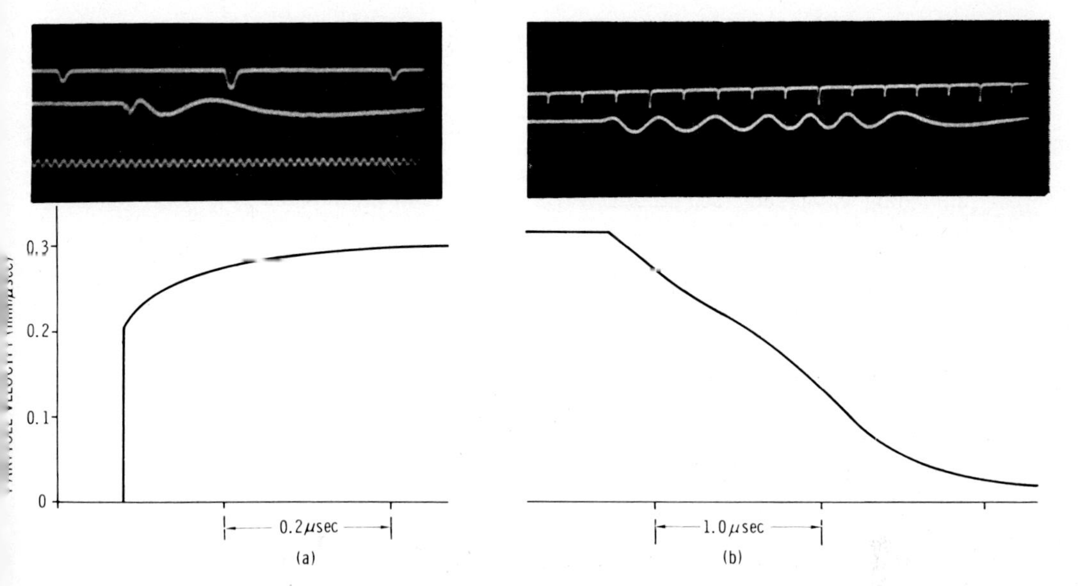

Fig. 10a and b. Velocity interferometer data. The upper traces are again the timing combs having marks at 0.2 μsec intervals with every fifth mark accented. The lowest trace in (a) is another timing signal with a period of 0.01 μsec. The center traces are the signals from the velocity interferometer. The velocity profiles at the bottom were obtained from the oscillograms.

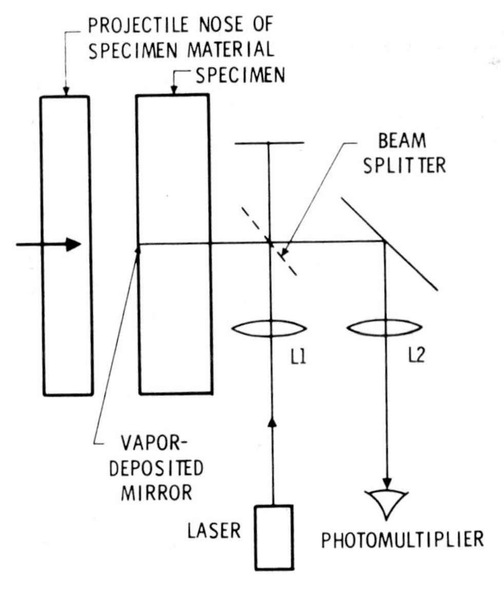

Fig. 11. Schematic of the index of refraction measuring experiment. Lens L1 brings the laser beam to focus at the two mirrors to minimize the sensitivity to tilt. Lens L2 recollimates the beam. Note that one beam penetrates through the specimen to the impact surface, upon which an extremely thin aluminum mirror is vapor deposited. (From BARKER and HOLLENBACH [1970, 2].)

regardless of the wave profile leading to the velocity v. This assumption is apparently valid for polymethyl methacrylate since both v and ν_m remained essentially constant while the shape of the compressive wavefront changed with time. A few observations involving unloading waves in the index-of-refraction shots further strengthened the assumption of single-valuedness. The maximum variation in ν_m was less than $\pm 0.2\%$ during the compressive wave phase of the shots while the unloading wave seemed to produce variations in ν_m of up to 0.5% at a given velocity v.

The use of the v-vs-ν_m measurements in reducing laser interferometer data was made more convenient by defining the frequency ν_d as the fringe frequency of an ordinary Michelson displacement interferometer with no window material involved, where one of the interferometer mirrors is stationary and the other is moving with the velocity v. Thus (cf. Eq. (5.4)),

$$\nu_d = \frac{2v}{\Lambda}. \tag{5.7}$$

It can be shown that, if the index of refraction of the specimen n varies with density ϱ according to the well-known Gladstone-Dale model,[30]

$$\frac{d\varrho}{\varrho} = \frac{dn}{(n-1)},$$

the frequency ν_m that will be observed in the experiment depicted in Fig. 11 will be equal to ν_d, regardless of the wave shape, wave velocity, or amplitude. Therefore, the fractional difference in frequency,

$$\frac{\Delta\nu}{\nu_d} = \frac{\nu_m - \nu_d}{\nu_d},$$

is a measure of the degree of departure of the index of refraction from the Gladstone-Dale model.

It was observed that $\Delta\nu/\nu_d$ is a slowly varying function of v in the stress range studied, and it is therefore convenient to use $\Delta\nu/\nu_d$ in correcting the interferometer data for the index-of-refraction effect. For example, consider a Michelson interferometer window experiment. The internal mirror velocity v at time t is given by

$$v(t) = \frac{\Lambda\nu_d(t)}{2}$$

by virtue of the definition of ν_d in (5.7). However, the velocity $v(t)$ leads to a measured fringe frequency of $\nu_m(t)$ instead of $\nu_d(t)$. Having the values of $\Delta\nu/\nu_d$-vs-v available for the window material and noting that

$$\nu_d = \frac{\nu_m}{1 + \frac{\Delta\nu}{\nu_d}},$$

we can find v from the equation

$$v(t) = \frac{\Lambda\nu_m(t)}{2\left(1 + \frac{\Delta\nu}{\nu_d}\right)}. \tag{5.8}$$

The solution of (5.8) may involve some iteration since $\Delta\nu/\nu_d$ is a function of v. A rapid convergence is assured, however, by the fact that the variation in $\Delta\nu/\nu_d$ is small compared to unity.

[30] Cf. GLADSTONE and DALE [1858, *1*], [1863, *1*]. In this regard, also see PRESTON [1928, *1*].

The velocity interferometer equation can be similarly derived to account for the variation in index of refraction. A Michelson interferometer produces a fringe frequency $\nu_m(t)$ which is equal to the total shift in frequency by beating the frequency-shifted light against the original light frequency. Thus, if ν_1^0 is the original light frequency and $\nu_1(t)$ is the shifted frequency, then $\nu_m(t) = \nu_1(t) - \nu_1^0$. The velocity interferometer, on the other hand, uses only light which undergoes the frequency shift ν_m and it produces a fringe frequency by beating the current light frequency $\nu_1(t)$ against the light frequency which was present a time τ earlier. Therefore, the fringe frequency produced by the velocity interferometer is

$$\nu_v(t) = \nu_1(t) - \nu_1(t-\tau).$$

Adding and substracting ν_1^0 on the right side shows that

$$\nu_v(t) = \nu_m(t) - \nu_m(t-\tau).$$

The fringe count $\mathscr{f}(t)$ is equal to the integral of the velocity interferometer frequency up to time t:

$$\mathscr{f}(t) = \int_0^t [\nu_m(s) - \nu_m(s-\tau)]\, ds. \tag{5.9}$$

In Eq. (5.9) it is assumed that $\nu_m(t) \equiv 0$ for $t < 0$, i.e., that the sample is initially at rest. Using this fact, (5.9) can be expressed as

$$\mathscr{f}(t) = \int_{t-\tau}^t \nu_m(s)\, ds. \tag{5.10}$$

Solving (5.8) for $\nu_m(t)$ and substituting into (5.10) yields

$$\mathscr{f}(t) = \frac{2}{\Lambda} \int_{t-\tau}^t v(s) \left(1 + \frac{\Delta\nu}{\nu_d}\right) ds. \tag{5.11}$$

As has already been noted, $(1 + \Delta\nu/\nu_d)$ is a slowly varying function of v. Moreover, the time interval τ between the limits of integration is very small (of the order of a few nanoseconds); thus $v(t)$ seldom changes appreciably during the period over which the integration is carried out. (In cases where $v(t)$ does change radically during times of the order of τ or less, different data-reduction techniques can often be used.) Therefore, we can remove the quantity $(1 + \Delta\nu/\nu_d)$ from the integral and set it equal to the value corresponding to the average value of v during the time interval $t-\tau$ to t. Then the average value of the velocity $v(t)$ during the interval of the integration is simply

$$\bar{v}(t) = \frac{1}{\tau} \int_{t-\tau}^t v(s)\, ds.$$

If a time must be assigned to $\bar{v}$, proper centering requires that it be $t - (\tau/2)$. Substituting for the integral in (5.11) leads to the velocity interferometer equation

$$v\left(t - \frac{\tau}{2}\right) = \frac{\Lambda}{2\tau} \frac{\mathscr{f}(t)}{\left(1 + \frac{\Delta\nu}{\nu_d}\right)}. \tag{5.12}$$

The solution of (5.12) may again require an iteration process. Because of the slowly varying nature of $\Delta\nu/\nu_d$, however, one can usually avoid iterating by making a sufficiently accurate first estimate of v. Note that the quantity $\Delta\nu/\nu_d$ in (5.8)

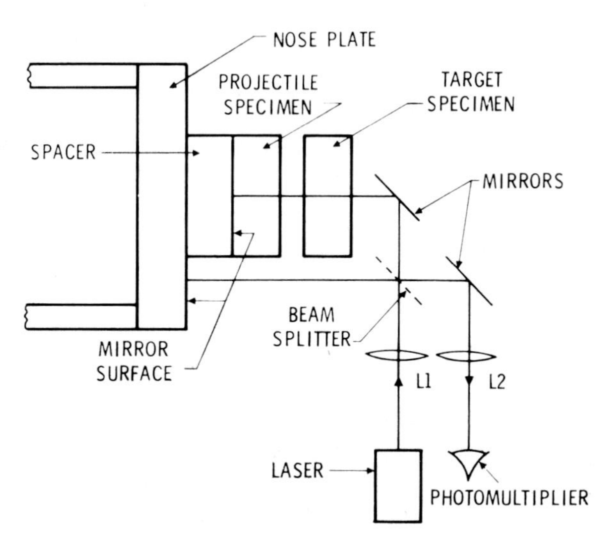

Fig. 12. The index of refraction measuring interferometer used in the higher impact velocity experiments. (From BARKER and HOLLENBACH [1970, *2*]).

and (5.12) can be either positive or negative, depending on the results of the $\Delta\nu/\nu_d$ measurements.

In practice, the method of measuring ν_m with the Michelson interferometer shown in Fig. 11 was useful only over the range of velocities up to 0.2 mm/μsec. At the higher velocities, the frequency ν_m exceeded the response capability of the photomultiplier tube. Therefore, the experimental configuration was modified to that shown in Fig. 12 where both of the mirrors of the interferometer were carried on the projectile noseplate. The interferometer was aligned while the projectile specimen rested at the muzzle of the gun barrel in contact with the target specimen, i.e., in the same position which was to be occupied at impact time. After the alignment, the projectile was drawn back to the breech end of the barrel, the barrel was evacuated, and the projectile was fired into the target.[31] No fringes were observed just before impact because both mirrors of the interferometer were traveling at the same speed and giving the same Doppler shift in light frequency. Just after impact, however, the waves traveling away from the impact surface into the projectile and target specimens gave an additional shift to the light frequency in that leg of the interferometer. The fringe frequency of the interferometer under these conditions was $\nu_i = 2|\Delta\nu| = 2|\nu_m - \nu_d|$, where, as before, ν_m is the fringe frequency that would have been present using the Michelson interferometer of Fig. 11 and $\nu_d = 2v/\Lambda$, where v is one-half the impact velocity. Since this type of experiment gives a direct measurement of $\Delta\nu$, it provides a particularly sensitive way of determining the effect of wave compression on the index of refraction. Note, however, that the sign of $\Delta\nu$ cannot be determined from the experiment depicted by Fig. 12. Therefore, the first type of experiment (Fig. 11) was used to establish the sign of $\Delta\nu$ for the material, and it was assumed that the sign did not change suddenly at the higher stress levels.

[31] Because of the precision normally required in interferometer alignment and because of the intrinsically violent acceleration of the projectile through the gun barrel, it was not obvious before the fact that the interferometer would return to its aligned condition at impact. It was found, surprisingly, that a number of these experiments gave consistently good results.

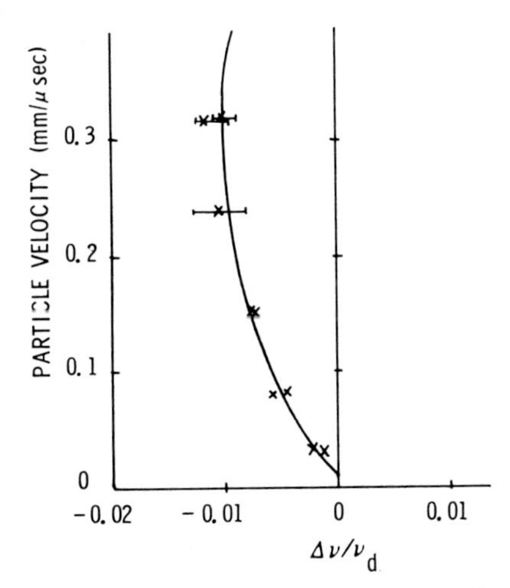

Fig. 13. Particle velocity vs. $\Delta\nu/\nu_d$ for polymethyl methacrylate. (From BARKER and HOLLENBACH [1970, *2*].)

The calibration of the variation of index of refraction with particle velocity in polymethyl methacrylate has been reported by BARKER and HOLLENBACH[32] and their results are shown in Fig. 13. The error flags have been placed only on the data points with the least precision. The duplicate experiments at nearly identical velocities serve to demonstrate the reproducibility of the polymer and of the experimental technique.

6. Determination of wave front velocities. The final aspect of the experimental techniques which we must consider is the determination of an average propagation velocity for the front of the transmitted wave profile. A high degree of accuracy is required for this measurement since it can be shown that the characterization of a specimen material is more sensitive to errors in wave velocity than to errors in particle velocity. The acceptable wave velocity uncertainty in the experiments discussed in this article was about $\pm 0.3\%$. This requirement translates to measuring distances to within a few hundredths of a millimeter and, more significantly, transit times to within a few nanoseconds.

In the absence of tilt and any deviation from planarity of the impacting surfaces, we may unambiguously define zero time as the time of impact. Then the average propagation velocity of a transmitted wave front would simply be the distance from the impact surface to the mirror divided by the time of arrival of the wave at the mirror. However, as was indicated in Sect. 4, the impact itself generally requires an appreciable time because the surfaces of the impacting plates cannot be made perfectly planar and because the tilt between the colliding surfaces cannot be completely eliminated. In fact, the time for the impacting surfaces to contact each other completely is rarely as small as a few nanoseconds and is usually of the order of 100 nsec or more. Thus, an impact or zero time cannot be clearly defined. In addition, if the transmitted wave profile is measured by a technique which averages the motion over a significant area in the plane of the measuring station, it is often impossible to accurately define an arrival time.

With laser interferometry, however, one brings the laser beam to focus on the mirror at the center line of the sample. Thus the lateral dimensions of the measuring

[32] BARKER and HOLLENBACH [1970, *2*]. They also calibrated two other window materials, fused silica and sapphire.

station are very small (about 0.1 mm) and hence an arrival time can be accurately determined. In addition a zero time can be defined as the time at which impact occurs at the center line of the specimen. If this time can be determined and if the wave in the specimen is not tilted with respect to the center line by more than a very few degrees, we can calculate the wave front propagation velocity accurately.[33]

The measurement of the time of impact at the center of the sample is complicated by the fact that one may not put a physical object there because a perturbation of the transmitted wave would result. A commonly employed method for determining impact time at the center utilizes the flush pin technique described in Sect. 4 which is used to measure tilt. If it is assumed that the two surfaces are plane, that the line of contact between the surfaces moves supersonically with respect to the target and projectile material sound velocities, and that the velocity of the uncontacted part of the projectile plate is constant, then the time when impact occurs at the center is the average of the times at which the flush pins close out. Uncertainties with this technique arise because the surfaces cannot be lapped perfectly flat and because the acceleration of the projectile can distort the flatness of the flyer plate attached to the projectile.

The technique used to determine the impact time at the center line for the experiments of this article made use of the fact that both the target and projectile plates were transparent. It consists of passing a light beam through the target and having it undergo total internal reflection at the center of the impact surface (Fig. 14a). When contact occurs at this point the light is no longer reflected but is transmitted through the projectile plate and reflected elsewhere (Fig. 14b). By monitoring the light beam reflected from the center of the target impact surface with a photomultiplier tube, the loss of intensity when impact occurs is easily detected. Thus, the impact time at the center is measured directly without distorting

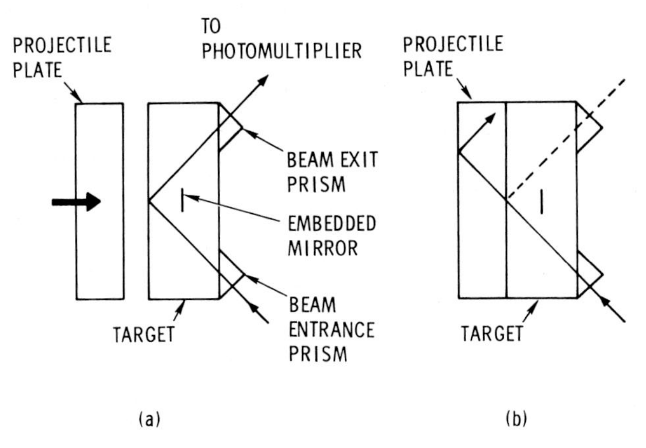

Fig. 14a and b. Experimental configuration for optical detection of impact time at the center of the specimen.

[33] It may occur to the reader that a tilt correction is still required even though the impact time and the wave profile are both measured along the center line of the target, for when the impacting surfaces are tilted the plane of the propagating wave in the specimen is also tilted. Thus, the measured wave profile originates not at the center of the impact surface, but rather at a lateral position depending on the wave tilt in the specimen. However, with impact tilts less than 0.002 radians per mm/μsec of the impact velocity (cf. Sect. 4) and with the wave speeds in polymethyl methacrylate of the order of 3 mm/μsec, it follows that using the impact time at the center caused an error of no more than a few thousandths of a percent.

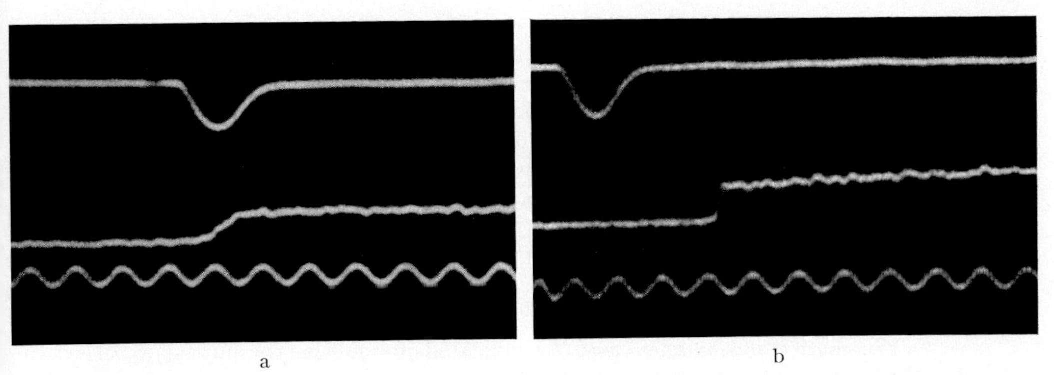

Fig. 15a and b. Oscillograms showing the loss of light at impact time. The top and bottom traces are for timing with the period on the top traces being 0.01 μsec. The center traces show the impact time. Decreasing light is upward on the oscillograms; hence, the upward jog in each center trace gives the impact time. The impact velocities in (a) and (b) were 0.061 and 0.640 mm/μsec, respectively.

the wave profile in the slightest. Fig. 15a and b shows the oscillograms resulting from the impact of polymethyl methacrylate plates at two different impact velocities.

The rise times of the signals in Fig. 15 can be attributed mainly to frustrated reflection and the response time of the photomultiplier. At low projectile velocities it appears that frustrated reflection accounts for most of the rise time. This well-known phenomenon[34] is caused by the fact that light undergoing total internal reflection penetrates slightly beyond the reflecting surface. The intensity falls off exponentially, being equal to 1% of the incident intensity at a distance of one wavelength from the boundary. As a second surface is brought to within a wavelength of the reflecting surface, frustrated reflection occurs and some of the incident light is transmitted across the gap. Since we resolve intensity changes of about 5%, the rise time of the signal can be no less than the time it takes the projectile to travel the last three-fourths to one wavelength to the impact surface. For a projectile velocity of 0.06 mm/μsec this would correspond to a rise time of between 7 and 10 nsec which appears to be in excellent agreement with the observed signal (cf. Fig. 15a).

At high projectile velocities the frustrated reflection effect contributes less to the signal rise time, and the response time of the photomultiplier tube becomes relatively more important. At the impact velocity of 0.64 mm/μsec of Fig. 15b, the frustrated reflection accounts for about a nanosecond of the rise time. The remainder of the observed rise time of about 2 nsec is attributed to the response time of the photomultiplier. From these considerations we can conclude that the time of impact is always 1 or 2 nsec before the minimum light level is reached. Therefore the technique of light beam reflection at the center of the impact surface detects the impact time to within about ± 2 nsec for the impact velocities of the experiments reported in this article.

Even after obtaining the signal at impact time, great care must be taken in establishing the time base if the time accuracy is to be maintained. The signal must be recorded on an oscillogram using a very fast sweep speed such that a time of 1 nsec can be easily distinguished. Likewise, the transmitted wave form must be similarly recorded. Timing combs must be recorded on each of the oscillograms in such a way that the sweep times of the different oscilloscopes can be very

[34] Cf. Wood [1934, *1*].

accurately related. To this end, any differences in the time mark transmission cable lengths must be accounted for since signals travel at a speed of about 20 cm/nsec through the cables. Also, the light path length difference between the impact light signal (sent directly to the photomultiplier) and the transmitted wave light signal (sent to the interferometer before going to the photomultiplier) must be taken into account since light travels at a speed of 30 cm/nsec. Although these details are not very difficult, they illustrate the care required to maintain nanosecond precision in relating the impact time to the time of arrival of the measured wave profile.

III. One-dimensional motions in materials with memory.

7. Kinematics and balance laws. As indicated previously, the impact experiments are designed so that the observation of the resulting wave profile along the center line of the target plate is completed before the arrival of unloading waves from the lateral surfaces of the target. For isotropic materials, symmetry requires the motion during the time of observation to be that of one-dimensional stretch and hence, *purely* longitudinal. Therefore, unless indicated otherwise, we shall confine our subsequent analysis to one-dimensional motions in viscoelastic solids. In this case, it is convenient to identify the body in its fixed reference configuration with an interval R of the real line and each material point with its position X in R. The *motion* of the material is described by the continuous function $\chi(\cdot,\cdot)$ which gives the spatial position

$$x=\chi(X,t) \tag{7.1}$$

at time t of the material point X. The *displacement* u of the material point X is

$$u=\chi(X,t)-X. \tag{7.2}$$

For suitably smooth motions,

$$F=\frac{\partial}{\partial X}\chi(X,t)$$

and

$$v=\dot{x}=\frac{\partial}{\partial t}\chi(X,t)=\frac{\partial}{\partial t}u(X,t) \tag{7.3}$$

define the *deformation gradient* F and *particle velocity* v of X at time t. The deformation gradient F is taken to be strictly positive; thus the motion $\chi(x,\cdot)$ is invertible. The *strain* ε of X at time t is given by

$$\varepsilon=-\frac{\partial}{\partial X}u(X,t)=1-F(X,t) \tag{7.4}$$

and has values in the interval $(-\infty, 1)$. Here the one-dimensional stress $\sigma(X,t)$ and strain $\varepsilon(X,t)$ are taken to be *positive in compression.*

The *history of the strain* ε^t up to time t of the material point X is defined by

$$\begin{aligned}\varepsilon^t(s)&=\varepsilon(X,t-s)\\&=1-F(X,t-s), \quad s\in[0,\infty).\end{aligned} \tag{7.5}$$

A knowledge of ε^t is equivalent to knowledge of the *present value* $\varepsilon=\varepsilon^t(0)$ and the *past history* ε^t_r, where ε^t_r is the history ε^t on the open interval $(0,\infty)$.

In one dimension, the balance of mass requires that the present density $\varrho(X,t)$ be related to the density ϱ_R in the reference configuration by

$$\varrho=\frac{\varrho_R}{1-\varepsilon(X,t)}. \tag{7.6}$$

Furthermore, in the absence of external body forces, the balance of linear momentum asserts that

$$\frac{d}{dt}\int_{X_1}^{X_2} \varrho_R v(X,t)\, dX = \sigma(X_1,t) - \sigma(X_2,t) \tag{7.7}$$

for every internal $[X_1, X_2]$ in R and for all times t. This balance law together with appropriate smoothness assumptions implies that

$$\varrho_R \dot{v} = -\frac{\partial \sigma}{\partial X} \tag{7.8}$$

for every (X, t) in $R \times (-\infty, \infty)$.

8. Classification of waves. By a *wave* we mean a motion which contains a propagating disturbance. Here we shall be particularly interested in waves involving jump discontinuities in certain kinematical fields.[1] The *intrinsic velocity* U of such a discontinuity (wave front) is defined by

$$U(t) = \frac{d}{dt} Y(t) \tag{8.1}$$

where $Y(t)$ is the material point in R at which the front is to be found at time t. Thus, the intrinsic velocity expresses the rate of advance of the front with respect to the material in the reference configuration and $Y(t)$, $t > 0$, gives the material trajectory of the front. Here we use the standard notation for the jump $[f]$ in a function $f(X, t)$ across the wave front at time t, i.e.,

$$[f](t) = f^-(t) - f^+(t) \tag{8.2}$$

where

$$f^\pm(t) = \lim_{X \to Y^\pm(t)} f(X,t). \tag{8.3}$$

With $U > 0$, f^+ and f^- are the limiting values of $f(X, t)$ immediately ahead of and behind the wave front. If

$$\dot{f}(X,t), \qquad \frac{\partial}{\partial X} f(X,t)$$

are also discontinuous at $X = Y(t)$ but continuous everywhere else, then

$$\frac{d[f]}{dt} = [\dot{f}] + U\left[\frac{\partial f}{\partial X}\right]. \tag{8.4}$$

This kinematical condition of compatibility is discussed in more detail by TRUESDELL and TOUPIN[2] and by CHEN.[3]

We assume throughout this article that the motion $\chi(\cdot,\cdot)$ is a continuous function of X, t for all X, t. A *shock wave* is a motion containing a wave front across

[1] Waves may also involve jump discontinuities in certain thermal fields. This aspect of wave propagation will be taken up in Sects. 25–27.

In three-dimensional motions the discontinuity is often termed a *singular surface*. A comprehensive treatment of the theory of singular surfaces has been given in the treatise by TRUESDELL and TOUPIN [1960, *7*]. In this regard, also see HILL [1961, *5*], THOMAS [1966, *13*], GURTIN [1972, *12*], and CHEN [1973, *4*]. The reader interested in the historical aspects of the theory are referred to TRUESDELL and TOUPIN [1960, *7*] and the treatise by GURTIN [1972, *12*]. The presentation of the theory given here closely follows that of COLEMAN, GURTIN, and HERRERA [1965, *4*], and COLEMAN and GURTIN [1965, *5*].

[2] TRUESDELL and TOUPIN [1960, *7*, Sect. 180].

[3] CHEN [1973, *4*, Sect. 4].

which the strain ε, the particle velocity v, and their derivatives are discontinuous. Thus, (8.4) with $f=u$ asserts that at $X=Y(t)$

$$[v]=U[\varepsilon]. \tag{8.5}$$

It is clear from this result that either $[v]$ or $[\varepsilon]$ can be taken as a measure of the *amplitude* of the shock. In general, the stress is also discontinuous at $X=Y(t)$ and the balance of linear momentum (7.7) implies that[4]

$$\begin{aligned} [\sigma] &= \varrho_R U[v], \\ \left[\frac{\partial\sigma}{\partial X}\right] &= -\varrho_R[\dot{v}]. \end{aligned} \tag{8.6}$$

The shock is said to be *compressive* if $[\varepsilon]>0$; *expansive* if $[\varepsilon]<0$.

An *acceleration wave* is a one-dimensional motion containing a wave front across which the strain ε and the particle velocity v are continuous but their derivatives are not. Thus, (8.4) with $f=v$ and $f=\varepsilon$ asserts that at $X=Y(t)$

$$[\dot{v}]=U[\dot{\varepsilon}]=U^2\left[\frac{\partial\varepsilon}{\partial X}\right]. \tag{8.7}$$

The jump in the particle acceleration is taken as the *amplitude* $a(t)$ of the wave front:

$$a(t)=[\dot{v}](t). \tag{8.8}$$

For a *compressive* wave, $a(t)>0$; while, on the other hand, for an *expansive* wave, $a(t)<0$. The balance relations (8.6) hold also for an acceleration wave.

In a *smooth structured wave*, all kinematical fields are continuous and have continuous derivatives.

9. Simple materials with fading memory. In this article we consider the propagation of waves in nonlinear homogeneous materials with long-range viscoelastic memory. The purely mechanical, one-dimensional response of such materials can be described in terms of a response functional for the stress, i.e., at a given material point the stress σ at time t is determined by the entire history of the strain ε^t:[5]

$$\sigma(t)=\mathscr{S}(\varepsilon^t)=\mathscr{S}(\varepsilon,\varepsilon_r^t). \tag{9.1}$$

It is further assumed that the material exhibits fading memory. This concept has been made mathematically precise by COLEMAN and NOLL[6] and implies that the stress functional $\mathscr{S}$ has certain smoothness properties. Let

$$\|\varepsilon^t\|^2=|\varepsilon(t)|^2+\int_0^\infty h^2(s)\,|\varepsilon^t(s)|^2\,ds \tag{9.2}$$

denote the norm of the history ε^t where $h(\cdot)$ is a fixed influence function, i.e., a continuous, monotone-decreasing, square-integrable function. The set of all histories with finite norm forms a Hilbert space. Then the constitutive functional $\mathscr{S}$ is assumed to be defined on this space and to be of class C^2. Notice that since $h(\cdot)$ is monotone decreasing, the assumption of fading memory implies that the

[4] Cf. TRUESDELL and TOUPIN [1960, *7*, Sect. 205] for the derivation of the three-dimensional counterpart of $(8.6)_1$. Also see JEFFREY [1964, *12*] and CHEN [1973, *4*, Sect. 5].

[5] Cf. NOLL [1958, *3*], COLEMAN [1964, *5*]. The formulation of the theory presented in this section is based on that given previously by COLEMAN and GURTIN [1965, *5*].

[6] COLEMAN and NOLL [1960, *1*], [1961, *1*], [1964, *6*]. More general formulations of the principle of fading memory have been given by COLEMAN and MIZEL [1966, *7*], [1968, *6*]. In this regard, also see PERZYNA [1967, *8*].

material response is influenced more by changes in strain in the recent past (small s) than by those occurring in the distant past (large s).

The assumed smoothness of the response functional $\mathcal{S}$ insures the existence of the ordinary partial derivatives

$$D_\varepsilon \mathcal{S}(\varepsilon^t) \equiv \frac{\partial}{\partial \varepsilon} \mathcal{S}(\varepsilon, \varepsilon_r^t), \qquad D_\varepsilon^2 \mathcal{S}(\varepsilon^t) \equiv \frac{\partial^2}{\partial \varepsilon^2} \mathcal{S}(\varepsilon, \varepsilon_r^t). \tag{9.3}$$

Since $\mathcal{S}$ can depend on the present value of strain for fixed past histories, the material is said to exhibit *instantaneous response* and we interpret

$$E_t = D_\varepsilon \mathcal{S}(\varepsilon^t), \qquad \tilde{E}_t = D_\varepsilon^2 \mathcal{S}(\varepsilon^t) \tag{9.4}$$

as instantaneous moduli. In particular, E_t is the *instantaneous tangent modulus corresponding to the history* ε^t and represents the initial slope of the stress-strain response corresponding to a sudden change in the strain history ε^t at time t. The modulus $\tilde{E}_t$ is the *instantaneous second-order modulus corresponding to the history* ε^t and represents the initial curvature of the instantaneous stress-strain response.[7] It will prove especially useful in what follows to specialize these definitions for the particular strain history

$$\varepsilon^t(s) = \varepsilon_{\mathrm{I}}(s) = \begin{cases} \varepsilon, & s = 0, \\ 0, & s \in (0, \infty) \end{cases} \tag{9.5}$$

which corresponds to a strain jump ε suddenly applied to a material point which has been unstrained for all past time. In this instance, the stress depends only on the strain jump ε and we call

$$\sigma = \mathcal{S}(\varepsilon_{\mathrm{I}}) = \sigma_{\mathrm{I}}(\varepsilon) \tag{9.6}$$

the *instantaneous stress-strain law*. Clearly, the function σ_{I} is of class C^2, and the instantaneous tangent modulus and second-order modulus become

$$E_{\mathrm{I}} = \frac{d}{d\varepsilon} \sigma_{\mathrm{I}}(\varepsilon), \qquad \tilde{E}_{\mathrm{I}} = \frac{d^2}{d\varepsilon^2} \sigma_{\mathrm{I}}(\varepsilon), \tag{9.7}$$

respectively.

Materials with fading memory are also known to exhibit *equilibrium response.* In particular, consider the equilibrium (constant) strain history

$$\varepsilon^t(s) = \varepsilon_{\mathrm{E}}(s) = \varepsilon, \qquad s \in [0, \infty). \tag{9.8}$$

The stress response to this history depends only on the value of strain ε and we call

$$\sigma = \mathcal{S}(\varepsilon_{\mathrm{E}}) = \sigma_{\mathrm{E}}(\varepsilon) \tag{9.9}$$

the *equilibrium stress-strain law*. The fading memory hypothesis insures that σ_{E} is also of class C^2, and the derivatives

$$E_{\mathrm{E}} = \frac{d}{d\varepsilon} \sigma_{\mathrm{E}}(\varepsilon), \qquad \tilde{E}_{\mathrm{E}} = \frac{d^2}{d\varepsilon^2} \sigma_{\mathrm{E}}(\varepsilon) \tag{9.10}$$

[7] The dependence of the instantaneous moduli E_t and $\tilde{E}_t$ on the strain history ε^t implies that they may change with time at a given material point; the subscript t serves to remind us of this possibility.

are called the *equilibrium tangent modulus* and *equilibrium second-order modulus*, respectively.

Another important property of nonlinear materials with memory is their response to small relative strain histories γ^t. By use of the smoothness of the functional $\mathcal{S}$, it can be shown that as $\|\gamma^t\| \to 0$ the stress is approximated by the constitutive equation of linear viscoelasticity,[8] i.e.,

$$\mathcal{S}(\varepsilon^t + \gamma^t) = \mathcal{S}(\varepsilon^t) + G_t(0)\,\gamma^t(0) + \int_0^\infty G_t'(s)\,\gamma^t(s)\,ds + o(\|\gamma^t\|). \tag{9.11}$$

The function $G_t(\cdot)$ is the *stress relaxation function corresponding to the strain history* ε^t and

$$G_t'(s) = \frac{d}{ds} G_t(s).$$

When it seems appropriate to emphasize the dependence of $G_t(\cdot)$ on the history ε^t, we retain the functional notation and write

$$G_t(s) = \mathcal{G}(\varepsilon^t; s).$$

Hereafter we assume that, for any fixed history ε^t,

$$\lim_{s\to\infty} G_t(s) = G_t(\infty)$$

exists and that $G_t'(s)$ and $G_t''(s)$ are both continuous, integrable functions on $[0, \infty)$. It should be noted that[9]

$$G_t(0) = E_t. \tag{9.12}$$

For our discussion of wave propagation, it is appropriate to impose certain curvature conditions on the stress functional $\mathcal{S}$ which are expected to hold for most viscoelastic materials. In particular, it will be assumed that for all $\varepsilon \in (0, 1)$ and all $s \in [0, \infty)$

$$\begin{gathered} \sigma_{\mathrm{I}}(\varepsilon) > \sigma_{\mathrm{E}}(\varepsilon) > 0, \\ E_{\mathrm{I}}(\varepsilon) > E_{\mathrm{E}}(\varepsilon) > 0, \\ \tilde{E}_{\mathrm{I}}(\varepsilon) > 0, \quad \tilde{E}_{\mathrm{E}}(\varepsilon) > 0, \end{gathered} \tag{9.13}$$

and

$$\mathcal{G}(\varepsilon^t; s) > 0, \quad \mathcal{G}'(\varepsilon^t; s) \leqq 0. \tag{9.14}$$

The inequalities (9.13) imply that the instantaneous and equilibrium stress-strain curves are strictly convex from below and that the instantaneous curve lies everywhere above the equlibrium curve. These curves are illustrated in Fig. 16, where it is assumed that $\varepsilon = 0$ corresponds to a natural equilibrium state, i.e., $\sigma_{\mathrm{I}}(0) = \sigma_{\mathrm{E}}(0) = 0$. The inequalities (9.14) assert that for all strain histories on (0,1) the stress relaxation function is positive and a monotone decreasing function of the elapsed time s.[10] BOWEN and CHEN[11] have shown that the inequality $(9.14)_2$, evaluated for $\varepsilon_t = \varepsilon_{\mathrm{E}}$ and $s = 0$, follows as a consequence of the second law of thermodynamics. The relation of this inequality to internal dissipation, in the context of linear viscoelasticity, has been discussed by GURTIN and HERRERA.[12]

[8] Cf. COLEMAN and NOLL [1961, *1*].

[9] See, for example, CHEN [1973, *4*, Sect. 12].

[10] The inequality $(9.14)_1$, along with (9.12), will insure that all wave velocities are real.

[11] BOWEN and CHEN [1973, *3*]. Also see Sect. 24 of this article.

[12] GURTIN and HERRERA [1965, *13*]. This inequality can also be related to work done by linear viscoelastic materials on certain strain paths; cf. DAY [1970, *9*].

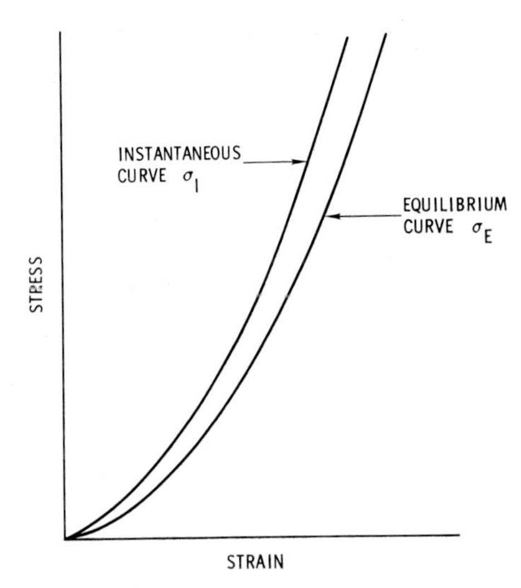

Fig. 16. Instantaneous and equilibrium stress-strain curves.

IV. Propagation of acoustic waves.

10. Infinitesimal sinusoidal progressive waves. Before turning to the study of waves which can be generated by impact loading conditions, we consider the propagation of infinitesimal progressive (acoustic) waves in nonlinear viscoelastic solids. The study of such waves provides important information with regard to the stress relaxation behavior of the material and, in fact, provides a method by which one can evaluate experimentally the longitudinal stress relaxation function corresponding to an equilibrium history. Moreover, as we shall see in a later section, the propagation of infinitesimal progressive waves can be related to the propagation of one-dimensional acceleration waves in the same material.

Here, following the analysis of COLEMAN and GURTIN,[1] we consider a simple material with memory characterized by (9.1) and suppose that the material is subjected to a small deformation w superimposed on a homogeneously deformed reference configuration. Then the motion is of the form

$$\chi(X, t-s) = (1-\mathring{\varepsilon})\, X + w(X, t-s), \quad s \in [0, \infty), \tag{10.1}$$

and the strain is given by

$$\varepsilon(X, t-s) = \mathring{\varepsilon} - \frac{\partial}{\partial X}\, w(X, t-s) \tag{10.2}$$

where $\mathring{\varepsilon}$ is the constant strain corresponding to the underlying deformed configuration. We identify the gradient in (10.2) as the small relative strain γ^t, and thus it follows from (7.8) and (9.11) that

$$\begin{aligned} \varrho_R\, \ddot{w}(X, t) &= G(0)\, \frac{\partial^2}{\partial X^2}\, w(X, t) \\ &\quad + \int_0^\infty G'(s)\, \frac{\partial^2}{\partial X^2}\, w(X, t-s)\, ds + o\left(\left\| \frac{\partial^2 w^t}{\partial X^2} \right\|\right). \end{aligned} \tag{10.3}$$

[1] COLEMAN and GURTIN [1965, 5]. The propagation of small-amplitude sinusoidal waves has also been studied by COLEMAN and GURTIN [1968, 3] for incompressible viscoelastic fluids undergoing steady shear.

Here $G(\cdot)$ is the stress relaxation function corresponding to the underlying equilibrium history $\mathring{\varepsilon}_E(\cdot)=\mathring{\varepsilon}$, i.e., $G(s)=\mathcal{G}(\mathring{\varepsilon}_E;s)$. By neglecting the remainder term in (10.3), we have the one-dimensional dynamical theory of linear viscoelasticity.[2]

We now seek the limitations which the dynamical theory of linear viscoelasticity imposes on a *damped infinitesimal sinusoidal progressive wave* given by

$$\begin{aligned} w(X,t) &= A\exp(-\alpha X)\cos\omega\left(t-\frac{X}{c}\right) \\ &= A\,\mathcal{R}e\left\{\exp\left[-\left(\alpha+\frac{i\omega}{c}\right)X\right]\exp(i\omega t)\right\} \end{aligned} \tag{10.4}$$

with *frequency* $\omega>0$, *attenuation* $\alpha>0$, *phase velocity* $c>0$, and *amplitude* A, all real numbers. The wave is called "infinitesimal" because it can be dynamically admissible in a general simple material with memory only if the remainder term in (10.3) is neglected. Clearly, (10.4) satisfies (10.3) if and only if

$$-\varrho_R\omega^2=\left(\alpha+\frac{i\omega}{c}\right)^2\{G(0)+\bar{G}'(\omega)\} \tag{10.5}$$

where $\bar{G}'(\omega)$ is the Fourier transform

$$\bar{G}'(\omega)=\int_0^\infty G'(s)\exp(-i\omega s)\,ds. \tag{10.6}$$

For each ω there is exactly one set of values $c=c(\omega)$, $\alpha=\alpha(\omega)$ satisfying (10.5):[3]

$$c^2(\omega)=\frac{1}{\varrho_R}|G(0)+\bar{G}'(\omega)|\sec^2\varphi(\omega),\qquad \alpha(\omega)=\frac{\omega}{c(\omega)}\tan\varphi(\omega) \tag{10.7}$$

where $0\leq\varphi<\pi/4$ and

$$\tan 2\varphi(\omega)=\frac{\mathcal{I}m\{G(0)+\bar{G}'(\omega)\}}{\mathcal{R}e\{G(0)+\bar{G}'(\omega)\}}. \tag{10.8}$$

By making use of some well-known theorems on Fourier transforms, one can easily show that Eqs. (10.7) imply that the limits

$$c_\infty=\lim_{\omega\to\infty}c(\omega),\qquad \alpha_\infty=\lim_{\omega\to\infty}\alpha(\omega), \tag{10.9}$$

which are called the *ultrasonic velocity* and *ultrasonic attenuation*, respectively, exist and are given by[4]

$$\begin{aligned} c_\infty^2 &= \frac{G(0)}{\varrho_R}, \\ c_\infty\,\alpha_\infty &= -\frac{G'(0)}{2G(0)}. \end{aligned} \tag{10.10}$$

The relation between the ultrasonic limits c_∞ and α_∞ and the material constants which govern the behavior of acceleration waves in homogeneously deformed regions will be considered in Sect. 20.

---

[2] Cf. LEITMAN and FISHER [1973, *10*, Chap. D] who also consider the propagation of sinusoidal progressive waves in three-dimensions.

[3] These equations have appeared extensively in the literature; see, for example, the survey by HUNTER [1960, *5*] and COLEMAN and GURTIN [1965, *5*]. Special forms of these results were derived earlier by HILLIER [1949, *1*] and SIPS [1951, *1*]. With the proper reinterpretation of the moduli, these results can also be applied to the analysis of stress pulse attenuation in linear viscoelastic rods; see, for example, KOLSKY [1965, *16*], STEVENS and MALVERN [1970, *21*], and WALSH [1971, *8*].

[4] This limiting behavior of the phase velocity was noted also by SIPS [1951, *1*] and HUNTER [1960, *5*]. Also see COLEMAN and GURTIN [1965, *5*] and LEITMAN and FISHER [1973, *10*, Sect. 43].

11. Determination of the stress relaxation function from acoustic wave experiments. It has long been recognized that, in principle, the stress relaxation function $G(\cdot)$ for a viscoelastic solid could be evaluated from experimental measurements of the frequency dependence of the phase velocity and attenuation of acoustic waves.[5] The application of this procedure, however, has been limited by the fact that good acoustic dispersion data have been available for only a very limited range of frequencies. Consequently, the corresponding relaxation function would be valid for only a very small time range and would have limited applicability. However, by noting that for many polymers the effects of time and temperature on the relaxation function superimpose,[6] one can overcome this difficulty and determine the velocity and attenuation of acoustic waves over a much greater range of frequencies than previously possible. Such results would then offer the opportunity to determine the stress relaxation function for a polymer which would be appropriate for a much wider time range.

There are, of course, several different experimental techniques which can be employed to obtain acoustic dispersion data for solid polymers. However, rather than give a detailed exposition of all such techniques,[7] we shall describe in this section a pulse-transmission method which has proven to give good results and which has been used by SUTHERLAND and LINGLE[8] in conjunction with time-temperature superposition to obtain dispersion data for the polymer, polymethyl methacrylate. Then, with these data, we follow the procedure outlined by NUNZIATO and SUTHERLAND[9] and evaluate the longitudinal stress relaxation function

$$G_0(s) = \mathscr{G}(0_{\mathrm{E}};\, s)$$

for the polymer which corresponds to the unstrained equilibrium reference configuration.[10]

Of special interest here is the form of the stress relaxation function which is appropriate for describing the one-dimensional shock wave response of viscoelastic materials. If the stress relaxation function $G_0(\cdot)$ is known for a polymer over a wide range of times, then by specifying the time-scale of a given experiment one can determine the stress relaxation function appropriate for that experiment. Here we show that by specifying the time-scale characteristic of plate impact experiments we can deduce the relaxation function appropriate for describing the wave propagational response of polymethyl methacrylate.[11] This result will prove to be especially useful in formulating a specific constitutive equation for this solid which is suitable for wave propagation studies.

As already indicated, the experimental determination of the phase velocity and attenuation over an extended range of frequencies employs a time-temperature

---

[5] See, for example, HILLIER [1949, *1*], GOTTENBERG and CHRISTENSEN [1964, *11*], KOLSKY [1965, *16*], MARVIN and MCKINNEY [1965, *17*], and FERRY [1970, *11*].

[6] The concept of time-temperature equivalence was originally proposed by LEADERMAN [1943, *1*] and subsequently introduced in a slightly different form by FERRY [1950, *1*]. Materials obeying time-temperature superposition are said to be "thermorheologically simple"; cf. SCHWARZL and STAVERMAN [1952, *1*]. A further discussion of linear viscoelastic materials of this type is contained in the expository article by STERNBERG [1964, *15*].

[7] There are several monographs on acoustics and experimental methods; see, for example, BARONE [1962, *2*] and TRUELL, ELBAUM, and CHICK [1969, *3*]. In this regard, also see MARVIN and MCKINNEY [1965, *17*] and FERRY [1970, *11*].

[8] SUTHERLAND and LINGLE [1972, *18*].

[9] NUNZIATO and SUTHERLAND [1973, *14*].

[10] Hereafter we denote the zero strain history $\varepsilon^t = \mathring{\varepsilon}_{\mathrm{E}} = 0$ by 0_{E}, for all $s \in [0, \infty)$.

[11] Cf. NUNZIATO and SUTHERLAND [1973, *14*].

superposition technique. This technique is based upon the observation that a change in the absolute temperature $\theta > 0$ shifts the time scale of the stress relaxation function by a factor $a(\theta)$:

$$G_0(s)\big|_{\theta=\theta_1} = G_0\left(\frac{s}{a(\theta)}\right)\bigg|_{\theta=\theta_0}, \tag{11.1}$$

for any two temperatures θ_0, θ_1. The shift factor $a(\theta)$ is related to the apparent activation energy of the polymer and depends strongly on the temperature, especially near the glass transition temperature θ_g.[12] By use of (11.1) and the results from Sect. 10, the effects of timetmperature superposition on the expressions for the phase velocity and attenuation are

$$\begin{aligned} c(\omega)\big|_{\theta=\theta_1} &= c(\omega\, a(\theta))\big|_{\theta=\theta_0}, \\ \alpha(\omega)\big|_{\theta=\theta_1} &= a(\theta)\, \alpha(\omega\, a(\theta))\big|_{\theta=\theta_0}. \end{aligned} \tag{11.2}$$

Thus, if the velocity and attenuation are measured experimentally as functions of temperature at some frequency, they are known for all frequencies within the range of applicability of (11.1).

To carry out this procedure experimentally, SUTHERLAND and LINGLE tested samples in a controlled temperature chamber with a temperature stability of $\pm 1\,^\circ$K. To ensure temperature stability within the sample, a 2 to 4 h wait period after a temperature change was used. Dispersion data were measured by using a modified version of the pulse-transmission technique introduced by MATTABONI and SCHREIBER.[13] Where MATTABONI and SCHREIBER obtained transit-time measurements by comparing the signal transmitted through a specimen to the signal used to drive the transmitting transducer, SUTHERLAND and LINGLE compared signals that were transmitted through samples of different lengths. This modification is advantageous in that both signals traverse the same electrical and acoustic path, except for the difference in length between the specimens. Thus, transit-time and attenuation measurements do not need to be corrected for different electrical paths.

Fig. 17 is a schematic diagram of the experimental configuration used. The output of a variable-frequency oscillator is divided by the binary divider to some submultiple and then used to supply a synchronized trigger to the pulse oscillator and the oscilloscope. The output of the pulse oscillator excites two identical transducers which are mounted to samples of the same material but of different lengths. After traversing the samples, the acoustic waves were monitored by a second set of transducers. The received signals are displayed on the oscilloscope with the output of the variable-frequency oscillator.

---

[12] For θ near θ_g the temperature dependence of $a(\theta)$ is fairly well represented for amorphous polymers by the Williams-Landel-Ferry formula

$$\log_{10} a(\theta) = -\frac{a_1(\theta - \theta_g)}{a_2 + \theta + \theta_g}$$

where a_1 and a_2 are empirical constants. However, for $\theta < \theta_g$, the temperature dependence of $a(\theta)$ is better represented by an equation of the simple Arrhenius form

$$\log_{10} a(\theta) = a_e\left(\frac{1}{\theta} - \frac{1}{\theta_0}\right).$$

Here the constant a_e is proportional to the apparent activation energy and $\theta_0 < \theta_g$ is some reference temperature. For a further discussion of the properties of $a(\theta)$, see FERRY [1970, *11*].

[13] MATTABONI and SCHREIBER [1967, *7*].

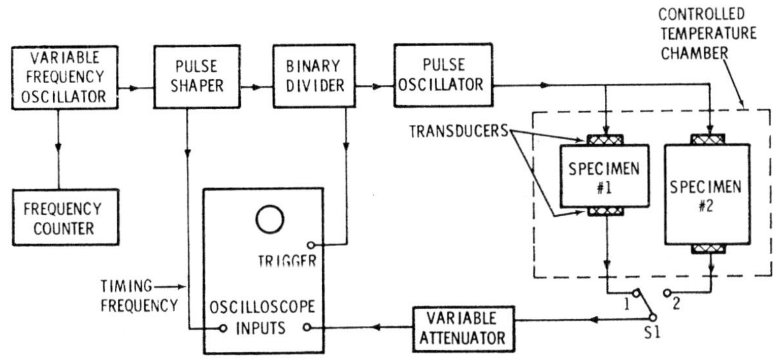

Fig. 17. Instrumentation schematic for acoustic measurements. (From SUTHERLAND and LINGLE [1972, *18*].)

The phase velocity c of the acoustic wave for any frequency ω is determined by

$$c = \frac{l_2 - l_1}{t_2 - t_1} = \frac{\Delta l}{\Delta t},$$

where t_i and l_i are the transit time and length of specimen i ($i = 1, 2$) corresponding to the temperature θ. The time Δt is measured directly by setting the period of the timing signal equal to an exact submultiple of the time delay to be measured and then using a frequency counter to determine the frequency of the timing signal. In this way, transit-time measurements are not restricted to the initial rise but may be made between any corresponding portion of the two transmitted waves. The errors in this technique, based on velocity measurements through several specimen thicknesses, are less than 1%.

Attenuation measurements were made by comparing the relative magnitudes of the signals transmitted through the two specimens. After the transmitted signal through one specimen is set to some selected level (as viewed on the oscilloscope), the signal transmitted through the other specimen is displayed, and a variable attenuator is used to attenuate this signal to the same level as the signal from the first specimen. The added attenuation in the electrical circuit, α_A, equals the difference in acoustic attenuation in dB between the two specimens. Thus,

$$\alpha = \frac{\alpha_A}{l_2 - l_1}.$$

The errors of this technique were estimated to be less than 10%.

Using this experimental method, SUTHERLAND and LINGLE[14] measured the phase velocity and attenuation of acoustic waves in polymethyl methacrylate at 1 MHz for a temperature range of 243 to 370 °K. These results, coupled with the relations (11.2) and the known dependence of the shift factor $a(\theta)$ on temperature,[15] serve to determine the phase velocity and attenuation curves shown in Figs. 18 and 19 corresponding to room temperature. The results obtained by this technique agree well with data on polymethyl methacrylate reported by other experimenters[16] for limited ranges of frequency.

[14] SUTHERLAND and LINGLE [1972, *18*].

[15] Since all measurements were made below the glass transition temperature of the polymer (for polymethyl methacrylate, $\theta_g = 378$ °K), the simple Arrhenius form for $a(\theta)$ was used with $a_e = -14.65 \times 10^3$ °K, $\theta_0 = 295$ °K.

[16] ASAY, DORR, ARNOLD, and GUENTHER [1965, *2*], ASAY, LAMBERSON, and GUENTHER [1969, *1*], ROMBERGER [1970, *18*], and R. LINGLE (unpublished).

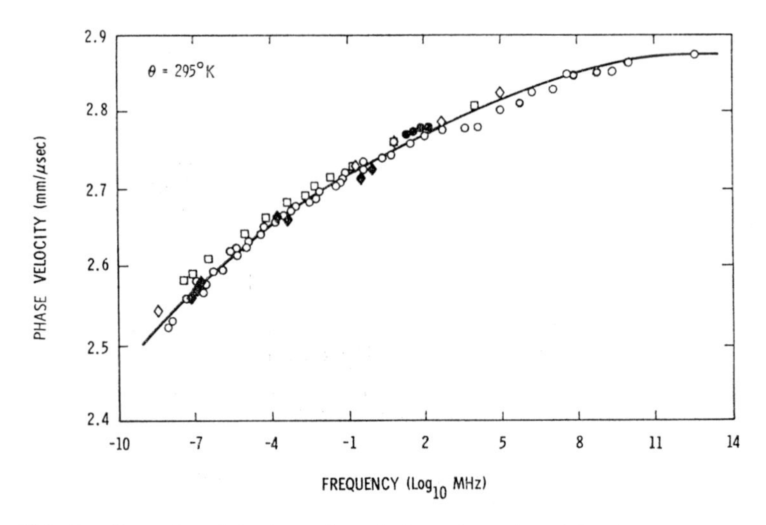

Fig. 18. Velocity dispersion data for polymethyl methacrylate. The data were obtained by ASAY, DORR, ARNOLD, and GUENTHER [1965, *2*] (◆), ASAY, LAMBERSON, and GUENTHER [1969, *1*] (●), ROMBERGER [1970, *18*] (◇), R. LINGLE (unpublished) (□), and SUTHERLAND and LINGLE [1972, *18*] (○).

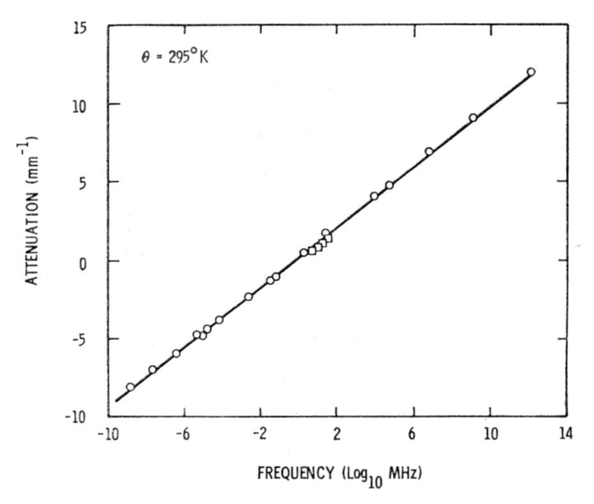

Fig. 19. Attenuation dispersion data for polymethyl methacrylate. The data were obtained by ASAY, LAMBERSON, and GUENTHER [1969, *1*] (□) and SUTHERLAND and LINGLE [1972, *18*] (○).

Clearly then, with the functions $c(\omega)$ and $\alpha(\omega)$ known, we can return to the results of Sect. 10 and evaluate the stress relaxation function $G_0(\cdot)$ corresponding to the underlying history 0_E. To this end, (10.7), (10.8), $(10.10)_1$, and the standard cosine inversion of a Fourier transform combine to yield

$$G_0(s) = \frac{2}{\pi}\int_0^s\int_0^\infty \left(\frac{\varrho_R c^2(\omega)}{\sec 2\varphi(\omega)\left\{\left[\frac{\alpha(\omega)\, c(\omega)}{\omega}\right]^2+1\right\}} - \varrho_R c_\infty^2\right)\cos(\omega s)\, d\omega\, ds + \varrho_R c_\infty^2 \tag{11.3}$$

where the phase angle $\varphi(\omega)$ is given by

$$\varphi(\omega)=\tan^{-1}\left[\frac{\alpha(\omega)\,c(\omega)}{\omega}\right]. \tag{11.4}$$

To evaluate the stress relaxation function from the experimental acoustic data, the integrand

$$\mathscr{I}(\omega)=\frac{\varrho_R c^2(\omega)}{\sec 2\varphi(\omega)\left\{\left[\frac{\alpha(\omega)\,c(\omega)}{\omega}\right]^2+1\right\}}-\varrho_R c_\infty^2$$

of (11.3) is evaluated pointwise and represented by the series[17]

$$\mathscr{I}(\omega)=\sum_{i=1}^{20} f_i \exp\left(-\frac{\omega}{W_i}\right). \tag{11.5}$$

The high-frequency limit $c_\infty=2.87$ mm/μsec was calculated by a method suggested by WALSH[18] in which one inspects the limiting behavior of the group velocity. Then using (11.5), the integration in Eq. (11.3) can be carried out to yield

$$G_0(s)=\frac{2}{\pi}\sum_{i=1}^{20} f_i \tan^{-1}(W_i s)+\varrho_R c_\infty^2. \tag{11.6}$$

Using this method, NUNZIATO and SUTHERLAND obtained the stress relaxation function for polymethyl methacrylate shown in Fig. 20. Although this relaxation function is shown for a time range of approximately 10^{-6} to 10^8 μsec, it should be realized that this result is only valid over times within the range of applicability of time-temperature superposition (cf. (11.1)). At very short times, i.e., times less than 10^{-6} μsec, where the molecular motions in the polymer occur within the time scale of acoustic experiments, the predictions of time-temperature superposition cannot be used.

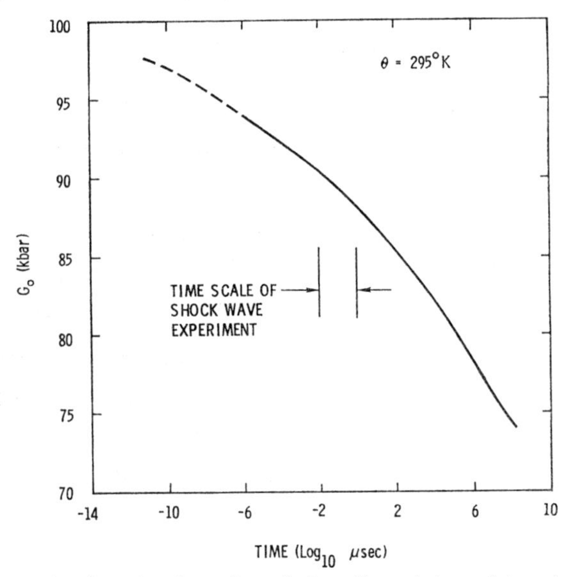

Fig. 20. Stress relaxation function for polymethyl methacrylate as determined from acoustic dispersion data. (From NUNZIATO and SUTHERLAND [1973, *14*].)

[17] A similar representation was suggested by SCHAPERY [1962, *8*]. We omit giving all the coefficients f_i and W_i in the present case.

[18] WALSH [1971, *8*].

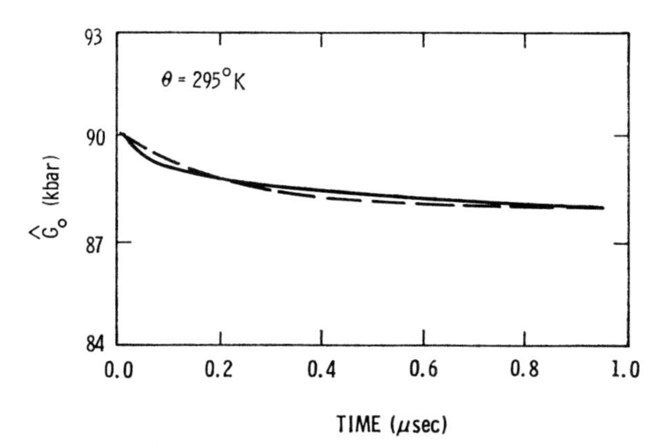

Fig. 21. Stress relaxation function for polymethyl methacrylate for the time range appropriate for shock wave studies. The dashed line is the single-exponential representation of the relaxation function.

To deduce the stress relaxation function appropriate for describing the response of a polymer in wave propagation experiments, we need to specify the time scale of the experiments. In the case of plate impact experiments, the characteristic time scale is 10^{-2} to 1 μsec, i.e., approximately the lower limit of experimental resolution to the time at which apparent equilibrium is achieved for wave profiles of infinite duration. Over this time scale, we denote the relaxation function by $\hat{G}_0(\cdot)$ where $\hat{G}_0(0)$ is equivalent to $G_0(10^{-2}\ \mu\text{sec})$ and $\hat{G}_0(\infty)$ is equivalent to $G_0(1\ \mu\text{sec})$. Then, from (11.6), the function $\hat{G}_0(\cdot)$ shown in Fig. 21 is obtained, in which

$$\hat{G}_0(0) = 90.1 \text{ kbars}, \qquad \hat{G}_0(\infty) = 88.0 \text{ kbars}. \tag{11.7}$$

The characteristic relaxation time τ is evaluated by the formula

$$\hat{G}_0(s)\big|_{s=\tau} = \frac{\hat{G}_0(0) - \hat{G}_0(\infty)}{e} + \hat{G}_0(\infty),$$

where e is the Naperian base. By use of (11.7), it follows that

$$\tau = 0.22\ \mu\text{sec}.$$

It is of particular interest to note that the relaxation function $\hat{G}_0(\cdot)$ can be well represented by the single exponential relation

$$\hat{G}_0(s) = [\hat{G}_0(0) - \hat{G}_0(\infty)] \exp\left(-\frac{s}{\tau}\right) + \hat{G}_0(\infty) \tag{11.8}$$

even though three decades of time (10^{-2} to 1 μsec) are involved (cf. Fig. 21).

Since the remainder of this article deals primarily with the propagation of waves generated by impact loading, we will drop the "^" notation on the stress relaxation function $\hat{G}_0(\cdot)$, recognizing, of course, that the interval $[0, \infty)$ of the argument corresponds to the characteristic time-scale of impact experiments.

V. Propagation of one-dimensional steady waves.

12. The existence of steady waves. In a series of papers, COLEMAN, GURTIN, HERRERA, and CHEN[1] have studied the propagation of shock and acceleration

[1] COLEMAN, GURTIN, and HERRERA [1965, *4*], COLEMAN and GURTIN [1965, *5*], CHEN and GURTIN [1970, *4*]. Also see CHEN [1973, *4*].

waves in general nonlinear materials with fading memory. Using the theory of singular surfaces, they obtained definitive results with regard to the qualitative behavior of the waves. However, their results are based on the assumption that solutions containing shock and acceleration discontinuities exist in such materials. For many years, it was believed that such surfaces of discontinuity were not possible in materials with internal (viscous) dissipation and that such discontinuities were replaced by sudden but smooth transitions. However, subsequent theoretical investigations by SIPS, PIPKIN, GREENBERG, and others [2] and the experimental studies of HALPIN and GRAHAM, BARKER and HOLLENBACH, and SCHULER [3] have shown that this notion is incorrect.

The existence of discontinuous solutions in nonlinear materials with memory was first demonstrated by PIPKIN. Using the specific constitutive assumption of finite linear viscoelasticity formulated by COLEMAN and NOLL,[4] PIPKIN obtained exact solutions to the one-dimensional steady field equation which exhibited smooth structured waves, acceleration waves, and shock waves. Subsequently, GREENBERG showed that steady shock wave solutions may exist for a large class of nonlinear materials with fading memory.[5] The theory of steady wave propagation in such dissipative materials is extremely important to the present discussion and it is worthwhile to consider the results of PIPKIN and GREENBERG in some detail.

A compressive motion is called *steady* if there exist functions $\hat{u}$, $\hat{v}$, and $\hat{\varepsilon}$ such that

$$u(X,t)=\hat{u}(x),$$
$$v(X,t)=\hat{v}(x)>0,$$
$$\varepsilon(X,t)=\hat{\varepsilon}(x)>0$$

where $x=\chi(X,t)$. Then from the kinematical definitions (7.2) to (7.4), it follows that

$$\hat{u}'(x)=-\frac{\hat{\varepsilon}(x)}{1-\hat{\varepsilon}(x)}.$$

[2] SIPS [1951, *1*] was the first to exhibit an explicit solution of the dynamical equations for linear viscoelastic solids showing a shock wave. Numerous other papers have dealt with solutions of dynamical linear viscoelasticity which contain shock and acceleration waves; see, for example, LEE and KANTER [1953, *1*], GLAUZ and LEE [1954, *1*], BERRY and HUNTER [1956, *1*], LEE and MORRISON [1956, *2*], BLAND [1957, *1*], LEE [1957, *4*], MORRISON [1957, *5*], DATTA [1959, *1*], ACHENBACH and CHAO [1962, *1*], CHU [1962, *3*], and VALANIS [1965, *21*], [1967, *11*].

PIPKIN [1966, *12*], and GREENBERG [1967, *3*] considered steady shock wave solutions in the nonlinear theory.

[3] HALPIN and GRAHAM [1965, *14*], BARKER and HOLLENBACH [1970, *2*] and SCHULER [1970, *19*] have observed shock discontinuities in the solid polymer, polymethyl methacrylate. HALPIN and GRAHAM used the quartz gauge instrumentation technique; while BARKER, HOLLENBACH, and SCHULER used laser interferometry.

[4] COLEMAN and NOLL [1961, *1*].

[5] GREENBERG [1967, *3*]. SKOBEEV [1967, *9*] and GREENBERG [1968, *9*] also established similar results for another class of viscoelastic materials, called *Maxwellian materials*, which in general may not be completely included in the constitutive assumption (9.1). Maxwellian materials obey the constitutive relation

$$\dot{\sigma}=\mathscr{E}(\varepsilon,\sigma)\,\dot{\varepsilon}+\mathscr{H}(\varepsilon,\sigma);$$

cf. SOKOLOVSKY [1948, *2*], MALVERN [1950, *2*], NOLL [1955, *2*], CRISTESCU [1964, *7*], LUBLINER [1964, *14*,], and DUNWOODY [1966, *8*]. A special case of this constitutive relation was first suggested by MAXWELL [1867, *1*] in his discussion of the kinetic theory of gases and was generalized to three dimensions by ZAREMBA [1903, *1*].

AMES, DAVY, and CHAND [1973, *1*] have also studied wave propagation in a class of Maxwellian materials and found that the only similarity solutions admissible were steady wave solutions. For a related study, see SULICIU, LEE, and AMES [1973, *19*].

Since for compressive motions $\hat{\varepsilon}$ is restricted to the interval (0, 1), $\hat{u}'(x)<0$ exists and is continuous except possibly at a countable set of points where it may suffer jump discontinuities. The monotonicity of $\hat{u}(x)$ insures the existence of the inverse

$$x=\tilde{x}(u).$$

Defining

$$\tilde{\varepsilon}(u)=\hat{\varepsilon}(\tilde{x}(u)), \tag{12.1}$$

it follows from (7.4) that

$$X_2-X_1=-\int_{u(X_1,t)}^{u(X_2,t)} \frac{du}{\tilde{\varepsilon}(u)} \tag{12.2}$$

for every interval $[X_1, X_2]$ in R. Differentiating (12.2) with respect to time and using $(7.3)_2$, we see that $\hat{v}$ and $\hat{\varepsilon}$ are not independent but must satisfy

$$\frac{\hat{v}(x_1)}{\hat{\varepsilon}(x_1)}=\frac{\hat{v}(x_2)}{\hat{\varepsilon}(x_2)}>0. \tag{12.3}$$

This result implies the existence of a number $V>0$ such that

$$\frac{\hat{v}(x)}{V}=\hat{\varepsilon}(x) \tag{12.4}$$

for all $x\in R$. Eqs. (7.3) and (12.4) further assert that the steady displacement field $u(X, t)$ obeys the linear partial differential equation

$$\frac{1}{V}\frac{\partial}{\partial t}u(X, t)+\frac{\partial}{\partial X}u(X, t)=0. \tag{12.5}$$

Every solution of this equation is expressible in the form

$$u(X, t)=u(\xi) \tag{12.6}$$

where

$$\xi=t-\frac{X}{V}.$$

The number V is called the *steady wave velocity*.[6] In view of (12.6), $(7.3)_2$, and (7.4), the corresponding particle velocity and strain become

$$\begin{aligned} v(\xi)&=u'(\xi), \\ \varepsilon(\xi)&=\frac{1}{V}u'(\xi)=\frac{1}{V}v(\xi). \end{aligned} \tag{12.7}$$

A steady strain field implies, through the constitutive assumption (9.1), a steady stress field

$$\sigma(X, t)=\sigma(\xi).$$

[6] In GREENBERG's formulation of steady motions [1967, *3*], the motion is expressible as

$$x=\chi(\xi), \quad \xi=t+\frac{X}{V}.$$

This results in a standing wave with the flow field passing through the wave with a free stream velocity V. The formulation we present here, which results in the representation (12.6), gives a somewhat different interpretation of V. In our case, the wave will be propagating into a region at rest with the steady wave velocity V. Since this is the situation encountered in wave propagation experiments involving solid materials, the latter formulation will prove to be the more useful.

Since the displacement is steady, (12.6) asserts that

$$\frac{d}{dt}\int\limits_{X_1}^{X_2} \varrho_R v(X, t)\, dX = \varrho_R V[v(\xi_2) - v(\xi_1)]$$

where $\xi_i = t - \frac{X_i}{V}$ $(i=1, 2)$ and, therefore, the balance of linear momentum (7.7) reduces to

$$\sigma(\xi_2) - \sigma(\xi_1) = \varrho_R V[v(\xi_2) - v(\xi_1)].$$

Hereafter, we shall assume that a natural equilibrium state exists far ahead of the wave, i.e.,

$$\lim_{\xi\to-\infty} v(\xi) = 0, \quad \lim_{\xi\to-\infty} \varepsilon(\xi) = 0, \quad \lim_{\xi\to-\infty} \sigma(\xi) = \sigma_E(0) = 0.$$

Then, by using (12.7), we can express the balance of linear momentum as

$$\sigma(\xi) = \varrho_R V\, v(\xi) = \varrho_R V^2 \varepsilon(\xi), \tag{12.8}$$

which holds for all $\xi \in (-\infty, \infty)$. Eqs. (12.8) and (9.1) combine to yield the functional equation

$$\varrho_R V^2 \varepsilon(\xi) = \mathscr{S}\big(\varepsilon(\xi - s)\big) \tag{12.9}$$

for the steady strain field $\varepsilon(\xi)$.

Here we wish to consider the existence of solutions $\varepsilon(\xi)$, $\varepsilon \in (0, 1)$ of (12.9) which are monotone increasing functions of ξ. GREENBERG[7] has given sufficient conditions for the existence of such solutions which are in essence the curvature conditions (9.13) and (9.14) and the requirement that for all monotone increasing functions $\varepsilon(\xi)$

$$\sigma_I(\varepsilon) \geqq \mathscr{S}\big(\varepsilon(\xi - s)\big) \geqq \sigma_E(\varepsilon), \tag{12.10}$$

and the *instantaneous longitudinal sound velocity* C_I,

$$C_I^2 = \frac{1}{\varrho_R} D_\varepsilon \mathscr{S}\big(\varepsilon(\xi - s)\big), \tag{12.11}$$

is a monotone increasing function of ξ. This latter requirement on the instantaneous sound velocity is to be expected on physical grounds. Intuitively, a steady wave in a nonlinear material with memory results from a balance being achieved between the tendency of the wave to decay due to internal dissipation (as manifested by the negative function $\mathscr{G}'(\varepsilon(\xi - s); s)$) and to grow due to nonlinear stress-strain response. Hence, in order to have a steady wave, each part of the wave must tend to overtake the part ahead of it. This will result if the instantaneous sound velocity C_I satisfies the monotonicity requirement. The restriction (12.10) merely asserts that for any value of strain $\varepsilon(\xi)$ achieved in a steady wave the corresponding value of stress $\sigma(\xi)$ must lie between the instantaneous and equilibrium stress-strain curves.

Under these conditions, steady waves do indeed exist and there are three types of solutions to consider which depend on the steady wave velocity V:[8]

(i) $(C_I)_0 > V > (C_E)_0$,

(ii) $(C_I)_0 = V$,

(iii) $(C_I)_0 < V$

[7] GREENBERG [1967, *3*].

[8] Cf. PIPKIN [1966, *12*], GREENBERG [1967, *3*], [1968, *9*]. We refer the reader to these studies for the proof of the results that follow.

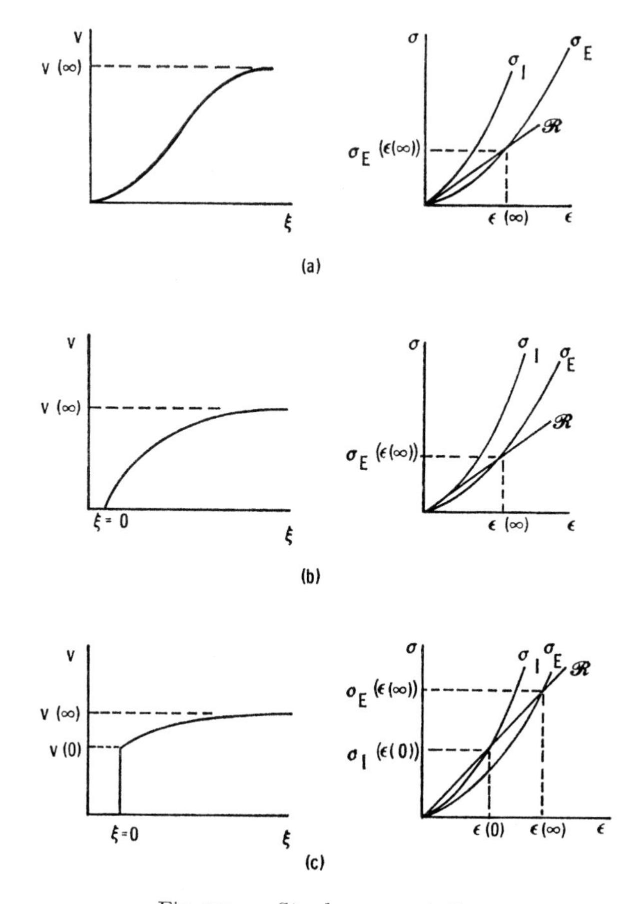

Fig. 22a–c. Steady wave solutions.

where $(C_\mathrm{I})_0$ and $(C_\mathrm{E})_0$ are the instantaneous and equilibrium sound velocities corresponding to the natural equilibrium state, i.e.,

$$(C_\mathrm{I})_0^2 = \frac{E_\mathrm{I}(0)}{\varrho_\mathrm{R}}, \qquad (C_\mathrm{E})_0^2 = \frac{E_\mathrm{E}(0)}{\varrho_\mathrm{R}}. \tag{12.12}$$

Now, it should be evident from (12.8) that each point in a steady wave lies on a secant (Rayleigh line) in the stress-strain plane connecting the initial and final states. The slope of the Rayleigh line $\mathscr{R}$ is $\varrho_\mathrm{R} V^2$. Thus, the type of solution depends upon the position of the Rayleigh line with respect to the instantaneous and equilibrium stress-strain curves, $\sigma_\mathrm{I}(\varepsilon)$ and $\sigma_\mathrm{E}(\varepsilon)$ (see Fig. 22). In case (i), the Rayleigh line lies below the instantaneous curve (Fig. 22a). Then there exists a solution satisfying the conditions far ahead of the wave and the solution is a smooth structured wave with the final equilibrium particle velocity $v(\infty)$ (and strain $\varepsilon(\infty)$) obeying

$$\varrho_\mathrm{R} V^2 \varepsilon(\infty) = \sigma_\mathrm{E}(\varepsilon(\infty)), \qquad v(\infty) = V\varepsilon(\infty). \tag{12.13}$$

Case (ii) corresponds to the Rayleigh line $\mathscr{R}$ tangent to the instantaneous curve (Fig. 22b). Thus, the solution is an acceleration wave with $\varepsilon(\infty)$, $v(\infty)$ satisfying

(12.13) and the steady wave velocity is the acceleration wave velocity $(C_{\text{I}})_0$.[9] Case (iii) is of particular importance in this article and corresponds to a Rayleigh line with a slope greater than the initial slope of the instantaneous curve (Fig. 22c). In this instance, the equilibrium conditions far ahead of the wave can be satisfied only if $v(\xi)\equiv 0$, $\varepsilon(\xi)\equiv 0$ for $\xi<0$ and there is a jump discontinuity in v (and ε) at $\xi=0$ with $v(0)$, $\varepsilon(0)$ satisfying

$$\varrho_{\text{R}} V^2 \varepsilon(0)=\sigma_{\text{I}}(\varepsilon(0)), \qquad v(0)=V\varepsilon(0). \tag{12.14}$$

This solution is a steady shock wave with the particle velocity (and strain) undergoing a smooth transition from $v(0)$ to $v(\infty)$ given by (12.13). Due to the monotonicity of the solution, $v(\infty)>v(0)$, $\varepsilon(\infty)>\varepsilon(0)$.

13. Experimental observation of steady shock waves. Using the plate impact configuration and the in-material, velocity interferometer instrumentation technique described in Chap. II (cf. Fig. 8), BARKER and HOLLENBACH[10] and SCHULER[11] have generated and observed steady shock waves in the solid polymer, polymethyl methacrylate. Their data are summarized in Table 1 and grouped according to nominal projectile velocities and increasing mirror distance.[12] In all shots except 320 and 321, the projectile noseplate was polymethyl methacrylate; hence, the resulting impact is symmetric. This symmetry implies that the boundary condition at the impact surface is one of constant velocity. Thus, in the absence of any unloading waves, the initial particle velocity jump at the impact surface as well as the maximum particle velocity behind the wave are one-half of the projectile velocity. Hereafter, these conditions of particle velocity at the impact surface will be referred to as impact conditions; the nominal projectile velocities will be used to designate various impact conditions. With the gun facility described in Sect. 4, the maximum projectile velocity is about 0.64 mm/μsec. This means that in a symmetric impact the maximum initial jump in particle velocity is 0.32 mm/μsec. To achieve a larger initial jump in particle velocity, a higher acoustic impedance projectile noseplate of fused silica was used on shots 320 and 321. The shots in the 300 series are pairwise duplicates of each other and serve to illustrate the reproducibility of the material.

In most of the experiments, the projectile noseplate, which was 6 mm thick, was mounted so that the face opposite the impact surface was a free surface. The compressive shock generated at impact reflects from this free surface as an unloading wave and propagates back into the projectile and target plates. In most cases, this unloading wave did not overtake the compressive wave front. However, at the higher projectile velocities and larger mirror distances, this did occur (as evidenced by the fact that the maximum particle velocity measured by the interferometer was less than one-half the projectile velocity, the value commensurate with a symmetric impact).

The tabulated shock velocities were obtained by dividing the distance between the impact surface and the plane of the mirror by the transit time of the wave as determined by the optical technique described in Sect. 6. Since the shock strength decays with distance from the impact surface, the shock velocity also decreases. Thus, the shock velocity given in Table 1 represents an average. However, only

[9] As we shall see in Sect. 20, the instantaneous sound velocity C_{I} is precisely the intrinsic velocity U of an acceleration wave.

[10] BARKER and HOLLENBACH [1970, *2*].

[11] SCHULER [1970, *19*].

[12] The data associated with shots numbered 312 to 322 are due to BARKER and HOLLENBACH. The remainder are due to SCHULER.

Table 1. *Shock wave measurements in polymethyl methacrylate.*

Nominal impact velocity (mm/μsec)	Shot number	Actual impact velocity v_m (mm/μsec)	Mirror distance X_m (mm)	Shock velocity (mm/μsec)	Unloading wave velocity (mm/μsec)	Particle velocity jump at shock v_0 (mm/μsec)	Maximum particle velocity $v_\infty = v_m/2$ (mm/μsec)
0.06	312	0.0609	6.349	2.843	2.97	0.0245	0.0305
	313	0.0613	6.284	2.844	2.96	0.0245	0.0306
	2109	0.0617	19.163	2.838	2.97	0.0240	0.0308
	2105	0.0617	37.452	2.834	2.95	0.0190	0.0306
0.15	314	0.1511	6.185	2.968	3.21	0.062	0.0756
	315	0.1516	6.388	2.959	3.22	0.065	0.0758
	2110	0.1511	12.824	2.974	3.22	0.061	0.0756
	2112	0.1547	12.898	2.958	3.22	0.057	0.0773
	2115	0.1514	37.539	2.991	3.23	0.060	0.0757
0.22	2119	0.2236	36.504	3.032	3.34	0.091	0.1118
0.30	316	0.3085	6.208	3.127	3.53	0.126	0.1542
	317	0.3090	6.160	3.130	3.54	0.127	0.1545
	2113	0.2999	25.252	3.113	3.51	0.122	0.1499
	2116	0.2985	37.283	3.106	3.49	0.122	0.1492
0.46	2106 *R*	0.4501	6.359	3.199	3.78	0.160	0.2250
	2107 *R*1	0.4604	25.314	3.178	3.76	0.150	0.2301
	2107 *R*2	0.4616	37.903	3.181	3.73	0.152	0.2307
0.64	318	0.6412	6.045	3.268	4.20	0.210	0.3206
	319	0.6391	6.116	3.268	4.18	0.205	0.3159
	2104	0.6431	18.782	3.249	—	0.175	0.3216
	2108	0.6217	25.189	3.199	4.07	0.162	0.3108
	2111	0.6165	37.490	3.203	—	0.161	0.3082
0.64	320	0.6092	6.269	3.349	4.84	0.300	—
	321	0.6147	6.350	3.342	4.87	0.315	—

at the 0.64 mm/μsec impact condition is a decrease in the average shock velocity clearly evident, and here the decrease between measurements at 6 and 37 mm from the impact surface amounts to 2%. This is consistent with the measurements of particle velocities which indicate that most of the shock decay occurs soon after impact in the region adjacent to the impact surface. At other impact conditions, only a slight decreasing trend of shock velocities is evident.

The column headed v_0 lists the jump in particle velocity at the shock.[13] The values obtained for these velocities are somewhat subjective, in that they resulted from plotting the velocity data points, drawing a smooth curve through them, and then reading the value of particle velocity at the shock arrival time. The accuracy of this procedure is estimated at $\pm 5\%$.

The measured particle velocity histories corresponding to the impact conditions of 0.06, 0.15, and 0.30 mm/μsec are shown in Fig. 23. The records of the 0.06 mm/μsec impact condition show a noticeable decay in the strength of the shock front and an accompanying smoothing of the wave shape with increasing distance of propagation. For the 0.15 mm/μsec impact condition, the particle velocity histories measured at nominally 6, 12, and 37 mm from the impact surface are nearly identical. Thus, it appears that within 6 mm from the impact surface, the wave generated at this impact condition becomes steady. The particle velocity records for the 0.30 mm/μsec impact condition also indicate that a steady wave profile has been achieved.

[13] The values v_0 and v_∞ correspond to $v(0)$ and $v(\infty)$, respectively, in Fig. 22.

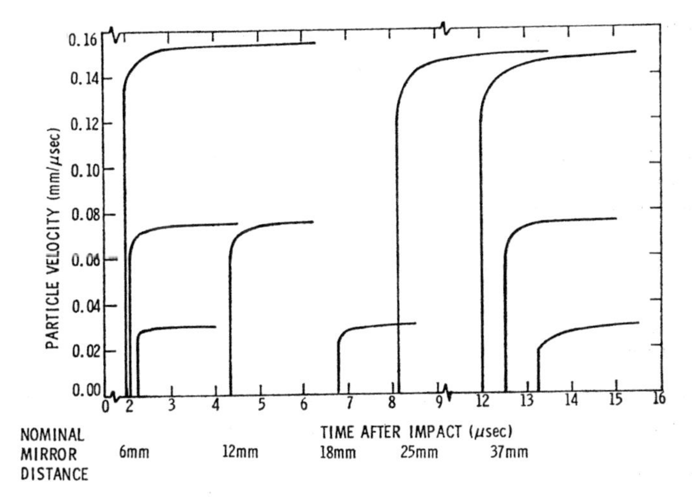

Fig. 23. Measured particle velocity histories for polymethyl methacrylate. (From BARKER and HOLLENBACH [1970, *2*], and SCHULER [1970, *19*].)

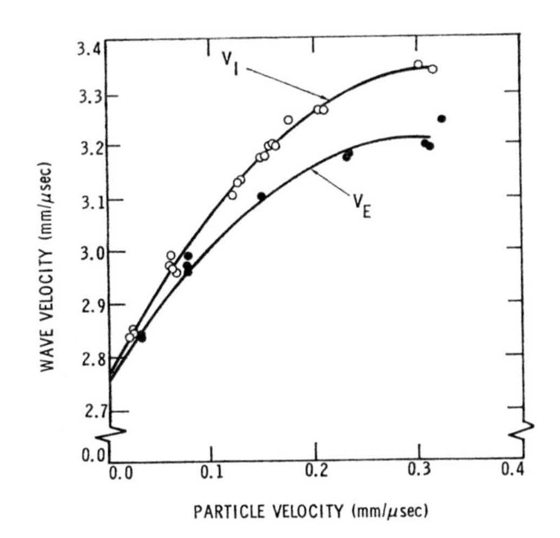

Fig. 24. Measured wave velocity vs. particle velocity for polymethyl methacrylate. (From SCHULER, NUNZIATO, and WALSH [1973, *18*].)

Now, from the analysis of steady waves, it is evident that measurement of the steady wave velocity V and the particle velocity jump at the front v_0 for various impact levels serves to determine the instantaneous response of the material, while measurement of V and the maximum equilibrium particle velocity v_∞ achieved for various impact levels serves to determine the equilibrium response of the material. Thus, from a series of steady wave experiments, two curves are generated in the wave velocity–particle velocity plane, "$V-v_0$" and "$V-v_\infty$." Consequently, for every value of particle velocity v, we can identify two wave velocities, $V_{\mathrm{I}}(v)$ and $V_{\mathrm{E}}(v)$. These curves may be represented by least squares, polynomial fits of the experimental data. In the case of polymethyl methacrylate[14]

$$V_{\mathrm{I}}(v)=(C_{\mathrm{I}})_0+b_{\mathrm{I}}v+a_{\mathrm{I}}v^2,\qquad V_{\mathrm{E}}(v)=(C_{\mathrm{E}})_0+b_{\mathrm{E}}v+a_{\mathrm{E}}v^2. \tag{13.1}$$

Fig. 24 shows the experimental data and the fits (13.1). The values of the instantaneous and equilibrium longitudinal sound velocities $(C_{\mathrm{I}})_0$ and $(C_{\mathrm{E}})_0$ in the fits (13.1) are related to the stress relaxation function $G_0(s)$ by

$$(C_{\mathrm{I}})_0^2=\frac{G_0(0)}{\varrho_{\mathrm{R}}},\qquad (C_{\mathrm{E}})_0^2=\frac{G_0(\infty)}{\varrho_{\mathrm{R}}}, \tag{13.2}$$

and they agree very well with the values determined from the acoustic dispersion data presented in Sect. 11.[15]

14. Finite linear viscoelasticity and the evaluation of material response functions. In order to analyze the steady wave profiles observed in polymethyl methacrylate, it is necessary to assume a specific constitutive model for the material. Following PIPKIN[16] and SCHULER,[17] we assume that the stress functional $\mathscr{S}$ can be represented by the one-dimensional counterpart of the constitutive equation of *finite linear viscoelasticity* developed by COLEMAN and NOLL:[18]

$$\begin{aligned}\sigma(t)&=\mathscr{S}(\varepsilon^t)\\&=\sigma_{\mathrm{I}}(\varepsilon(t))+\int_0^\infty K'(\varepsilon;s)\{1-(1-\varepsilon^t(s))^2\}\,ds\end{aligned} \tag{14.1}$$

for all compressive strain histories $\varepsilon^t(\cdot)$.[19] The relaxation function $K'(\cdot;\cdot)$ has the property

$$\lim_{s\to\infty}K'(\varepsilon,s)=0,\qquad \varepsilon\in(0,1).$$

Evaluating (14.1) for the equilibrium history ε_{E} shows that the instantaneous and equilibrium response functions are related by

$$\sigma_{\mathrm{I}}(\varepsilon)=\sigma_{\mathrm{E}}(\varepsilon)+[K(\varepsilon;0)-K(\varepsilon;\infty)]\{1-(1-\varepsilon)^2\}. \tag{14.2}$$

Now, from the measurements of the steady wave velocity and the particle velocity at the head and at the tail of the steady wave the instantaneous and equilibrium response functions, $\sigma_{\mathrm{I}}(\varepsilon)$ and $\sigma_{\mathrm{E}}(\varepsilon)$, can be uniquely determined. Combining (13.1) with $(12.13)_2$ and $(12.14)_2$, the velocities V_{I} and V_{E} can alternatively be expressed in terms of strain:

$$\begin{aligned}V_{\mathrm{I}}(\varepsilon)&=\frac{[1-b_{\mathrm{I}}\varepsilon]-\{[1-b_{\mathrm{I}}\varepsilon]^2-4(C_{\mathrm{I}})_0a_{\mathrm{I}}\varepsilon^2\}^{\frac12}}{2a_{\mathrm{I}}\varepsilon^2},\\ V_{\mathrm{E}}(\varepsilon)&=\frac{[1-b_{\mathrm{E}}\varepsilon]-\{[1-b_{\mathrm{E}}\varepsilon]^2-4(C_{\mathrm{E}})_0a_{\mathrm{E}}\varepsilon^2\}^{\frac12}}{2a_{\mathrm{E}}\varepsilon^2}.\end{aligned} \tag{14.3}$$

[14] The coefficients for polymethyl methacrylate for $v<0.3$ mm/µsec are: $(C_{\mathrm{I}})_0=2.76$ mm/µsec, $b_{\mathrm{I}}=3.62$, $a_{\mathrm{I}}=-5.64$ µsec/mm, $(C_{\mathrm{E}})_0=2.74$ mm/µsec, $b_{\mathrm{E}}=3.50$, $a_{\mathrm{E}}=-6.89$ µsec/mm. Cf. SCHULER [1970, *19*].

[15] From the acoustic dispersion data, $(C_{\mathrm{I}})_0=2.76$ mm/µsec and $(C_{\mathrm{E}})_0=2.73$ mm/µsec. Cf. NUNZIATO and SUTHERLAND [1973, *14*].

[16] PIPKIN [1966, *12*].

[17] SCHULER [1970, *19*].

[18] COLEMAN and NOLL [1961, *1*]. This assumption for the stress functional $\mathscr{S}$ is motivated in part by the fact that the general three-dimensional formulation of finite linear viscoelasticity (cf. Sect. 23 of this article) enjoys the correct invariance required by material objectivity and thus is meaningful in situations involving finite deformations.

[19] The constitutive equation of finite linear viscoelasticity has also been used to analyze wave propagation in nonlinear viscoelastic rods by YUAN and LIANIS [1974, *4*].

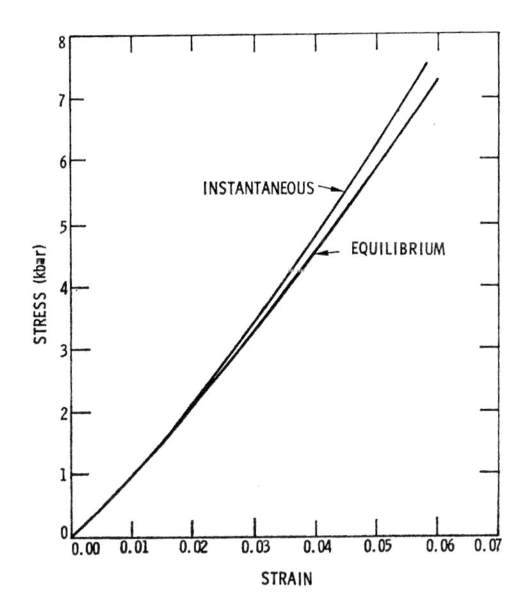

Fig. 25. Instantaneous and equilibrium stress-strain curves for polymethyl methacrylate. (From SCHULER and WALSH [1971, *6*].)

Then, it follows from $(12.13)_1$ and $(12.14)_1$ that

$$\begin{aligned}\sigma_{\mathrm{I}}(\varepsilon) &= \varrho_{\mathrm{R}} V_{\mathrm{I}}^2(\varepsilon)\,\varepsilon,\\ \sigma_{\mathrm{E}}(\varepsilon) &= \varrho_{\mathrm{R}} V_{\mathrm{E}}^2(\varepsilon)\,\varepsilon.\end{aligned} \tag{14.4}$$

The instantaneous and equilibrium stress-strain curves for polymethyl methacrylate are shown in Fig. 25.

In order to proceed further, some assumption must be made with regard to the form of the stress relaxation function $K(\varepsilon; s)$. Consistently with the assumptions made by PIPKIN[20] and GREENBERG[21] to insure the existence of steady waves, we assume that

$$K(\varepsilon; s) = [K(\varepsilon; 0) - K(\varepsilon; \infty)] \exp\left(-\frac{s}{\tau}\right) + K(\varepsilon; \infty) \tag{14.5}$$

where $\tau > 0$ is a constant relaxation time.[22] Our analysis of acoustic dispersion data has shown that a single exponential relaxation function of this type is indeed appropriate for shock wave propagation in polymethyl methacrylate even though three decades of time are involved in such experiments.[23]

15. Steady shock wave solutions. It is significant that by using the assumptions of the previous section along with steady wave and acoustic dispersion data we are able to construct a complete constitutive model for a nonlinear viscoelastic solid. Here we use this model to obtain steady shock wave solutions of (12.9) and compare the calculated wave profiles with experimental observations.

[20] PIPKIN [1966, *12*].

[21] GREENBERG [1967, *3*].

[22] SCHULER [1970, *19*] used a similar form for $K(\varepsilon; s)$ in his analysis of steady shock waves. It is interesting to note that finite linear viscoelastic materials with the relaxation function (14.5) are also Maxwellian; cf. HERRMANN and NUNZIATO [1973, *6*].

[23] Cf. Sect. 11.

It follows from (12.9), (14.1), (14.2), and (14.5) that the steady strain field $\varepsilon(\xi)$ in the nonlinear viscoelastic material is the solution of the integral equation

$$\frac{\sigma_{\mathrm{I}}(\varepsilon)-\varrho_{\mathrm{R}} V^2 \varepsilon}{\sigma_{\mathrm{I}}(\varepsilon)-\sigma_{\mathrm{E}}(\varepsilon)}=\frac{1}{\tau}\int_0^\infty \exp\left(-\frac{s}{\tau}\right)\frac{\{1-(1-\varepsilon(\xi-s))^2\}}{\{1-(1-\varepsilon(\xi))^2\}}\,ds \tag{15.1}$$

where $\sigma_{\mathrm{I}}(\varepsilon)$ and $\sigma_{\mathrm{E}}(\varepsilon)$ are represented by (14.4). This expression can be solved by differentiating it with respect to ξ and then integrating the resulting first-order differential equation by quadrature.[24] Using (12.7) to express the result in terms of particle velocity, we find that the solution of (15.1) is

$$\begin{aligned}\frac{\xi-\xi_0}{\tau}=&-\ln\left\{\frac{V-(C_{\mathrm{E}})_0-b_{\mathrm{E}}v-a_{\mathrm{E}}v^2}{(C_{\mathrm{I}})_0-(C_{\mathrm{E}})_0+(b_{\mathrm{I}}-b_{\mathrm{E}})\,v+(a_{\mathrm{I}}-a_{\mathrm{E}})\,v^2}\right\}\\&+\xi_1\ln|v-v_1|+\xi_2\ln|v-v_2|+\xi_3\ln|v|+\xi_4\ln|2V-v|\end{aligned} \tag{15.2}$$

where v_1 and v_2 are the solutions of the quadratic

$$a_{\mathrm{E}}v^2+b_{\mathrm{E}}v+(C_{\mathrm{E}})_0-V=0, \tag{15.3}$$

and ξ_1, ξ_2, ξ_3, ξ_4 are constants given by

$$\xi_1=-\frac{2[(C_{\mathrm{I}})_0-V+b_{\mathrm{I}}v_1+a_{\mathrm{I}}v_1^2]\,(V-v_1)}{a_{\mathrm{E}}(v_2-v_1)\,v_1(2V-v_1)},$$

$$\xi_2=-\frac{2[(C_{\mathrm{I}})_0-V+b_{\mathrm{I}}v_2+a_{\mathrm{I}}v_2^2]\,(V-v_2)}{a_{\mathrm{E}}(v_2-v_1)\,v_2(2V-v_2)},$$

$$\xi_3=-\frac{V-(C_{\mathrm{I}})_0}{V-(C_{\mathrm{E}})_0},\quad \xi_4=\frac{[(C_{\mathrm{I}})_0-V+2b_{\mathrm{I}}V+4a_{\mathrm{I}}V^2]}{a_{\mathrm{E}}(2V-v_1)\,(2V-v_2)}.$$

Now, for the solution to be that of a steady shock wave, $V>(C_{\mathrm{I}})_0$. In this instance, $\xi_3<0$ and the equilibrium conditions far ahead of the wave (i.e., v vanishes as $\xi\to-\infty$) can be satisfied only if $v\equiv 0$ for $\xi<0$ and there is a jump discontinuity in v at $\xi=0$. The constant ξ_0 is chosen so as to render $v(0)$ the smallest positive root of

$$a_{\mathrm{I}}v^2+b_{\mathrm{I}}v+(C_{\mathrm{I}})_0-V=0. \tag{15.4}$$

The maximum particle velocity achieved, $v(\infty)$, is the smallest positive root of (15.3).

To compare the predicted steady shock wave solutions (15.2) with the experimentally observed wave profiles, it remains to specify the relaxation time τ. Certainly, one could use the value of τ determined from the acoustic analysis of Sect. 11. However, SCHULER[25] has suggested an alternative approach which is based on the experimentally observed steady wave profile. For each specified impact condition, the steady wave velocity V is known and the wave profile can be calculated as a function of the non-dimensional parameter $(\xi-\xi_0)/\tau$. Thus, in each case a relaxation time can be determined such that the particle velocity halfway between the jump at the shock v_0 and the maximum v_∞ occurred at a time commensurate with the experimental data. Using this approach, SCHULER obtained an average relaxation time for polymethyl methacrylate of $\tau=0.25$ µsec. This result compares favorably with the relaxation time of 0.22 µsec deduced from acoustic data.

Generally, it is difficult to make good comparisons of calculated wave profiles with observed wave profiles. This is primarily due to the fact that for a given

[24] Cf. PIPKIN [1966, *12*], SCHULER [1970, *19*].
[25] SCHULER [1970, *19*].

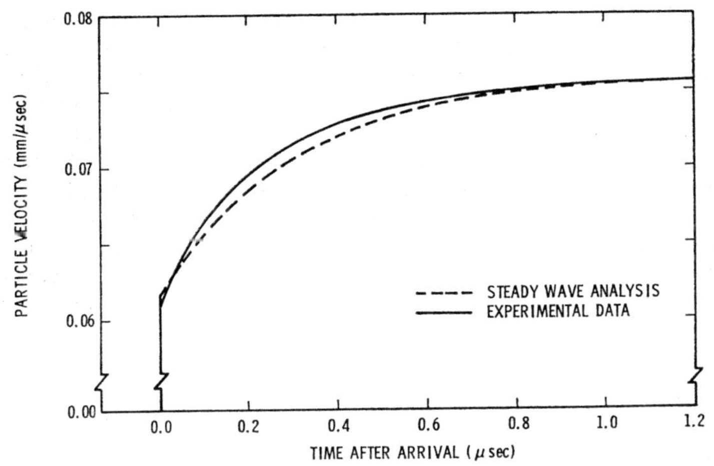

Fig. 26. Comparison of measured and calculated steady wave profiles for polymethyl methacrylate; impact velocity of 0.15 mm/μsec; mechanical model. (From SCHULER [1970, *19*].)

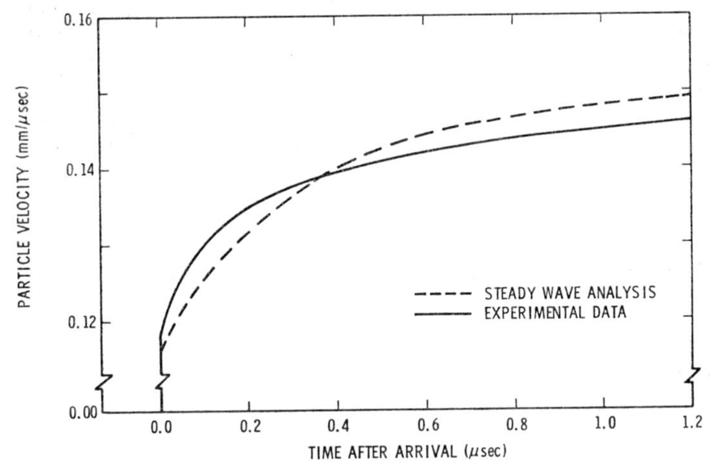

Fig. 27. Comparison of measured and calculated steady wave profiles for polymethyl methacrylate; impact velocity of 0.30 mm/μsec; mechanical model. (From SCHULER [1970, *19*].)

steady wave velocity the values of the particle velocity jump at the shock and the maximum particle velocity calculated using the polynomial fits (13.1) will not in general be commensurate with those actually observed. Thus, in making the calculations shown in Figs. 26 and 27 for polymethyl methacrylate, SCHULER[25] made slight adjustments in the steady wave velocity V to give the best agreement at the wave front and at the tail of the wave. The comparisons shown correspond to wave profiles experimentally observed at propagation distances greater than 37 mm. The profile in Fig. 26 corresponds to an impact condition of 0.15 mm/μsec and a measured steady wave velocity of 2.991 mm/μsec; while the profile in Fig. 27 corresponds to an impact condition of 0.30 mm/μsec and a shock velocity of 3.106 mm/μsec. The steady wave velocities used by SCHULER in the calculations were 2.965 and 3.106 mm/μsec, respectively. It is seen that good agreement was obtained, particularly for the 0.15 mm/μsec impact condition.

The calculated profiles shown in Figs. 26 and 27 were obtained by use of (15.2), which is a necessary condition for a steady wave. However, the wave profiles must also satisfy the sufficiency requirement based on the sound velocity criterion discussed in Sect. 12. SCHULER[26] has shown that this criterion is indeed satisfied for the profiles shown. However, for impact conditions greater than 0.30 mm/μsec (about 7.5 kbar), this criterion is not satisfied. At this stress level, SCHULER[26] and BARKER and HOLLENBACH[27] have detected an inflection in the equilibrium stress-strain curve of polymethyl methacrylate. This observation suggests that the finite linear viscoelastic constitutive assumption used here is not valid for the material for stresses above this level. A characterization of the dynamic viscoelastic response of polymethyl methacrylate above 7.5 kbar has been considered by SCHULER and NUNZIATO[28] in terms of a Maxwellian constitutive model.

VI. Growth and decay of one-dimensional shock waves.

16. The shock amplitude equation. The general theory of the propagation of one-dimensional shock waves in nonlinear materials with fading memory characterized by (9.1) has been developed by COLEMAN, GURTIN, and HERRERA[1] and by CHEN and GURTIN.[2] In particular, CHEN and GURTIN showed that the growth and decay behavior of a shock discontinuity is governed by a differential equation relating the strain ε^- to the strain gradient $(\partial\varepsilon/\partial X)^-$ immediately behind the front.[3] Here we present the main results of their analysis.

Consider a compressive shock wave propagating into a region at rest and unstrained for all past times, i.e., for $X \geqq Y(t)$, $\varepsilon^t(s)=0$ for all $s \in [0,\infty)$ and

$$[\varepsilon]=\varepsilon^- > 0, \quad \left[\frac{\partial\varepsilon}{\partial X}\right]=\left(\frac{\partial\varepsilon}{\partial X}\right)^- . \tag{16.1}$$

It follows from the definition of the instantaneous response function σ_{I} that

$$[\sigma]=\sigma_{\mathrm{I}}(\varepsilon^-).$$

Thus (8.5) and $(8.6)_1$ imply that the intrinsic velocity U is given by[4]

$$\varrho_{\mathrm{R}}\, U^2 = \frac{\sigma_{\mathrm{I}}(\varepsilon^-)}{\varepsilon^-} \tag{16.2}$$

and by the convexity condition $(9.13)_3$,

$$\frac{\varrho_{\mathrm{R}}\, U^2}{E_{\mathrm{I}}^-} \equiv \mu^* < 1 \tag{16.3}$$

where $E_{\mathrm{I}}^- = E_{\mathrm{I}}(\varepsilon^-)$. The inequality (16.3) asserts that *the shock velocity is always subsonic with respect to the material behind the wave front.* Clearly, by (16.2), the

[26] SCHULER [1970, *19*], [1970, *20*].

[27] BARKER and HOLLENBACH [1970, *2*].

[28] SCHULER and NUNZIATO [1974, *3*].

[1] COLEMAN, GURTIN, and HERRERA [1965, *4*].

[2] CHEN and GURTIN [1970, *4*].

[3] A similar relation has been derived by DUVALL and ALVERSON [1963, *1*], AHRENS and DUVALL [1966, *1*], and LUBLINER and GREEN [1970, *16*] for Maxwellian materials; by LUBLINER and SECOR [1966, *10*] for viscoelastic materials with the stress functional represented by a multi-integral expansion (cf. GREEN and RIVLIN [1957, *2*]); and by HUILGOL [1973, *7*] for plane waves in three-dimensional materials with memory. Also see the review article by CHEN [1973, *4*, Sect. 16].

[4] Cf. COLEMAN, GURTIN, and HERRERA [1965, *4*].

shock velocity depends on the strain amplitude ε^- and in fact

$$\frac{dU}{dt} = \frac{(1-\mu^*)\,E_{\mathrm{I}}}{2\varrho_{\mathrm{R}}\,U\,\varepsilon^-}\,\frac{d\varepsilon^-}{dt}\,. \tag{16.4}$$

Thus, the shock velocity increases, decreases, or remains the same according to whether the shock amplitude ε^- increases, decreases, or remains the same.

Using the properties of shock waves and the assumed smoothness of the stress functional $\mathscr{S}$, CHEN and GURTIN[5] derived the *shock amplitude equation*

$$\frac{d\varepsilon^-}{dt} = U\,\frac{(1-\mu^*)}{(1+3\mu^*)}\left\{\lambda^* - \left(\frac{\partial\varepsilon}{\partial X}\right)^-\right\} \tag{16.5}$$

where

$$\lambda^* = \frac{G'(\varepsilon;\,0)\,\varepsilon^-}{U E_I\,(1-\mu^*)}\,. \tag{16.6}$$

Here $G'(\varepsilon^-;\,0) = \mathscr{G}'(\varepsilon_{\mathrm{I}};\,0)$ is the initial slope of the stress relaxation function corresponding to the jump strain history (9.5); thus, by the curvature assumption $(9.14)_2$, $\lambda^* \leq 0$. It is immediately evident from (16.3) and the amplitude equation (16.5) that the growth or decay behavior of the wave front depends on the strain gradient immediately behind the front, i.e.,

(i) *if* $\lambda^* < \left(\frac{\partial\varepsilon}{\partial X}\right)^-$, *then* $\frac{d\varepsilon^-}{dt} < 0$,

(ii) *if* $\lambda^* = \left(\frac{\partial\varepsilon}{\partial X}\right)^-$, *then* $\frac{d\varepsilon^-}{dt} = 0$, *or*

(iii) *if* $\lambda^* > \left(\frac{\partial\varepsilon}{\partial X}\right)^-$, *then* $\frac{d\varepsilon^-}{dt} > 0$.

In view of the steady wave condition (ii), λ^* is called the *critical jump in strain gradient*.

Experimental shock wave studies involve the measurement of particle velocity histories at given material points in a sample, and thus it is often more convenient to express the amplitude equation (16.5) in terms of particle velocity:[6]

$$\frac{dv^-}{dt} = \frac{(1-(\mu^*)^2)}{(1+3\mu^*)}\,\{(\dot{v})^- - U^2|\lambda^*|\}. \tag{16.7}$$

Here $(\dot{v})^-$ is the particle acceleration immediately behind the shock front and $U^2|\lambda^*|$ is the *critical acceleration*. Clearly, the front grows if $(\dot{v})^- > U^2|\lambda^*|$, decays if $(\dot{v})^- < U^2|\lambda^*|$, or is steady if $(\dot{v})^- = U^2|\lambda^*|$.

It should be noted that, in general, it is not possible to determine the particle velocity amplitude v^- of a shock wave as a function of time from the differential equation (16.7) because the particle acceleration $(\dot{v})^-$ is not known in advance. The value of this quantity at each instant reflects the boundary conditions associated with the generation and subsequent propagation of the shock wave. However, there are certain shock wave problems in which approximate solutions of (16.7) can be obtained (we shall discuss one such problem in Sect. 18).

In the case of weak shock discontinuities, it can be shown that the shock amplitude equation (16.7) reduces to

$$\frac{dv^-}{dt} = \frac{G_0'(0)}{2(E_{\mathrm{I}})_0}\,v^- \tag{16.8}$$

[5] CHEN and GURTIN [1970, *4*].

[6] In view of (8.5), the shock velocity U is also expressible as a function of particle velocity.

where $G_0'(0)$ and $(E_{\mathrm{I}})_0$ are the initial slobe of the stress relaxation function and the instantaneous tangent modulus evaluated at $\varepsilon^t = O_{\mathrm{E}}$. Eq. (16.8) has the solution

$$v^-(t) = v^-(0) \exp\left\{\frac{G_0'(0)}{2(E_{\mathrm{I}})_0} t\right\}, \tag{16.9}$$

which implies that the amplitude of a weak shock front decays exponentially to zero as $t \to \infty$.[7]

It is also of interest to note that as $v^- \to 0\,(\varepsilon^- \to 0)$

$$U \to U_0 \tag{16.10}$$

and

$$U^2 |\lambda^*| \to 2\lambda_0 \tag{16.11}$$

where

$$U_0^2 = \frac{(E_{\mathrm{I}})_0}{\varrho_{\mathrm{R}}}, \qquad \lambda_0 = -\frac{G_0'(0)\, U_0}{(\tilde{E}_{\mathrm{I}})_0}, \tag{16.12}$$

and $(\tilde{E}_{\mathrm{I}})_0$ is the instantaneous second-order modulus evaluated for the history $\varepsilon^t = O_{\mathrm{E}}$. As we shall see in a later section,[8] U_0 is the intrinsic velocity of an acceleration wave propagating into an undeformed region at rest and the constant λ_0 is called the critical amplitude. Thus, as the amplitude of the shock discontinuity tends to zero, the intrinsic shock velocity is approximated by the intrinsic acceleration wave velocity U_0 and the critical acceleration has as its limit twice the critical amplitude λ_0 of an acceleration wave propagating into an unstrained equilibrium region.[9]

In the next three sections we shall apply the results presented here to a particular viscoelastic solid and consider several one-dimensional problems involving the behavior of compressive shock waves.

17. The critical acceleration. As we have seen, the growth and decay of a shock wave propagating into a region unstrained for all past times depends on the relative magnitude of the particle acceleration immediately behind the shock $(\dot{v})^-$ and the critical acceleration

$$U^2 |\lambda^*| = \frac{U |G'(\varepsilon^-; 0)|\, \varepsilon^-}{E_{\mathrm{I}}^- (1 - \mu^*)} \tag{17.1}$$

where $\mu^* < 1$ is defined by (16.3). The critical acceleration depends upon the strain (or particle velocity) both explicitly and through the strain dependence of the shock velocity U, the instantaneous tangent modulus E_{I}, and the initial slope of the relaxation function $G'(\varepsilon^-; 0)$. Hence, one method of obtaining $\lambda^*(\varepsilon^-)$ for a particular material is to use the constitutive model developed from the analysis of experimentally observed steady shock waves to determine the strain dependence of these functions. Another method of evaluating $\lambda^*(\varepsilon^-)$ is suggested by observing from (8.5), (16.4) and (16.7) that, for a steady shock wave, U is constant and

$$|\lambda^*| = \frac{(\dot{v})^-}{U^2}.$$

[7] Cf. CHEN and GURTIN [1970, *4*]. This is the type of behavior predicted also by the linear theory of viscoelasticity; cf. FISHER [1965, *10*], VALANIS [1965, *20*], and LEITMAN and FISHER [1973, *10*, Sect. 41]. Also see footnote 2, p. 32. Exponential decay of a weak shock wave has been observed by K. W. SCHULER (unpublished) in polymethyl methacrylate and by DUVALL [1964, *9*] in Sioux quartzite.

[8] Cf. Sect. 22.

[9] Cf. CHEN and GURTIN [1970, *4*].

Thus, $\lambda^*(\varepsilon^-)$ can be obtained directly by measuring $(\dot{v})^-$ for steady shocks of varying strengths.[10] Here, following the study of SCHULER and WALSH[11] we evaluate the critical acceleration from experimental shock propagation studies involving the solid polymer polymethyl methacrylate using both methods and contrast the results.

As described in Sect. 14, the strain dependence of the shock velocity is obtained directly from the experimental measurements which yield the shock velocity-particle velocity dependence, i.e.,

$$U(\varepsilon^-) = V_{\mathrm{I}}(\varepsilon^-). \tag{17.2}$$

Further, from the results of the steady wave studies and $(14.3)_1$, $(14.4)_1$, and (17.2), the strain dependence of the instantaneous tangent modulus can be obtained:

$$E_{\mathrm{I}}(\varepsilon^-) = \varrho_{\mathrm{R}} U^2(\varepsilon^-) \left\{ \frac{1 + b_{\mathrm{I}}\varepsilon^- + 2a_{\mathrm{I}}(\varepsilon^-)^2 U(\varepsilon^-)}{1 - b_{\mathrm{I}}\varepsilon^- - 2a_{\mathrm{I}}(\varepsilon^-)^2 U(\varepsilon^-)} \right\}. \tag{17.3}$$

This leaves the function $G'(\varepsilon^-; 0)$ to be determined. As one would expect from the analysis of steady waves, this function can be determined from the measured instantaneous and equilibrium response functions and the relaxation time τ.

To show this, we consider the constitutive relation (14.1) for finite linear viscoelastic materials with $K(\varepsilon; s)$ represented by (14.5):

$$\begin{aligned} \sigma(t) &= \mathscr{S}(\varepsilon^t) \\ &= \sigma_{\mathrm{I}}(\varepsilon(t)) + \int_0^\infty K'(\varepsilon; s)\,\{1 - (1 - \varepsilon^t(s))^2\}\, ds. \end{aligned} \tag{17.4}$$

The response functions involved in the definition of the critical acceleration have been associated with the constitutive assumption (9.1) for materials with memory. In order to relate (17.4) to this general formulation, it is necessary to note first that, for small relative strain histories γ^t, (9.11) can be written alternatively as

$$\mathscr{S}(\varepsilon^t + \gamma^t) = \mathscr{S}(\varepsilon^t) + \frac{d}{d\alpha}\, \mathscr{S}(\varepsilon^t + \alpha\gamma^t)\big|_{\alpha=0} + o(\|\gamma^t\|)$$

where

$$\frac{d}{d\alpha}\, \mathscr{S}(\varepsilon^t + \alpha\gamma^t)\big|_{\alpha=0} = G_t(0)\gamma^t(0) + \int_0^\infty G_t'(s)\, \gamma^t(s)\, ds. \tag{17.5}$$

Thus, by computing the derivative (17.5) of the stress functional (17.4) for an arbitrary strain history and evaluating the result for the appropriate underlying

---

[10] Cf. CHEN and GURTIN [1970, *4*], CHEN [1973, *4*, Sects. 16 and 17]. Of course, it is not necessary to have steady shock waves in order to evaluate $\lambda^*(\varepsilon^-)$. For an unsteady wave, (16.7) implies that

$$|\lambda^*| = \frac{1}{U^2} \left\{ (\dot{v})^- + \frac{(1 + 3\mu^*)}{(1 - (\mu^*)^2)} \frac{dv^-}{dt} \right\}.$$

From a series of impact experiments, we can directly compute U and μ^* as functions of particle velocity (or strain) by a procedure similar to that described in Sect. 14 (cf. Eqs. (17.2) and (17.3)). Furthermore, by observing the growth or decay of the wave for a given impact condition, $dv^-(t)/dt$ and $(\dot{v})^-$ can be evaluated for various values of the wave amplitude v^- (or ε^-). Thus λ^* is determined. Using a similar procedure and the shock amplitude equation for Maxwellian materials, ASAY, FOWLES, and GUPTA [1972, *1*] have determined the material relaxation properties of lithium fluoride from precursor decay studies.

[11] SCHULER and WALSH [1971, *6*].

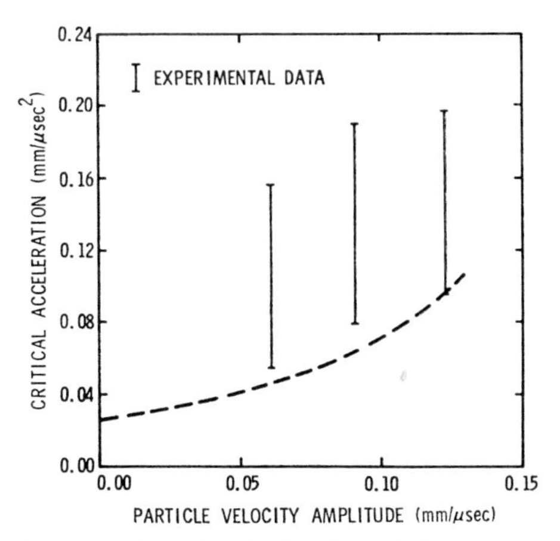

Fig. 28. Comparison of measured and calculated critical acceleration vs. particle velocity behind the shock wave for polymethyl methacrylate; mechanical model. (From SCHULER, NUNZIATO, and WALSH [1973, *18*].)

history, i.e., in the case of shock wave studies, for the jump history ε_{I}, we can identify the functions

$$G_t(0) = \mathcal{G}(\varepsilon_{\mathrm{I}}; 0),$$
$$G'_t(s) = \mathcal{G}'(\varepsilon_{\mathrm{I}}; s) = G'(\varepsilon; s).$$

Following this through, we find that

$$\begin{aligned} G_t(0) &= E_{\mathrm{I}}(\varepsilon^-), \\ G'_t(s) &= G'(\varepsilon^-; s) = 2K'(\varepsilon^-; s). \end{aligned} \tag{17.6}$$

Combining $(17.6)_2$ with (14.5) and using (14.2) yields

$$G'(\varepsilon^-; 0) = -\frac{2[\sigma_{\mathrm{I}}(\varepsilon^-) - \sigma_{\mathrm{E}}(\varepsilon^-)]}{\varepsilon^-(2-\varepsilon^-)\,\tau} \tag{17.7}$$

when evaluated at $s=0$.

With $U(\varepsilon^-)$, $E_{\mathrm{I}}(\varepsilon^-)$, and $G'(\varepsilon^-; 0)$ known, the calculation of $U^2(\varepsilon^-)|\lambda^*(\varepsilon^-)|$ can be carried out using (17.1). Actually, in presenting the results for polymethyl methacrylate we follow the lead indicated earlier and determine the critical acceleration $U^2|\lambda^*|$ as a function of the particle velocity behind the shock. Fig. 28 shows this dependence along with the ranges of values of $U^2|\lambda^*| = (\dot{v})^-$ determined graphically from the experimental steady wave profiles at three impact conditions. The error brackets reflect not only the difference between the results for several experiments at the same nominal impact condition, but also the uncertainties in determining the slopes. It should be noted that the calculated values of $U^2|\lambda^*|$ are extremely sensitive to variations in the fits for σ_{I} and σ_{E} and the relaxation time τ. Nevertheless, the correlation in Fig. 28 appears to give at least a qualitative verification of the concept of a critical acceleration and of the methods of obtaining material response functions from experimental shock wave studies.

Of course, the continued growth or decay of the shock front cannot be predicted solely from these results. That is, as the jump in particle velocity varies, the particle acceleration immediately behind the front will in general change. From the dependence of λ^* on ε^- (and hence on v^-), the value of the critical acceleration

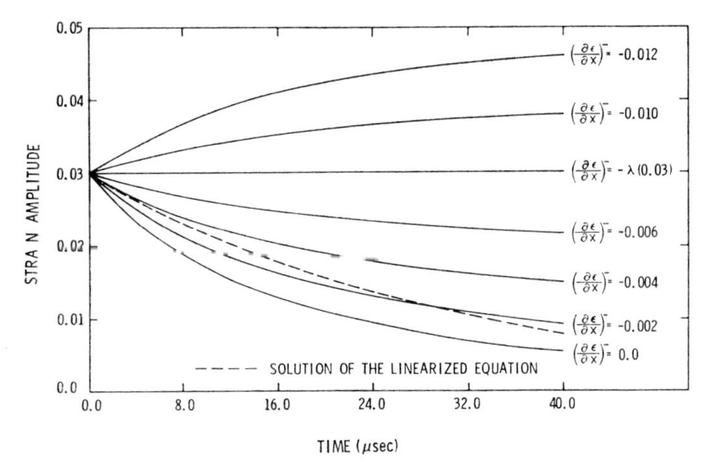

Fig. 29. Strain behind the shock wave vs. time for fixed values of strain gradient for polymethyl methacrylate; mechanical model. (From CHEN, GURTIN, and WALSH [1970, *5*].)

changes and, in turn, affects the rate of growth or decay of the wave through (16.7). This has been considered by CHEN, GURTIN, and WALSH[12] in a study of the long-range amplitude behavior. Assuming $(\partial\varepsilon/\partial X)^-$ is a fixed constant, they calculated the variation of the strain amplitude $\varepsilon^-(t)$ in polymethyl methacrylate for an initial amplitude $\varepsilon^-(0)=0.03$. Their results for several fixed values of the strain gradient are shown in Fig. 29. Note that the solution of the linearized equation (16.9) does *not* furnish a lower bound for the behavior of the amplitude $\varepsilon^-(t)$.

18. Shock pulse attenuation. In considering the shock amplitude equation (16.7) it was pointed out that this equation can not in general be used to determine $v^-(t)$ since the particle acceleration $(\dot{v})^-(t)$ is not known in advance. However, NUNZIATO and SCHULER[13] have shown that (16.7) can be used to evaluate $v^-(t)$ for a certain type of shock wave problem, called the *thin pulse problem*, by approximating $(\dot{v})^-(t)$. The problem is one of calculating the attenuation of a pulse which consists of a shock front propagating into an unstrained region at rest followed immediately by an unloading wave of arbitrary shape which unloads the material to its original reference configuration (cf. Fig. 30). Here we follow the approach of NUNZIATO and SCHULER to calculate the attenuation of a thin shock pulse in polymethyl methacrylate and compare the results with experimental observations.[14]

For viscoelastic materials, the shock loads the material along a Rayleigh line to a point ε^- on the instantaneous stress-strain curve $\sigma_1(\varepsilon)$ and the unloading wave unloads it along some unknown stress-strain path. This unloading path will always lie between the instantaneous and equilibrium stress-strain curves. However, in a

[12] CHEN, GURTIN, and WALSH [1970, *5*].

[13] NUNZIATO and SCHULER [1973, *13*].

[14] Various other approximate methods have been developed to analyze the propagation of unsteady shock waves in viscoelastic solids. For example, ACHENBACH and REDDY [1967, *1*], SUN [1970, *22*], and VOGT and SCHAPERY [1971, *7*] have used wave front expansions to obtain approximate solutions in linear and nonlinear materials. LUBLINER and SECOR [1967, *6*] have used power series to obtain approximate solutions for semilinear materials with the stress function represented by a multi-integral expansion. Finally, LUBLINER and GREEN [1970, *16*] used a perturbation analysis to obtain approximate solutions for Maxwellian materials which are only slightly nonlinear. The predictions of these approximate analyses apparently have not been compared with experimental observations.

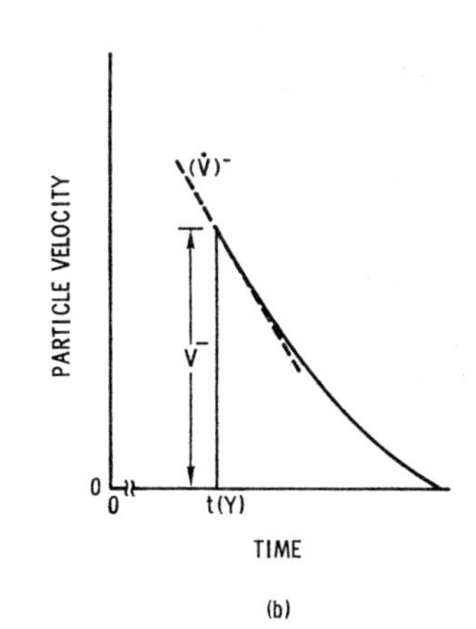

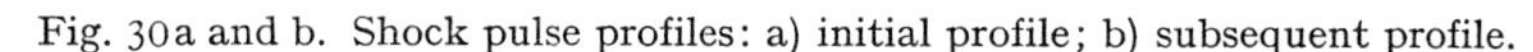

Fig. 30a and b. Shock pulse profiles: a) initial profile; b) subsequent profile.

study of the unloading behavior of polymethyl methacrylate from precompressed equilibrium states, SCHULER[15] found that unloading waves always unload the material *initially* down the instantaneous stress-strain curve through that point on the equilibrium curve σ_E corresponding to the initially deformed state.[16] Physically, this is due to the fact that in the initial portion of the unloading wave the strain rates are very high. In the case of very thin shock pulses, high strain rates are achieved throughout the pulse profile, and thus it is reasonable to assume that in this case the material unloads along the instantaneous curve going through the point to which the shock loaded the material. In other words, *we assume that the unloading stress-strain path is the instantaneous stress-strain curve* $\sigma_I(\varepsilon)$.

This assumption means that the effects of rate-dependence on the unloading wave are neglected and the unloading wave is treated as a wave propagating in a nonlinear elastic material whose stress-strain curve is $\sigma_I(\varepsilon)$.[17] Furthermore the following simplifying assumptions will be made concerning the unloading wave:

(i) *the unloading wave is a simple wave, and*

(ii) *the unloading path in the stress-particle velocity plane is the instantaneous stress-particle velocity curve,* $\sigma_I(v)$:

$$\sigma_I(v) = \varrho_R V_I(v)\, v$$

where $V_I(v)$ *is defined by* $(13.1)_1$.

Assumption (ii) is an approximation which eliminates the reflection which results from the unloading wave—shock wave interaction and hence insures that the unloading wave remains a simple wave for all time. This approximation should prove to be especially good for shocks of weak to moderate strength.

Assumption (ii) also gives a one-to-one correspondence between points on the unloading wave and points on the instantaneous stress-particle velocity curve over the corresponding range of particle velocity. Furthermore, from assumption (i)

[15] SCHULER [1970, *20*].

[16] As we shall see in Sect. 21, unloading waves can be interpreted as expansive acceleration waves. Since such waves propagate with an instantaneous sound velocity, they must initially unload the material along the appropriate instantaneous stress-strain curve.

[17] Nevertheless, the effects of internal dissipation will still enter the attenuation calculations through the critical acceleration $U^2|\lambda^*|$ which appears in the shock amplitude equation (16.7).

each point on the unloading wave propagates at a constant level of particle velocity with the corresponding intrinsic instantaneous sound velocity

$$C_I^2 = \frac{E_I}{\varrho_R}. \tag{18.1}$$

Let the smooth function $t_0(v)$ represent the initial unloading wave profile at the material point X_0 (Fig. 30a). Then the unloading portion of the subsequent profile (Fig. 30b) can also be represented by the smooth function $t(v)$ which is related to the initial profile $t_0(v)$ through the simple wave relation[18]

$$t(v) = t_0(v) + \frac{X - X_0}{C_I(v)} \tag{18.2}$$

for each $v \leqq v^-$ and for all material points $X \geqq X_0$.

Differentiating the simple wave description (18.2) with respect to v, inverting, and evaluating the result at the shock front yields the following expression for the particle acceleration immediately behind the front:

$$(\dot{v})^- = \left[\frac{1}{\dot{v}_0(v^-)} - \alpha^*(v^-)\,[Y(t) - X_0]\right]^{-1}. \tag{18.3}$$

Here, $Y(t) \geqq X_0$ is the location of the shock front at time t, $\dot{v}_0(v^-) < 0$ is the slope of the initial unloading wave profile at the particle velocity v^-, and

$$\alpha^*(v^-) = \frac{1}{C_I^2(v^-)}\left[\frac{d C_I(v)}{dv}\right]_{v=v^-}. \tag{18.4}$$

Substitution of (18.3) in the shock amplitude equation (16.7) yields the ordinary differential equation

$$\frac{dv^-}{dt} = \frac{1 - (\mu^*(v^-))^2}{1 + 3\mu^*(v^-)}\left\{\frac{\dot{v}_0(v^-)}{1 - \dot{v}_0(v^-)\,\alpha^*(v^-)\,[Y(t) - X_0]} - U^2(v^-)\,|\lambda^*(v^-)|\right\}, \tag{18.5}$$

where $\mu^* < 1$ and λ^* are given by (16.3) and (16.6), respectively. Since $\sigma_I(\varepsilon)$ is convex, it can be shown that $\alpha^*(v^-) > 0$ for all $v^- \neq 0$. As we have already noted, $\dot{v}_0(v^-) < 0$. Thus, it follows from (8.1) and (18.5) that *the shock pulse amplitude v^- is monotone decreasing and will approach a limiting value (zero) in an infinite time.*

Using (13.1) along with (8.5), (14.3), (14.4), (17.3), (17.6), and (18.1), we can evaluate $\mu^*(v^-)$, $\alpha^*(v^-)$, and $\lambda^*(v^-)$. Furthermore, by specifying the initial shock pulse profile, we know $\dot{v}_0(v^-)$. Thus, we can integrate the coupled set of differential equations (8.1) and (18.5) to determine the attenuation curve $v^-(X)$. Once the shock wave amplitude v^- is known at any material point $X \geqq X_0$, the corresponding unloading wave profile can then be determined through (18.2).

The experimental technique which was employed by NUNZIATO and SCHULER[19] to generate and observe shock pulse propagation in viscoelastic materials is shown schematically in Fig. 31. The specimen through which the pulse propagation is to be observed is attached as the noseplate of a projectile which is accelerated by a light gas gun.[20] The pulse is produced by the impact of the specimen against a thin (0.0216 mm) gold foil target which was tautly stretched over a hole in an aluminum support plate and retained by an "O" ring. In order to insure that the motion which results at impact is one-dimensional, great care was taken to maintain the

[18] Cf. COURANT and FRIEDRICHS [1948, *1*, Sect. 42]. A similar representation of the unloading wave was also employed by FOWLES [1960, *2*] and by BERTHOLF and OLIVER [1970, *3*] in studies of attenuation in elastic materials. Their approach to the problem differs from that discussed here.

[19] NUNZIATO and SCHULER [1973, *13*].

[20] Cf. Sect. 4.

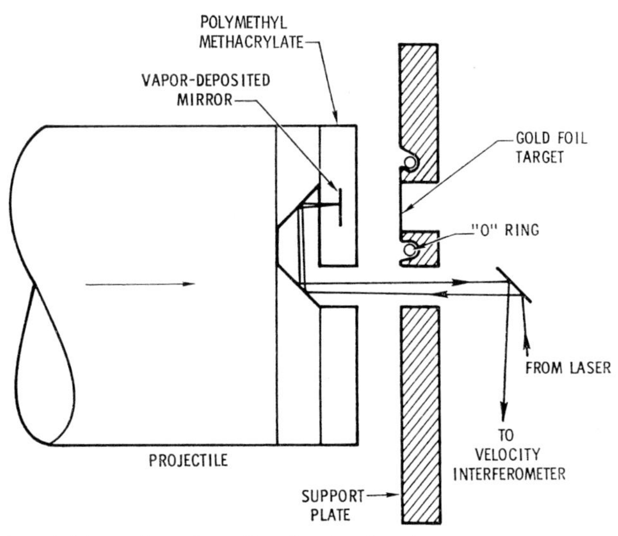

Fig. 31. Experimental impact configuration for shock pulse attenuation studies. (From NUNZIATO and SCHULER [1973, *13*].)

Table 2. *Shock pulse attenuation measurements in polymethyl methacrylate.*

Shot number	Impact velocity v_m (mm/μsec)	Mirror distance X_m (mm)	Particle velocity amplitude v^- (mm/μsec)
2400	0.3081	0.480	0.2600
2401	0.3034	1.257	0.2150
2402	0.3111	3.200	0.1655
2403	0.3094	5.969	0.1144

flatness of both the foil and the impact surface of the projectile and to minimize the tilt between them. The projectile velocity was measured by the electrically charged pin technique described in Sect. 4.

The specimen again consisted in two plates of polymethyl methacrylate bonded together with an embedded mirror located at various distances from the impact surface. The motion of the mirror was monitored by the velocity interferometer instrumentation technique described in Sect. 5. The data from a series of four experiments with a nominal impact condition of 0.31 mm/μsec are given in Table 2 and grouped according to increasing mirror distance (from the impact surface).[21]

Due to the impedance mismatch between polymethyl methacrylate and gold, an unloading wave from the free surface of the foil reflects from the impact surface and only partially unloads the polymethyl methacrylate. The wave reflected back into the gold subsequently re-reflects at the free surface of the gold foil, and thus establishes a reverberation pattern. Due to this reverberation the polymethyl methacrylate unloads in a series of ≈ 14 nsec wide steps of decreasing amplitude.

[21] KELLER [1968, *13*] has also observed thin pulse propagation in polymethyl methacrylate using a streak camera. The pulses were generated using an exploding foil technique and in all the experiments the particle velocity was greater than 0.40 mm/μsec. His data are not included since they are outside the range of applicability of the viscoelastic constitutive model used for polymethyl methacrylate.

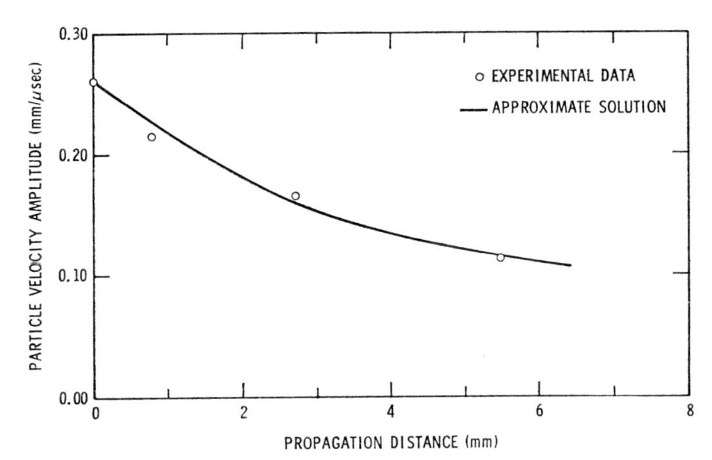

Fig. 32. Comparison of measured and calculated shock pulse amplitudes in polymethyl methacrylate. (From NUNZIATO and SCHULER [1973, *13*].)

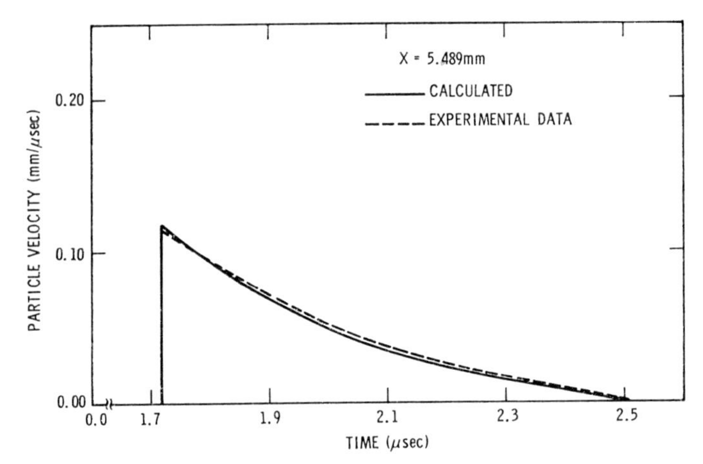

Fig. 33. Comparison of measured and calculated shock pulse profiles in polymethyl methacrylate. (From NUNZIATO and SCHULER [1973, *13*].)

These steps quickly spread out and, indeed, even after a very short propagation distance of 0.48 mm, these steps were indistinguishable, having merged into a very steep unloading profile, which was exponential in character.

The reverberation pattern makes it difficult to determine the particle velocity-time profile at the point where attenuation begins. Therefore, the particle velocity history measured at the mirror location $X_{\mathrm{m}}=0.48$ mm was used as the initial pulse shape. From these data the initial unloading profile was fitted as[22]

$$t_0(v) = -k \ln\left(\frac{v}{v_0^-}\right), \qquad v \leqq v_0^- . \tag{18.6}$$

It then follows that

$$\dot{v}_0(v^-) = -\frac{v^-}{k} .$$

With the initial pulse shape specified and the material response functions known, we can calculate the attenuation curve $v^-(X)$ by integrating the differential

[22] Here $k=0.134$ μsec and $v_0^- = 0.26$ mm/μsec.

equations (8.1) and (18.5) numerically. The resulting solution is compared with the experimental data points in Fig. 32. Here the propagation distance X has been normalized with respect to the mirror location closest to the impact surface (shot 2400). It is evident that the approximate solution compares well ($\pm 5\%$) with the experimentally observed attenuation of the shock amplitude. Since the impact velocity in the experiments was relatively low, reflections of the unloading wave from the shock wave were not expected to influence the attenuation and this is borne out in Fig. 32. By assuming unloading along the instantaneous curve, one would expect the predicted attenuation to be less than that observed. However, the good comparison obtained seems to indicate that in these experiments the unloading stress-strain path does lie very close to the instantaneous curve.

Finally, from the simple wave description (18.2) and the initial pulse shape (18.6), we can determine the wave shape at any other point. The pulse shape was calculated for $X=5.489$ mm and compared with the particle velocity-time history measured at that point. The results are shown in Fig. 33 and are in good agreement.

19. Some remarks on the evolution of steady shock waves. The observation of steady shock waves in a nonlinear viscoelastic solid has been shown to play an important role in the characterization of the dynamic response of the material. In view of this, a better understanding of the development of steady shock waves in plate impact experiments is desirable. However, the problem is very difficult to analyze and a complete solution to the problem requires the use of numerical techniques.[23] Nevertheless, there are some general qualitative features of steady wave evolution which should be mentioned. Furthermore, using the shock amplitude equation, we can construct a lower bound on the propagation distance required for the development of steady wave profiles in polymethyl methacrylate which illustrates these features.[24]

In a symmetric impact experiment, $v^-(0)=v_0^-$ and $(\dot{v})^-(0)=(\dot{v})_0^-=0$. After impact ($t>0$, $X>0$), there is a noticeable decay in the strength of the shock front and an accompanying smoothing of the wave shape with increasing propagation distance. For example, this evolutionary behavior in polymethyl methacrylate can be seen in Fig. 23 for the nominal impact condition of 0.06 mm/μsec ($v_0^-=0.03$ mm/μsec). The important aspect to note here is that the strength of the shock front decays with propagation distance and that the particle acceleration $(\dot{v})^-$ immediately behind the shock front, after impact, is always positive. Thus, we can conclude from the shock amplitude equation (16.7) that at any instant $t>0$

$$U^2|\lambda^*| \geqq (\dot{v})^- > 0$$

and, consequently,

$$\left|\frac{dv^-}{dt}\right| \leqq \frac{(1-(\mu^*)^2)}{(1+3\mu^*)} U^2|\lambda^*|. \tag{19.1}$$

This inequality gives an *upper bound* on the magnitude of the rate of decay of a shock front in symmetric impact experiments. Thus, treating (19.1) as an equality and integrating it along with (8.1), we can determine the *lower bound* on the attenua-

[23] Algorithms which numerically compute solutions to the initial-boundary value problems associated with shock wave propagation in nonlinear viscoelastic solids have been developed by HICKS and HOLDRIDGE [1972, *13*] and by LAWRENCE [1972, *14*]. The code developed by HICKS and HOLDRIDGE is based on the method of characteristics; while that due to LAWRENCE employs a finite-difference scheme. Both algorithms are based on the constitutive assumption of Maxwellian materials. For a further discussion of the application of numerical methods to wave propagation problems, see the review article by HERRMANN and HICKS [1973, *5*].

[24] Cf. NUNZIATO and SCHULER [1973, *12*].

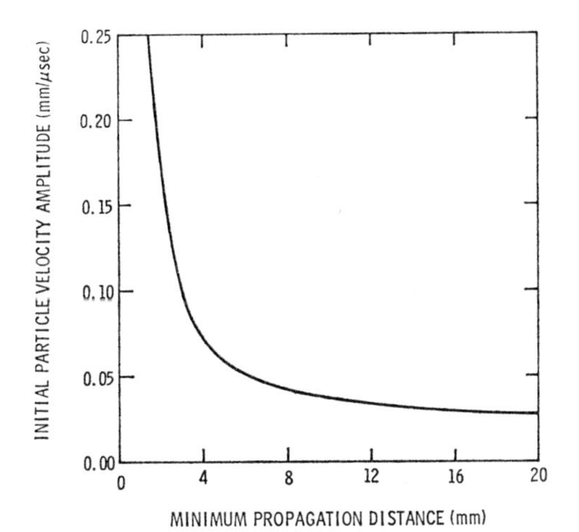

Fig. 34. Minimum propagation distance for the evolution of a steady shock wave as a function of initial particle velocity behind the wave for polymethyl methacrylate. (From NUNZIATO and SCHULER [1973, *12*].)

tion curve $v^-(X)$. Comparison of this calculated curve with the experimentally measured value of v^- corresponding to a steady wave (with the same initial value v_0^-) determines the minimum distance δ required for the steady wave to develop.

Using the experimental data for polymethyl methacrylate and the previously determined material functions μ^*, λ^*, and U, NUNZIATO and SCHULER[25] determined the resulting dependence of δ on the initial amplitude v_0^-, which is shown in Fig. 34. The results illustrate two general qualitative features of steady wave evolution which are consistent with experimental observations:

(i) *the greater the initial amplitude* v_0^-, *the faster the wave evolves toward the steady wave shape, and*

(ii) *below a certain initial amplitude* v_0^-, *a steady shock wave is not possible.*

The first observation suggests that there is greater stress relaxation in the material at the higher impact conditions. The second observation is to be expected from our previous discussion of weak shock waves and the fact that they always decay. In the case of polymethyl methacrylate, $v_0^- \geqq 0.03$ mm/μsec is necessary for the development of a steady shock wave.

The curve in Fig. 34 represents only a lower bound. While data corresponding to points which are to the right of the curve may represent steady shock waves, there is no guarantee. In fact it appears from the experimental data for polymethyl methacrylate that the actual distance for steady wave evolution is approximately three times the minimum distance given in Fig. 34.[26]

VII. Growth and decay of one-dimensional acceleration waves

20. The amplitude of waves in homogeneously deformed regions. The propagation of one-dimensional acceleration waves in nonlinear materials with fading

[25] NUNZIATO and SCHULER [1973, *12*].

[26] A numerical solution for the evolution of a steady shock wave in polymethyl methacrylate has been obtained using the characteristic computer code developed by HICKS and HOLDRIDGE [1972, *13*]. The results are commensurate with the experimental observations and show that the wave tends *asymptotically* to the steady wave shape.

memory characterized by (9.1) has been considered in detail by COLEMAN, GURTIN, and HERRERA.[1] In particular, they showed that the intrinsic velocity U of an acceleration wave propagating into a region with an arbitrary strain history ε^t satisfies[2]

$$U^2 = \frac{E_t}{\varrho_R}. \tag{20.1}$$

Notice that in view of (12.11) and (9.4) this is precisely the instantaneous longitudinal sound velocity of the material. COLEMAN and GURTIN[3] further showed that if the acceleration wave amplitude $a(t) = [\dot{v}]\ (t)$ is given at $t=0$, then it is possible to derive an expression for $a(t)$ at later times, the form of the expression depending upon the strain history of the region traversed by the wave front up to time t. Here it will suffice to give only their results for waves propagating into homogeneously deformed regions at rest.

Since $U(t) > 0$, we assume that for $t \geqq 0$ and $X \geqq Y(t)$ the motion is of the form

$$\chi(X, t-s) = (1-\mathring{\varepsilon})X + \mathring{X}, \quad s \in [0, \infty), \tag{20.2}$$

where $\mathring{\varepsilon}$ and $\mathring{X}$ are constants. In this case, the moduli E_t, $\tilde{E}_t$, and $G_t(s)$ are time-independent and are given by

$$\begin{aligned} E_t &= D_\varepsilon \mathscr{S}(\mathring{\varepsilon}_E) = E, \\ \tilde{E}_t &= D_\varepsilon^2 \mathscr{S}(\mathring{\varepsilon}_E) = \tilde{E}, \\ G_t(s) &= \mathscr{G}(\mathring{\varepsilon}_E; s) = G(s). \end{aligned} \tag{20.3}$$

Using (20.3), COLEMAN and GURTIN have shown that the intrinsic velocity U is a constant:

$$U^2 = \frac{E}{\varrho_R}, \tag{20.4}$$

and that the acceleration wave amplitude $a(t)$ satisfies the differential equation

$$\frac{da}{dt} = -\mu a + \beta a^2, \tag{20.5}$$

where μ and β are constants given by

$$\mu = -\frac{G'(0)}{2G(0)}, \quad \beta = \frac{\tilde{E}}{2G(0)U}. \tag{20.6}$$

Clearly, U, μ, and β depend only on the underlying equilibrium strain $\mathring{\varepsilon}$. The differential equation (20.5) is a Bernoulli equation[4], and it can be readily integrated

[1] COLEMAN, GURTIN, and HERRERA [1965, *4*], COLEMAN and GURTIN [1965, *5*], COLEMAN and GURTIN [1966, *4*]. Also see the review by CHEN [1973, *4*, Sect. 14].

[2] This result has also been obtained by COLEMAN, GREENBERG, and GURTIN [1966, *3*], and by DUNWOODY [1966, *8*] for certain classes of Maxwellian materials. The general propagation condition for acceleration waves has been derived for linear viscoelastic materials by HERRERA and GURTIN [1965, *15*], for materials of the integral type by VARLEY [1965, *22*] and WATERSTON [1969, *5*], for certain Maxwellian materials by DUNWOODY and DUNWOODY [1965, *8*], and for materials with memory by COLEMAN and GURTIN [1966, *6*] and HUILGOL [1973, *8*]. FISHER and GURTIN [1965, *11*] have also shown that, in the context of linear viscoelasticity, shocks and all higher-order waves obey the same propagation condition as acceleration waves; also see LEITMAN and FISHER [1973, *10*, Sect. 41].

[3] COLEMAN and GURTIN [1965, *5*].

[4] The general properties of the Bernoulli equation (20.5) with time-dependent coefficients $\mu(t)$, $\beta(t)$, have been studied by BAILEY and CHEN [1971, *1*], [1972, *2*], and BOWEN and CHEN [1972, *5*]. Also see CHEN [1973, *4*, Sect. 13].

to yield the solution[5]

$$a(t)=\frac{\lambda}{\left(\frac{\lambda}{a(0)}-1\right)\exp(\mu t)+1} \tag{20.7}$$

where

$$\lambda\equiv\frac{\mu}{\beta}=-\frac{G'(0)\,U}{\tilde{E}}. \tag{20.8}$$

Thus, contrary to the case of shock waves, an exact closed-form expression for the time-dependent behavior of the acceleration wave amplitude can be obtained which depends only on the properties of the material.

From the curvature assumptions imposed in Sect. 9 and (20.3), it is clear that the wave velocity is real and that

$$\mu>0, \quad \beta>0, \quad \lambda>0.$$

Thus, a straightforward analysis of Eq. (20.7) yields the following results:[6]

(i) *if* $a(0)<\lambda$, *then* $a(t)\to 0$ *monotonically as* $t\to\infty$;
(ii) *if* $a(0)=\lambda$, *then* $a(t)\equiv a(0)$; *or*
(iii) *if* $a(0)>\lambda$, *then* $a(t)\to\infty$ *monotonically in a finite time* t_∞,

$$t_\infty=-\frac{1}{\mu}\ln\left(1-\frac{\lambda}{a(0)}\right). \tag{20.9}$$

Notice that if the wave is expansive ($a<0$), then $a(0)$ is always less than λ, and the wave front decays. However, if the wave is compressive ($a>0$), then λ plays the role of a *critical amplitude*. If the amplitude is less than the critical amplitude, then the internal dissipation of the material (manifested by a strictly negative value of $G'(0)$) is the governing effect, and the wave front decays. However, if the amplitude is greater than the critical value, then the nonlinearity of the instantaneous stress-strain curve is the dominating effect and the acceleration wave front achieves an infinite amplitude in a finite time. It has been conjectured that the approach of $a(t)$ to ∞ as $t\to t_\infty$ is indicative of the formation of a shock wave at time t_∞. In Sect. 22 we shall discuss some experimental results reported by WALSH and SCHULER[7] which support this conjecture.

Finally, it is of interest to note the relation between acceleration waves and infinitesimal sinusoidal progressive waves which were discussed in Sect. 10. Setting the results (10.10) alongside (20.4) and (20.7), we see that an acceleration wave entering a homogeneously deformed region at rest propagates with the velocity

$$U=c_\infty$$

---

[5] Cf. COLEMAN and GURTIN [1965, *5*]. In this regard, also see VARLEY [1965, *22*], DUNWOODY and DUNWOODY [1965, *8*], COLEMAN, GREENBERG, and GURTIN [1966, *3*], DUNWOODY [1966, *8*], and COLEMAN and GURTIN [1968, *3*]. LUBLINER [1967, *5*] has obtained an approximate solution of the dynamical field equations for a nonlinear viscoelastic solid which contains an acceleration wave. He found that the amplitude of the wave also obeys (20.7).

In the linear theory of viscoelasticity, the amplitude of an acceleration wave obeys

$$a(t)=a(0)\exp(-\mu t) \tag{*}$$

with μ given by $(20.6)_1$; cf. COLEMAN and GURTIN [1965, *5*], FISHER [1965, *10*], and COLEMAN and GURTIN [1966, *4*]. Eq. (*) also governs the propagation of all higher-order waves (cf. FISHER [1965, *10*] and COLEMAN, GREENBERG, and GURTIN [1966, *3*]), and, as already indicated, also governs the decay of weak shock waves (cf. Sect. 16). Also see LEITMAN and FISHER [1973, *10*, Sect. 41].

[6] Cf. COLEMAN and GURTIN [1965, *5*].

[7] WALSH and SCHULER [1973, *20*].

and has an amplitude $a(t)$ given by (20.7) with

$$\mu = c_\infty \alpha_\infty, \qquad \lambda = \frac{2 c_\infty^4 \alpha_\infty \varrho_R}{\tilde{E}},$$

where c_∞ and α_∞ are the ultrasonic velocity and attenuation for progressive waves in the same material.[8] This result extends to materials with memory TRUESDELL's "second theorem of equivalence" for velocities of acceleration waves and plane infinitesimal progressive waves in perfectly elastic materials subject to homogeneous strain.[9]

21. Expansive waves in precompressed regions. It is evident from Fig. 10 that an unloading wave may be considered to be an expansive acceleration wave propagating into a deformed region. In this section we consider the propagation of such unloading waves in regions which have been subjected to the passage of steady shock waves. In each case the shock wave is assumed to reach equilibrium before the arrival of the unloading wave, and thus we have an example of an expansive acceleration wave propagating into a homogeneously deformed region at equilibrium. Applying the constitutive model developed in Sect. 14 and the results of the previous section, SCHULER[10] determined the velocity of such waves in polymethyl methacrylate as a function of the initial equilibrium strain and compared the results with experimentally measured unloading wave velocities. Here we present his results and, in addition, calculate the amplitude decay of an expansive wave for a particular value of initial strain. These results are compared with measured values of the wave amplitude.

α) *Acceleration wave velocities.* The velocity U of an acceleration wave propagating in a finite linear viscoelastic material is given by (20.1) where, by $(9.4)_1$ and (14.1),

$$E_t = E_{\mathrm{I}}(\varepsilon) + \int_0^\infty \left[\frac{d}{d\varepsilon} K'(\varepsilon; s)\right]\left\{1 - \left(1 - \varepsilon^t(s)\right)^2\right\} ds. \tag{21.1}$$

We note that in general the intrinsic velocity U is a function of the current strain state and the past history of the strain. However, as we have just indicated, we wish to calculate the acceleration wave velocity at a material particle which has been subjected to the passage of a steady shock wave and has had a relatively long time to reach equilibrium prior to the arrival of the unloading wave.[11] Thus, the appropriate history to employ in (21.1) is $\mathring{\varepsilon}_{\mathrm{E}}(s) = \mathring{\varepsilon}$, $s \in [0, \infty)$, and it follows from (21.1), (14.2), and (20.1) that

$$U^2 = \frac{E}{\varrho_R} = \frac{1}{\varrho_R}\left\{E_{\mathrm{E}}(\mathring{\varepsilon}) + \frac{2(1-\mathring{\varepsilon})}{(2-\mathring{\varepsilon})\,\mathring{\varepsilon}}\left[\sigma_{\mathrm{I}}(\mathring{\varepsilon}) - \sigma_{\mathrm{E}}(\mathring{\varepsilon})\right]\right\}. \tag{21.2}$$

By use of the representations (14.4) for the instantaneous and equilibrium response functions, σ_{I} and σ_{E}, the velocity of an acceleration wave predicted by (21.2) can be calculated as a function of the initial strain $\mathring{\varepsilon}$. The curve shown in Fig. 35 is the result of such a calculation for polymethyl methacrylate.

[8] This result is due to COLEMAN and GURTIN [1965, *5*].

[9] TRUESDELL [1961, *8*].

[10] SCHULER [1970, *20*].

[11] For all the experimental data reported here, the difference in time at the embedded mirror between the arrival of the shock wave and the arrival of the unloading wave is about an order of magnitude greater than the relaxation time of the material.

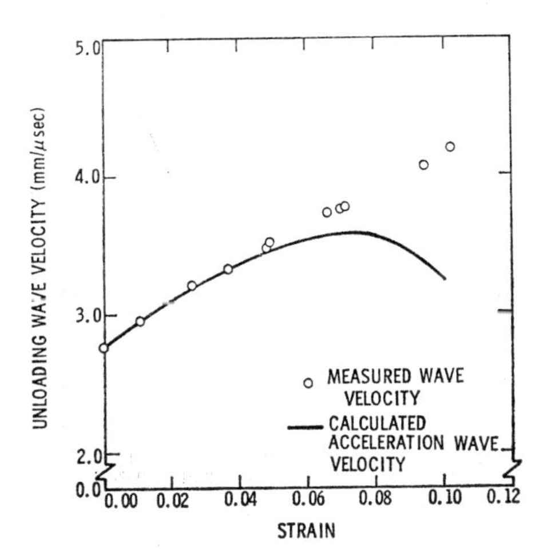

Fig. 35. Comparison of measured and calculated unloading wave velocities in polymethyl methacrylate as a function of initial equilibrium strain.

Expansive acceleration (unloading) waves were produced in polymethyl methacrylate by performing symmetric impact experiments in which the center region of the back of the flyer plate was unsupported. The initial shock wave traveling from the impact plane into the flyer plate thus reflects from the rear free surface of the flyer as an expansive acceleration (unloading) wave of infinite amplitude. In each experiment, the thickness of the flyer plate X_p and the distance from the impact surface to the embedded mirror X_m were such that the material particle at the mirror achieved equilibrium prior to the arrival of the unloading wave. The unloading wave velocities shown in Table 1 are average velocities calculated from

$$U = \frac{X_m + X_p}{t_a - \dfrac{X_p}{V}}$$

where t_a is the time after impact at which the unloading wave arrives at the mirror and V is the measured steady shock velocity. For a given impact condition no significant variation of unloading wave velocity was observed in polymethyl methacrylate over the range of propagation distances employed in the experiments (typically from 6 to 37 mm). The measured unloading wave velocity and corresponding equilibrium strain values are also shown in Fig. 35. Clearly, for strains less than about 6% good agreement exists between the calculated acceleration wave velocity corresponding to the equilibrium history and the observed unloading wave velocity. The lack of agreement above about 6% strain (7.5 kbar) is indicative of the inflection in the equilibrium behavior of polymethyl methacrylate reported by SCHULER[12] and BARKER and HOLLENBACH[13] and is due to the fact that the existence of a steady wave in polymethyl methacrylate traveling faster than 3.1 mm/μsec is not consistent with the constitutive assumption of finite linear viscoelasticity.

β) *Expansive wave decay.*[14] It was shown in Sect. 20 that the growth or decay of an acceleration wave propagating in a homogeneously deformed region is

[12] SCHULER [1970, *19*].
[13] BARKER and HOLLENBACH [1970, *2*].
[14] The results presented here have not been reported previously.

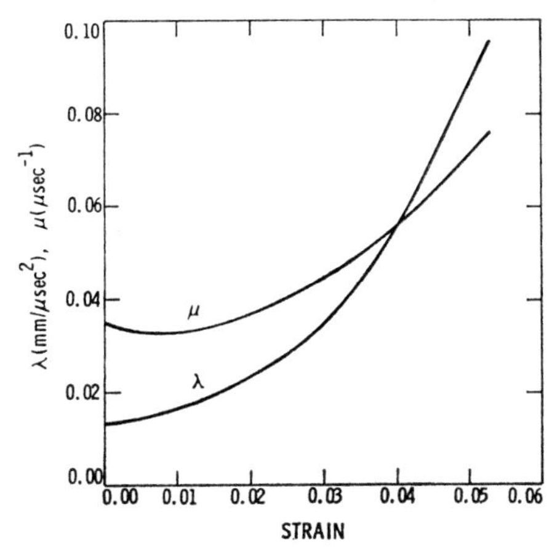

Fig. 36. Calculated critical amplitude λ and μ for polymethyl methacrylate as a function of initial equilibrium strain.

governed by a critical amplitude λ and a parameter μ, both of which depend on the acceleration wave velocity U, the initial value of the stress relaxation function $G(0)$, the instantaneous second-order modulus $\tilde{E}$, and the initial slope of the stress relaxation function $G'(0)$. Here we wish to apply these results to the case of an expansive wave propagating into a region with a known initial equilibrium strain $\mathring{\varepsilon}$.

For a finite linear viscoelastic solid, the intrinsic velocity U is given by (21.2) and it follows that

$$E = G(0) = \varrho_{\mathrm{R}} U^2. \tag{21.3}$$

Furthermore, by $(9.4)_2$ and (14.1), the instantaneous second-order modulus $\tilde{E}_t$ can be expressed as

$$\tilde{E}_t = \tilde{E}_{\mathrm{I}}(\varepsilon) + \int_0^\infty \left[\frac{d^2}{d\varepsilon^2} K'(\varepsilon; s)\right] \{1 - (1 - \varepsilon^t(s))^2\} ds. \tag{21.4}$$

Evaluating this result for the equilibrium history (9.8) and using (14.2) yields

$$\begin{aligned} \tilde{E} = \tilde{E}_{\mathrm{E}}(\mathring{\varepsilon}) + \frac{2}{(2-\mathring{\varepsilon})\mathring{\varepsilon}} \Big\{ & 2[E_{\mathrm{I}}(\mathring{\varepsilon}) - E_{\mathrm{E}}(\mathring{\varepsilon})](1-\mathring{\varepsilon}) \\ & - [\sigma_{\mathrm{I}}(\mathring{\varepsilon}) - \sigma_{\mathrm{E}}(\mathring{\varepsilon})] \left[1 + \frac{4(1-\mathring{\varepsilon})^2}{(2-\mathring{\varepsilon})\mathring{\varepsilon}}\right] \Big\}. \end{aligned} \tag{21.5}$$

Finally, to determine the initial slope of the stress relaxation function $G'(0)$, we carry out a calculation analogous to that in Sect. 17 except that the identity (17.5) is now evaluated for the equilibrium history (9.8). This yields

$$G'(0) = -\frac{2[\sigma_{\mathrm{I}}(\mathring{\varepsilon}) - \sigma_{\mathrm{E}}(\mathring{\varepsilon})]}{\mathring{\varepsilon}(2-\mathring{\varepsilon})\tau}. \tag{21.6}$$

Using the instantaneous and equilibrium stress-strain curves given in Sect. 14 and the relaxation time τ obtained in Sect. 15, we can calculate, from $(20.6)_1$ and (20.8), the variation of the critical amplitude λ and the parameter μ with the initial state of equilibrium strain for polymethyl methacrylate. The results are shown in Fig. 36 and, in view of (20.7), enable us to determine the amplitude behavior of an acceleration wave for any known value of initial strain. It is of interest to note that the critical amplitude λ *increases* with initial strain.

The experimental observation of the decay of an expansive acceleration wave requires that we observe the unloading wave in the material sample at various distances from the impact surface. Furthermore, in each experiment, the impact condition must be nominally the same and, most important of all, the impact must be symmetric with the thickness of the flyer plate X_p nominally the same. The flyer plate thickness is the quantity which determines the initial amplitude of the acceleration wave introduced into the material sample after the passage of the shock front. The impact condition determines the equilibrium state of strain into which the unloading wave propagates.

Shots 315, 2110, and 2115 on polymethyl methacrylate (cf. Table 1) for the nominal impact condition of 0.15 mm/μsec all had essentially the same flyer thickness (for 315, $X_p=5.936$ mm; for 2110, $X_p=6.038$ mm; and for 2115, $X_p=6.040$ mm) and provide measurements at approximately 6, 12, and 37 mm. Thus, the observation of the unloading wave at each mirror location should provide an observation of expansive wave decay. These unloading profiles are shown in Fig. 37 and confirm the decrease in the amplitude of the wave with propagation distance (or time after impact). To correlate these results with the theoretical predictions of acceleration wave behavior, we take the first profile as the initial condition $a(0)$ and then calculate the amplitude of the two later profiles, $a(2.085\ \mu\text{sec})$ and $a(9.690\ \mu\text{sec})$, using (20.7). Thus, the initial amplitude $a(0)$ of the acceleration wave is -0.105 mm/μsec² and for the 0.15 mm/μsec impact condition, $\mathring{\varepsilon}=0.0254$. For this initial strain, the calculated velocity $U=3.19$ mm/μsec compares very well with the observed unloading wave velocity of 3.22 mm/μsec, and from Fig. 36 we have a critical amplitude $\lambda=0.028$ mm/μsec² and $\mu=0.040$/μsec. Using these values, we calculate $a(2.085\ \mu\text{sec})=-0.074$ mm/μsec² and $a(9.690\ \mu\text{sec})=-0.032$ mm/μsec². The slopes corresponding to these amplitudes are also shown in Fig. 37 and are in good agreement with the measured values of -0.070 mm/μsec² and -0.035 mm/μsec², respectively.

It should be noted at this point that the velocity and amplitude behavior of unloading waves in polymethyl methacrylate for stresses below 7.5 kbar can also be accurately predicted by assuming that the material has been subjected to the jump history (9.5). This appears to be due to the fact that over this stress range the instantaneous and equilibrium stress-strain curves for the material are very close together. Consequently, it would appear that rate effects have only a small influence on the unloading behavior of the material.[15] This gives further support

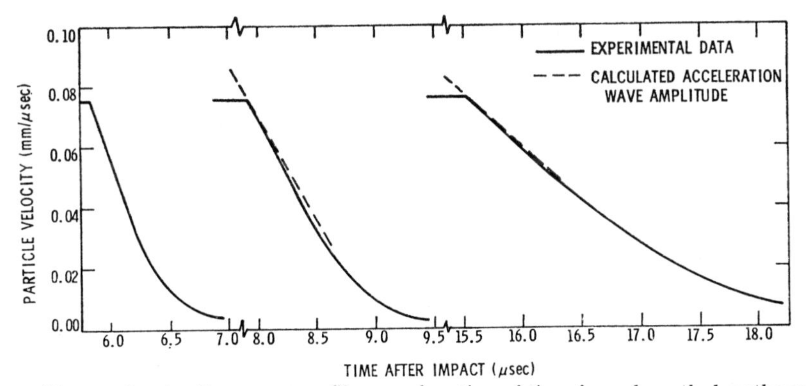

Fig. 37. Measured unloading wave profiles as a function of time for polymethyl methacrylate.

[15] Cf. SCHULER [1970, *20*]. In this regard, also see SCHULER and NUNZIATO [1974, *3*] who make use of this observation to construct a Maxwellian constitutive model which describes the response of polymethyl methacrylate to 60 kbar.

to the assumption in our discussion of shock pulse attenuation that the unloading response can be treated as nonlinear elastic.[16]

22. Compressive wave growth in undeformed regions at rest. Here we consider the case of an acceleration wave propagating into an unstrained region at rest and define the moduli

$$E_t = D_\varepsilon \mathcal{S}(O_\mathrm{E}) = (E_\mathrm{I})_0, \quad \tilde{E}_t = D_\varepsilon^2 \mathcal{S}(O_\mathrm{E}) = (\tilde{E}_\mathrm{I})_0, \quad G_t(s) = \mathcal{G}(O_\mathrm{E};\, s) = G_0(s), \tag{22.1}$$

where we recall the definition of the rest history $\mathring{\varepsilon}_\mathrm{E}(s) = O_\mathrm{E}(s) = 0$, $s \in [0, \infty)$. Thus the intrinsic velocity U is given by

$$U^2 = \frac{(E_\mathrm{I})_0}{\varrho_\mathrm{R}} \equiv U_0^2, \tag{22.2}$$

and λ and μ become

$$\lambda = -\frac{G_0'(0)\, U_0}{(\tilde{E}_\mathrm{I})_0} \equiv \lambda_0, \qquad \mu = -\frac{G_0'(0)}{2G_0(0)} \equiv \mu_0 \tag{22.3}$$

with $G_0(0)$ and $G_0'(0)$ the initial value and slope of the stress relaxation function $G_0(\cdot)$. Methods for determining these constants for use in predicting the growth and decay of acceleration waves in undeformed regions from the results of steady shock wave experiments have been suggested by CHEN and GURTIN[17] and by WALSH and SCHULER.[18]

CHEN and GURTIN have observed that λ_0 and μ_0 can be determined directly from experimental steady wave data without appealing to an explicit constitutive relation. That is, in a steady shock wave, $(\dot{v})^- = U^2 |\lambda^*|$ (the critical acceleration). As shown in Sect. 17, this quantity can be evaluated graphically from steady wave profiles for various amplitudes v^-. Then extrapolating this data to $v^- = 0$ and using (16.11) yields the critical amplitude λ_0. Finally, by use of (9.7), the constants $(E_\mathrm{I})_0 = G_0(0)$ and $(\tilde{E}_\mathrm{I})_0$ can be evaluated directly from the instantaneous stress-strain curve $\sigma_\mathrm{I}(\varepsilon)$ and μ_0 can be calculated from (22.3), i.e.,

$$\mu_0 = \frac{(\tilde{E}_\mathrm{I})_0\, \lambda_0}{2(E_\mathrm{I})_0\, U_0}. \tag{22.4}$$

Although this approach is satisfactory in principle, the usually limited number of experimentally determined particle accelerations $(\dot{v})^-$ and their large uncertainty make the value λ_0 obtained by an extrapolation to zero particle velocity questionable (e.g., Fig. 28). WALSH and SCHULER, on the other hand, have used an alternative technique to evaluate λ_0 and μ_0 which in essence makes use of more experimental data. They evaluated (21.2), (21.3), (21.5), and (21.6), which are based on the constitutive model of finite linear viscoelasticity, at $\varepsilon = 0$ to obtain the constants necessary to calculate λ_0 and μ_0 from (22.3). This is equivalent to extrapolating the curves for λ and μ in Fig. 36 to zero strain. By use of the values of λ_0 and μ_0 obtained in this manner,[19] the critical time t_∞ at which a shock may form can be determined from (20.9) as a function of the initial amplitude $a(0)$. This result, again for the polymer, polymethyl methacrylate, is shown in Fig. 38. Also shown in this figure is the $t_\infty(a(0))$ behavior for a nonlinear elastic material[20]

---

[16] Cf. Sect. 18.

[17] CHEN and GURTIN [1972, *6*]. Also see CHEN [1973, 4, Sect. 17].

[18] WALSH and SCHULER [1973, *20*].

[19] For polymethyl methacrylate, $\lambda_0 = 0.013$ mm/μsec² and $\mu_0 = 0.035$/μsec; cf. WALSH and SCHULER [1973, *20*].

[20] Cf. CHEN [1973, *4*, Sect. 14].

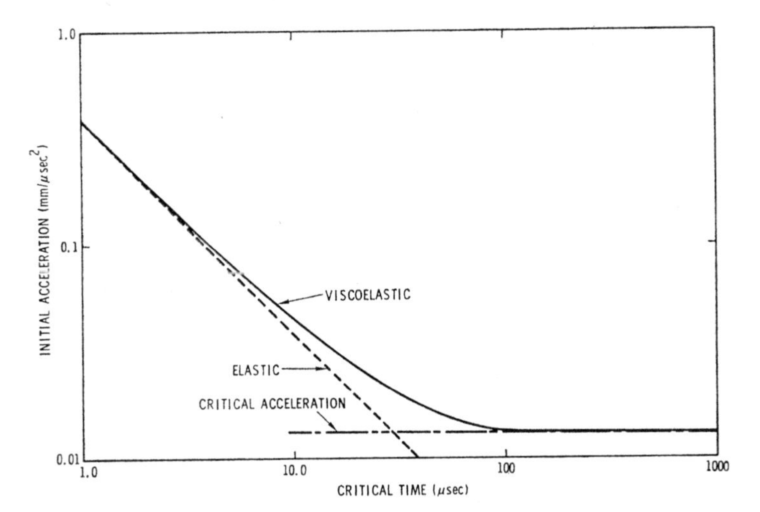

Fig. 38. Critical time vs. initial amplitude of compressive acceleration waves for polymethyl methacrylate. (From WALSH and SCHULER [1973, *20*].)

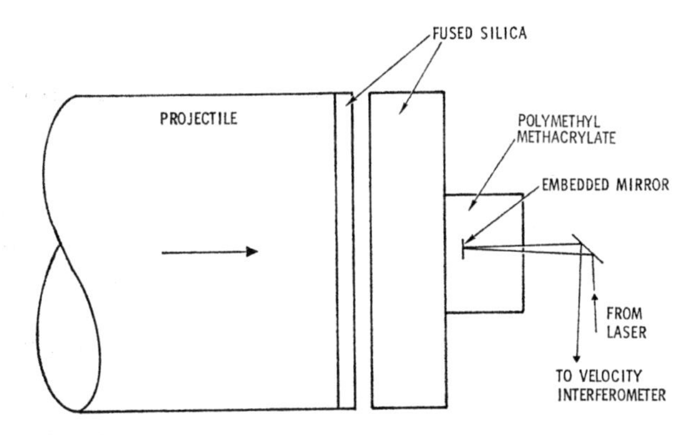

Fig. 39. Experimental plate impact configuration for acceleration wave studies. (From WALSH and SCHULER [1973, *20*].)

which has the same values for the tangent and second-order moduli as polymethyl methacrylate (curve labeled elastic).[21]

WALSH and SCHULER[22] also compared the predictions of Fig. 38 with experimental observations of the growth of compressive acceleration waves in polymethyl methacrylate. The experimental configuration used is shown in Fig. 39. The generation of the acceleration wave is due to the unique properties of fused silica.[23] Unlike most materials, the one-dimensional stress-strain curve of this elastic material (for stresses less than about 35 kbar) has a curvature such that compressive discontinuities tend to spread, i.e., $\tilde{E} < 0$ (cf. Fig. 40). Thus, as the shock discontinuity produced at the impact surface propagates into the fused silica, it spreads continuously, forming an acceleration wave which is introduced into the polymethyl methacrylate sample. It is not difficult to show using a singular surface

[21] The values of $(E_{\mathrm{I}})_0 = 90.3$ kbar and $(\tilde{E}_{\mathrm{I}})_0 = 1312$ kbar were used in the calculation.
[22] WALSH and SCHULER [1973, *20*].
[23] Cf. BARKER and HOLLENBACH [1970, *2*].

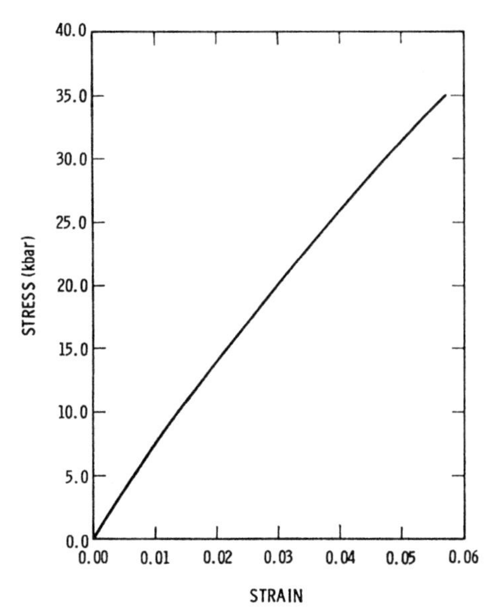

Fig. 40. Stress-strain curve for fused silica.

analysis that the amplitude of the acceleration discontinuity introduced varies inversely with the thickness of the fused silica.[24] In order to maintain a one-dimensional motion for the duration of the experiment, the thickness of the fused silica is limited to about 4 cm. This corresponds to a lower limit on the acceleration discontinuity producible with this technique of approximately 0.27 mm/μsec². Unfortunately, this value of $a(0)$ is greater than the value of $\lambda_0 = 0.013$ mm/μsec² calculated previously and thus, as expected, only acceleration waves which grow were observed.

The data from six experiments were available for correlation with the predicted behavior of acceleration waves in polymethyl methacrylate. These data, which are shown in Fig. 41 (curves A1, A2, A3 and B1, B2, B3),[25] can be grouped into two series depending on the thickness of the fused silica employed.

The first series (input A) employed a 25.5 mm thick piece of fused silica to obtain an initial acceleration wave amplitude of 0.41 mm/μsec². From Fig. 38 this input wave amplitude corresponds to a critical time $t_\infty = 0.94$ μsec. The transit time from the fused silica-polymethyl methacrylate interface to the mirror was 1.06, 1.63, and 2.25 μsec for shots A1, A2, A3, respectively.[26] Thus, in each of these experiments a shock discontinuity is expected to have formed. The wave profiles measured at the mirror are shown in Fig. 41, where it is clear that the shocks have formed.

The second series (input B) used a 38.1 mm thick piece of fused silica which yields an initial acceleration level of $a(0) = 0.27$ mm/μsec². From Fig. 38 the

---

[24] This can be seen from the acceleration wave amplitude equation for nonlinear elastic materials (cf. CHEN [1973, *4*, Sect. 14]) by noting that the propagation distance X obeys $X = U_0 t$ and that at the impact plane the wave is a shock, i.e., $a(0) = \infty$.

[25] Two of these experiments (A3 and B3) were performed by BARKER and HOLLENBACH [1970, *2*]. The remainder are due to WALSH and SCHULER [1973, *20*].

[26] Here transit times are calculated from $t = X_m / U_0$ where the acceleration wave velocity U_0 is the lead coefficient of the shock velocity-particle velocity fit $V_I(v^-)$ given in Sect. 13, i.e., $U_0 = (C_I)_0 = 2.76$ mm/μsec. As was noted there, this value agrees very well with that deduced from acoustic dispersion data (cf. Sect. 11).

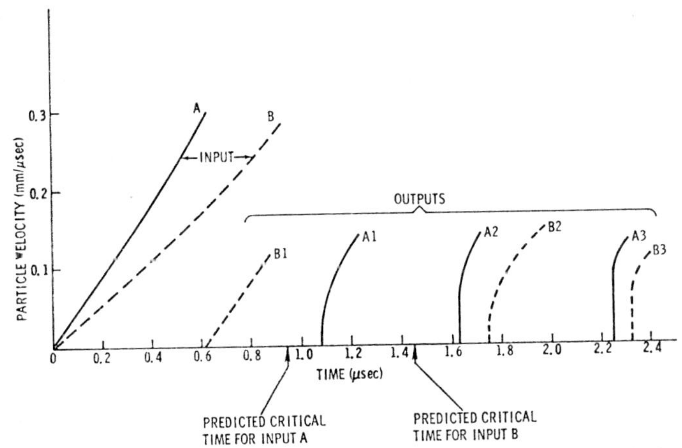

Fig. 41. Experimental results for acceleration wave propagation in polymethyl methacrylate (From WALSH and SCHULER [1973, *20*].)

corresponding critical time is $t_\infty = 1.45$ μsec. Two of the shots in this series (B2, B3) had propagation times (1.74, 2.32 μsec, respectively) greater than t_∞. Thus, again, for B2, B3, a shock wave is expected to form and this is borne out in Fig. 41. In shot B1 a wave was observed after propagating for 0.63 μsec. Since this time is less than t_∞, the formation of the shock discontinuity is not predicted. The results are shown in Fig. 41 where, as expected, no shock is evident. In addition, since $a(0) > \lambda_0$, the acceleration amplitude is increasing and can be predicted from (20.7). Considering $a(0) = 0.27$ mm/μsec², we calculate $a(0.63\ \mu\text{sec}) = 0.47$ mm/μsec². The measured level of 0.46 mm/μsec² is in good agreement with this predicted value.

It is interesting to note from Fig. 38 that, at these levels of initial acceleration, the predicted t_∞ for polymethyl methacrylate is close to that for a nonlinear elastic solid with the same tangent and second-order moduli, E and $\tilde{E}$. Of course for elastic materials there is no critical acceleration. Nevertheless, an acceleration wave whose initial amplitude is an order of magnitude greater than the critical acceleration will propagate in nonlinear polymeric solids (such as polymethyl methacrylate) as if the material were elastic.[27] This observation suggests that in certain cases the simpler elastic theory can be used to interpret the results of acceleration wave experiments conducted on nonlinear viscoelastic materials.

23. Toward a three-dimensional characterization of nonlinear viscoelastic solids. As we have seen in this chapter and in the previous chapter on shock waves, the growth and decay of one-dimensional waves in nonlinear viscoelastic solids can be accurately predicted using the one-dimensional constitutive equation developed from acoustic and steady wave studies. However, in order to predict the dynamic response of viscoelastic solids in loading configurations more general than those associated with one-dimensional strain, it is necessary to consider a specific, nonlinear, three-dimensional constitutive assumption, such as that of finite linear viscoelasticity. Furthermore, one must be able to evaluate the relevant material response functions. In the case of isotropic finite linear viscoelasticity there are 15 independent response functions to determine.[28] Thus, while the consideration

[27] By considering the asymptotic behavior of the solutions of the Bernoulli equation (20.5) which governs acceleration wave propagation, BAILEY and CHEN [1972, *2*] reach a similar conclusion. See also CHEN [1973, *4*, Sect. 14].

[28] Cf. COLEMAN and NOLL [1961, *1*].

of one-dimensional strain (longitudinal motion) provides some information with regard to these functions, it is obviously not sufficient to determine the general constitutive model completely. In view of this, it is desirable to develop a program by which one can evaluate the material functions necessary to characterize the dynamic, three-dimensional response of viscoelastic solids from experimental data.[29] A step in this direction has been taken by NUNZIATO, SCHULER, and WALSH[30] who considered the dilatational response of finite linear viscoelastic solids. Motivated by the one-dimensional studies, they characterized the dilatational response in terms of an instantaneous and equilibrium response, i.e., the response to a suddenly applied dilatational volume change and the response to a similar volume change held for long times. Then, using an acceleration wave analysis and the equivalence of acceleration waves and acoustic waves, they give a procedure for determining the instantaneous pressure-density curve from the equilibrium pressure-density curve and results of acoustic wave experiments performed at various equilibrium states of hydrostatic pressure.[31] Here we outline the analysis by NUNZIATO, SCHULER, and WALSH and give their results for the polymer, polymethyl methacrylate.

α) *Dilatational response of isotropic finite linear viscoelastic solids.* Here we consider a three-dimensional body $\mathscr{B}$ which is identified with a region in Euclidean space. The motion of $\mathscr{B}$ is described by the function $\boldsymbol{\chi}$ such that

$$\boldsymbol{x}=\boldsymbol{\chi}(\boldsymbol{X}, t)$$

gives the position $\boldsymbol{x}$ at time t of the material particle which occupied the position $\boldsymbol{X}$ in some homogeneous reference configuration. Assuming sufficient smoothness of $\boldsymbol{\chi}(\cdot, \cdot)$, we define

$$\dot{\boldsymbol{x}}=\frac{\partial}{\partial t}\boldsymbol{\chi}(\boldsymbol{X}, t), \qquad \ddot{\boldsymbol{x}}=\frac{\partial^2}{\partial t^2}\boldsymbol{\chi}(\boldsymbol{X}, t)$$

as the particle velocity and acceleration, and

$$\boldsymbol{F}=\nabla_{\boldsymbol{X}}\boldsymbol{\chi}(\boldsymbol{X}, t)$$

as the deformation gradient. The tensors

$$\boldsymbol{C}=\boldsymbol{F}^{\mathrm{T}}\boldsymbol{F}, \qquad \boldsymbol{B}=\boldsymbol{F}\boldsymbol{F}^{\mathrm{T}}$$

are called the *right* and *left Cauchy-Green tensors*, respectively.[32] Also their histories $\boldsymbol{C}^t(\cdot)$ and $\boldsymbol{B}^t(\cdot)$ are defined in the same manner as the one-dimensional strain history $\varepsilon^t(\cdot)$ in Sect. 7.

Through the balance of mass, $\boldsymbol{F}$, $\boldsymbol{C}$, and $\boldsymbol{B}$ are related to the present material density ϱ and the reference density ϱ_{R} by

$$\frac{\varrho_{\mathrm{R}}}{\varrho}=\det\boldsymbol{F}=(\det\boldsymbol{C})^{\frac{1}{2}}=(\det\boldsymbol{B})^{\frac{1}{2}}.$$

The balance of linear momentum, neglecting body forces, is

$$\mathrm{div}_{\boldsymbol{x}}\,\boldsymbol{t}=\varrho\ddot{\boldsymbol{x}} \tag{23.1}$$

where $\boldsymbol{t}=\boldsymbol{t}^{\mathrm{T}}$ is the *Cauchy stress.*

[29] Such a program has been discussed for elastic materials by TRUESDELL [1961, *8*].

[30] NUNZIATO, SCHULER, and WALSH [1972, *16*].

[31] In examining existing experimental techniques, it was found that, while several techniques lead to a determination of the (near) equilibrium dilatational response, none seemed to provide sufficient information to determine directly the instantaneous response. A discussion of the various techniques is given by MARVIN and MCKINNEY [1965, *17*].

[32] The usual direct tensor notation is followed: $\boldsymbol{A}^{\mathrm{T}}$ is the transpose of $\boldsymbol{A}$, $\boldsymbol{A}^{-1}$ the inverse, det $\boldsymbol{A}$ the determinant and tr $\boldsymbol{A}$ the trace; $\mathbf{1}$ is the unit or identity tensor.

In terms of the deformation history[33]

$$\boldsymbol{J}^t(s) = \mathbf{1}_E(s) - (\boldsymbol{F}^T)^{-1}\, \boldsymbol{C}^t(s)\, \boldsymbol{F}^{-1} \tag{23.2}$$

and a tensor-valued stress relaxation function $\boldsymbol{\Gamma}(\boldsymbol{B}; s)$, the class of homogeneous, finite linear viscoelastic solids is defined by the constitutive equation

$$\boldsymbol{t} = \boldsymbol{H}(\boldsymbol{B}) + \int_0^\infty \boldsymbol{\Gamma}(\boldsymbol{B}; s)\{\boldsymbol{J}^t(s)\}\, ds. \tag{23.3}$$

The tensor-valued function $\boldsymbol{H}(\boldsymbol{B})$ will be shown to relate to the equilibrium response; also, it is assumed that $\boldsymbol{\Gamma}(\boldsymbol{B}; s)$ has the property $\lim_{s\to\infty} \boldsymbol{\Gamma}(\boldsymbol{B}; s) = 0$. For isotropic materials, $\boldsymbol{H}(\boldsymbol{B})$ and $\boldsymbol{\Gamma}(\boldsymbol{B}; s)$ have the representations

$$\boldsymbol{H}(\boldsymbol{B}) = h_0 \mathbf{1} + h_1 \boldsymbol{B} + h_2 \boldsymbol{B}^2, \tag{23.4}$$

$$\begin{aligned}\boldsymbol{\Gamma}(\boldsymbol{B}; s)\{\boldsymbol{J}^t(s)\} = {} & \boldsymbol{M}_1(\boldsymbol{B}; s)\boldsymbol{J}^t(s) + \boldsymbol{J}^t(s)\boldsymbol{M}_1(\boldsymbol{B}; s) \\ & + \operatorname{tr}[\boldsymbol{J}^t(s)\boldsymbol{M}_2(\boldsymbol{B}; s)]\mathbf{1} + \operatorname{tr}[\boldsymbol{J}^t(s)\boldsymbol{M}_3(B; s)]\boldsymbol{B} \\ & + \operatorname{tr}[\boldsymbol{J}^t(s)\boldsymbol{M}_4(\boldsymbol{B}; s)]\boldsymbol{B}^2.\end{aligned} \tag{23.5}$$

The quantities h_0, h_1, and h_2 are functions of the principal invariants of $\boldsymbol{B}$, and the tensor functions $\boldsymbol{M}_i(\boldsymbol{B}; s)$ $(i = 1, 2, 3, 4)$ are isotropic for each s, $s \in [0, \infty)$, with representations of the form (23.4).[34]

Here we are interested in the dilatational response of an isotropic solid to hydrostatic loads. In this case, the deformation can be represented by

$$\boldsymbol{F} = v\mathbf{1}$$

where $v \in (0, 1)$, is the *compression ratio*. Thus, it follows that

$$\boldsymbol{C} = \boldsymbol{B} = v^2 \mathbf{1}$$

and

$$\frac{\varrho_R}{\varrho} = v^3. \tag{23.6}$$

The equilibrium dilatational response of the solid is its response to a constant hydrostatic deformation for all time, i.e.,

$$\boldsymbol{F}^t(s) = v\mathbf{1}_E(s), \qquad s \in [0, \infty).$$

Then $\boldsymbol{C}^t(s) = v^2 \mathbf{1}_E(s)$, $s \in [0, \infty)$, and hence

$$\boldsymbol{J}^t(s) = 0.$$

The stress response (23.3) to this deformation is

$$\boldsymbol{t} = \boldsymbol{t}_{EH} = \boldsymbol{H}(\boldsymbol{B})\big|_{\boldsymbol{B} = v^2\mathbf{1}} \tag{23.7}$$

and, in view of (23.4) and (23.6), serves to define the *equilibrium pressure-density curve* π_E:

$$\pi_E(\varrho) = \tfrac{1}{3} \operatorname{tr}(\boldsymbol{t}_{EH}) = h_0 + h_1 v^2 + h_2 v^4. \tag{23.8}$$

---

[33] The function $\mathbf{1}_E(s)$, $s \in [0, \infty)$ is the constant tensor history with unity as its diagonal components.

[34] Cf. COLEMAN and NOLL [1961, *1*]. It is a simple matter to show that in the case of one-dimensional strain, i.e.,

$$[\boldsymbol{F}]_{ij} = \begin{bmatrix} 1-\varepsilon & 0 & 0 \\ 0 & 1 & 0 \\ 0 & 0 & 1 \end{bmatrix},$$

the longitudinal component of stress $t_{11} = \sigma$ can be written in the form (14.1).

The *equilibrium bulk modulus* B_E is defined by[35]

$$B_E(\varrho) = \frac{\varrho^2}{\varrho_R} \frac{d}{d\varrho} \pi_E(\varrho). \tag{23.9}$$

The instantaneous dilatational response of an isotropic viscoelastic solid is its response to a hydrostatic compression suddenly applied at time t, i.e.,

$$\boldsymbol{F}^t(s) = \begin{cases} v\mathbf{1}, & s=0, \\ \mathbf{1}, & s \in (0, \infty). \end{cases}$$

Then $\boldsymbol{C}^t(s) = \mathbf{1}_E(s)$, and hence

$$\boldsymbol{J}^t(s) = \left(\frac{v^2-1}{v^2}\right) \mathbf{1}_E(s).$$

The stress response to this deformation is

$$\boldsymbol{t} = \boldsymbol{t}_{IH} = \boldsymbol{H}(\boldsymbol{B})\big|_{\boldsymbol{B}=v^2\mathbf{1}} + \left(\frac{v^2-1}{v^2}\right)\left[\int_0^\infty \boldsymbol{\Gamma}(\boldsymbol{B}; s)\big|_{\boldsymbol{B}=v^2\mathbf{1}}\{\mathbf{1}_E(s)\}\, ds\right].$$

Thus, by (23.4) to (23.8), we can define the *instantaneous pressure-density curve* π_I by

$$\pi_I(\varrho) = -\frac{1}{3}\operatorname{tr}(\boldsymbol{t}_{IH}) = \pi_E(\varrho) + \left[1 - \left(\frac{\varrho}{\varrho_R}\right)^{\frac{2}{3}}\right]\gamma(\varrho) \tag{23.10}$$

where

$$\gamma(\varrho) = \int_0^\infty \{2m_1(\varrho; s) + 3\,[m_2(\varrho; s) + m_3(\varrho; s)\, v^2 + m_4(\varrho; s)\, v^4]\}\, ds, \tag{23.11}$$

and, for $i = 1, 2, 3, 4$,

$$m_i(\varrho; s) = \tfrac{1}{3}\operatorname{tr}[\boldsymbol{M}_i(\boldsymbol{B}; s)\big|_{\boldsymbol{B}=v^2\mathbf{1}}]. \tag{23.12}$$

In view of (23.10), we define the *instantaneous bulk modulus* by

$$B_I(\varrho) = \frac{\varrho^2}{\varrho_R} \frac{d}{d\varrho} \pi_I(\varrho). \tag{23.13}$$

For a body subject to compressive dilatation we would expect $\pi_I \geqq \pi_E$. Thus, since $v \in (0, 1)$, we conclude that $\gamma(\varrho) < 0$. Consequently, if the body were loaded instantaneously by a hydrostatic compression and then held at fixed density, (23.10) implies that the pressure would relax to the equilibrium curve by an amount equal to $[1 - (\varrho/\varrho_R)^{\frac{2}{3}}]\,\gamma(\varrho)$. The quantity $\gamma(\varrho)$ is a measure of the material's bulk relaxation. However, such bulk stress relaxation experiments apparently have not been carried out,[36] and therefore, to determine the instantaneous and equilibrium curves, π_I and π_E, we must turn to other experimental procedures. One such procedure involves results from wave propagation experiments, in particular acceleration waves, which we consider next.

β) The velocity of acceleration waves in isotropic solids subject to hydrostatic pressure. In a three-dimensional body, a wave front is represented by a smoothly propagating surface $\mathscr{S}(t)$, $t \in (0, t_0)$, across which certain kinematical quantities suffer jump discontinuities. The hypersurface

$$\mathscr{W} = \{(\boldsymbol{x}, t) : \boldsymbol{x} \in \mathscr{S}(t),\ t \in (0, t_0)\}$$

[35] The determination of the equilibrium isothermal bulk modulus for a given solid will be shown to be particularly useful in the more general considerations of thermoviscoelastic response (cf. Sect. 28).

[36] The corresponding hydrostatic creep experiments have been performed for some polymers; cf. FINDLEY, REED, and STERN [1967, *2*] and FERRY [1970, *11*].

is also assumed to be smooth and is called the *local trajectory* of the wave front. At each point $(\boldsymbol{x}, t)$ on $\mathscr{W}$ there exists a normal $(\boldsymbol{n}, -U)$ where $\boldsymbol{n}(|\boldsymbol{n}|=1)$ represents its *local direction of propagation* and $U(t)>0$ its *local speed of propagation*.[37]

Analogous to one-dimensional waves, the surface $\mathscr{W}$ is called an *acceleration wave* if the motion $\boldsymbol{\chi}(\cdot, \cdot)$ and its first derivatives are continuous everywhere on $\mathscr{B}\times(-\infty, \infty)$ while its second-order derivatives are continuous everywhere on $\mathscr{B}\times(-\infty, \infty)$ except at $\mathscr{W}$ where they may suffer, at most, jump discontinuities. The *amplitude* $\boldsymbol{a}(t)$ of the wave front is then defined through the jump in the acceleration vector $\ddot{\boldsymbol{x}}$:

$$[\ddot{\boldsymbol{x}}](t)=\boldsymbol{a}(t).$$

The wave is said to be *longitudinal* if and only if its local direction is parallel to its amplitude vector, i.e.,

$$\boldsymbol{a}\times\boldsymbol{n}=0;$$

transverse if and only if its local direction is perpendicular to its amplitude vector, i.e.,

$$\boldsymbol{a}\cdot\boldsymbol{n}=0.$$

NUNZIATO, SCHULER, and WALSH[38] have considered the propagation of acceleration waves in isotropic finite linear viscoelastic solids subjected to a uniform state of hydrostatic pressure and derived explicit formulae for the velocities of propagation. In this particular case every wave is either longitudinal or transverse. The longitudinal wave velocity U_{L} obeys[39]

$$\begin{aligned} U_{\mathrm{L}}^2= & \frac{4v^2}{3\varrho}(h_1+2h_2v^2)+\frac{d}{d\varrho}\pi_{\mathrm{E}}(\varrho) \\ & -\frac{2}{\varrho}\int_0^\infty[2m_1(\varrho;s)+m_2(\varrho;s)+m_3(\varrho;s)v^2+m_4(\varrho;s)v^4]\,ds \end{aligned} \tag{23.14}$$

and the transverse wave velocity U_{T} is given by

$$U_{\mathrm{T}}^2=\frac{v^2}{\varrho}(h_1+2h_2v^2)-\frac{2}{\varrho}\int_0^\infty m_1(\varrho;s)\,ds. \tag{23.15}$$

Here h_i $(i=0, 1, 2)$ and m_i $(i=1, 2, 3, 4)$ are defined through (23.4) and (23.12), respectively, v is the compression ratio corresponding to the underlying state of dilatation, and $\pi_{\mathrm{E}}(\varrho)$ is the equilibrium pressure-density curve.

γ) *Experimental determination of the instantaneous and equilibrium pressure-density curves.* Several experimental studies have considered the quasi-static response of polymers to hydrostatic-type loading.[40] In particular, FINDLEY, REED, and STERN[41] have investigated the hydrostatic creep of solid polymer samples immersed in a pressurized liquid and GIELESSEN and KOPPELMANN, HEYDEMANN and GUICKING, and SPETZLER,[42] using a similar hydrostatic environment, have

[37] Cf. GURTIN [1972, *12*, Sect. 72]. In this regard, also see TRUESDELL and TOUPIN [1960, *7*, Sect. 173] and CHEN [1973, *4*, Sect. 4].

[38] NUNZIATO, SCHULER, and WALSH [1972, *16*].

[39] Cf. TRUESDELL [1961, *8*], who gives the corresponding results for elastic materials. Also see CHEN [1973, *4*, Sect. 10].

[40] A comprehensive review of hydrostatic loading experiments and the bulk properties of polymers has been given by CURRO [1974, *1*]. In this regard, also see the excellent discussions by MARVIN and MCKINNEY [1965, *17*] and FERRY [1970, *11*, Chap. 18].

[41] FINDLEY, REED, and STERN [1967, *2*].

[42] GIELESSEN and KOPPELMANN [1960, *3*], HEYDEMANN and GUICKING [1963, *2*], and H. A. SPETZLER (unpublished).

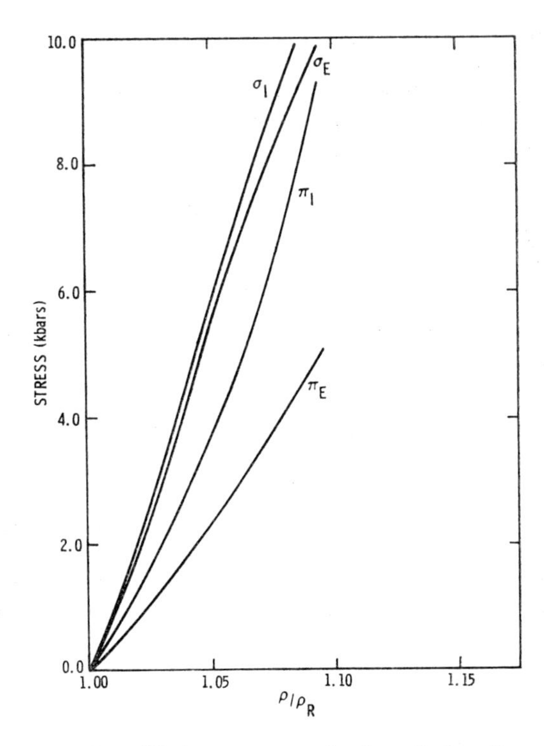

Fig. 42. Instantaneous and equilibrium response of polymethyl methacrylate to hydrostatic compression (π_I and π_E) and to one-dimensional compressive strain (σ_I and σ_E). (From NUNZIATO, SCHULER, and WALSH [1972, *16*].)

determined the pressure-density response of polymethyl methacrylate. In each of these studies sufficient time was allowed to elapse after each pressure change to ensure that mechanical equilibrium was achieved before making the measurements. Since these experiments were performed using purely hydrostatic loading conditions and are in essence equilibrium experiments, we interpret the resulting pressure-density curves as an accurate description of the equilibrium response of polymethyl methacrylate.[43] The composite of the results of these studies is shown in Fig. 42 as the curve π_E.

The determination of the instantaneous pressure-density curve π_I is not, however, as straightforward. That is, as we have already observed, the dilatational response to a suddenly applied hydrostatic load has not been measured directly. However, in view of (23.10), π_I can be obtained if, in addition to the equilibrium curve π_E, the bulk relaxation parameter $\gamma(\varrho)$ is also known. It follows from (23.11), (23.14), and (23.15) that this function γ is directly related to the velocities of longitudinal and transverse acceleration waves in a solid subjected to hydrostatic pressure, i.e.,

$$\gamma(\varrho) = \frac{3}{2}\varrho\left\{\frac{d}{d\varrho}\pi_E(\varrho) - \left[U_L^2(\varrho) - \frac{4}{3}U_T^2(\varrho)\right]\right\}.$$

[43] The use of this equilibrium response data as the equilibrium data for the relatively short time scale appropriate for wave propagation studies is a result of the fact that the modulus associated with the bulk response of viscoelastic solids exhibits a narrow range of magnitudes when measured over a wide range of time (cf. FERRY [1970, *11*]).

Thus, (23.10) becomes

$$\pi_{\mathrm{I}}(\varrho)=\pi_{\mathrm{E}}(\varrho)+\frac{3}{2}\,\varrho\left\{1-\left(\frac{\varrho}{\varrho_{\mathrm{R}}}\right)^{\frac{2}{3}}\right\}\left\{\frac{d}{d\varrho}\,\pi_{\mathrm{E}}(\varrho)-\left(U_{\mathrm{L}}^{2}(\varrho)-\frac{4}{3}\,U_{\mathrm{T}}^{2}(\varrho)\right)\right\}. \qquad (23.16)$$

As a consequence of the equivalence of acceleration wave and high-frequency acoustic wave velocities, (23.16) can be viewed as a relation between π_{I}, π_{E}, and high-frequency acoustic wave velocities corresponding to various levels of equilibrium hydrostatic pressure. Such acoustic velocities can be readily measured and have been reported extensively in the literature.[44] In particular, ASAY, LAMBERSON, and GUENTHER[45] have measured the longitudinal and transverse acoustic wave velocities in a pressurized sample of polymethyl methacrylate using an echo-pulse transmission technique. Using this data along with the previously determined equilibrium curve π_{E} and Eq. (23.16), we generate the instantaneous pressure-density curve π_{I} shown in Fig. 42.[46]

Thus, we see that from wave propagation experiments it is possible to evaluate, in addition to the one-dimensional (longitudinal) stress-strain response, the dilatational pressure-density response of a nonlinear viscoelastic solid. The instantaneous and equilibrium curves σ_{I} and σ_{E} for polymethyl methacrylate[47] are also shown in Fig. 42 and provide some indication of the shear strength of the solid at these high stress levels.

VIII. Thermodynamic influences on viscoelastic wave propagation.

24. Thermodynamics of materials with memory.[1] In all of the discussions to this point the consideration of thermodynamic influences on the propagation of shock and acceleration waves in viscoelastic solids has been neglected. In this chapter we consider such influences and thus permit the stress to be affected not only by the history of the strain, but also by the history of a thermodynamic variable such as the absolute temperature or entropy. No restriction is placed on the influence of the history of strain and temperature (or entropy) on the stress other than to suppose that it be consistent with the principle of fading memory and be compatible with the laws of thermodynamics.[2]

Throughout our discussion, we confine our attention to homogeneous viscoelastic solids which do not conduct heat. Furthermore, the kinematical concepts and the forms of the balance laws will again be those associated with one dimension.[3] That is, the material is assumed to occupy an interval of the real line R in

[44] See, for example, MARVIN and MCKINNEY [1965, *17*] and FERRY [1970, *11*].

[45] ASAY, LAMBERSON, and GUENTHER [1969, *1*].

[46] It should be noted that the acoustic data used in this calculation was obtained at the frequency which yields a longitudinal wave velocity consistent with the velocity used in the one-dimensional acceleration wave study (cf. Sect. 22).

[47] Cf. Sect. 14.

[1] Many of the results discussed in this section and used in subsequent sections are due to COLEMAN [1964, *4*, *5*]. A comprehensive review of the thermodynamics of materials with fading memory has been given by DAY [1972, *8*].

[2] A thermodynamically consistent constitutive theory of nonlinear materials with memory may be formulated using either temperature or entropy as the independent thermodynamic variable (cf. COLEMAN [1964, *4*]). Both formulations have their advantages. By using temperature as the independent variable, the construction of a specific constitutive model becomes much simpler and makes more economical use of equilibrium thermophysical data such as specific heat and thermal expansion data. On the other hand, use of the entropy as the independent variable permits a more concise discussion of the general properties of shock and acceleration wave propagation. Since, in subsequent sections, we shall be concerned with the general theory of wave propagation as well as the application of a specific thermoviscoelastic constitutive model, we present both formulations of the constitutive theory.

[3] Cf. Sect. 7.

its homogeneous reference configuration. To the balance laws of mass and linear momentum given in Sect. 7, we adjoin the balance of energy and the second law of thermodynamics. In the absence of heat conduction and external heat supplies, the balance of energy takes the form

$$\frac{d}{dt}\int_{X_1}^{X_2}\left[\frac{1}{2}\varrho_R v^2(X,t)+e(X,t)\right]dX=\sigma(X_1,t)\,v(X_1,t)-\sigma(X_2,t)\,v(X_2,t) \tag{24.1}$$

and the second law of thermodynamics (Clausius-Duhem inequality) asserts that

$$\frac{d}{dt}\int_{X_1}^{X_2}\varrho_R\eta(X,t)\,dX\geqq 0 \tag{24.2}$$

for all times t and for every interval $[X_1, X_2]$ in R. Here e is the *internal energy* (in units of energy per unit volume) and η is the *entropy*. With appropriate smoothness assumptions, (24.1) and (24.2) imply that

$$\dot{e}=\sigma\dot{\varepsilon}$$

and (24.3)

$$\dot{\eta}\geqq 0$$

for every (X, t) in $R\times(-\infty, \infty)$.

α) *Constitutive assumption with temperature as the independent thermodynamic variable.* When using the absolute temperature $\theta>0$ as the independent thermodynamic variable, the one-dimensional response of nonlinear materials with memory is characterized in terms of response functionals for the *free energy*

$$\psi=e-\theta\eta, \tag{24.4}$$

the stress σ, and the entropy η, i.e., at a material point X, ψ, σ, and η at time t are determined by the entire history of the strain ε^t and the temperature θ^t:

$$\begin{aligned}\psi(t)&=\hat{\mathscr{P}}(\varepsilon^t, \theta^t),\\ \sigma(t)&=\hat{\mathscr{S}}(\varepsilon^t, \theta^t),\\ \eta(t)&=\hat{\mathscr{N}}(\varepsilon^t, \theta^t).\end{aligned} \tag{24.5}$$

Since a knowledge of the histories ε^t and θ^t defined by

$$\begin{aligned}\varepsilon^t(s)&=\varepsilon(X, t-s),\\ \theta^t(s)&=\theta(X, t-s),\end{aligned}$$

$s\in[0, \infty)$, is equivalent to a knowledge of the present values $\varepsilon=\varepsilon^t(0)$, $\theta=\theta^t(0)$ and the past histories

$$\varepsilon_r^t(t)=\varepsilon^t(s),\qquad \theta_r^t(t)=\theta^t(s),\qquad s\in(0, \infty),$$

the constitutive assumption (24.5) can alternatively be written as[4]

$$\begin{aligned}\psi(t)&=\hat{\mathscr{P}}(\varepsilon, \theta, \varepsilon_r^t, \theta_r^t),\\ \sigma(t)&=\hat{\mathscr{S}}(\varepsilon, \theta, \varepsilon_r^t, \theta_r^t),\\ \eta(t)&=\hat{\mathscr{N}}(\varepsilon, \theta, \varepsilon_r^t, \theta_r^t).\end{aligned} \tag{24.6}$$

[4] Cf. COLEMAN [1964, *5*], COLEMAN and GURTIN [1965, *6*].

In view of (24.4), the internal energy e can be defined in terms of the free energy, temperature, and entropy and hence also depends on the strain and temperature histories:

$$e(t)=\psi(t)+\theta(t)\,\eta(t)=\hat{\mathscr{E}}(\varepsilon,\theta,\varepsilon_r^t,\theta_r^t). \tag{24.7}$$

It is further assumed that the material exhibits fading memory in the sense of COLEMAN and NOLL.[5] This concept has been extended to functionals of the type (24.6) by COLEMAN[6] and by MIZEL and WANG,[7] and, in essence implies that the functionals $\hat{\mathscr{P}}$, $\hat{\mathscr{S}}$, $\hat{\mathscr{N}}$, and $\hat{\mathscr{E}}$ are continuously differentiable on the space of all histories with finite norm.[8] Here the free energy functional $\hat{\mathscr{P}}$ is assumed to be of class C^3 and the functionals $\hat{\mathscr{S}}$, $\hat{\mathscr{N}}$, and $\hat{\mathscr{E}}$ are assumed to be of class C^2.

Let $\hat{\mathscr{F}}$ stand for the functionals $\hat{\mathscr{P}}$, $\hat{\mathscr{S}}$, $\hat{\mathscr{N}}$, and $\hat{\mathscr{E}}$. Then, the principle of fading memory implies the existence of the ordinary partial derivatives

$$\begin{aligned} D_\varepsilon\hat{\mathscr{F}}(\varepsilon^t,\theta^t) &\equiv \frac{\partial}{\partial\varepsilon}\hat{\mathscr{F}}(\varepsilon,\theta,\varepsilon_r^t,\theta_r^t),\\ D_\theta\hat{\mathscr{F}}(\varepsilon^t,\theta^t) &\equiv \frac{\partial}{\partial\theta}\hat{\mathscr{F}}(\varepsilon,\theta,\varepsilon_r^t,\theta_r^t) \end{aligned} \tag{24.8}$$

as well as of higher derivatives such as $D_\varepsilon^2\hat{\mathscr{F}}$, $D_\varepsilon D_\theta\hat{\mathscr{F}}=D_\theta D_\varepsilon\hat{\mathscr{F}}$, $D_\theta^2\hat{\mathscr{F}}$. The derivatives $D_\varepsilon\hat{\mathscr{F}}$ and $D_\theta\hat{\mathscr{F}}$ defined by (24.8) represent instantaneous moduli. In particular,

$$E_t^{\mathrm{T}}=D_\varepsilon\hat{\mathscr{S}}(\varepsilon^t,\theta^t) \tag{24.9}$$

is the *instantaneous isothermal tangent modulus*; it represents the initial slope of the stress-strain response to a sudden isothermal change in the present value of strain history ε^t at time t and is taken to be strictly positive, $E_t^{\mathrm{T}}>0$. The modulus A_t defined by

$$A_t=D_\theta\hat{\mathscr{S}}(\varepsilon^t,\theta^t) \tag{24.10}$$

is the *instantaneous stress-temperature modulus*. The modulus

$$\tilde{E}_t^{\mathrm{T}}=D_\varepsilon^2\hat{\mathscr{S}}(\varepsilon^t,\theta^t), \tag{24.11}$$

which is called the *instantaneous isothermal second-order modulus*, represents the initial curvature of the isothermal stress-strain response. Finally,

$$\varkappa_t=D_\theta\hat{\mathscr{E}}(\varepsilon^t,\theta^t) \tag{24.12}$$

is the *instantaneous specific heat* and, following a customary procedure in thermodynamics, we assume that $\varkappa_t>0$.

The assumed smoothness of the stress functional $\hat{\mathscr{S}}$ also insures that for small relative strain and temperature histories, γ^t and φ^t, the material response can be approximated by linear viscoelasticity, i.e.,

$$\begin{aligned} \hat{\mathscr{S}}(\varepsilon^t+\gamma^t,\theta^t+\varphi^t) &= \hat{\mathscr{S}}(\varepsilon^t,\theta^t)+G_t(0)\gamma^t(0)+L_t(0)\,\varphi^t(0)\\ &\quad+\int_0^\infty G_t'(s)\gamma^t(s)\,ds+\int_0^\infty L_t'(s)\,\varphi^t(s)\,ds+o(\|\gamma^t\|+\|\varphi^t\|) \end{aligned} \tag{24.13}$$

---

[5] Cf. Sect. 9.

[6] COLEMAN [1964, *4*]. Also see COLEMAN and GURTIN [1965, *6*].

[7] MIZEL and WANG [1966, *11*].

[8] Cf. Eq. (9.2).

as $(\|\gamma^t\|+\|\varphi^t\|)\to 0$. The quantities $G_t(s)=\hat{\mathscr{G}}(\varepsilon^t,\theta^t;s)$ and $L_t(s)=\hat{\mathscr{L}}(\varepsilon^t,\theta^t;s)$ are called the *stress-strain relaxation function* and the *stress-temperature relaxation function*, respectively, and

$$G_t'(s)=\frac{d}{ds}G_t(s),\qquad L_t'(s)=\frac{d}{ds}L_t(s).$$

An analogous approximation holds for $\hat{\mathscr{E}}(\varepsilon^t+\gamma^t,\theta^t+\varphi^t)$ which involves an *energy-strain relaxation function* $H_t(s)=\hat{\mathscr{H}}(\varepsilon^t,\theta^t;s)$ and an *energy-temperature relaxation function* $J_t(s)=\hat{\mathscr{J}}(\varepsilon^t,\theta^t;s)$. The relaxation functions $G_t(s)$, $L_t(s)$, and $J_t(s)$ are related to the instantaneous moduli E_t^{T}, A_t, and $\varkappa_t$ through the expressions

$$G_t(0)=E_t^{\mathrm{T}},\qquad L_t(0)=A_t,\qquad J_t(0)=\varkappa_t. \tag{24.14}$$

Again we have used the subscript t as a reminder that all of these functions depend, in general, on the underlying histories ε^t, θ^t.

As in the previous considerations with the mechanical theory it is useful to consider the instantaneous and equilibrium behavior of the solid. The instantaneous response of the solid is its response to the jump history

$$\big(\varepsilon^t(s),\theta^t(s)\big)=\big(\varepsilon_{\mathrm{I}}(s),\theta_{\mathrm{I}}(s)\big)=\begin{cases}(\varepsilon,\theta), & s=0,\\ (0,\theta_0), & s\in(0,\infty)\end{cases} \tag{24.15}$$

with θ_0 the reference temperature corresponding to the unstrained state. For such a history, the functionals $\hat{\mathscr{P}}$, $\hat{\mathscr{S}}$, $\hat{\mathscr{N}}$, and $\hat{\mathscr{E}}$ depend only on the present values ε, θ and thus

$$\begin{aligned}\psi&=\hat{\mathscr{P}}(\varepsilon_{\mathrm{I}},\theta_{\mathrm{I}})=\hat{\psi}_{\mathrm{I}}(\varepsilon,\theta),\\ \sigma&=\hat{\mathscr{S}}(\varepsilon_{\mathrm{I}},\theta_{\mathrm{I}})=\hat{\sigma}_{\mathrm{I}}(\varepsilon,\theta),\\ \eta&=\hat{\mathscr{N}}(\varepsilon_{\mathrm{I}},\theta_{\mathrm{I}})=\hat{\eta}_{\mathrm{I}}(\varepsilon,\theta),\\ e&=\hat{\mathscr{E}}(\varepsilon_{\mathrm{I}},\theta_{\mathrm{I}})=\hat{e}_{\mathrm{I}}(\varepsilon,\theta).\end{aligned} \tag{24.16}$$

These functions are called the *instantaneous response functions* and it follows from the smoothness hypothesis that $\hat{\psi}_{\mathrm{I}}$ is of class C^3 and that $\hat{\sigma}_{\mathrm{I}}$, $\hat{\eta}_{\mathrm{I}}$ and $\hat{e}_{\mathrm{I}}$ are of class C^2. Clearly, by (24.9) to (24.12) and (24.15), the *instantaneous isothermal moduli* become

$$\begin{aligned}E_{\mathrm{I}}^{\mathrm{T}}&=\frac{\partial}{\partial\varepsilon}\hat{\sigma}_{\mathrm{I}}(\varepsilon,\theta),\\ A_{\mathrm{I}}&=\frac{\partial}{\partial\theta}\hat{\sigma}_{\mathrm{I}}(\varepsilon,\theta),\\ \tilde{E}_{\mathrm{I}}^{\mathrm{T}}&=\frac{\partial^2}{\partial\varepsilon^2}\hat{\sigma}_{\mathrm{I}}(\varepsilon,\theta),\\ \varkappa_{\mathrm{I}}&=\frac{\partial}{\partial\theta}\hat{e}_{\mathrm{I}}(\varepsilon,\theta).\end{aligned} \tag{24.17}$$

It will prove useful in what follows to also define the *instantaneous stress-energy modulus* [9]

$$\Gamma_{\mathrm{I}}=\frac{A_{\mathrm{I}}}{\varkappa_{\mathrm{I}}}. \tag{24.18}$$

[9] In the physics literature, this modulus is frequently called the "Grüneisen parameter" after E. GRÜNEISEN [1926, *1*].

The equilibrium response of a nonlinear material with memory is its response to the constant history

$$\left(\varepsilon^t(s),\, \theta^t(s)\right) = \left(\varepsilon_{\mathrm{E}}(s),\, \theta_{\mathrm{E}}(s)\right) = (\varepsilon,\, \theta), \qquad s \in [0,\, \infty). \tag{24.19}$$

In this instance, the functionals $\widehat{\mathscr{P}}$, $\widehat{\mathscr{S}}$, $\widehat{\mathscr{N}}$, and $\widehat{\mathscr{E}}$ depend only upon the values of strain ε and temperature θ, and we call

$$\begin{aligned} \psi &= \widehat{\mathscr{P}}(\varepsilon_{\mathrm{E}},\, \theta_{\mathrm{E}}) = \hat{\psi}_{\mathrm{E}}(\varepsilon,\, \theta), \\ \sigma &= \widehat{\mathscr{S}}(\varepsilon_{\mathrm{E}},\, \theta_{\mathrm{E}}) = \hat{\sigma}_{\mathrm{E}}(\varepsilon,\, \theta), \\ \eta &= \widehat{\mathscr{N}}(\varepsilon_{\mathrm{E}},\, \theta_{\mathrm{E}}) = \hat{\eta}_{\mathrm{E}}(\varepsilon,\, \theta), \\ e &= \widehat{\mathscr{E}}(\varepsilon_{\mathrm{E}},\, \theta_{\mathrm{E}}) = \hat{e}_{\mathrm{E}}(\varepsilon,\, \theta) \end{aligned} \tag{24.20}$$

the *equilibrium response functions*. The fading memory assumption also insures that $\hat{\psi}_{\mathrm{E}}$ is of class C^3 and that $\hat{\sigma}_{\mathrm{E}}$, $\hat{\eta}_{\mathrm{E}}$, and $\hat{e}_{\mathrm{E}}$ are of class C^2. The derivatives

$$\begin{aligned} E_{\mathrm{E}}^{\mathrm{T}} &= \frac{\partial}{\partial \varepsilon}\, \hat{\sigma}_{\mathrm{E}}(\varepsilon,\, \theta), \\ A_{\mathrm{E}} &= \frac{\partial}{\partial \theta}\, \hat{\sigma}_{\mathrm{E}}(\varepsilon,\, \theta), \\ \tilde{E}_{\mathrm{E}}^{\mathrm{T}} &= \frac{\partial^2}{\partial \varepsilon^2}\, \hat{\sigma}_{\mathrm{E}}(\varepsilon,\, \theta), \\ \varkappa_{\mathrm{E}} &= \frac{\partial}{\partial \theta}\, \hat{e}_{\mathrm{E}}(\varepsilon,\, \theta) \end{aligned} \tag{24.21}$$

are the *equilibrium isothermal moduli.*[10]

In considering the thermodynamics of nonlinear materials with fading memory described by (24.6), Coleman[11] showed that the second law imposes certain restrictions on the response functionals $\widehat{\mathscr{P}}$, $\widehat{\mathscr{S}}$, $\widehat{\mathscr{N}}$. In particular, he showed that these functionals are not independent but rather are related through the *stress* and *entropy relations*

$$\begin{aligned} \widehat{\mathscr{S}}(\varepsilon^t,\, \theta^t) &= D_\varepsilon \widehat{\mathscr{P}}(\varepsilon^t,\, \theta^t), \\ \widehat{\mathscr{N}}(\varepsilon^t,\, \theta^t) &= -D_\theta \widehat{\mathscr{P}}(\varepsilon^t,\, \theta^t). \end{aligned} \tag{24.22}$$

Furthermore, the free energy functional $\widehat{\mathscr{P}}$ must be such that the *internal dissipation* $\mathfrak{d}$, defined by

$$\mathfrak{d}(t) = \frac{1}{\theta}\left\{\widehat{\mathscr{S}}(\varepsilon^t,\, \theta^t)\, \dot{\varepsilon} - \widehat{\mathscr{N}}(\varepsilon^t,\, \theta^t)\, \dot{\theta} - \frac{\partial}{\partial t}\, \widehat{\mathscr{P}}(\varepsilon^t,\, \theta^t)\right\}, \tag{24.23}$$

is non-negative:

$$\mathfrak{d} \geqq 0. \tag{24.24}$$

In exploring the implications of this dissipation inequality, Coleman[12] established that the free energy is a minimum at equilibrium, i.e., for all histories with the present values $(\varepsilon,\, \theta)$,

$$\widehat{\mathscr{P}}(\varepsilon^t,\, \theta^t) \geqq \hat{\psi}_{\mathrm{E}}(\varepsilon,\, \theta). \tag{24.25}$$

---

[10] We can also define an equilibrium stress-energy modulus $\Gamma_{\mathrm{E}} = A_{\mathrm{E}}/\varkappa_{\mathrm{E}}$.

[11] Coleman [1964, *4*]. Also see Coleman and Mizel [1968, *5*], Gurtin [1968, *10*], Coleman and Owen [1970, *7*], and the excellent surveys by Truesdell [1969, *4*] and Day [1972, *8*].

[12] Coleman [1964, *4*].

An immediate consequence of this minimal property is that[13]

$$D_\varepsilon \hat{\mathscr{S}}(\varepsilon_E, \theta_E) \geqq E_E^T(\varepsilon, \theta), \qquad D_\theta \hat{\mathscr{E}}(\varepsilon_E, \theta_E) \leqq \varkappa_E(\varepsilon, \theta). \tag{24.26}$$

BOWEN and CHEN[14] have further explored the consequences of (24.24) and proved that

$$\hat{\mathscr{G}}'(\varepsilon_E, \theta_E; 0) \leqq 0, \qquad \hat{\mathscr{J}}'(\varepsilon_E, \theta_E; 0) \geqq 0. \tag{24.27}$$

As in the mechanical theory, it is appropriate at this point to impose certain curvature restrictions on the isothermal stress-strain response of the material. Motivated by $(24.26)_1$, we assume that for all $\varepsilon \in (0, 1)$ and $\theta > 0$

$$\begin{gathered} E_I^T(\varepsilon, \theta) > E_E^T(\varepsilon, \theta) > 0, \\ \tilde{E}_I^T(\varepsilon, \theta) > 0, \qquad \tilde{E}_E^T(\varepsilon, \theta) > 0. \end{gathered} \tag{24.28}$$

This implies that the instantaneous and equilibrium isothermal stress-strain curves, $\hat{\sigma}_I(\varepsilon, \theta)$ and $\hat{\sigma}_E(\varepsilon, \theta)$, are strictly convex from below. As seems reasonable for most materials, we also assume that

$$A_I(\varepsilon, \theta) > 0, \qquad A_E(\varepsilon, \theta) > 0 \tag{24.29}$$

for all (ε, θ) and that the state $(0, \theta_0)$ corresponds to a *natural equilibrium state* of the material, i.e.,[15]

$$\begin{gathered} (\hat{\sigma}_I)_0 = (\hat{\sigma}_E)_0 = 0, \\ (\hat{\eta}_I)_0 = (\hat{\eta}_E)_0 = \eta_0, \\ (\hat{e}_I)_0 = (\hat{e}_E)_0 = e_0. \end{gathered}$$

β) Constitutive assumption with entropy as the independent thermodynamic variable. The formulation of the constitutive theory for materials with memory using entropy as the independent variable closely parallels the formulation in terms of temperature. In this case the one-dimensional response of such materials can be described in terms of constitutive equations for the internal energy e, the stress σ, and the temperature θ:[16]

$$\begin{aligned} e(t) &= \tilde{\mathscr{E}}(\varepsilon^t, \eta^t) = \tilde{\mathscr{E}}(\varepsilon, \eta, \varepsilon_r^t, \eta_r^t), \\ \sigma(t) &= \tilde{\mathscr{S}}(\varepsilon^t, \eta^t) = \tilde{\mathscr{S}}(\varepsilon, \eta, \varepsilon_r^t, \eta_r^t), \\ \theta(t) &= \tilde{\mathscr{T}}(\varepsilon^t, \eta^t) = \tilde{\mathscr{T}}(\varepsilon, \eta, \varepsilon_r^t, \eta_r^t). \end{aligned} \tag{24.30}$$

The response functionals are assumed to have all the smoothness properties associated with the fading memory hypothesis, and in the present context $\tilde{\mathscr{E}}$ is of class C^3 and $\tilde{\mathscr{S}}$ and $\tilde{\mathscr{T}}$ are of class C^2. Then, if $\tilde{\mathscr{F}}$ stands for $\tilde{\mathscr{E}}$, $\tilde{\mathscr{S}}$, or $\tilde{\mathscr{T}}$, we can define the differential operators $D_\varepsilon \tilde{\mathscr{F}}$ and $D_\eta \tilde{\mathscr{F}}$ in a manner similar to (24.8) and it follows from the second law that[17]

$$\begin{aligned} \tilde{\mathscr{S}}(\varepsilon^t, \eta^t) &= D_\varepsilon \tilde{\mathscr{E}}(\varepsilon^t, \eta^t), \\ \tilde{\mathscr{T}}(\varepsilon^t, \eta^t) &= D_\eta \tilde{\mathscr{E}}(\varepsilon^t, \eta^t). \end{aligned} \tag{24.31}$$

[13] Cf. COLEMAN [1964, *5*], COLEMAN and GURTIN [1965, *7*].

[14] BOWEN and CHEN [1973, *3*]. They only give the proof of $(24.27)_1$; an analogous argument will establish $(24.27)_2$.

[15] The notation $(\cdot)_0$ is used to designate evaluation at $(\varepsilon, \theta) = (0, \theta_0)$, i.e., the equilibrium state.

[16] The smoothness assumption on the constitutive functionals necessary to ensure the existence of the functional transformation $(\varepsilon^t, \theta^t) \to (\varepsilon^t, \eta^t)$ is discussed in detail by COLEMAN [1964, *4*]. Also see COLEMAN and GURTIN [1965, *6*].

[17] Cf. COLEMAN [1964, *4*].

As before, we need to introduce certain instantaneous moduli. We call

$$E_t^{\mathrm{N}} = D_\varepsilon \tilde{\mathscr{S}}(\varepsilon^t, \eta^t),$$
$$M_t = D_\eta \tilde{\mathscr{S}}(\varepsilon^t, \eta^t), \tag{24.32}$$
$$\tilde{E}_t^{\mathrm{N}} = D_\varepsilon^2 \tilde{\mathscr{S}}(\varepsilon^t, \eta^t)$$

the *instantaneous isentropic tangent modulus*, the *instantaneous stress-entropy modulus*, and the *instantaneous isentropic second-order modulus*, respectively. A straightforward application of the chain rule for functionals yields the useful relations[18]

$$E_t^{\mathrm{N}} = E_t^{\mathrm{T}} + \frac{\theta A_t^2}{\varkappa_t},$$
$$M_t = \frac{\theta}{\varkappa_t} A_t. \tag{24.33}$$

With $\varkappa_t > 0$, $(24.33)_1$ implies that

$$E_t^{\mathrm{N}} \geqq E_t^{\mathrm{T}}. \tag{24.34}$$

Analogous to (24.5), we now consider the jump history[19]

$$(\varepsilon^t(s), \eta^t(s)) = (\varepsilon_{\mathrm{I}}(s), \eta_{\mathrm{I}}(s)) = \begin{cases} (\varepsilon, \eta), & s = 0, \\ (0, \eta_0), & s \in (0, \infty) \end{cases} \tag{24.35}$$

and define the *instantaneous response functions*

$$e = \tilde{\mathscr{E}}(\varepsilon_{\mathrm{I}}, \eta_{\mathrm{I}}) = \tilde{e}_{\mathrm{I}}(\varepsilon, \eta),$$
$$\sigma = \tilde{\mathscr{S}}(\varepsilon_{\mathrm{I}}, \eta_{\mathrm{I}}) = \tilde{\sigma}_{\mathrm{I}}(\varepsilon, \eta), \tag{24.36}$$
$$\theta = \tilde{\mathscr{T}}(\varepsilon_{\mathrm{I}}, \eta_{\mathrm{I}}) = \tilde{\theta}_{\mathrm{I}}(\varepsilon, \eta).$$

The fading memory hypothesis laid down for the functionals $\tilde{\mathscr{E}}$, $\tilde{\mathscr{S}}$, and $\tilde{\mathscr{T}}$ ensures that $\tilde{e}_{\mathrm{I}}$, $\tilde{\sigma}_{\mathrm{I}}$, and $\tilde{\theta}_{\mathrm{I}}$ are smooth functions and it follows from (24.32) that

$$E_{\mathrm{I}}^{\mathrm{N}} = \frac{\partial}{\partial \varepsilon} \tilde{\sigma}_{\mathrm{I}}(\varepsilon, \eta),$$
$$M_{\mathrm{I}} = \frac{\partial}{\partial \eta} \tilde{\sigma}_{\mathrm{I}}(\varepsilon, \eta), \tag{24.37}$$
$$\tilde{E}_{\mathrm{I}}^{\mathrm{N}} = \frac{\partial^2}{\partial \varepsilon^2} \tilde{\sigma}_{\mathrm{I}}(\varepsilon, \eta)$$

represent the *instantaneous isentropic moduli*.

In a similar manner, we can also consider the constant (equilibrium) history $(\varepsilon^t(s), \eta^t(s)) = (\varepsilon_{\mathrm{E}}(s), \eta_{\mathrm{E}}(s)) = (\varepsilon, \eta)$, $s \in [0, \infty)$, and define the *equilibrium response functions* $\tilde{e}_{\mathrm{E}}$, $\tilde{\sigma}_{\mathrm{E}}$, $\tilde{\theta}_{\mathrm{E}}$ and the corresponding *equilibrium isentropic moduli* $E_{\mathrm{E}}^{\mathrm{N}}$, M_{E}, $\tilde{E}_{\mathrm{E}}^{\mathrm{N}}$.

25. Propagation of steady shock waves. Although a formal proof of the existence of steady waves in general nonconducting materials with memory is lacking, steady shock wave solutions have been exhibited by NUNZIATO and WALSH[20] for

[18] Cf. COLEMAN and GURTIN [1965, *6*].

[19] It is, of course, tacitly assumed that the state $(0, \eta_0)$ corresponds to the state $(0, \theta_0)$. Thus, the designation $(\cdot)_0$ can imply evaluation at either $(0, \theta_0)$ or $(0, \eta_0)$, whichever is appropriate.

[20] NUNZIATO and WALSH [1973, *16*].

a particular thermoviscoelastic constitutive assumption and compared with the steady wave profiles experimentally observed by BARKER and HOLLENBACH and by SCHULER in polymethyl methacrylate.[21] On the basis of their results, one would expect that for most materials the overall properties of steady wave solutions in the thermoviscoelastic case are qualitatively the same as in the purely mechanical case[22] provided we interpret the sound velocities as isentropic sound velocities and interpret the stress-strain curves as Hugoniot stress-strain curves.

α) *Steady waves in thermoviscoelastic solids.* In considering steady compressive motions in thermoviscoelastic media, we recall from Sect. 12 that the displacement $u(X,t)$ is expressible in the form

$$u(X,t)=u(\xi), \quad \xi=t-\frac{X}{V}, \tag{25.1}$$

where $V>0$ is the steady wave velocity, and that the corresponding particle velocity and strain are given by

$$\begin{aligned} v(\xi)&=u'(\xi),\\ \varepsilon(\xi)&=\frac{1}{V}u'(\xi)=\frac{1}{V}v(\xi). \end{aligned} \tag{25.2}$$

In the present case, the temperature field is also assumed to be steady:

$$\theta(X,t)=\theta(\xi).$$

In view of the constitutive equations $(24.6)_2$ and (24.7), steady strain and temperature fields imply steady stress and internal energy fields, and thus

$$\sigma(X,t)=\sigma(\xi), \quad e(X,t)=e(\xi).$$

If it is assumed that a natural equilibrium state exists far ahead of the wave, i.e., if

$$\begin{aligned} &\lim_{\xi\to-\infty}\varepsilon(\xi)=0, &&\lim_{\xi\to-\infty}\theta(\xi)=\theta_0,\\ &\lim_{\xi\to-\infty}\sigma(\xi)=(\hat{\sigma}_{\mathrm{E}})_0=0, &&\lim_{\xi\to-\infty}e(\xi)=(\hat{e}_{\mathrm{E}})_0=e_0, \end{aligned} \tag{25.3}$$

then the balance of linear momentum becomes[23]

$$\sigma(\xi)=\varrho_{\mathrm{R}}V\,v(\xi)=\varrho_{\mathrm{R}}V^2\varepsilon(\xi). \tag{25.4}$$

Furthermore, for steady displacement and internal energy fields,

$$\frac{d}{dt}\int_{X_1}^{X_2}\left[\frac{1}{2}\varrho_{\mathrm{R}}v^2(X,t)+e(X,t)\right]dX=V\left\{\frac{1}{2}\varrho_{\mathrm{R}}[v^2(\xi_2)-v^2(\xi_1)]+[e(\xi_2)-e(\xi_1)]\right\}.$$

Thus, by (25.3), and (25.4), the balance of energy (24.1) asserts that

$$e(\xi)=e_0+\tfrac{1}{2}\varrho_{\mathrm{R}}V^2\varepsilon^2(\xi)=e_0+\tfrac{1}{2}\sigma(\xi)\,\varepsilon(\xi). \tag{25.5}$$

Eqs. (25.4) and (25.5) are required to hold for all $\xi\in(-\infty,\infty)$ and combine with $(24.6)_2$ and (24.7) to yield the coupled functional equations

$$\begin{aligned} \varrho_{\mathrm{R}}V^2\varepsilon(\xi)&=\hat{\mathscr{S}}(\varepsilon(\xi-s),\theta(\xi-s)),\\ e_0+\tfrac{1}{2}\varrho_{\mathrm{R}}V^2\varepsilon^2(\xi)&=\hat{\mathscr{E}}(\varepsilon(\xi-s),\theta(\xi-s)) \end{aligned} \tag{25.6}$$

for the steady strain and temperature fields, $\varepsilon(\xi)$ and $\theta(\xi)$.

---

[21] BARKER and HOLLENBACH [1970, *2*] and SCHULER [1970, *19*]. Also see Sect. 13 of this article.

[22] Cf. Sect. 12.

[23] Cf. Sect. 12.

Now, as we have already indicated, the existence of steady wave solutions $\varepsilon(\xi)>0$, $\theta(\xi)>\theta_0$ of the functional equations (25.6), which are monotone increasing functions of ξ, has not been formally established. Nevertheless, by making a specific constitutive assumption, it is possible to exhibit steady wave solutions. In the next two subsections, we formulate such a specific constitutive model and, using (25.6), calculate steady shock wave solutions.

β) *A specific thermoviscoelastic constitutive assumption and the evaluation of material response functions.* Since the free energy functional $\hat{\mathscr{P}}$ serves to determine the stress σ and the entropy η through the relations (24.22) (and hence the internal energy e through (24.7)), any formulation of a specific constitutive model should be based on an explicit assumption for the free energy. In view of this and the minimal property (24.25), the free energy functional $\hat{\mathscr{P}}$ is assumed to be the following quadratic form:[24]

$$\psi(t)=\hat{\mathscr{P}}(\varepsilon^t,\theta^t)=\hat{\psi}_E(\varepsilon,\theta)+\tfrac{1}{2}\Omega(\varepsilon)\{\hat{\mathscr{P}}_1(\varepsilon^t)+\hat{\mathscr{P}}_2(\theta^t)\}^2 \tag{25.7}$$

where $\Omega(\varepsilon)>0$ for all $\varepsilon\in[0,1)$ and the functionals $\hat{\mathscr{P}}_1(\varepsilon^t)$ and $\hat{\mathscr{P}}_2(\theta^t)$ have the representations

$$\hat{\mathscr{P}}_1(\varepsilon^t)=\varepsilon+\int_0^\infty f'(s)\,\varepsilon^t(s)\,ds, \tag{25.8}$$

$$\hat{\mathscr{P}}_2(\theta^t)=-\frac{1}{\theta_0}\left[g_0\,\theta+\int_0^\infty g'(s)\,\theta^t(s)\,ds\right]. \tag{25.9}$$

Here $f(\cdot)$ and $g(\cdot)$ are mechanical and thermal relaxation functions with the properties $f(0)=1$, $g(0)=g_0>0$ and $f(s)$ and $g(s)\to 0$ as $s\to\infty$. The specific constitutive equations for the stress and the internal energy can be computed from (25.7) using (24.22) and (24.7):

$$\begin{aligned}\sigma(t)&=\hat{\mathscr{S}}(\varepsilon^t,\theta^t)\\&=\hat{\sigma}_E(\varepsilon,\theta)+\tfrac{1}{2}\Omega'(\varepsilon)\{\hat{\mathscr{P}}_1(\varepsilon^t)+\hat{\mathscr{P}}_2(\theta^t)\}^2+\Omega(\varepsilon)\{\hat{\mathscr{P}}_1(\varepsilon^t)+\hat{\mathscr{P}}_2(\theta^t)\},\end{aligned} \tag{25.10}$$

$$\begin{aligned}e(t)&=\hat{\mathscr{E}}(\varepsilon^t,\theta^t)\\&=\hat{e}_E(\varepsilon,\theta)+\tfrac{1}{2}\Omega(\varepsilon)\{\hat{\mathscr{P}}_1(\varepsilon^t)+\hat{\mathscr{P}}_2(\theta^t)\}^2+\Omega(\varepsilon)\,g_0\frac{\theta}{\theta_0}\{\hat{\mathscr{P}}_1(\varepsilon^t)+\hat{\mathscr{P}}_2(\theta^t)\}.\end{aligned} \tag{25.11}$$

Evaluating (25.10) and (25.11) for the jump history (24.15), we obtain the relations

$$\hat{\sigma}_I(\varepsilon,\theta)=\hat{\sigma}_E(\varepsilon,\theta)+\tfrac{1}{2}\Omega'(\varepsilon)\,h^2(\varepsilon,\theta)+\Omega(\varepsilon)\,h(\varepsilon,\theta), \tag{25.12}$$

$$\hat{e}_I(\varepsilon,\theta)=\hat{e}_E(\varepsilon,\theta)+\tfrac{1}{2}\Omega(\varepsilon)\,h(\varepsilon,\theta)\left[h(\varepsilon,\theta)+2g_0\frac{\theta}{\theta_0}\right] \tag{25.13}$$

where

$$h(\varepsilon,\theta)=\varepsilon-g_0\left(\frac{\theta}{\theta_0}-1\right). \tag{25.14}$$

For this particular model, the instantaneous and equilibrium moduli defined in $(24.17)_{1,2,4}$ and $(24.21)_{1,2,4}$ are related by

$$\begin{aligned}E_I^T(\varepsilon,\theta)&=E_E^T(\varepsilon,\theta)+\tfrac{1}{2}\Omega''(\varepsilon)\,h^2(\varepsilon,\theta)+2\Omega'(\varepsilon)\,h(\varepsilon,\theta)+\Omega(\varepsilon),\\A_I(\varepsilon,\theta)&=A_E(\varepsilon,\theta)-\frac{g_0}{\theta_0}\,[\Omega'(\varepsilon)\,h(\varepsilon,\theta)+\Omega(\varepsilon)],\\\varkappa_I(\varepsilon,\theta)&=\varkappa_E(\varepsilon,\theta)-\Omega(\varepsilon)\,g_0^2\,\frac{\theta}{\theta_0^2}.\end{aligned} \tag{25.15}$$

[24] Cf. Nunziato and Walsh [1973, *16*]. A similar form for the free energy was used by Coleman and Mizel [1968, *7*] in a study of the stability of viscoelastic filaments.

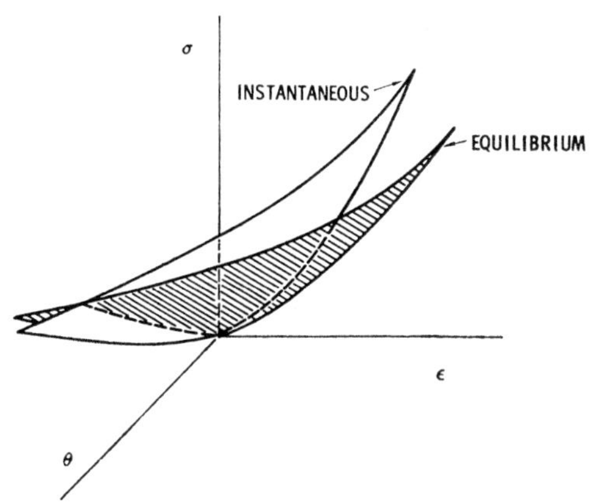

Fig. 43. Instantaneous and equilibrium surfaces in stress-strain-temperature space.

Notice that with $\Omega(0)>0$ $(25.15)_{1,3}$ are consistent with the inequalities (24.26) at the natural equilibrium state $(0, \theta_0)$. It is also interesting to note that with $g_0>0$ and $\Omega(0)>0$,

$$(A_{\mathrm{I}})_0 \leqq (A_{\mathrm{E}})_0 .$$

This inequality has been found to hold for polymethyl methacrylate by NUNZIATO and WALSH[25] and, on physical grounds, it is expected to hold for most polymers. This observation, along with the curvature conditions (24.28), suggests that the instantaneous and equilibrium surfaces in stress-strain-temperature space intersect and Fig. 43 is an illustration of this. It is clear from (25.12) that the projection of the line of intersection is given in the temperature-strain plane by $h(\varepsilon, \theta)=0$. A figure similar to Fig. 43 could also be drawn for the internal energy and the plane of intersection of the instantaneous and equilibrium surfaces in this case is also given by $h(\varepsilon, \theta)=0$.

Now, in order to use this constitutive formulation to calculate steady shock wave solutions, it is necessary to evaluate the material response functions involved; namely, the equilibrium functions $\hat{\sigma}_{\mathrm{E}}(\varepsilon, \theta)$, $\hat{e}_{\mathrm{E}}(\varepsilon, \theta)$, the function $\Omega(\varepsilon)$, and the relaxation functions $f(s)$ and $g(s)$. A procedure for evaluating these from experimental steady shock wave, thermophysical, and bulk response data has been developed by NUNZIATO and WALSH[26] and applied to the polymer, polymethyl methacrylate.

As we have already seen in Sect. 14, measurements of the steady wave velocity V and the particle velocity v at the head and tail of a steady shock wave enable one to determine the instantaneous and equilibrium response functions, $\sigma_{\mathrm{I}}(\varepsilon)$ and $\sigma_{\mathrm{E}}(\varepsilon)$. In the present context, these functions represent the *instantaneous* and *equilibrium Hugoniot stress-strain curves* for the material and can be represented in terms of least-squares polynomial fits in strain,

$$\begin{aligned} \sigma_{\mathrm{I}}(\varepsilon) &= l_{\mathrm{I}}\,\varepsilon + m_{\mathrm{I}}\,\varepsilon^2 + n_{\mathrm{I}}\,\varepsilon^3, \\ \sigma_{\mathrm{E}}(\varepsilon) &= l_{\mathrm{E}}\,\varepsilon + m_{\mathrm{E}}\,\varepsilon^2 + n_{\mathrm{E}}\,\varepsilon^3 \end{aligned} \tag{25.16}$$

with the coefficients corresponding to the equilibrium state $(0, \theta_0)$ ahead of the wave.[27] The coefficients l_{I} and l_{E} are related to the instantaneous and equilibrium

[25] NUNZIATO and WALSH [1973, *15*]. Also see Sect. 28 of this article.

[26] NUNZIATO and WALSH [1973, *16*].

[27] For polymethyl methacrylate, the values of the coefficients are: $l_{\mathrm{I}}=90.44$ kbar, $m_{\mathrm{I}}=681.1$ kbar, $n_{\mathrm{I}}=-1396.0$ kbar, $l_{\mathrm{E}}=88.38$ kbar, $m_{\mathrm{E}}=680.7$ kbar, $n_{\mathrm{E}}=-2720.0$ kbar. The curves are shown in Fig. 25. For the experimental data cited in Sect. 13, $\theta_0=295\,^{\circ}$K.

isentropic sound velocities by[28]

$$l_{\mathrm{I}}=\varrho_{\mathrm{R}}(C_{\mathrm{I}})_0^2=(E_{\mathrm{I}}^{\mathrm{N}})_0, \quad l_{\mathrm{E}}=\varrho_{\mathrm{R}}(C_{\mathrm{E}})_0^2=(E_{\mathrm{E}}^{\mathrm{N}})_0. \tag{25.17}$$

Since the energy equation (25.5) must hold everywhere in a steady wave, evaluation of this expression at the shock front and at the tail of the wave yields

$$\begin{aligned} \hat{H}_{\mathrm{I}}(\varepsilon,\theta) &= \hat{e}_{\mathrm{I}}(\varepsilon,\theta)-e_0-\tfrac{1}{2}\hat{\sigma}_{\mathrm{I}}(\varepsilon,\theta)\,\varepsilon=0,\\ \hat{H}_{\mathrm{E}}(\varepsilon,\theta) &= \hat{e}_{\mathrm{E}}(\varepsilon,\theta)-e_0-\tfrac{1}{2}\hat{\sigma}_{\mathrm{E}}(\varepsilon,\theta)\,\varepsilon=0. \end{aligned} \tag{25.18}$$

The functions $\hat{H}_{\mathrm{I}}(\cdot,\cdot)$ and $\hat{H}_{\mathrm{E}}(\cdot,\cdot)$ are called the *instantaneous* and *equilibrium Hugoniot temperature-strain curves*. Since the Hugoniot stress-strain curves (25.16) are one-to-one, the Hugoniot temperature-strain curves must also be one-to-one,[29] and we assume they have the representations[30]

$$\begin{aligned} \Theta_{\mathrm{I}}(\varepsilon) &= \theta_0\{1+\mathfrak{a}_{\mathrm{I}}\,\varepsilon+\mathfrak{b}_{\mathrm{I}}\,\varepsilon^2\},\\ \Theta_{\mathrm{E}}(\varepsilon) &= \theta_0\{1+\mathfrak{a}_{\mathrm{E}}\,\varepsilon+\mathfrak{b}_{\mathrm{E}}\,\varepsilon^2\}. \end{aligned} \tag{25.19}$$

The coefficients $\mathfrak{a}_{\mathrm{I}}$, $\mathfrak{b}_{\mathrm{I}}$, $\mathfrak{a}_{\mathrm{E}}$, $\mathfrak{b}_{\mathrm{E}}$ also depend upon the reference state $(0,\theta_0)$ and, after some lengthy calculations, can be evaluated from thermophysical data.[31] The Hugoniot temperature-strain curves for polymethyl methacrylate are shown in Fig. 44.

The material function $\Omega(\varepsilon)$ is evaluated using (25.13) and steady shock wave data:

$$\Omega(\varepsilon)=\frac{2\{e_{\mathrm{I}}(\varepsilon)-\hat{e}_{\mathrm{E}}(e,\Theta_{\mathrm{I}}(\varepsilon))\}}{h(\varepsilon,\Theta_{\mathrm{I}}(\varepsilon))\left\{h(\varepsilon,\Theta_{\mathrm{I}}(\varepsilon))+2g_0\dfrac{\Theta_{\mathrm{I}}(\varepsilon)}{\theta_0}\right\}}, \tag{25.20}$$

where

$$e_{\mathrm{I}}(\varepsilon)=\hat{e}_{\mathrm{I}}(\varepsilon,\Theta_{\mathrm{I}}(\varepsilon))=e_0+\tfrac{1}{2}\sigma_{\mathrm{I}}(\varepsilon)\,\varepsilon,$$

and the constant g_0, which is a measure of the thermal relaxation, is calculated from thermophysical data, the sound velocities, and the instantaneous stress-energy modulus $(\Gamma_{\mathrm{I}})_0$.[32]

In view of (25.7), it would be natural to formulate the equilibrium model by making an explicit assumption for the equilibrium free energy function. However, in order to make more efficient use of the thermophysical property data, the formulation is presented here in terms of the equilibrium internal energy function. That is, we assume that

$$\hat{e}_{\mathrm{E}}(\varepsilon,\theta)=e_0+\int_{\theta_0}^{\theta}\varkappa_{\mathrm{E}}(0,\omega)\,d\omega+D(\varepsilon)\,W(\theta), \tag{25.21}$$

---

[28] In the present case, $(C_{\mathrm{I}})_0=2.76$ mm/μsec, $(C_{\mathrm{E}})_0=2.73$ mm/μsec. The difference in the value of $(C_{\mathrm{E}})_0$ given here and that value given previously is a result of the experimental data fits.

[29] Cf. NUNZIATO and HERRMANN [1972, *15*].

[30] It should be noted that the experimental determination of a single set of Hugoniot stress-strain curves is not sufficient to determine completely a thermomechanical constitutive model for a material. In such instances, additional assumptions with regard to the material response are required. However, from Hugoniot data at several initial temperatures, it is possible to construct a constitutive model, with no *a priori* assumptions, for the limited range over which the experimental data is valid. The general procedure for constructing such a constitutive model is outlined by LYSNE and HARDESTY [1973, *11*].

[31] Cf. NUNZIATO and WALSH [1973, *16*]. For polymethyl methacrylate, the coefficients are: $\mathfrak{a}_{\mathrm{E}}=(\Gamma_{\mathrm{E}})_0=0.66$, $\mathfrak{b}_{\mathrm{E}}=-1.50$, $\mathfrak{a}_{\mathrm{I}}=(\Gamma_{\mathrm{I}})_0=0.55$, $\mathfrak{b}_{\mathrm{I}}=-0.94$.

[32] Cf. NUNZIATO and WALSH [1973, *16*]. The value of $(\Gamma_{\mathrm{I}})_0$ is evaluated from energy deposition experiments involving a very small temperature rise. Such situations can be achieved in electron beam experiments (cf. Sect. 4).

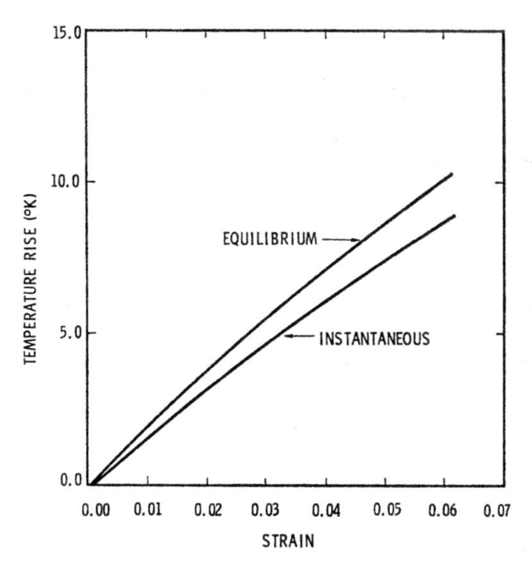

Fig. 44. Instantaneous and equilibrium Hugoniot temperature-strain curves calculated for polymethyl methacrylate. (From NUNZIATO and WALSH [1973, *16*].)

where $D(0)=0$, $W(\theta_0)=1$. Then, by use of the equilibrium counterparts of (24.22) and (24.7) along with (25.21), the equilibrium stress function is found to be given by

$$\hat{\sigma}_E(\varepsilon, \theta) = M(\varepsilon)\frac{\theta}{\theta_0} - \theta\, D'(\varepsilon)\int_{\theta_0}^{\theta}\frac{W(\omega)}{\omega^2}\,d\omega. \tag{25.22}$$

The strain-dependent functions $D(\varepsilon)$ and $M(\varepsilon)$ are evaluated from the results of steady wave experiments, i.e., using (25.21) and (25.22) along with (25.19) and the fact that $\sigma_E(\varepsilon) = \hat{\sigma}_E(\varepsilon, \Theta_E(\varepsilon))$:

$$D(\varepsilon) = \frac{1}{W(\Theta_E(\varepsilon))}\left\{\tfrac{1}{2}\sigma_E(\varepsilon)\,\varepsilon - \int_{\theta_0}^{\Theta_E(\varepsilon)}\varkappa_E(0,\omega)\,d\omega\right\},$$

$$M(\varepsilon) = \frac{\theta_0}{\Theta_E(\varepsilon)}\left\{\sigma_E(\varepsilon) + \Theta_E(\varepsilon)\,D'(\varepsilon)\int_{\theta_0}^{\Theta_E(\varepsilon)}\frac{W(\omega)}{\omega^2}\,d\omega\right\}.$$

It follows from (25.22) that

$$W(\theta) = 1 + \int_{\theta_0}^{\theta}\frac{\omega}{\theta_0 (A_E)_0}\left[\frac{\partial}{\partial\theta}A_E(0,\omega)\right]d\omega.$$

To evaluate the equilibrium response functions, we need, in addition to the steady shock wave data, the temperature dependence of the equilibrium stress-temperature modulus $A_E(0, \theta)$ and the equilibrium specific heat $\varkappa_E(0, \theta)$ at zero strain. For a particular viscoelastic material, the modulus $A_E(0, \theta)$ can be calculated from measured values of the equilibrium coefficient of linear thermal expansion β_E and the equilibrium isothermal bulk modulus B_E:[33]

$$A_E(0, \theta) = 3\beta_E(\theta)\, B_E(\theta). \tag{25.23}$$

[33] Cf. Sect. 23 for a consideration of the bulk response of nonlinear viscoelastic solids.

Then, the function $\varkappa_E(0, \theta)$ is computed from the measured equilibrium specific heat at zero pressure $\tilde{\varkappa}_E$,

$$\varkappa_E(0, \theta) = \tilde{\varkappa}_E(\theta) - 3\theta A_E(0, \theta)\beta_E(\theta). \tag{25.24}$$

By use of data for polymethyl methacrylate obtained by TOULOUKIAN[34] and by HEYDEMANN and GUICKING,[35] A_E and $\varkappa_E$ can be represented by[36]

$$\begin{aligned} A_E(0, \theta) &= A_0 + A_1\theta + A_2\theta^2 + A_3\theta^3, \\ \varkappa_E(0, \theta) &= \varkappa_0 + \varkappa_1\theta + \varkappa_2\theta^2 + \varkappa_3\theta^3. \end{aligned} \tag{25.25}$$

Finally, the calculation of steady shock wave solutions requires that the relaxation functions $f(\cdot)$ and $g(\cdot)$ be specified. Motivated by the successful use of an exponential relaxation function in the purely mechanical theory,[37] it is assumed that

$$f(s) = \exp\left(\frac{-s}{\tau_\varepsilon}\right), \qquad g(s) = g_0 \exp\left(\frac{-s}{\tau_\theta}\right). \tag{25.26}$$

The mechanical relaxation time τ_ε to be used for polymethyl methacrylate is that determined from acoustic dispersion data by NUNZIATO and SUTHERLAND;[38] while the thermal relaxation time τ_θ is evaluated by obtaining the best overall agreement between calculated and experimentally observed steady wave profiles. This completes the determination of the relevant material functions for the steady wave problem and, in essence, serves to characterize completely the dynamic thermoviscoelastic response of polymethyl methacrylate.

γ) *Steady shock wave solutions.* The constitutive equations (25.10) and (25.11) combine with the balance equations (25.6) to form a coupled set of integral equations for the steady strain and temperature fields $\varepsilon(\xi)$ and $\theta(\xi)$:

$$\begin{aligned} \varrho_R V^2 \varepsilon(\xi) &= \hat{\sigma}_E(\varepsilon(\xi), \theta(\xi)) + \tfrac{1}{2}\Omega'(\varepsilon(\xi))\{\widehat{\mathscr{P}}_1(\varepsilon(\xi - s)) + \widehat{\mathscr{P}}_2(\theta(\xi - s))\}^2 \\ &\quad + \Omega(\varepsilon(\xi))\{\widehat{\mathscr{P}}_1(\varepsilon(\xi - s)) + \widehat{\mathscr{P}}_2(\theta(\xi - s))\}, \\ e_0 + \tfrac{1}{2}\varrho_R V^2 \varepsilon^2(\xi) &= \hat{e}_E(\varepsilon(\xi), \theta(\xi)) + \tfrac{1}{2}\Omega(\varepsilon(\xi))\{\widehat{\mathscr{P}}_1(\varepsilon(\xi - s)) + \widehat{\mathscr{P}}_2(\theta(\xi - s))\}^2 \\ &\quad + \Omega(\varepsilon(\xi))\, g_0 \frac{\theta(\xi)}{\theta_0}\{\widehat{\mathscr{P}}_1(\varepsilon(\xi - s)) + \widehat{\mathscr{P}}_2(\theta(\xi - s))\}. \end{aligned} \tag{25.27}$$

Here we will confine our attention to steady shock wave solutions, i.e., solutions for given values of the steady wave velocity V with

$$V^2 > (C_I)_0^2 = \frac{(E_I^N)_0}{\varrho_R}.$$

Such solutions of (25.27) can satisfy the equilibrium conditions (25.3) far ahead of the wave only if $\varepsilon(\xi) = 0$, $\theta(\xi) = \theta_0$ for $\xi < 0$ and there is a jump discontinuity in strain ε and temperature θ at $\xi = 0$. The value $\varepsilon(0)$ is determined by (12.14) and $(25.16)_1$, i.e., $\varepsilon(0)$ is the smallest positive root of the quadratic

$$n_I \varepsilon^2 + m_I \varepsilon + \varrho_R[(C_I)_0^2 - V^2] = 0.$$

---

[34] TOULOUKIAN [1967, *10*].

[35] HEYDEMANN and GUICKING [1963, *2*].

[36] The values of the coefficients are: $A_0 = 0.762 \times 10^{-2}$ kbar/°K, $A_1 = 0.163 \times 10^{-4}$ kbar/°K², $A_2 = -1.14 \times 10^{-8}$ kbar/°K³, $A_3 = -0.727 \times 10^{-10}$ kbar/°K⁴, $\varkappa_0 = -0.460 \times 10^{-2}$ kbar/°K, $\varkappa_1 = 1.75 \times 10^{-4}$ kbar/°K², $\varkappa_2 = -0.688 \times 10^{-6}$ kbar/°K³, $\varkappa_3 = 0.106 \times 10^{-8}$ kbar/°K⁴.

[37] Cf. Sect. 15.

[38] NUNZIATO and SUTHERLAND [1973, *14*]. Also see Sect. 11 of this article.

Similarly, by (12.13) and $(25.16)_2$, the maximum strain $\varepsilon(\infty)$ achieved for a given steady wave velocity V is the smallest positive root of

$$n_E\,\varepsilon^2 + m_E\,\varepsilon + \varrho_R\,[(C_E)_0^2 - V^2] = 0.$$

The corresponding temperatures, $\theta(0)$ and $\theta(\infty)$, are calculated by (25.19):

$$\theta(0) = \Theta_I(\varepsilon(0)), \qquad \theta(\infty) = \Theta_E(\varepsilon(\infty)).$$

To obtain steady shock wave solutions, NUNZIATO and WALSH[39] solved the coupled integral equations (25.27) numerically. In particular, (25.27) can be converted to a coupled set of ordinary differential equations which, with the material functions known, can be integrated for the strain $\varepsilon(\xi)$ and the temperature $\theta(\xi)$ by a Runge-Kutta integration technique. The corresponding particle velocity profile $v(\xi)$ follows from $(25.2)_2$. In Figs. 45 and 46 the comparison between calculated steady wave profiles and wave profiles experimentally observed by BARKER and HOLLENBACH and by SCHULER[40] are shown for the impact conditions of 0.15 mm/µsec (Fig. 45) and 0.30 mm/µsec (Fig. 46). These are the same profiles considered for the mechanical case,[41] and again we see that good agreement is obtained by using slightly adjusted steady wave velocities of 2.963 and 3.106 mm/µsec, respectively. The thermal relaxation time providing the best agreement was $\tau_\theta = 0.10$ µsec. It is of interest to point out that the profiles are indeed sensitive to the relaxation time τ_θ which may seem to be somewhat surprising since steady wave propagation is essentially a mechanical problem. As one would expect, the corresponding temperature profiles show the same overall structure as the particle velocity profiles.

Finally, it should be noted that due to the observed inflection in the equilibrium response of polymethyl methacrylate the thermoviscoelastic model constructed here is only expected to be valid for stresses up to approximately 7.5 kbar.

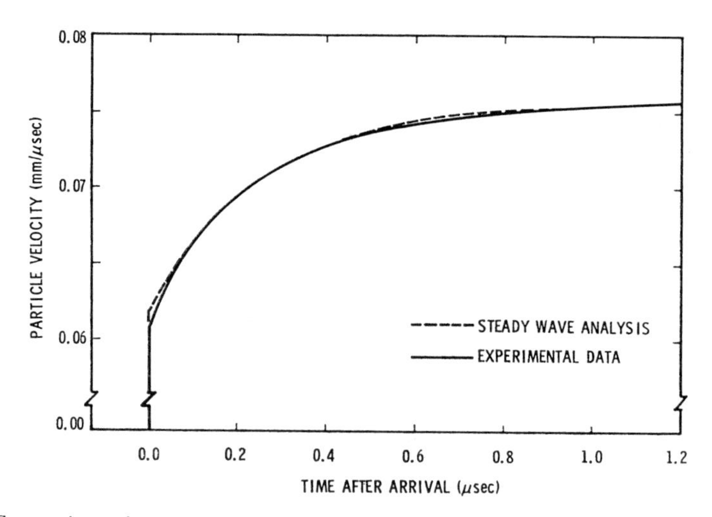

Fig. 45. Comparison of measured and calculated steady wave profiles for polymethyl methacrylate; impact velocity of 0.15 mm/µsec; thermomechanical model. (From NUNZIATO and WALSH [1973, *16*].)

[39] NUNZIATO and WALSH [1973, *16*].
[40] BARKER and HOLLENBACH [1970, *2*] and SCHULER [1970, *19*].
[41] Cf. Sects. 13 and 15.

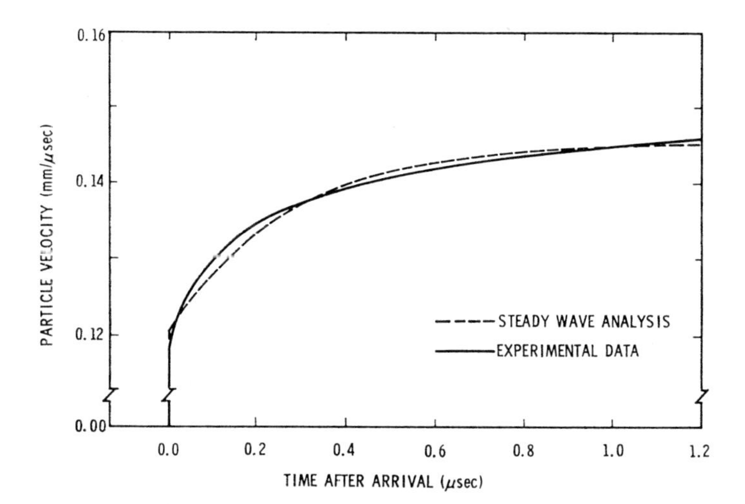

Fig. 46. Comparison of measured and calculated steady wave profiles for polymethyl methacrylate; impact velocity of 0.30 mm/μsec; thermomechanical model. (From NUNZIATO and WALSH [1973, *16*].)

26. The growth and decay of shock waves. Thermodynamic influences on the propagation of one-dimensional shock waves in nonconducting materials with memory have been considered by COLEMAN and GURTIN and by CHEN and GURTIN.[42] Using the general constitutive assumption (24.30), COLEMAN and GURTIN showed that most of the classical theory of Hugoniot curves, as developed by BETHE and WEYL,[43] could be extended to solids with long-range viscoelastic memory. Subsequently, CHEN and GURTIN made use of this extension to consider the growth and decay of shock waves. In particular, they derived a differential equation analogous to (16.5) which relates the strain ε^- and the strain gradient $(\partial\varepsilon/\partial X)^-$ behind the wave when the region ahead of the wave has been in an unstrained equilibrium state for all past time. This shock amplitude equation serves to establish the existence of a critical jump in strain gradient which, along with another parameter ζ relating the temperature θ^- behind the wave to the strain ε^-, governs the local behavior of the wave.

Fundamental to the classical theory of Hugoniot curves are certain curvature assumptions on the isentropic stress-strain law and the assumption that there exists for the material a Hugoniot stress-strain law of the form $\sigma^-=\sigma_{\mathrm{I}}(\varepsilon^-)$. These assumptions can be well motivated from physical considerations. However, NUNZIATO and HERRMANN[44] have shown that the requirement that the Hugoniot curve have the representation $\sigma^-=\sigma_{\mathrm{I}}(\varepsilon^-)$ is not necessary in order to discuss the

[42] COLEMAN and GURTIN [1966, *5*], [1968, *4*] and CHEN and GURTIN [1972, *7*]. Shock wave propagation in materials with memory which do conduct heat has been studied by ACHENBACH, VOGEL, and HERRMANN [1968, *1*], ACHENBACH and HERRMANN [1970, *1*], and DUNWOODY [1972, *9*, *10*].

[43] BETHE [1942, *1*] and WEYL [1949, *2*]. Expositions of the classical theory of Hugoniot curves have also been given by COURANT and FRIEDRICHS [1948, *1*, Sect. 65], SERRIN [1959, *2*, Sect. 56], and HAYES [1960, *4*]. The interested reader will find references to the original contributions of RANKINE, HUGONIOT, DUHEM, ZEMPLEN, JOUGUET, and others in SERRIN'S article and in the treatise by TRUESDELL and TOUPIN [1960, *7*, Sects. 189 and 241].

[44] NUNZIATO and HERRMANN [1972, *15*]. Though they considered elastic nonconductors, their results apply directly to nonlinear viscoelastic materials which exhibit instantaneous elasticity.

existence and behavior of shock waves. In this section, we consider the growth and decay of shock waves in nonconducting materials with memory in the context of the general theory of Hugoniot curves developed by NUNZIATO and HERRMANN and present some results obtained by NUNZIATO and WALSH[45] concerning a verification of the concept of the critical jump in strain gradient for polymethyl methacrylate.

α) *General properties of shock transition.* In the consideration of thermodynamic influences, the temperature θ and the entropy η, as well as the strain ε, are discontinuous across a shock front. If $Y(t)$ is the location of the front in the reference configuration R at time t, then at $X=Y(t)$ the stress σ and internal energy e are also discontinuous and it follows from the balance of momentum (7.7) and energy (24.1) that[46]

$$[\sigma]=\varrho_{\mathrm{R}}U[v], \qquad \left[\frac{\partial\sigma}{\partial X}\right]=-\varrho_{\mathrm{R}}[\dot{v}], \tag{26.1}$$

$$U[e+\tfrac{1}{2}\varrho_{\mathrm{R}}v^2]=[\sigma v], \qquad [\dot{e}]=[\sigma\dot{\varepsilon}]. \tag{26.2}$$

The second law requires that[47]

$$[\eta]\geqq 0. \tag{26.3}$$

Combining $(26.1)_1$ and $(26.2)_1$ with the compatibility condition (8.5) yields the well-known results[48]

$$\varrho_{\mathrm{R}}U^2=\frac{[\sigma]}{[\varepsilon]}, \tag{26.4}$$

$$[e]=\tfrac{1}{2}(\sigma^-+\sigma^+)[\varepsilon]. \tag{26.5}$$

Eq. (26.5) is called the *Hugoniot relation*.

The general theory of shock waves can best be presented using the constitutive formulation based on entropy as the independent variable. Here we assume that the wave is entering a region which has been unstrained and at constant and uniform entropy for all past times,[49] i.e., for all $X\geqq Y(t)$,

$$\varepsilon^t(s)=0, \qquad \eta^t(s)=\eta_0, \qquad s\in[0,\infty).$$

Therefore

$$[\varepsilon]=\varepsilon^-, \qquad \left[\frac{\partial\varepsilon}{\partial X}\right]=\left(\frac{\partial\varepsilon}{\partial X}\right)^-, \qquad [\eta]=\eta^--\eta_0,$$

and the history immediately behind the shock is the jump history (24.35). Thus, the definitions of the instantaneous response functions (24.36) and of the natural equilibrium state imply that

$$\begin{aligned} e^-&=\tilde{e}_{\mathrm{I}}(\varepsilon^-,\eta^-), & e^+&=\tilde{e}_{\mathrm{I}}(0,\eta_0)=e_0,\\ \sigma^-&=\tilde{\sigma}_{\mathrm{I}}(\varepsilon^-,\eta^-), & \sigma^+&=\tilde{\sigma}_{\mathrm{I}}(0,\eta_0)=0,\\ \theta^-&=\tilde{\theta}_{\mathrm{I}}(\varepsilon^-,\eta^-), & \theta^+&=\tilde{\theta}_{\mathrm{I}}(0,\eta_0)=\theta_0. \end{aligned} \tag{26.6}$$

In view of these relations and (26.5), it is evident that the strain ε^- and the entropy η^- cannot be arbitrary. If we define the function

$$\tilde{H}_{\mathrm{I}}(\varepsilon,\eta)=\tilde{e}_{\mathrm{I}}(\varepsilon,\eta)-e_0-\tfrac{1}{2}\tilde{\sigma}_{\mathrm{I}}(\varepsilon,\eta)\,\varepsilon, \tag{26.7}$$

---

[45] NUNZIATO and WALSH [1973, *17*].

[46] Cf. TRUESDELL and TOUPIN [1960, *7*, Sects. 205 and 241]. Also see CHEN [1973, *4*, Sect. 5].

[47] Cf. TRUESDELL and TOUPIN [1960, *7*, Sect. 258].

[48] TRUESDELL and TOUPIN [1960, *7*, Sect. 297] refer to (26.4) as the Rankine-Hugoniot equation. Also see CHEN [1973, *4*, Sect. 22].

[49] The results to be presented here apply equally well to arbitrarily deformed states; cf. NUNZIATO and HERRMANN [1972, *15*].

then by (26.5) the curve

$$\tilde{H}_{\mathrm{I}}(\varepsilon^{-},\eta^{-})=0, \tag{26.8}$$

called the *instantaneous Hugoniot entropy-strain curve*, represents all entropy-strain states $(\varepsilon^{-},\eta^{-})$ attainable in a shock jump from the initial state $(0,\eta_0)$. It follows that if the instantaneous stress-entropy modulus $M_{\mathrm{I}}\neq 0$, then the inverse function

$$\eta=\tilde{\sigma}_{\mathrm{I}}^{-1}(\varepsilon,\sigma)$$

exists, and consequently (26.7) may be rewritten as

$$\tilde{H}_{\mathrm{I}}\left(\varepsilon,\tilde{\sigma}_{\mathrm{I}}^{-1}(\varepsilon,\sigma)\right)=H_{\mathrm{I}}(\varepsilon,\sigma). \tag{26.9}$$

The curve

$$H_{\mathrm{I}}(\varepsilon^{-},\sigma^{-})=0 \tag{26.10}$$

defines the *instantaneous Hugoniot stress-strain curve* $\mathscr{H}$ for the material. Clearly, from (26.4) the shock velocity is determined by the secant (Rayleigh line) connecting the initial state (0, 0) and the final state $(\varepsilon^{-},\sigma^{-})$ on the Hugoniot curve $\mathscr{H}$. It should be emphasized that in general the Hugoniot relation (26.10) *cannot* be solved for one of its arguments in terms of the other. Thus, the possibility that the Hugoniot curve $\mathscr{H}$ may exhibit stress or strain extrema or inflection points must be considered.[50]

From (24.30), $(24.31)_1$, (26.7), (26.9), and the smoothness of the response functional $\tilde{\mathscr{E}}$, it follows that the functions $\tilde{H}_{\mathrm{I}}(\varepsilon,\eta)$ and $\mathrm{H}_{\mathrm{I}}(\varepsilon,\sigma)$ are single-valued, continuous, and differentiable functions of their arguments. This insures that the Hugoniot curve $\mathscr{H}$ is not closed and is fundamental to establishing the general properties of the curve (26.10). In particular, assuming that the instantaneous isentropic stress-strain curve is strictly convex from below and thus satisfies the inequalities

$$E_{\mathrm{I}}^{\mathrm{N}}(\varepsilon,\eta)>0,\qquad \tilde{E}_{\mathrm{I}}^{\mathrm{N}}(\varepsilon,\eta)>0,\qquad M_{\mathrm{I}}(\varepsilon,\eta)>0, \tag{26.11}$$

we have the following results:[51]

(i) *the shock wave is compressive, i.e.,* $\varepsilon^{-}>0$;[52]

(ii) *the entropy* η^{-} *and the shock velocity* U *increase monotonically as the Hugoniot stress-strain curve* $\mathscr{H}$ *is traversed outward from the origin* (0, 0); *and*

(iii) *the shock velocity is supersonic with respect to the material ahead of the wave and subsonic with respect to the material behind the wave, i.e.,*

$$(E_{\mathrm{I}}^{\mathrm{N}})_0<\varrho_{\mathrm{R}}U^2<(E_{\mathrm{I}}^{\mathrm{N}})^{-}$$

where $(E_{\mathrm{I}}^{\mathrm{N}})^{-}=E_{\mathrm{I}}^{\mathrm{N}}(\varepsilon^{-},\eta^{-})$.[53]

[50] The possible existence of such critical points was first discussed by BETHE [1942, *1*]; however, his study is based on assumptions which differ from those employed here. A detailed description of critical stress and strain points in the context of the present assumptions has been given by NUNZIATO and HERRMANN [1972, *15*].

[51] Cf. NUNZIATO and HERRMANN [1972, *15*]. These results are also valid for $M_{\mathrm{I}}<0$.

[52] If $\tilde{E}_{\mathrm{I}}^{\mathrm{N}}<0$, then the shock is expansive $(\varepsilon^{-}<0)$. The properties of waves in materials for which the isentropic stress-strain curve is not strictly convex or concave have been considered by COWPERTHWAITE [1968, *8*] and DE BELLOY [1970, *10*].

[53] DUNWOODY [1972, *10*] has shown that a similar result holds also for materials with memory which do conduct heat. In that case, the instantaneous Hugoniot stress-strain curve is the instantaneous isothermal stress-strain curve, and thus (iii) is an immediate consequence of curvature conditions such as (24.28).

Statements (i) and (iii) generalize results previously obtained by COLEMAN and GURTIN for the case when (26.10) has the solution $\sigma^- = \sigma_{\mathrm{I}}(\varepsilon^-)$.[54] The proof of (i) to (iii) in the present, more general, context is quite lengthy. However, it involves a rather straightforward extension of some geometrical arguments given by COWAN and SERRIN[55] and of some parameterizations suggested by WEYL.[56]

COLEMAN and GURTIN also examined the properties of weak shock waves and extended many of the classical results to materials with memory.[57] In particular, for small ε^- the jump in entropy is third-order in the amplitude, i.e.,

$$[\eta] = \frac{(\tilde{E}_{\mathrm{I}}^{\mathrm{N}})_0}{12\varrho_{\mathrm{R}}\theta_0}(\varepsilon^-)^3 + O\left((\varepsilon^-)^4\right).$$

This implies that the instantaneous Hugoniot stress-strain curve and the instantaneous isentropic stress-strain curve not only touch at $\varepsilon^- = 0$ but also have there the same slope and curvature; in fact, as $\varepsilon^- \to 0$ we have

$$\sigma^- = (E_{\mathrm{I}}^{\mathrm{N}})_0 \varepsilon^- + \tfrac{1}{2}(\tilde{E}_{\mathrm{I}}^{\mathrm{N}})_0 (\varepsilon^-)^2 + O\left((\varepsilon^-)^3\right). \tag{26.12}$$

It follows from this result and (26.1), that as $\varepsilon^- \to 0$ the shock velocity U is approximated by

$$U^2 = U_0^2 + O(\varepsilon^-) \tag{26.13}$$

where

$$U_0^2 = (C_{\mathrm{I}})_0^2 = \frac{(E_{\mathrm{I}}^{\mathrm{N}})_0}{\varrho_{\mathrm{R}}}$$

is the instantaneous isentropic sound velocity.

β) *The shock amplitude equation.* It is clear from (26.8), (26.6) and (26.4) that the entropy η^-, the internal energy e^-, the stress σ^-, the temperature θ^- and the shock velocity U all depend on the shock amplitude ε^-. In particular, differentiating (26.5) and $(26.6)_{1,2}$ with respect to t and combining the results with (26.4), we arrive at[58]

$$\frac{d\eta^-}{dt} = \frac{(E_{\mathrm{I}}^{\mathrm{N}})^-(1-\mu)}{M_{\mathrm{I}}^-(2\zeta - 1)} \frac{d\varepsilon^-}{dt} \tag{26.14}$$

where

$$\mu = \frac{\varrho_{\mathrm{R}} U^2}{(E_{\mathrm{I}}^{\mathrm{N}})^-}, \qquad \zeta = \frac{\theta^-}{M_{\mathrm{I}}^- \varepsilon^-}. \tag{26.15}$$

The parameters μ and ζ play an important role in shock wave theory, and it follows from (i) and (iii) of the previous subsection and the curvature condition (26.11) that

$$0 < \mu < 1, \qquad \zeta > 0. \tag{26.16}$$

[54] COLEMAN and GURTIN [1966, *5*], [1968, *4*]. As we indicated previously, their results are based on the earlier studies of Hugoniot curves by BETHE [1942, *1*] and WEYL [1949, *2*]. In this regard, also see COURANT and FRIEDRICHS [1948, *1*], SERRIN [1959, *2*, Sect. 56], HAYES [1960, *4*], THOMAS [1969, *2*], and CHEN [1973, *4*, Sect. 22].

[55] COWAN [1958, *1*] and SERRIN [1959, *2*, Sect. 56]. Maintaining the generality considered here, COWAN established (i) and (ii) for elastic fluids.

[56] WEYL [1949, *2*].

[57] COLEMAN and GURTIN [1966, *5*], [1968, *4*]. WATERSTON [1969, *5*] considered some properties of weak shocks in transversely isotropic and isotropic viscoelastic materials whose response functionals are represented by multi-integral expansions. For a discussion of the classical theory of weak waves, see COURANT and FRIEDRICHS [1948, *1*], SERRIN [1959, *2*], and HAYES [1960, *4*].

[58] Cf. CHEN and GURTIN [1972, *7*].

By (26.6), (26.14), and (26.15), we also have[59]

$$\begin{aligned}
\frac{de^-}{dt} &= \left\{ \frac{(2\zeta-1)\,M_{\mathrm{I}}^-\sigma^- + (1-\mu)(E_{\mathrm{I}}^{\mathrm{N}})^-\,\theta^-}{M_{\mathrm{I}}^-\,(2\zeta-1)} \right\} \frac{d\varepsilon^-}{dt}, \\
\frac{d\sigma^-}{dt} &= (E_{\mathrm{I}}^{\mathrm{N}})^- \left\{ \frac{2\zeta-\mu}{2\zeta-1} \right\} \frac{d\varepsilon^-}{dt}, \qquad (26.17) \\
\frac{d\theta^-}{dt} &= M_{\mathrm{I}}^- \left\{ \frac{(2+\gamma)\,\zeta-1}{2\zeta-1} \right\} \frac{d\varepsilon^-}{dt}
\end{aligned}$$

where

$$\gamma = \frac{(E_{\mathrm{I}}^{\mathrm{N}})^-\,(1-\mu)\,\varepsilon^-}{\varkappa_{\mathrm{I}}^-\,M_{\mathrm{I}}^-}.$$

Furthermore, differentiating (26.4) with respect to t and using $(26.17)_2$ yields[60]

$$\frac{dU}{dt} = \frac{\zeta\mu\,(E_{\mathrm{I}}^{\mathrm{N}})^-}{\varrho_{\mathrm{R}}\,U(2\zeta-1)\,\varepsilon^-}\,\frac{d\varepsilon^-}{dt}. \qquad (26.18)$$

Clearly, the variation of η^-, e^-, σ^-, θ^-, and U with the amplitude ε^- will depend on the magnitude of the parameter ζ. This quantity also influences the growth or decay of the amplitude ε^-.

Using these results along with general properties of shock waves and the assumed smoothness of the response functionals $\tilde{\mathscr{E}}$ and $\tilde{\mathscr{S}}$, CHEN and GURTIN[61] have derived the shock amplitude equation

$$\frac{d\varepsilon^-}{dt} = \frac{U(1-\mu)\,(2\zeta-1)}{(3\mu+1)\,\zeta-(3\mu-1)} \left\{ \hat{\lambda} - \left(\frac{\partial\varepsilon}{\partial X} \right)^- \right\} \qquad (26.19)$$

where $\hat{\lambda}$ is related to the stress and energy relaxation properties of the material.[62] A straightforward analysis of this equation, noting (26.16), yields the following results with regard to the growth or decay of the wave:[63]

(i) *If* $\zeta > \frac{1}{2}$ *or* $0 < \zeta < (3\mu-1)/(3\mu+1)$, *then*

$$\hat{\lambda} < \left(\frac{\partial\varepsilon}{\partial X} \right)^- \Leftrightarrow \frac{d\varepsilon^-}{dt} < 0,$$

$$\hat{\lambda} > \left(\frac{\partial\varepsilon}{\partial X} \right)^- \Leftrightarrow \frac{d\varepsilon^-}{dt} > 0.$$

(ii) *If* $\frac{1}{2} > \zeta > (3\mu-1)/(3\mu+1)$, *then*

$$\hat{\lambda} < \left(\frac{\partial\varepsilon}{\partial X} \right)^- \Leftrightarrow \frac{d\varepsilon^-}{dt} > 0,$$

$$\hat{\lambda} > \left(\frac{\partial\varepsilon}{\partial X} \right)^- \Leftrightarrow \frac{d\varepsilon^-}{dt} < 0.$$

(iii) *If* $\zeta = \frac{1}{2}$ *or* $\hat{\lambda} = (\partial\varepsilon/\partial X)^-$, *then* $d\varepsilon^-/dt = 0$.

In view of (i) to (iii), $\hat{\lambda}$ is called the *critical jump in strain gradient.*

[59] Cf. CHEN and GURTIN [1972, *7*].

[60] Cf. CHEN and GURTIN [1972, *7*].

[61] CHEN and GURTIN [1972, *7*].

[62] The precise form of $\hat{\lambda}$ is not necessary in order to examine the qualitative behavior of the wave amplitude. However, for specific applications, an expression for $\hat{\lambda}$ will be required and is given in (26.23) using temperature as the independent variable.

[63] Recall that external body forces and heat supplies have been neglected. However, such external fields can influence the wave behavior through the quantity $\hat{\lambda}$; see for example, WALSH [1967, *12*] and CHEN [1971, *3*] who studied the influence of discontinuous external fields, and NUNZIATO and WALSH [1972, *17*], [1974, *2*], who examined the influence of continuous external fields. These studies all pertain to elastic nonconductors; however, the results can easily be generalized to materials with memory.

Now, if the Hugoniot stress-strain curve is one-to-one and hence has the representation

$$\sigma^- = \sigma_I(\varepsilon^-), \tag{26.20}$$

then the monotonicity requirement on the entropy along the Hugoniot and (26.14) imply that $\zeta > \frac{1}{2}$.[64] In this case, the behavior of the wave is as described in (i) above and is the same as in the purely mechanical theory.[65] On the other hand, if the Hugoniot stress-strain curve $\mathscr{H}$ cannot be represented by (26.20) (and hence is multi-valued), then there exists at least one point $(\varepsilon^-, \sigma^-)$ on the curve $\mathscr{H}$, as the curve is traversed outward from the origin (0, 0), at which the strain ε^- is a relative maximum. At this point, the slope of the Hugoniot curve is infinite, and it follows from $(26.17)_2$ that $\zeta = \frac{1}{2}$.[66] Thus, by (iii) above, the shock is steady and independent of the strain gradient behind the wave. For an amplitude ε^- corresponding to a point $(\varepsilon^-, \sigma^-)$ on the curve $\mathscr{H}$ which is beyond the relative strain maximum, as the curve is traversed outward from (0, 0), (26.14) and the monotonicity of the entropy shows that $\zeta < \frac{1}{2}$. Thus, the behavior is given by (ii) above and is exactly opposite to that in the purely mechanical theory.

CHEN and GURTIN[67] considered also the behavior of weak waves and showed that the results (16.9) to (16.11), obtained in the purely mechanical case, apply in the present more general case provided the instantaneous tangent modulus, the instantaneous second-order modulus, and the stress relaxation function are taken *at fixed entropy*. That is, all weak waves decay exponentially. Furthermore, as the shock amplitude tends to zero, the critical acceleration[68] $U^2|\hat{\lambda}|$ has as its limit twice the critical amplitude of an acceleration wave propagating into an unstrained region in equilibrium.[69]

γ) *Shock wave behavior in a particular thermoviscoelastic solid.* NUNZIATO and WALSH[70] have used the above results to examine the behavior of shock waves in a particular thermoviscoelastic solid. Here we follow their analysis and use the constitutive model described in Sect. 25 to evaluate the relevant material response functions, including the critical jump in strain gradient, and compare the predictions with experimental results obtained from the observation of steady waves in polymethyl methacrylate.

Since the constitutive model was formulated using temperature as the independent variable, we need to first rewrite the parameters appearing in the shock amplitude equation in terms of the temperature. In this case, a shock wave will grow or decay according to (26.19) where, by (26.15), (24.33), and (24.18),

$$\mu = \frac{\varrho_R U^2}{(E_I^N)^-}, \quad \zeta = \frac{\varkappa_I^-}{\varepsilon^- A_I^-} = \frac{1}{\varepsilon^- \Gamma_I^-}, \tag{26.21}$$

$$(E_I^N)^- = (E_I^T)^- + \frac{\theta^- (A_I^-)^2}{\varkappa_I^-}, \tag{26.22}$$

[64] Cf. NUNZIATO and HERRMANN [1972, *15*]. CHEN and GURTIN [1972, *7*] consider only this case and follow a similar argument.

[65] Cf. Sect. 16.

[66] Cf. NUNZIATO and HERRMANN [1972, *15*]. Points at which ε^- correspond to relative strain maxima on the Hugoniot curve $\mathscr{H}$ are called *critical strain points*.

[67] CHEN and GURTIN [1972, *7*].

[68] Naturally, a shock amplitude equation similar to (26.19) could be given in terms of the particle velocity v^-, the particle acceleration $(\dot{v})^-$, and the critical acceleration $U^2|\hat{\lambda}|$; cf. Sect. 16.

[69] Cf. Sect. 27.

[70] NUNZIATO and WALSH [1973, *17*].

and

$$\hat{\lambda} = \frac{1}{(1-\mu)\, U(E_{\mathrm{I}}^{\mathrm{N}})^{-}} \{G'(\varepsilon^{-}, \theta^{-}; 0)\, \varepsilon^{-} - L'(\varepsilon^{-}, \theta^{-}; 0)\, (\theta^{-} - \theta_0) - \Gamma_{\mathrm{I}}^{-} [H'(\varepsilon^{-}, \theta^{-}; 0)\, \varepsilon^{-} - J'(\varepsilon^{-}, \theta^{-}; 0)\, (\theta^{-} - \theta_0)]\}. \tag{26.23}$$

Here $G'(\varepsilon^{-}, \theta^{-}; 0) = \hat{\mathscr{G}}'(\varepsilon_{\mathrm{I}}, \theta_{\mathrm{I}}, 0)$, etc., are the initial slopes of the relaxation functions corresponding to the jump history (24.15). Notice that the quantities U, μ, ζ, and $\hat{\lambda}$ are functions only of the strain amplitude since immediately behind the shock wave the temperature and strain are related through the Hugoniot relation (26.5) which, in the present context, takes the form[71]

$$\hat{H}_{\mathrm{I}}(\varepsilon^{-}, \theta^{-}) = \hat{e}_{\mathrm{I}}(\varepsilon^{-}, \theta^{-}) - e_0 - \tfrac{1}{2}\hat{\sigma}_{\mathrm{I}}(\varepsilon^{-}, \theta^{-})\, \varepsilon^{-} = 0. \tag{26.24}$$

For a reasonable range of strain amplitudes, we expect that $\hat{\lambda} < 0$.[72]

The determination of the critical jump in strain gradient follows a procedure similar to that outlined in Sect. 17 for the mechanical theory. We first note that for polymethyl methacrylate the instantaneous Hugoniot stress-strain curve is one-to-one for stresses less than 7.5 kbar and thus can be represented by (26.20) with $\sigma_{\mathrm{I}}(\varepsilon)$ given by $(25.16)_1$. Then it follows from (26.4) that the shock velocity obeys

$$\varrho_{\mathrm{R}}\, U^2(\varepsilon^{-}) = \frac{\sigma_{\mathrm{I}}(\varepsilon^{-})}{\varepsilon^{-}}.$$

As we also showed in Sect. 25, the instantaneous Hugoniot temperature-strain curve (26.24) has the representation

$$\theta^{-} = \Theta_{\mathrm{I}}(\varepsilon^{-})$$

with $\Theta_{\mathrm{I}}(\varepsilon)$ given by $(25.19)_1$. The amplitude dependence of the moduli $(E_{\mathrm{I}}^{\mathrm{N}})^{-}$ and Γ_{I}^{-} can now be evaluated using (26.22), (25.15), and the response functions $\hat{\sigma}_{\mathrm{E}}(\varepsilon, \theta)$, $\hat{e}_{\mathrm{E}}(\varepsilon, \theta)$, and $\Omega(\varepsilon)$ determined in Sect. 25. Fig. 47 shows this dependence of the moduli $(E_{\mathrm{I}}^{\mathrm{N}})^{-}$ and Γ_{I}^{-} as well as that of the shock velocity U and the temperature θ^{-}.

To evaluate the relaxation functions involved in the expression for $\hat{\lambda}$, it is necessary to relate the response functions of the thermoviscoelastic model (25.10) and (25.11) to those defined in the more general formulation $(24.6)_2$ and (24.7) in terms of which $\hat{\lambda}$ is given. To this end, we compute the derivatives of the stress (25.10) and the internal energy (25.11) with respect to the relative strain and temperature history and evaluate these for the appropriate underlying history. That is, we calculate

$$\frac{d}{d\alpha} \mathscr{S}(\varepsilon^t + \alpha\gamma^t, \theta^t + \alpha\varphi^t)\big|_{\alpha=0} = G_t(0)\, \gamma^t(0) + L_t(0)\, \varphi^t(0) + \int_0^\infty G_t'(s)\, \gamma^t(s)\, ds + \int_0^\infty L_t'(s)\, \varphi^t(s)\, ds$$

and evaluate the result for the jump history, $\varepsilon^t = \varepsilon_{\mathrm{I}}$, $\theta^t = \theta_{\mathrm{I}}$. A similar procedure applies for the internal energy. Carrying this out yields the following identification

[71] Cf. Eq. $(25.18)_1$.

[72] As we shall see shortly, $\hat{\lambda} < 0$ for $\varepsilon^{-} < 0.04$ in polymethyl methacrylate.

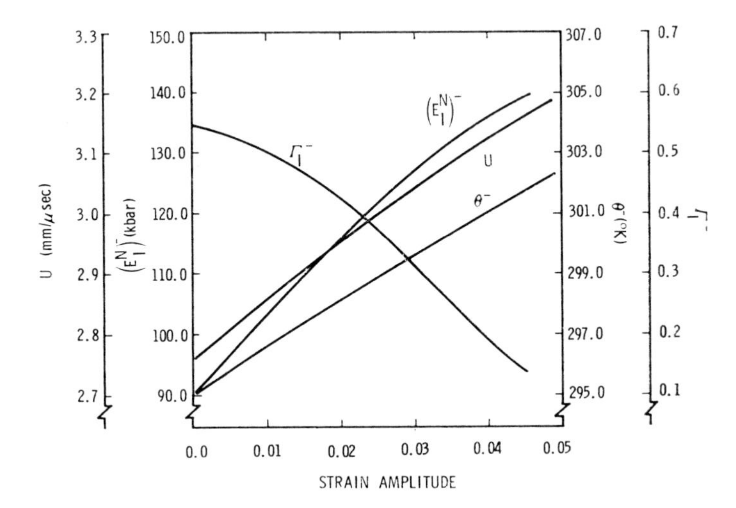

Fig. 47. The instantaneous isentropic stress-strain modulus $(E_I^N)^-$, the instantaneous stress-energy modulus Γ_I^-, and the temperature θ^- evaluated behind the shock along with the shock velocity U as functions of the strain amplitude for polymethyl methacrylate. (From NUNZIATO and WALSH [1973, *17*].)

of the relaxation functions:

$$
\begin{aligned}
G'(\varepsilon,\theta;0) &= -\frac{1}{\tau_\varepsilon}\left[\Omega'(\varepsilon)\,h(\varepsilon,\theta)+\Omega(\varepsilon)\right],\\
L'(\varepsilon,\theta;0) &= \frac{g_0}{\theta_0\,\tau_\theta}\left[\Omega'(\varepsilon)\,h(\varepsilon,\theta)+\Omega(\varepsilon)\right],\\
H'(\varepsilon,\theta,0) &= -\frac{\Omega(\varepsilon)}{\tau_\varepsilon}\,(\varepsilon+g_0),\\
J'(\varepsilon,\theta;0) &= \frac{g_0\Omega(\varepsilon)}{\theta_0\tau_\theta}\,(\varepsilon+g_0).
\end{aligned}
\tag{26.25}
$$

Here, as before, τ_ε and τ_θ are the relaxation times and $h(\varepsilon,\theta)$ is given by (25.14).

Thus, with all the material response functions known, we can calculate $|\hat{\lambda}(\varepsilon^-)|$ and compare the results with experimentally determined values. For a steady wave,

$$\hat{\lambda}(\varepsilon^-) = \left(\frac{\partial\varepsilon}{\partial X}\right)^- = -\frac{(\dot{v})^-}{U^2}$$

where $(\dot{v})^-$ is the value of the particle acceleration immediately behind the shock front. Thus, as in the mechanical case, we give the results in terms of the critical acceleration $U^2|\hat{\lambda}|$ and its dependence on the particle velocity amplitude v^-. Fig. 48 indicates the amplitude dependence of $U^2|\hat{\lambda}|$ for polymethyl methacrylate using the thermoviscoelastic constitutive model which was developed in Sect. 25 from the results of steady wave experiments and thermophysical and acoustic data. The correlation between the values of particle acceleration $(\dot{v})^-$ determined directly from steady wave profiles[73] and the calculated curve is good. It should be noted that the curve of $U^2|\hat{\lambda}|$ is somewhat higher than the one based on the mechanical model (cf. Fig. 28) and is believed to indicate partially the influence of the thermodynamic effects on shock propagation in nonlinear polymeric materials.

[73] Cf. Sect. 13.

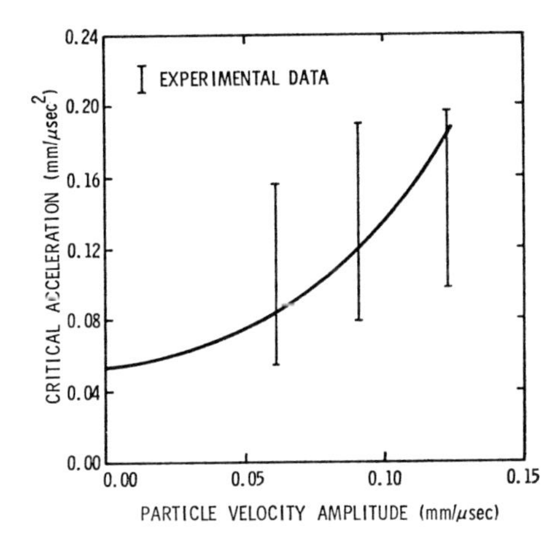

Fig. 48. Comparison of measured and calculated critical acceleration vs. particle velocity for polymethyl methacrylate; thermomechanical model. (From NUNZIATO and WALSH [1973, *17*].)

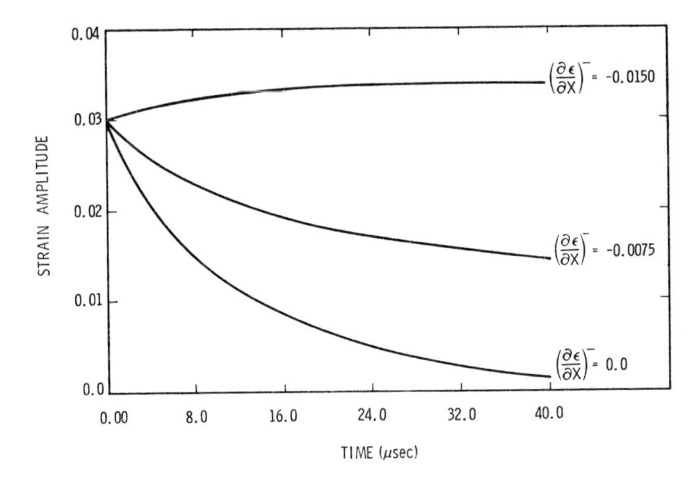

Fig. 49. Strain behind the shock wave vs. time for fixed values of strain gradient for polymethyl methacrylate; thermomechanical model. (From NUNZIATO and WALSH [1973, *17*].)

Since in general the time variation of the particle acceleration (and hence the strain gradient) behind the wave is not known in advance, it is not possible to determine $\varepsilon^-(t)$ completely from (26.19). Nevertheless, by making certain assumptions on $(\partial\varepsilon/\partial X)^-$ it is possible to gain some insight into the amplitude behavior of a shock discontinuity which might be expected. NUNZIATO and WALSH[74] assumed $(\partial\varepsilon/\partial X)^-$ to be constant at a value (i) greater than and (ii) less than $\hat{\lambda}$ and showed the resulting growth and decay behavior of the wave (cf. Fig. 49).

27. The growth and decay of acceleration waves. In their study of thermodynamic influences on the growth and decay of acceleration waves in materials with memory, COLEMAN and GURTIN[75] showed that every acceleration wave with intrinsic

[74] NUNZIATO and WALSH [1973, *17*].

[75] COLEMAN and GURTIN [1965, *6*], COLEMAN and GURTIN [1968, *4*].

velocity $U(t) \neq 0$ propagating in a nonconductor is homentropic.[76] Further, if the wave is entering a region at rest in a state of constant and uniform entropy, then (20.1) and (20.7), which give the velocity U_0 and amplitude $a(t)$ of an acceleration wave neglecting thermal effects, apply in the present more general case provided that the relaxation function and second-order modulus used in Sect. 20 are taken at fixed entropy.[77] In view of this, it would again seem possible to use the results from experimental shock wave studies and the results from the theory of weak waves given in Sect. 26 to evaluate the material constants necessary to predict the critical time t_∞ at which an acceleration wave would (appear to) form a shock discontinuity in a nonconductor. NUNZIATO and WALSH[78] have carried out just such a calculation for polymethyl methacrylate. The values of the constants $(E_\mathrm{I}^\mathrm{N})_0$, $(\tilde{E}_\mathrm{I}^\mathrm{N})_0$, and U_0 follow directly from (26.20), $(25.16)_1$, and (26.12). The critical amplitude λ_0 is one-half the zero-amplitude value of the calculated critical acceleration shown in Fig. 48, and furthermore, by (20.6),

$$\mu_0 = \frac{(\tilde{E}_\mathrm{I}^\mathrm{N})_0 \lambda_0}{2(E_\mathrm{I}^\mathrm{N})_0 U_0}.$$

Using these values of μ_0 and λ_0, NUNZIATO and WALSH calculated critical times from (20.9) which were commensurate with the experimental results presented in Sect. 22. These results give further support to the observation that if the assumption that the material is a thermal nonconductor is valid (which seems reasonable for polymeric solids), then the study of acceleration waves provides no new information about the thermodynamic effects on the material response.

28. Stress-energy response. Our entire discussion up to this point has dealt with one-dimensional wave propagation in nonlinear viscoelastic materials which can be produced by mechanical impact. However, in the context of thermovisco-

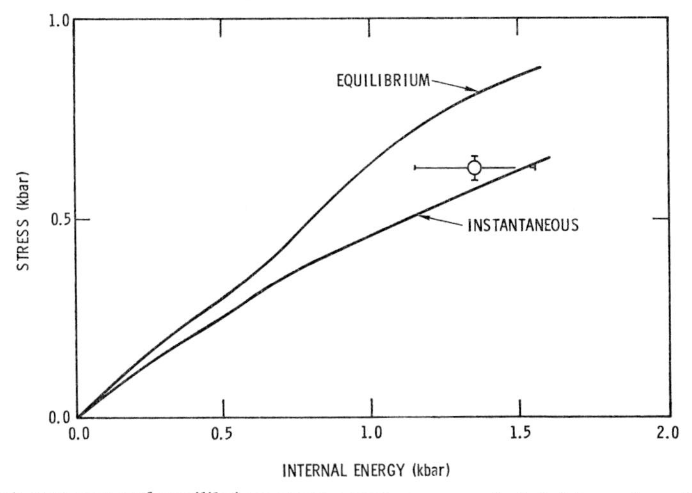

Fig. 50. Instantaneous and equilibrium stress-energy curves calculated for polymethyl methacrylate. (From NUNZIATO and WALSH [1973, *15*].)

[76] Cf. COLEMAN and GURTIN [1965, *7*], [1966, *6*] who also derived the general propagation condition for acceleration waves in the context of the three-dimensional theory of materials with memory. In this regard, also see MANDEL and BRUN [1968, *14*] and a related study by WANG and BOWEN [1966, *14*].

[77] Under this same condition, the equivalence of acceleration waves and infinitesimal sinusoidal progressive waves described in Sect. 20 also holds.

[78] NUNZIATO and WALSH [1973, *17*].

elastic response it is also worthwhile to consider the one-dimensional response of the material to thermal impact, i.e., the response to rapid deposition of energy. Important factors in predicting such response are the instantaneous and equilibrium stress-energy curves at zero strain. Using the thermoviscoelastic model constructed in Sect. 25, NUNZIATO and WALSH[79] have calculated the stress-energy curves for polymethyl methacrylate and their results are shown in Fig. 50. Also shown here is the experimental result obtained by BARKER[80] for a polymethyl methacrylate sample which experienced an instantaneous deposition of energy corresponding to a significant temperature rise. The good agreement with the instantaneous curve gives further support to the usefulness of the model constructed.

It is interesting to note that, unlike the mechanical stress-strain response (Fig. 25), the instantaneous stress-energy response is *less* than the equilibrium response. This appears to be due to the manner in which energy is absorbed and transferred in a polymer. In the case of rapid energy deposition, the energy is initially absorbed by the high frequency modes and then is transferred over a period of tens of nanoseconds to the translational modes. Since it is the translational modes which contribute to the stress, the stress increases with time in a constant volume situation. Thus, $(\Gamma_{\mathrm{I}})_0 \leqq (\Gamma_{\mathrm{E}})_0$ and it is for this reason, along with the inequality $(24.26)_2$, that we previously observed that $(A_{\mathrm{I}})_0 \leqq (A_{\mathrm{E}})_0$.[81]

References.

Italic numbers in parentheses following each reference indicate the sections in which it is cited.

1858 *1.* DALE, the Rev. T. P., and J. H. GLADSTONE: On the influence of temperature on the refraction of light. Phil. Trans. Roy. Soc. (London), Ser. A **148**, 887–894. (*5*)

1863 *1.* GLADSTONE, J. H., and the Rev. T. P. DALE: Researches on the refraction, dispersion, and sensitiveness of liquids. Phil. Trans. Roy. Soc. (London), Ser. A **153**, 317–343. (*5*)

1867 *1.* MAXWELL, J. C.: On the dynamical theory of gases. Phil. Trans. Roy. Soc. (London), Ser. A **157**, 49–88. Reprinted in Maxwell's Scientific Papers. Cambridge University Press 1890. (*12*)

1903 *1.* ZAREMBA, S.: Sur une forme perfectionnée de la théorie de la relaxation. Bull. Int. Acad. Sci. Cracovie, 594–614. (*12*)

1926 *1.* GRÜNEISEN, E.: Zustand des festen Körpers. In: Handbuch der Physik, vol. X. Berlin: Springer. (*24*)

1928 *1.* PRESTON, T.: Theory of Light. New York: Macmillan. (*5*)

1934 *1.* WOOD, R. W.: Physical Optics. New York: Macmillan. (*6*)

1942 *1.* BETHE, H.: The theory of shock waves for an arbitrary equation of state. Office of Scientific Research and Development, Report No. 545. (*26*)

1943 *1.* LEADERMAN, H.: Elastic and creep properties of filamentous materials. Washington, D. C.: Textile Foundation. (*11*)

1948 *1.* COURANT, R., and K. O. FRIEDRICHS: Supersonic Flow and Shock Waves. New York: Interscience. (*18, 26*)

2. SOKOLOVSKY, V. V.: The propagation of elastic-viscous-plastic waves in bars [in Russian]. Prikl. Mat. i Mek. **12**, 261–280. (*12*)

[79] NUNZIATO and WALSH [1973, *15*].

[80] L. M. BARKER (unpublished)

[81] This difference between $(\Gamma_{\mathrm{I}})_0$ and $(\Gamma_{\mathrm{E}})_0$, or for that matter $\Gamma_{\mathrm{I}}(0, \theta)$ and $\Gamma_{\mathrm{E}}(0, \theta)$, which is a result of thermal rate effects, is significant and may serve to explain in part the apparent disparity between reported values of the stress-energy modulus (Grüneisen parameter) for polymeric solids determined by different techniques (cf. NUNZIATO and WALSH [1973, *15*]).

1949 *1.* HILLIER, K. W.: A method of measuring some dynamic elastic constants and its application to the study of high polymers. Proc. Phys. Soc. (London) B **62**, 701–713. (*10, 11*)

2. WEYL, H.: Shock waves in arbitrary fluids. Comm. Pure Appl. Math. **2**, 103–122. (*26*)

1950 *1.* FERRY, J. D.: Mechanical properties of substances of high molecular weight. VI. J. Am. Chem. Soc. **72**, 3746–3752. (*11*)

2. MALVERN, L. E.: Plastic wave propagation in a bar of material exhibiting a strain rate effect. Quart. Appl. Math. **8**, 405–411. (*12*)

1951 *1.* SIPS, R.: Propagation phenomena in elastic-viscous media. J. Polymer Sci. **6**, 285–293. (*10, 12*)

1952 *1.* SCHWARZL, F., and A. J. STAVERMAN: Time-temperature dependence of linear viscoelastic behavior. J. Appl. Phys. **23**, 838–843. (*11*)

1953 *1.* LEE, E. H., and I. KANTER: Wave propagation in finite rods of viscoelastic material. J. Appl. Phys. **24**, 1115–1122. (*12*)

1954 *1.* GLAUZ, R. D., and E. H. LEE: Transient wave analysis in a linear time-dependent material. J. Appl. Phys. **25**, 947–953. (*12*)

1955 *1.* MINSHALL, S.: Properties of elastic and plastic waves determined by pin contactors and crystals. J. Appl. Phys. **26**, 463–469. (*5*)

2. NOLL, W.: On the continuity of the solid and fluid states. J. Rational Mech. Anal. **4**, 3–81. Reprinted in Continuum Mechanics II: Rational Mechanics of Materials (ed. by C. TRUESDELL). Intl. Sci. Rev. Ser. New York: Gordon and Breach 1965. (*12*)

1956 *1.* BERRY, D. S., and S. C. HUNTER: The propagation of dynamic stresses in viscoelastic rods. J. Mech. Phys. Solids **4**, 72–95. (*12*)

2. LEE, E. H., and J. A. MORRISON: A comparison of the propagation of longitudinal waves in rods of viscoelastic materials. J. Polymer Sci. **19**, 93–110. (*12*)

1957 *1.* BLAND, D. R.: Application of the one-sided Fourier transform to the stress analysis of linear viscoelastic materials. In: Proceedings of the Conference on the Properties of Materials at High Rates of Strain. London: The Institution of Mechanical Engineers. (*12*)

2. GREEN, A. E., and R. S. RIVLIN: The mechanics of non-linear materials with memory. Arch. Rational Mech. Anal. **1**, 1–21. Reprinted in Continuum Mechanics II: Rational Mechanics of Materials (ed. by C. TRUESDELL). Intl. Sci. Rev. Ser. New York: Gordon and Breach 1965. (*16*)

3. HUGHES, D. S., W. M. OTTO, E. B. BLANKENSHIP, M. F. GOURLEY, and L. E. GOURLEY: Shock-waves in metals generated by projectile impacts. Bull. Am. Phys. Soc. **2**, 214. (*4*)

4. LEE, E. H.: Stress analysis for viscoelastic materials. In: Proceedings of the Conference on the Properties of Materials at High Rates of Strain. London: The Institution of Mechanical Engineers. (*12*)

5. MORRISON, J. A.: Wave propagation in rods of Voigt material and visco-elastic materials with three-parameter models. Quart. Appl. Math. **14**, 153–169. (*12*)

1958 *1.* COWAN, R. D.: Properties of the Hugoniot function. J. Fluid Mech. **3**, 531–545. (*26*)

2. KOLSKY, H.: The propagation of stress waves in viscoelastic solids. Appl. Mech. Rev. **11**, 465–468. (*1*)

3. NOLL, W.: A mathematical theory of the mechanical behavior of continuous media. Arch. Rational Mech. Anal. **2**, 197–226. Reprinted in Continuum Mechanics II: Rational Mechanics of Materials (ed. by C. TRUESDELL). Intl. Sci. Rev. Ser. New York: Gordon and Breach, 1965, and in Continuum Theory of Inhomogeneities in Simple Bodies. Berlin-Heidelberg-New York: Springer 1968. (*9*)

4. RICE, M. H., R. G. MCQUEEN, and J. M. WALSH: Compression of solids by strong shock waves. In: Solid State Physics, vol. 6 (ed. by F. SEITZ and D. TURNBULL). New York: Academic Press. (*4*)

1959 *1.* DATTA, S.: A note on stress waves in viscoelastic rods. J. Technol. Calcutta **4**, 41–46. (*12*)

2. SERRIN, J.: Mathematical principles of classical fluid mechanics. In: Handbuch der Physik, vol. VIII/1 (ed. by S. FLÜGGE and C. TRUESDELL). Berlin-Göttingen-Heidelberg: Springer. (*26*)

1960 *1.* COLEMAN, B. D., and W. NOLL: An approximation theorem for functionals with applications in continuum mechanics. Arch. Rational Mech. Anal. **6**, 355–370. Reprinted in Continuum Mechanics II: Rational Mechanics of Materials. (ed. by C. TRUESDELL). Intl. Sci. Rev. Ser. New York: Gordon and Breach 1965. (*9*)

2. FOWLES, G. R.: Attenuation of the shock wave produced in a solid by a flying plate. J. Appl. Phys. **31**, 655–661. (*18*)

3. GIELESSEN, J., u. J. KOPPELMANN: Über die Druckabhängigkeit der elastischen Moduln von Polymethacrylsäuremethylester (PMMA) bei Raumtemperatur. Kolloid-Z. **172**, 162–166. (*23*)
4. HAYES, W. D.: Gasdynamic Discontinuities. In: Fundamentals of Gas Dynamics. Princeton, N. J.: Princeton University Press. (*26*)
5. HUNTER, S. C.: Viscoelastic waves. In: Progress in Solids Mechanics, vol. I (ed. by I. N. SNEDDON and R. HILL). Amsterdam: North Holland Publishing Co. (*10*)
6. KOLSKY, H.: Viscoelastic waves. In: International Symposium on Stress Wave Propagation in Materials (ed. by N. DAVIDS). New York: Interscience. (*1*)
7. TRUESDELL, C., and R. A. TOUPIN: The classical field theories. In: Handbuch der Physik, vol. III/1 (ed. by S. FLÜGGE). Berlin-Göttingen-Heidelberg: Springer. (*8, 23, 26*)

1961 1. COLEMAN, B. D., and W. NOLL: Foundations of linear viscoelasticity. Rev. Mod. Phys. **33**, 239–249. Erratum ibid., **36**, 1103. Reprinted in Continuum Mechanics III: Foundations of Elasticity Theory (ed. by C. TRUESDELL). Intl. Sci. Rev. Ser. New York: Gordon and Breach 1965. (*9, 12, 14, 23*)
2. DAVIS, W. C., and B. G. CRAIG: Smear camera technique for free-surface velocity measurement. Rev. Sci. Instr. **32**, 579–581. (*5*)
3. DUVALL, G. E.: Properties and application of shock waves. In: Response of Metals to High Velocity Deformations (ed. by P. G. SHEWMON and V. F. ZACKAY). New York: Interscience. (*4*)
4. FOWLES, G. R.: Shock wave compression of hardened and annealed 2024 aluminum. J. Appl. Phys. **32**, 1475–1487. (*5*)
5. HILL, R.: Discontinuity relations in mechanics of solids. In: Progress in Solid Mechanics, vol. II (ed. by I. N. SNEDDON and R. HILL). Amsterdam: North Holland Publishing Co. (*8*)
6. HUGHES, D. S., L. E. GOURLEY, and M. F. GOURLEY: Shock wave compression of iron and bismuth. J. Appl. Phys. **32**, 624–629. (*5*)
7. RICE, M. H,: Capacitor technique for measuring the velocity of a plane conducting surface. Rev. Sci. Instr. **32**, 449–451. (*5*)
8. TRUESDELL, C.: General and exact theory of waves in finite elastic strain. Arch. Rational Mech. Anal. **8**, 263–296. Reprinted in Continuum Mechanics III: Foundations of Elasticity Theory (ed. by C. TRUESDELL). Intl. Sci. Rev. Ser. New York: Gordon and Breach 1965, and in Wave Propagation in Dissipative Materials. Berlin-Heidelberg-New York: Springer 1965. (*20, 23*)

1962 1. ACHENBACH, J. D., and C. C. CHAO: A three-parameter viscoelastic model particularly suited for dynamic problems. J. Mech. Phys. Solids **10**, 245–252. (*12*)
2. BARONE, A.: Generation, detection, and measurement of ultrasound. In: Handbuch der Physik, vol. XI/2 (ed. by S. FLÜGGE). Berlin-Göttingen-Heidelberg: Springer. (*11*)
3. CHU, B. T.: Stress waves in isotropic linear viscoelastic materials. J. Mécanique **1**, 439–462. (*12*)
4. GUENTHER, A. H., D. C. WUNSCH, and T. D. SOAPES: Acceleration of thin plates by exploding foil techniques. In: Exploding Wires, vol. 2 (ed. by W. G. CHACE and H. K. MOORE). New York: Plenum Press. (*4*)
5. KELLER, D. V., and J. R. PENNING: Exploding foils—The production of plane shock waves and the acceleration of thin plates. In: Exploding Wires, vol. 2 (ed. by W. G. CHACE and H. K. MOORE). New York: Plenum Press. (*4*)
6. KOLSKY, H.: The detection and measurement of stress waves. In: Experimental Techniques in Shock and Vibration. New York: American Society of Mechanical Engineers. (*1*)
7. NEILSON, F. W., W. B. BENEDICK, W. P. BROOKS, R. A. GRAHAM, and G. W. ANDERSON: Electrical and optical effects of shock waves in crystalline quartz. In: Les Ondes de Détonation. Paris: Centre National de La Recherche Scientifique. (*5*)
8. SCHAPERY, R. A.: Approximate methods of transform inversion for viscoelastic stress analysis. In: Proceedings of the Fourth U. S. National Congress of Applied Mechanics (ed. by R. M. ROSENBERG). New York: American Society of Mechanical Engineers. (*11*)

1963 1. DUVALL, G. E., and R. C. ALVERSON: Fundamental research in support of Vela-Uniform. Tech. Summary Rept. 4. Menlo Park, Calif.: Stanford Research Institute (*16*)
2. HEYDEMANN, P., and H. D. GUICKING: Specific volume of polymers as a function of temperature and pressure. Kolloid-Z.-Z. für Poly. **193**, 16–25. (*23, 25*)
3. TAYLOR, J. W., and M. H. RICE: Elastic-plastic properties of iron. J. Appl. Phys. **34**, 364–371. (*5*)

1964 *1.* BARKER, L. M., and R. E. HOLLENBACH: System for measuring the dynamic properties of materials. Rev. Sci. Instr. **35**, 742–746. (*4*)

2. BERNSTEIN, D., and D. D. KEOUGH: Piezoresistivity of manganin. J. Appl. Phys. **35**, 1471–1474. (*5*)

3. BRIDGMAN, P. W.: Collected Experimental Papers. Cambridge: Harvard University Press. (*5*)

4. COLEMAN, B. D.: Thermodynamics of materials with memory. Arch. Rational Mech. Anal. **17**, 1–46. (*24*)

5. COLEMAN, B. D.: Thermodynamics, strain impulses, and viscoelasticity. Arch. Rational Mech. Anal. **17**, 230–254. (*9, 24*)

6. COLEMAN, B. D., and W. NOLL: Simple fluids with fading memory. In: Proceedings of the IUTAM Symposium on Second-Order Effects in Elasticity, Plasticity, and Fluid Dynamics (ed. by M. REINER and D. ABIR). New York: Macmillan. (*9*)

7. CRISTESCU, N.: Some problems of the mechanics of extensible strings. In: Proceedings of the IUTAM Symposium on Stress Waves in Anelastic Solids. Berlin-Göttingen-Heidelberg-New York: Springer. (*12*)

8. DREMIN, A. N., and G. A. ADADUROV: The behavior of glass under dynamic loading [in Russian]. Fiz. Tverd. Tela. **6**, 1757–1764. (*5*)

9. DUVALL, G. E.: Propagation of plane waves in a stress-relaxing medium. In: Proceedings of the IUTAM Symposium on Stress Waves in Anelastic Solids. Berlin-Göttingen-Heidelberg-New York: Springer. (*16*)

10. FULLER, P. J. A., and J. H. PRICE: Dynamic pressure measurements to 300 kilobars with a resistance transducer. Brit. J. Appl. Phys. **15**, 751–758. (*5*)

11. GOTTENBERG, W. G., and R. M. CHRISTENSEN: An experiment for determination of the mechanical property in shear for a linear, isotropic viscoelastic solid. Int. J. Egng. Sci. **2**, 45–57. (*11*)

12. JEFFERY, A.: A note on the derivation of the discontinuity conditions across contact discontinuities, shocks, and phase fronts. Z. Angew. Math. Phys. **15**, 68–71. (*8*)

13. KOLSKY, H.: Stress waves in solids. J. Sound Vib. **1**, 88–110. (*1*)

14. LUBLINER, J.: A generalized theory of strain-rate dependent plastic wave propagation in bars. J. Mech. Phys. Solids **12**, 59–65. (*12*)

15. STERNBERG, E.: On the analysis of thermal stresses in viscoelastic solids. In: Proceedings of the Third Symposium on Naval Structural Mechanics (ed. by A. M. FREUDENTHAL, B. A. BOLEY, and H. LIEBOWITZ). New York: Macmillan. (*11*)

16. THUNBORG, Jr., S., G. E. INGRAM, and R. A. GRAHAM: Compressed gas gun for controlled planar impacts over a wide velocity range. Rev. Sci. Instr. **35**, 11–14. (*4*)

1965 *1.* AL'TSHULER, L. V.: Use of shock waves in high-pressure physics [in Russian]. Usp. Fiz. Nauk. **85**, 197–258. (*4*)

2. ASAY, J. R., A. J. DORR, N. D. ARNOLD, and A. H. GUENTHER: Ultrasonic wave velocity-temperature studies in several plastics, plastic foams, and nose-cone materials. Rept. AFWL-TR-65-188. Albuquerque, N. M.: Air Force Weapons Laboratory. (*11*)

3. BARKER, L. M., and R. E. HOLLENBACH: Interferometer technique for measuring the dynamic mechanical properties of materials. Rev. Sci. Instr. **36**, 1617–1620. (*4, 5*)

4. COLEMAN, B. D., M. E. GURTIN, and I. HERRERA R.: Waves in materials with memory I. The velocity of one-dimensional shock and acceleration waves. Arch. Rational Mech. Anal. **19**, 1–19. Reprinted in Wave Propagation in Dissipative Materials. Berlin-Heidelberg-New York: Springer 1965. (*8, 12, 16, 20*)

5. COLEMAN, B. D., and M. E. GURTIN: Waves in materials with memory II. On the growth and decay of one-dimensional acceleration waves. Arch. Rational Mech. Anal. **19**, 239–265. Reprinted in Wave Propagation in Dissipative Materials. Berlin-Heidelberg-New York: Springer 1965. (*8, 9, 10, 12, 20*)

6. COLEMAN, B. D., and M. E. GURTIN: Waves in materials with memory III. Thermodynamic influences on the growth and decay of acceleration waves. Arch. Rational Mech. Anal. **19**, 266–298. Reprinted in Wave Propagation in Dissipative Materials. Berlin-Heidelberg-New York: Springer 1965. (*24, 27*)

7. COLEMAN, B. D., and M. E. GURTIN: Waves in materials with memory IV. Thermodynamics and the velocity of general acceleration waves. Arch. Rational Mech. Anal. **19**, 317–338. Reprinted in Wave Propagation in Dissipative Materials. Berlin-Heidelberg-New York: Springer 1965. (*24, 27*)

8. DUNWOODY, J., and N. T. DUNWOODY: Acceleration waves in viscoelastic media of Maxwell-Zaremba type. Int. J. Engng. Sci. **3**, 417–427. (*20*)

9. EDEN, G., and P. W. WRIGHT: A technique for the precise measurement of the motion of a plane free surface. In: Proceedings of the Fourth Symposium on Detonation. Washington, D. C.: U. S. Government Printing Office. (*5*)

10. FISHER, G. M. C.: The decay of plane waves in the linear theory of viscoelasticity. Division of Applied Mathematics Rept. No. 2. Providence, R. I.: Brown University. (*16, 20*)

11. FISHER, G. M. C., and M. E. GURTIN: Wave propagation in the linear theory of viscoelasticity. Quart. Appl. Math. **23**, 257–263. (*20*)

12. GRAHAM, R. A., F. W. NEILSON, and W. B. BENEDICK: Piezoelectric current from shock-loaded quartz—a submicrosecond stress gauge. J. Appl. Phys. **36**, 1775–1783. (*5*)

13. GURTIN, M. E., and I. HERRERA R.: On dissipation inequalities and linear viscoelasticity. Quart. Appl. Math. **23**, 235–245. (*9*)

14. HALPIN, W. J., and R. A. GRAHAM: Shock wave compression of plexiglas from 3 to 20 kilobars. In: Proceedings of the Fourth Symposium on Detonation. Washington, D. C.: U. S. Government Printing Office. (*12*)

15. HERRERA R., I., and M. E. GURTIN: A correspondence principle for viscoelastic wave propagation. Quart. Appl. Math. **22**, 360–364. (*20*)

16. KOLSKY, H.: Experimental studies of the mechanical behavior of linear viscoelastic solids. In: Proceedings of the Fourth Symposium on Naval Structural Mechanics (ed. by A. C. ERINGEN, H. LIEBOWITZ, S. L. KOH, and J. M. CROWLEY). New York: Pergamon Press. (*1, 10, 11*)

17. MARVIN, R. S., and J. E. MCKINNEY: Volume relaxation in amorphous polymers. In: Physical Acoustics, vol. II B (ed. by W. P. MASON). New York: Academic Press. (*11, 23*)

18. PEYRE, C., J. PUJOL, and J. THOUVENIN: Experimental method for analysis of the structure of a shock wave in a solid. In: Proceedings of the Fourth Symposium on Detonation. Washington, D. C.: U. S. Government Printing Office. (*5*)

19. TRUESDELL, C., and W. NOLL: The non-linear field theories of mechanics. In: Handbuch der Physik, vol. III/3 (ed. by S. FLÜGGE). Berlin-Heidelberg-New York: Springer. (*2*)

20. VALANIS, K. C.: Propagation and attenuation of waves in linear viscoelastic solids. J. Math. Phys. **44**, 227–239. (*16*)

21. VALANIS, K. C.: Wave propagation in viscoelastic solids with measured relaxation or creep functions. In: Proceedings of the Fourth International Congress of Rheology (ed. by E. H. LEE). New York: Interscience. (*12*)

22. VARLEY, E.: Acceleration fronts in viscoelastic materials. Arch. Rational Mech. Anal. **19**, 215–225. (*20*)

1966 *1.* AHRENS, T. J., and G. E. DUVALL: Stress relaxation behind elastic shock waves in rocks. J. Geophys. Res. **71**, 4349–4360. (*16*)

2. AHRENS, T. J., and M. H. RUDERMAN: Immersed-foil method for measuring shock wave profiles in solids. J. Appl. Phys. **37**, 4758–4765. (*5*)

3. COLEMAN, B. D., J. M. GREENBERG, and M. E. GURTIN: Waves in materials with memory V. On the amplitude of acceleration waves and mild discontinuities. Arch. Rational Mech. Anal. **22**, 333–354. (*20*)

4. COLEMAN, B. D., and M. E. GURTIN: Acceleration waves in nonlinear materials. In: Modern Developments in the Mechanics of Continua; Proceedings of an International Conference on Rheology (ed. by S. ESKINAZI). New York: Academic Press. (*20*)

5. COLEMAN, B. D., and M. E. GURTIN: Thermodynamics and one dimensional shock waves in materials with memory. Proc. Roy. Soc. (London), Ser. A **292**, 562–574. (*26*)

6. COLEMAN, B. D., and M. E. GURTIN: Thermodynamics and wave propagation. Quart. Appl. Math. **24**, 257–262. (*20, 27*)

7. COLEMAN, B. D., and V. J. MIZEL: Norms and semi-groups in the theory of fading memory. Arch. Rational Mech. Anal. **23**, 87–123. (*9*)

8. DUNWOODY, J.: Longitudinal wave propagation in a rate dependent material. Int. J. Engng. Sci. **4**, 277–287. (*12, 20*)

9. JONES, A. H., W. M. ISBELL, and C. I. MAIDEN: Measurement of the very-high pressure properties of materials using a light gas gun. J. Appl. Phys. **37**, 3493–3499. (*4*)

10. LUBLINER, J., and G. A. SECOR: The propagation of shock waves in nonlinear viscoelastic materials. Structural Engineering Laboratory Rept. 66-11. Berkeley, Calif.: University of California. (*16*)

11. MIZEL, V. J., and C.-C. WANG: A fading memory hypothesis which suffices for chain rules. Arch. Rational Mech. Anal. **28**, 124–134. (*24*)

12. PIPKIN, A. C.: Shock structure in a viscoelastic fluid. Quart. Appl. Math. **23**, 297–303 (*12, 14, 15*)

13. Thomas, T. Y.: The general theory of compatibility conditions. Int. J. Engng. Sci. **4**, 207–233. (*8*)

14. Wang, C.-C., and R. M. Bowen: On the thermodynamics of nonlinear materials with quasi-elastic response. Arch. Rational Mech. Anal. **22**, 79–99. (*27*)

1967 *1.* Achenbach, J. D., and D. P. Reddy: Note on wave propagation in linearly viscoelastic media. Z. Angew. Math. Phys. **18**, 141–144. (*18*)

2. Findley, W. N., R. M. Reed, and P. Stern: Hydrostatic creep of plastics. J. Appl. Mech. **34**, 895–904. (*23*)

3. Greenberg, J. M.: The existence of steady shock waves in nonlinear materials with memory. Arch. Rational Mech. Anal. **24**, 1–21. (*12, 14*)

4. Jacobson, R. S.: Magnetic acceleration of flyer plates for shock wave testing of materials. Rept. SCL DR-67-60. Livermore, Calif.: Sandia Laboratories. (*4*)

5. Lubliner, J.: On the calculation of acceleration waves in viscoelastic materials. J. Mécanique **6**, 243–250. (*20*)

6. Lubliner, J., and G. A. Secor: Wave propagation in semilinear viscoelastic materials. Int. J. Engng. Sci. **5**, 755–772. (*18*)

7. Mattaboni, P., and E. Schreiber: Method of pulse transmission measurements for determining sound velocities. J. Geophys. Res. **72**, 5160–5163. (*11*)

8. Perzyna, P.: On fading memory of a material. Arch. Mech. Stos. **19**, 537–547. (*9*)

9. Skobeev, A. M.: Stress waves in a viscoplastic medium [in Russian]. Inzh. Zh. Mekh. Tverd. Tela. **2**, 139–142. (*12*)

10. Touloukian, Y. S.: Thermophysical properties of high temperature solid materials, vol. 6, part II. New York: Macmillan. (*25*)

11. Valanis, K. C.: Some exact wave propagation solutions in viscoelastic materials. Acta Mechanica **4**, 170–190. (*12*)

12. Walsh, E. K.: Induced one-dimensional waves in elastic nonconductors. J. Appl. Mech. **34**, 937–941. (*26*)

1968 *1.* Achenbach, J. D., S. M. Vogel, and G. Herrmann: On stress waves in viscoelastic media conducting heat. In: Proceedings of the IUTAM Symposium on the Irreversible Aspect of Continuum Mechanics and Transfer of Physical Characteristics in Moving Fluids (ed. by H. Parkus and L. I. Sedov). Wien-New York: Springer. (*26*)

2. Barker, L. M.: The fine structure of compressive and release wave shapes in aluminum measured by the velocity interferometer technique. In: Proceedings of the IUTAM Symposium on the Behavior of Dense Media Under High Dynamic Pressures. New York: Gordon and Breach. (*5*)

3. Coleman, B. D., and M. E. Gurtin: On the stability against shear waves of steady flows of non-linear viscoelastic fluids. J. Fluid Mech. **33**, 165–181. (*10, 20*)

4. Coleman, B. D., and M. E. Gurtin: Thermodynamics and wave propagation in non-linear materials with memory. In: Proceedings of the IUTAM Symposium on the Irreversible Aspect of Continuum Mechanics and Transfer of Physical Characteristics in Moving Fluids (ed. by H. Parkus and L. I. Sedov). Wien-New York: Springer. (*26, 27*)

5. Coleman, B. D., and V. J. Mizel: A general theory of dissipation in materials with memory. Arch. Rational. Mech. Anal. **27**, 255–274. (*24*)

6. Coleman, B. D., and V. J. Mizel: On the general theory of fading memory. Arch. Rational Mech. Anal. **29**, 18–31. (*9*)

7. Coleman, B. D., and V. J. Mizel: On the stability of solutions of functional-differential equations. Arch. Rational Mech. Anal. **30**, 173–196. (*25*)

8. Cowperthwaite, M.: Properties of some Hugoniot curves associated with shock instability. J. Franklin Inst. **285**, 275–284. (*26*)

9. Greenberg, J. M.: Existence of steady waves for a class of nonlinear dissipative materials. Quart. Appl. Math. **26**, 27–34. (*12*)

10. Gurtin, M. E.: On the thermodynamics of materials with memory. Arch. Rational Mech. Anal. **28**, 40–50. (*24*)

11. Isbell, W. M., F. H. Shipman, and A. H. Jones: Use of a light gas gun in studying material behavior at megabar pressures. In: Proceedings of the IUTAM Symposium on the Behavior of Dense Media Under High Dynamic Pressures. New York: Gordon and Breach. (*4*)

12. Karnes, C. H.: The plate impact configuration for determining mechanical properties of materials at high strain rates. In: Mechanical Behavior of Materials Under Dynamic Loads (ed. by U. S. Lindholm). Berlin-Heidelberg-New York: Springer. (*4*)

13. Keller, D. V.: Behavior of short duration shocks in plexiglas. In: Proceedings of the IUTAM Symposium on the Behavior of Dense Media Under High Dynamic Pressures. New York: Gordon and Breach. (*18*)

14. Mandel, J., and L. Brun: Thermodynamique et ondes dans les milieux viscoelastiques. J. Mech. Phys. Solids **16**, 33–58. (*27*)

1969 *1.* Asay, J. R., D. L. Lamberson, and A. H. Guenther: Pressure and temperature dependence of the acoustic velocities in polymethyl methacrylate. J. Appl. Phys. **40**, 1768–1783. (*11, 23*)

2. Thomas, T. Y.: Mathematical foundations of the theory of shocks in gases. J. Math. Anal. Appl. **26**, 595–629. (*26*)

3. Truell, R., C. Elbaum, and B. B. Chick: Ultrasonic Methods in Solid State Physics. New York-London: Academic Press. (*11*)

4. Truesdell, C.: Rational Thermodynamics. New York: McGraw-Hill. (*24*)

5. Waterston, R. J.: One-dimensional shock waves and acceleration fronts in nonlinear viscoelastic materials. Quart. J. Mech. and Appl. Math. **22**, 261–281. (*20, 26*)

1970 *1.* Achenbach, J. D., and G. Herrmann: Propagation of second-order thermomechanical disturbances in viscoelastic solids. In: Proceedings of the IUTAM Symposium on Thermoinelasticity (ed. by B. A. Boley). New York-Wien: Springer. (*26*)

2. Barker, L. M., and R. E. Hollenbach: Shock-wave studies of PMMA, fused silica, and sapphire. J. Appl. Phys. **41**, 4208–4226. (*5, 12, 13, 15, 21, 22, 25*)

3. Bertholf, L. D., and M. L. Oliver: Approximate analytic expressions for the attenuation of a triangular pressure pulse with distance. Rept. SC RR-69-596. Albuquerque, N. M.: Sandia Laboratories. (*18*)

4. Chen, P. J., and M. E. Gurtin: On the growth of one-dimensional shock waves in materials with memory. Arch. Rational Mech. Anal. **36**, 33–46. (*12, 16, 17*)

5. Chen, P. J., M. E. Gurtin, and E. K. Walsh: Shock amplitude variation in polymethyl methacrylate for fixed values of the strain gradient. J. Appl. Phys. **41**, 3557–3558. (*17*)

6. Clifton, R. J.: On the analysis of the laser velocity-interferometer. J. Appl. Phys. **41**, 5335–5337. (*5*)

7. Coleman, B. D., and D. R. Owen: On the thermodynamics of materials with memory. Arch. Rational Mech. Anal. **36**, 245–269. (*24*)

8. Commerford, G. L., and J. S. Whittier: Uniaxial-strain wave-propagation experiments using shock-tube loading. Exp. Mech. **10**, 120–126. (*4*)

9. Day, W. A.: On the monotonicity of the relaxation functions of viscoelastic materials. Proc. Cambridge Phil. Soc. **67**, 503–508. (*9*)

10. de Belloy, J. B.: Étude des ondes de choc longitudinales dans les matériaux élastiques non linéaires à partir du réseau des isentropiques. J. Mécanique **9**, 277–307. (*26*)

11. Ferry, J. D.: Viscoelastic Properties of Polymers. New York: Wiley. (*11, 23*)

12. Fowles, G. R., G. E. Duvall, J. Asay, P. Bellamy, F. Feistman, D. Grady, T. Michaels, and R. Mitchell: Gas gun for impact studies. Rev. Sci. Instr. **41**, 984–996. (*4*)

13. Ingram, G. E., and R. A. Graham: Quartz gauge technique for impact experiments. In: Proceedings of the Fifth Symposium on Detonation. Washington, D. C.: U. S. Government Printing Office. (*5*)

14. Jacobs, S. J., and D. J. Edwards: Experimental study of the electromagnetic velocity-gauge technique. In: Proceedings of the Fifth Symposium on Detonation. Washington, D. C.: U. S. Government Printing Office. (*5*)

15. Keough, D. D., and J. Y. Wong: Variation of the shock piezoresistance coefficient of manganin as a function of deformation. J. Appl. Phys. **41**, 3508–3515. (*5*)

16. Lubliner, J., and R. J. Green: Shock-wave propagation in instantaneously nonlinear materials. J. Mécanique **9**, 507–522. (*16, 18*)

17. Perry, F. C.: Thermoelastic response of polycrystalline metals to relativistic electron beam absorption. J. Appl. Phys. **41**, 5017–5022. (*4*)

18. Romberger, A. B.: The temperature dependence of the index of refraction and hypersonic speed, and an investigation of the Landau-Placzek ratio in three polymethacrylates. Thesis, Pennsylvania State University. (*11*)

19. Schuler, K. W.: Propagation of steady shock waves in polymethyl methacrylate. J. Mech. Phys. Solids **18**, 277–293. (*12–15, 21, 25*)

20. Schuler, K. W.: The speed of release waves in polymethyl methacrylate. Proceedings of the Fifth Symposium on Detonation. Washington, D. C.: U. S. Government Printing Office. (*15, 18, 21*)

21. Stevens, A. L., and L. E. Malvern: Wave propagation in prestrained polyethylene rods. Exp. Mech. **10**, 1–7. (*10*)

22. Sun, C. T.: Transient wave propagation in a viscoelastic rod. J. Appl. Mech. **37**, 1141–1144. (*18*)

1971 *1.* BAILEY, P., and P. J. CHEN: On the local and global behavior of acceleration waves. Arch. Rational Mech. Anal. **41**, 121–131. (*20*)

2. BARKER, L. M.: Velocity interferometer data reduction. Rev. Sci. Instr. **42**, 276–278. (*5*)

3. CHEN, P. J.: External influences on the growth and decay of one-dimensional shock waves in elastic non-conductors. Int. J. Solids Struct. **7**, 1697–1703. (*26*)

4. GAUSTER, W. B.: Effect of heating rate in thermoelastic stress production. J. Mech. Phys. Solids **19**, 137–145. (*4*)

5. PERCIVAL, C. M.: A quantitative measurement of thermally induced stress waves. AIAA J. **9**, 347–349. (*4*)

6. SCHULER, K. W., and E. K. WALSH: Critical-induced acceleration for shock propagation in polymethyl methacrylate. J. Appl. Mech. **38**, 641–645. (*14, 17*)

7. VOGT, J. P., and R. A. SCHAPERY: Uniaxial shock wave propagation in a viscoelastic material. Int. J. Solids Struct. **7**, 505–521. (*18*)

8. WALSH, E. K.: The decay of stress waves in one-dimensional polymer rods. Trans. Soc. Rheol. **15**, 345–353. (*10, 11*)

1972 *1.* ASAY, J. R., G. R. FOWLES, and Y. GUPTA: Determination of material relaxation properties from measurements on decaying elastic shock fronts. J. Appl. Phys. **43**, 744–746. (*17*)

2. BAILEY, P., and P. J. CHEN: On the local and global behavior of acceleration waves. Addendum: Asymptotic behavior. Arch. Rational. Mech. Anal. **44**, 212–216. (*20, 22*)

3. BARKER, L. M.: Laser interferometry in shock-wave research. Exp. Mech. **12**, 209–215. (*5*)

4. BARKER, L. M., and R. E. HOLLENBACH: Laser interferometer for measuring high velocities of any reflecting surface. J. Appl. Phys. **43**, 4669–4675. (*5*)

5. BOWEN, R. M., and P. J. CHEN: Some comments on the behavior of acceleration waves of arbitrary shape. J. Math. Phys. **13**, 948–950. (*20*)

6. CHEN, P. J., and M. E. GURTIN: On the use of experimental results concerning steady shock waves to predict the acceleration wave response of nonlinear viscoelastic materials. J. Appl. Mech. **39**, 295–296. (*22*)

7. CHEN, P. J., and M. E. GURTIN: Thermodynamic influences on the growth of one-dimensional shock waves in materials with memory. Z. Angew. Math. Phys. **23**, 69–79. (*26*)

8. DAY, W. A.: The Thermodynamics of Simple Materials with Fading Memory. Berlin-Heidelberg-New York: Springer. (*24*)

9. DUNWOODY, J.: One-dimensional shock waves in heat conducting materials with memory. 1. Thermodynamics. Arch. Rational Mech. Anal. **47**, 117–148. (*26*)

10. DUNWOODY, J.: One-dimensional shock waves in heat conducting materials with memory. 2. Shock analysis. Arch. Rational Mech. Anal. **48**, 192–204. (*26*)

11. GRAHAM, R. A.: Strain dependence of longitudinal piezoelectric, elastic, and dielectric constants of X-cut quartz. Phys. Rev. B **6**, 4779–4792. (*5*)

12. GURTIN, M. E.: The linear theory of elasticity. In: Handbuch der Physik, vol. VIa/2 (ed. by C. TRUESDELL). Berlin-Heidelberg-New York: Springer. (*8, 23*)

13. HICKS, D. L., and D. B. HOLDRIDGE: The CONCHAS wavecode. Rept. SC RR-72-0451. Albuquerque, N. M.: Sandia Laboratories. (*19*)

14. LAWRENCE, R. J.: A general viscoelastic constitutive relation for use in wave propagation calculations. Rept. SC RR-72-0114. Albuquerque, N. M.: Sandia Laboratories. (*19*)

15. NUNZIATO, J. W., and W. HERRMANN: The general theory of shock waves in elastic nonconductors. Arch. Rational Mech. Anal. **47**, 272–287. (*25, 26*)

16. NUNZIATO, J. W., K. W. SCHULER, and E. K. WALSH: The bulk response of viscoelastic solids. Trans. Soc. Rheol. **16**, 15–32. (*23*)

17. NUNZIATO, J. W., and E. K. WALSH: Propagation and growth of shock waves in inhomogeneous fluids. Phys. Fluids **15**, 1397–1402. (*26*)

18. SUTHERLAND, H. J., and R. LINGLE: An acoustic characterization of polymethyl methacrylate and three epoxy formulations. J. Appl. Phys. **43**, 4022–4026. (*11*)

1973 *1.* AMES, W. F., D. T. DAVY, and R. CHAND: Closed form similarity solutions for the nonlinear viscoelastic wave problem. In: Proceedings of the Fourth Canadian Congress of Applied Mechanics (ed. by A. BAZERGUI). Montreal: Ecole Polytechnique (abstract only). (*12*)

2. BELL, J. F.: The experimental foundations of solid mechanics. In: Handbuch der Physik, vol. VIa/1 (ed. by C. TRUESDELL). Berlin-Heidelberg-New York: Springer. (*1*)

3. BOWEN, R. M., and P. J. CHEN: Thermodynamic restriction on the initial slope of the stress relaxation function. Arch. Rational Mech. Anal. **51**, 278–284. (*9, 24*)

4. CHEN, P. J.: Growth and decay of waves in solids. In: Handbuch der Physik, vol. VIa/3 (ed. by C. TRUESDELL). Berlin-Heidelberg-New York: Springer. (*1, 2, 8, 9, 12, 16, 17, 20, 22, 23, 26*)
5. HERRMANN, W., and D. L. HICKS: Numerical analysis methods. In: Metallurgical Effects at High Strain Rates (ed. by R. W. ROHDE, B. M. BUTCHER, J. R. HOLLAND, and C. H. KARNES). New York: Plenum Press. (*19*)
6. HERRMANN, W., and J. W. NUNZIATO: Nonlinear constitutive equations. In: Dynamic Response of Materials to Intense Impulsive Loading (ed. by P. C. CHOU and A. K. HOPKINS). Dayton, Ohio: Air Force Materials Laboratory. (*14*)
7. HUILGOL, R. R.: Growth of plane shock waves in materials with memory. Int. J. Engng. Sci. **11**, 75–86. (*16*)
8. HUILGOL, R. R.: Acceleration waves in isotropic simple materials. Arch. Mech. Stos. 25, 365–376. (*20*)
9. LEE, L. M.: Nonlinearity in the piezoresistance coefficient of impact loaded manganin. J. Appl. Phys. **44**, 4017–4022. (*5*)
10. LEITMAN, M. J., and G. M. C. FISHER: The linear theory of viscoelasticity. In: Handbuch der Physik, vol. VIa/3 (ed. by C. TRUESDELL). Berlin-Heidelberg-New York: Springer. (*10, 16, 20*)
11. LYSNE, P. C., and D. R. HARDESTY: The fundamental equation of state of liquid nitromethane to 100 kbar. J. Chem. Phys. **59**, 6512–6523. (*25*)
12. NUNZIATO, J. W., and K. W. SCHULER: Evolution of steady shock waves in polymethyl methacrylate. J. Appl. Phys. **44**, 4774–4775. (*19*)
13. NUNZIATO, J. W., and K. W. SCHULER: Shock pulse attenuation in a nonlinear viscoelastic solid. J. Mech. Phys. Solids **21**, 447–457. (*18*)
14. NUNZIATO, J. W., and H. J. SUTHERLAND: Acoustical determination of stress relaxation functions for polymers. J. Appl. Phys. **44**, 184–187. (*11, 13, 25*)
15. NUNZIATO, J. W., and E. K. WALSH: Instantaneous and equilibrium Grüneisen parameters for a nonlinear viscoelastic polymer. J. Appl. Phys. **44**, 1207–1211. (*25, 28*)
16. NUNZIATO, J. W., and E. K. WALSH: Propagation of steady shock waves in nonlinear thermoviscoelastic solids. J. Mech. Phys. Solids **21**, 317–335. (*25*)
17. NUNZIATO, J. W., and E. K. WALSH: Amplitude behavior of shock waves in a thermoviscoelastic solid. Int. J. Solids Struct. **9**, 1373–1383. (*26, 27*)
18. SCHULER, K. W., J. W. NUNZIATO, and E. K. WALSH: Recent results in nonlinear viscoelastic wave propagation. Int. J. Solids Struct. **9**, 1237–1281. (*13, 17*)
19. SULICIU, I., S. Y. LEE, and W. F. AMES: Nonlinear traveling waves for a class of rate-type materials. J. Math. Anal. Appl. **42**, 313–322. (*12*)
20. WALSH, E. K., and K. W. SCHULER: Acceleration wave propagation in a nonlinear viscoelastic solid. J. Appl. Mech. **40**, 705–710. (*20, 22*)
1974 *1.* CURRO, J. G.: Hydrostatic equations of state for polymers. J. Macro. Sci.-Rev. in Macro. Chem. (in press). (*23*)
2. NUNZIATO, J. W., and E. K. WALSH: Shock wave propagation in elastic nonconductors with continuous external radiation. Acta Mechanica (in press). (*26*)
3. SCHULER, K. W., and J. W. NUNZIATO: The dynamic mechanical behavior of polymethyl methacrylate. Rheologica Acta **13**, 265–273. (*15, 21*)
4. YUAN, H. L., and G. LIANIS: Experimental investigation of wave propagation in nonlinear viscoelastic materials. Rheologica Acta (in press). (*14*)

Addendum

The following additional papers bear on the subject of this article. The italic numbers in parentheses indicate the sections in which reference to the paper would have been made, had it been available in time.

1973 *A.* DUNWOODY, J.: One-dimensional shock waves in heat conducting materials with memory. 3. Evolutionary behavior. Arch. Rational Mech. Anal. **50**, 278–289. (*26*)
B. FOWLES, G. R.: Experimental technique and instrumentation. In: Dynamic Response of Materials to Intense Impulsive Loading (ed. by P. C. CHOU and A. K. HOPKINS). Dayton, Ohio: Air Force Material Laboratory. (*4, 5*)
C. SADD, M. H.: A note on shear waves in BKZ fluids. Trans. Soc. Rheol. **17**, 647–657. (*20*)
D. SPENCE, D. A.: Nonlinear wave propagation in viscoelastic materials. In: Nonlinear Elasticity (ed. by R. W. DICKEY). New York: Academic Press. (*12, 20*)

E. SWAN, G. W., G. E. DUVALL, and C. K. THORNHILL: On steady wave profiles in solids. J. Mech. Phys. Solids **21**, 215–227. (*12*)

1974 *A.* BLAKE, T. R., and J. F. WILSON: A numerical study of plane longitudinal waves in a nonlinear viscoelastic material. J. Appl. Mech. **41**, 111–116. (*19*),

B. BOWEN, R. M., and P. J. CHEN: Acceleration waves in a mixture of chemically reacting materials with memory. Acta Mechanica (in press). (*20*)

C. BOWEN, R. M., P. J. CHEN, and J. W. NUNZIATO: Shock waves in a mixture of chemically reacting materials with memory. Acta Mechanica (in press). (*26*)

D. NACHLINGER, R. R., J. W. NUNZIATO, and L. T. WHEELER: Theorems on wave propagation and uniqueness for a class of nonlinear dissipative materials. J. Math. Anal. Appl. (in press). (*12, 15*)

E. NACHLINGER, R. R., and L. T. WHEELER: Wave propagation and uniqueness in one-dimensional simple materials. J. Math. Anal. Appl. (in press). (*12, 15*)

F. NUNZIATO, J. W., K. W. SCHULER, and D. B. HAYES: Numerical wave propagation calculations for nonlinear viscoelastic solids. In: Proceedings of the First International Conference on Computational Methods in Nonlinear Mechanics. Austin, Texas: University of Texas. (*19*)

G. TING, T. C. T.: Propagation of discontinuities of all orders in nonlinear media. In: Proceedings of the Tenth Anniversary Meeting of the Society of Engineering Science (in press). (*20*)

Waves in Solids.

By

R. N. Thurston.

With 3 Figures.

A. Introduction.

1. Scope.

This article discusses the theoretical basis for understanding mechanical waves in solids and the interpretation of experiments on mechanical waves. There are many mechanical wave phenomena in solids that can be adequately described within pure mechanics, or at least within mechanics and thermostatics. However, some of the most interesting and fruitful things about mechanical waves are the nonmechanical effects that accompany them. This is especially true in many of the present-day applications.

We shall be concerned mainly with small-amplitude waves that are only slightly damped. In typical experiments, the travel time and attenuation of essentially sinusoidal waves are measured as functions of temperature, frequency, applied static stress, impurity concentration, magnetic field intensity, or other variables. Besides revealing the dependence of effective elastic coefficients on the varied quantities, such experiments have led to a better understanding of the causes of damping, and have given direction to the search for low-loss materials needed in certain applications.

In these experiments, the wave constitutes only a small perturbation from an initial state. After passage of the wave, the material returns to a state that may be slightly different from the initial one because of dissipative heating, but the difference is usually kept imperceptibly small in order to simplify the interpretation of the experiments. In other important experiments, the disturbance is deliberately made very large in order to investigate the behavior of the material under extremely high stresses and very fast deformations. Here the final state may be grossly different from the initial one because of permanent deformation, appreciable temperature rise, etc. The latter experiments are described in James Bell's article, "The Experimental Foundations in Solid Mechanics" in Volume VIa/1. Such experiments are ordinarily harder to interpret than the former, but they furnish information that cannot be obtained any other way. The goal is often to determine whether a proposed constitutive assumption is appropriate, or to develop materials and structures that can withstand violent shocks.

Some material on waves in solids may be found in other volumes of this encyclopedia. For example, in Volume XI/2, Acoustics II (1962) are the articles "Generation, Detection, and Measurement of Ultrasound" by A. Barone, and "High Frequency Ultrasonic Stress Waves in Solids" by R. Truell and C. Elbaum. Seismological applications are discussed in the articles by K. E. Bullen and by W. M. Ewing and Frank Press in Volume XLVII, Geophysics I (1956).

This introduction is followed by 11 chapters, entitled Foundations, Equilibrium states, Electromechanical interactions, Material symmetry, Exponentially damped plane waves, Linear viscoelastic interactions, Thermoviscoelastic media, Small-amplitude waves that are sinusoidal in time, Ultrasonic measurements as a function of static initial stress, Analysis of ultrasonic measurements as a function of temperature, and Examples. The examples include elastic waves in cubic crystals, thermoviscoelastic waves in cubic crystals, piezoelectrically excited vibrations, radial motion of thin circular piezoelectric ceramic disks, waves of finite amplitude in elastic media, longitudinal shock waves in elastic solids, reflection of longitudinal waves at normal incidence, and longitudinal motion of a piezoelectric material. The reader is referred to the table of contents for a further indication of the coverage.

2. Notation.

The notation follows the general scheme common in modern continuum mechanics with vectors and tensors in bold face type when considered as entities. We sometimes also indicate a vector by its general component, considering the "velocity v_i" to be synonymous with the "velocity $\boldsymbol{v}$" unless i has been given a specific meaning. In general, $\boldsymbol{X}$ indicates a point in the material by giving its position vector in the reference configuration, while $\boldsymbol{x}$ in the function $\boldsymbol{x}(\boldsymbol{X}, t)$ is its position at time t. The function $\boldsymbol{x}(\boldsymbol{X}, t)$ describes the motion.

All symbols are defined where first used. Most quantities relating to the geometry of deformation and motion are defined in Sect. 3. We use the familiar J for the Jacobian of the deformation gradient matrix, but V_{ij} for the strain components [cf. Eqs. (3.2) and (3.5)]. Many of the other symbols may be quickly identified from the tables, especially Tables 10.1 and 11.1 for thermomechanical quantities, Tables 14.1 and 14.2 when electrical variables are present, and Table 28.1 and the surrounding text for quantities associated with different reference states in the discussion of wave propagation in an initially stressed medium.

Attention is directed to the referral of thermodynamic functions, such as the internal energy U, to unit volume in the reference state in Sect. 10 and 11, in order to free several equations from the reference density $\bar{\varrho}$. However, in Sect. 28 it is more advantageous to use the internal energy per unit mass, which is there denoted by $\mathscr{U}$ (script) [cf. Eq. (28.4)].

In Sects. 12 through 14, it is important to note that the conventional electromagnetic fields are denoted by script capitals while ordinary capitals of the same letters are used for deformation-transformed quantities that have some advantages in a material description [cf. Eqs. (12.8) and (12.9)].

The notation becomes particularly burdensome in discussions of "effective" and "thermodynamic" elastic coefficients and their derivatives with respect to pressure and temperature because of the many different functions that are involved. Therefore, the reader is encouraged to be persistent in following the distinctions that are made.

B. Foundations.

3. Motion and deformation.

3.1. Geometry of Deformation. Strain. Let coordinates be referred to a fixed rectangular Cartesian coordinate system. The motion of a continuous medium can be specified by three functions $x_i(X_1, X_2, X_3, t)$, $i=1, 2, 3$, or, in vector form,

$\boldsymbol{x}(\boldsymbol{X}, t)$. For each time t, these functions map the medium from a fixed reference configuration, in which the particles have coordinates (X_1, X_2, X_3), into the present configuration in which the corresponding coordinates are (x_1, x_2, x_3). For fixed t, the three functions define a deformation from the configuration $\boldsymbol{X}$ to $\boldsymbol{x}$. The particle *displacement* $\boldsymbol{u}$ from the reference configuration to the present one is defined by

$$\boldsymbol{u} = \boldsymbol{x} - \boldsymbol{X}. \tag{3.1}$$

The local properties of a deformation are characterized by the nine deformation gradients $\partial x_i/\partial X_j$. A tensor C_{ij}, called the right Cauchy-Green tensor [TRUESDELL and NOLL, 1965, NFT*, p. 53], and a strain tensor V_{ij} may be defined by

$$C_{ij} = \frac{\partial x_k}{\partial X_i} \frac{\partial x_k}{\partial X_j} = \delta_{ij} + 2 V_{ij}. \tag{3.2}$$

Except where otherwise noted, summation over repeated subscripts is to be understood. The symbol δ_{ij} is the Kronecker delta, equal to 1 if $i=j$, and otherwise 0. Purely rigid motions are characterized by $C_{ij} = \delta_{ij}$, or $V_{ij} = 0$. In a strain, some of the particles of the medium must have moved either closer together or farther apart.

From (3.1),

$$x_k = X_k + u_k, \tag{3.3}$$

$$\frac{\partial x_k}{\partial X_i} = \delta_{ki} + \frac{\partial u_k}{\partial X_i}. \tag{3.4}$$

From (3.2) and (3.4),

$$V_{ij} = \frac{1}{2} \left(\frac{\partial u_i}{\partial X_j} + \frac{\partial u_j}{\partial X_i} + \frac{\partial u_k}{\partial u_i} \frac{\partial u_k}{\partial X_j} \right). \tag{3.5}$$

The word *right* in the name right Cauchy-Green tensor is motivated by the fact that C_{ij} is the square of the right stretch tensor. Right and left stretch tensors U and L are defined [TRUESDELL and NOLL, 1965, NFT, pp. 52–53] with the help of the polar decomposition theorem

$$F = RU = LR \tag{3.6}$$

where F is the matrix of the deformation gradients $\partial x_i/\partial X_j$, i.e.

$$F = \left\| \begin{matrix} \frac{\partial x_1}{\partial X_1} & \frac{\partial x_1}{\partial X_2} & \frac{\partial x_1}{\partial X_3} \\ \frac{\partial x_2}{\partial X_1} & \frac{\partial x_2}{\partial X_2} & \frac{\partial x_2}{\partial X_3} \\ \frac{\partial x_3}{\partial X_1} & \frac{\partial x_3}{\partial X_2} & \frac{\partial x_3}{\partial X_2} \end{matrix} \right\|, \tag{3.7}$$

R is the orthogonal rotation tensor, and U and L are symmetric. From the above, the matrix of the tensor C_{ij} is

$$C = F^T F = (RU)^T RU = UR^T RU = U^2. \tag{3.8}$$

Here the superscript T denotes the transpose. The last equality in (3.8) follows because R is orthogonal ($R^{-1} = R^T$).

After a deformation has taken place, an additional deformation that carries x_k to x'_i can be described by a matrix of deformation gradients $F' = \|\partial x'_i/\partial x_k\|$, and the resultant deformation by the matrix product $F'F$, whose elements are $(\partial x'_i/\partial x_k)(\partial x_k/\partial X_j) = \partial x'_i/\partial X_j$. If F' represents a rigid rotation, then it is an orthog-

* NFT is an abbreviation for "The non-linear field theories of mechanics."

onal matrix, and $(F'F)^T(F'F)$ reduces to $F^T F$ by the same operations as in (3.8). Thus, an additional rigid rotation leaves C_{ij} (and V_{ij}) unchanged.

In an actual motion, the deformation gradients are of course functions of position and time. However, to discuss the rotation and stretch of material line and surface elements at a point, it is convenient to imagine a spatially homogeneous deformation in which the deformation gradients are constants. Aside from a translation of the point at the origin, such a deformation brings the point X_i to the new place $x_i = (\partial x_i/\partial X_j) X_j$. Consequently, also, $X_i = (\partial X_i/\partial x_j) x_j$. A material plane, of unit normal $\boldsymbol{N}$ in the reference configuration, represented by $N_i X_i = 0$, has the new representation $N_i(\partial X_i/\partial x_j) x_j = 0$, with new direction numbers $N_i(\partial X_i/\partial x_j)$. The new direction cosines are

$$n_j = N_i(\partial X_i/\partial x_j)/f_{\boldsymbol{N}}, \tag{3.9}$$

where the normalization factor $f_{\boldsymbol{N}}$, obtained by setting $n_j n_j = 1$, satisfies

$$f_{\boldsymbol{N}}^2 = G_{ik} N_i N_k \tag{3.10}$$

with

$$G_{ik} \equiv \frac{\partial X_i}{\partial x_j} \frac{\partial X_k}{\partial x_j} = C_{ik}^{-1}. \tag{3.11}$$

By repeated application of the formula

$$\frac{\partial x_i}{\partial X_j} \frac{\partial X_j}{\partial x_k} = \delta_{ik}, \tag{3.12}$$

it is readily verified that, as indicated in (3.11), the tensor G is the inverse of C:

$$C_{ik} G_{kj} = \frac{\partial x_s}{\partial X_i} \frac{\partial x_s}{\partial X_k} \frac{\partial X_k}{\partial x_m} \frac{\partial X_j}{\partial x_m} = \frac{\partial x_s}{\partial X_i} \delta_{sm} \frac{\partial X_j}{\partial x_m} = \delta_{ij}. \tag{3.13}$$

Now consider a pair of parallel planes a unit distance apart in the reference configuration. The initially normal material line segment $\boldsymbol{N}$ from one plane to the other is, in general, rotated and stretched by the homogeneous deformation into a new direction given by a unit vector $\boldsymbol{m}$ and a new length $\lambda_{\boldsymbol{N}}$ such that its new components are

$$\lambda_{\boldsymbol{N}} m_j = (\partial x_j/\partial X_k) N_k. \tag{3.14}$$

The new separation of the planes is obtained by projecting this vector onto the new normal $\boldsymbol{n}$. Thus, the ratio of the new to the old separation of the planes is

$$(l/\bar{l})_{\boldsymbol{N}} = \lambda_{\boldsymbol{N}} \boldsymbol{m} \cdot \boldsymbol{n} = \frac{N_i N_k}{f_{\boldsymbol{N}}} \frac{\partial x_j}{\partial X_k} \frac{\partial X_i}{\partial x_j} = \frac{1}{f_{\boldsymbol{N}}}. \tag{3.15}$$

The factor $\lambda_{\boldsymbol{N}}$ in (3.14) is called the stretch [TRUESDELL and TOUPIN, 1960, CFT*, p. 255] of the material line of reference direction $\boldsymbol{N}$. By setting $m_j m_j = 1$ in (3.14), we obtain

$$\lambda_N^2 = C_{ik} N_i N_k. \tag{3.16}$$

By putting $\boldsymbol{N}$ in (3.14) along each of the coordinate axes in turn, we find that the corresponding material line elements get rotated and stretched into $\partial \boldsymbol{x}/\partial X_1$, $\partial \boldsymbol{x}/\partial X_2$, $\partial \boldsymbol{x}/\partial X_3$, respectively. By considering these line elements as three intersecting edges of an initially rectangular parallelepiped, it can be seen that the ratio of new to old volume is the triple scalar product

$$\frac{V}{\bar{V}} = \frac{\partial \boldsymbol{x}}{\partial X_1} \cdot \frac{\partial \boldsymbol{x}}{\partial X_2} \times \frac{\partial \boldsymbol{x}}{\partial X_3} = \det F, \tag{3.17}$$

* CFT is an abbreviation for "The classical field theories."

where F is the matrix of the deformation gradients displayed in (3.7). We adopt the widely used symbol J for the determinant of F:

$$J \equiv \det F. \tag{3.18}$$

J is the functional determinant, or Jacobian of (x_1, x_2, x_3) with respect to (X_1, X_2, X_3). Now the ratio $A/\bar{A}$ of new to old cross sectional areas initially perpendicular to $\boldsymbol{N}$ must satisfy $A/\bar{A} = V\,\bar{l}/\bar{V}\,l$, and hence

$$(A/\bar{A})_{\boldsymbol{N}} = J f_{\boldsymbol{N}}. \tag{3.19}$$

The volume ratio in (3.17) is named the cubical dilatation [LOVE, 1927, Sect. 25' p. 61]. If names are needed for the ratios in (3.15) and (3.19), I suggest *thickness swell* for (3.15) and *surface stretch* for (3.19). $f_{\boldsymbol{N}}$ could be called the *thickness shrink* of the direction $\boldsymbol{N}$.

3.2. Derivatives with respect to time. The material derivative of a quantity, denoted by a dot over the symbol, is the time derivative following the motion of a particle. Thus, $\dot{\boldsymbol{x}}$ is the velocity and $\ddot{\boldsymbol{x}}$ the acceleration.

Since the reference configuration $\boldsymbol{X}$ is here considered fixed, the time derivative following a particle is the derivative at constant $\boldsymbol{X}$. In the case of a reference configuration $\boldsymbol{X}$ that is not strictly stationary, the functions $x_i(\boldsymbol{X}, t)$ do not suffice to specify the motion. The specification could be completed by giving $X_k(\boldsymbol{a}, t)$, the motion of the reference configuration with respect to a fixed master configuration $\boldsymbol{a}$. The material derivative is then the derivative at constant $\boldsymbol{a}$. For example,

$$\dot{x}_i = \frac{\partial x_i}{\partial t} + \dot{X}_k \frac{\partial x_i}{\partial X_k}, \tag{3.20}$$

$$\frac{\partial \dot{x}_i}{\partial X_r} = \frac{\partial^2 x_i}{\partial X_r \partial t} + \frac{\partial}{\partial X_r}\left(\dot{X}_k \frac{\partial x_i}{\partial X_k}\right), \tag{3.21}$$

$$\dot{\overline{\left(\frac{\partial x_i}{\partial X_r}\right)}} = \frac{\partial^2 x_i}{\partial X_r \partial t} + \dot{X}_k \frac{\partial^2 x_i}{\partial X_k \partial X_r} = \frac{\partial \dot{x}_i}{\partial X_r} - \frac{\partial \dot{X}_k}{\partial X_r}\frac{\partial x_i}{\partial X_k}, \tag{3.22}$$

$$\dot{C}_{rs} = \frac{\partial x_i}{\partial X_s}\frac{\partial \dot{x}_i}{\partial X_r} - C_{sk}\frac{\partial \dot{X}_k}{\partial X_r} + \frac{\partial x_i}{\partial X_r}\frac{\partial \dot{x}_i}{\partial X_s} - C_{rk}\frac{\partial \dot{X}_k}{\partial X_s}. \tag{3.23}$$

By multiplying (3.23) by $(\partial X_s/\partial x_q)\,(\partial X_r/\partial x_p)$ and summing over r and s, we show that

$$2d_{pq} = \frac{\partial X_s}{\partial x_q}\frac{\partial X_r}{\partial x_p}\dot{C}_{rs} + C_{sk}\frac{\partial \dot{X}_k}{\partial x_p}\frac{\partial X_s}{\partial x_q} + C_{rk}\frac{\partial \dot{X}_k}{\partial x_q}\frac{\partial X_r}{\partial x_p} \tag{3.24}$$

where d_{pq} is the *deformation rate* or *stretching tensor*

$$d_{pq} = \frac{1}{2}\left(\frac{\partial \dot{x}_p}{\partial x_q} + \frac{\partial \dot{x}_q}{\partial x_p}\right). \tag{3.25}$$

For a fixed reference configuration $\boldsymbol{X}$, these formulas can be simplified immediately by setting $\dot{\boldsymbol{X}} = 0$. From (3.2), $\dot{C}_{ij} = 2\dot{V}_{ij}$. Hence, for fixed $\boldsymbol{X}$, (3.24) yields the important formula

$$d_{ij} = \frac{\partial X_r}{\partial x_i}\frac{\partial X_s}{\partial x_j}\dot{V}_{rs}. \tag{3.26}$$

Unless otherwise stated the configuration $\boldsymbol{X}$ is assumed to be fixed.

A description in which the independent variables are $(\boldsymbol{X}, t)$ is called a material description, and one in which the independent variables are $(\boldsymbol{x}, t)$ is called a spatial description [TRUESDELL and TOUPIN, 1960, CFT, p. 326]. By the chain rule, the material time derivative of a function $w(\boldsymbol{x}, t)$ is

$$\dot{w} = \frac{\partial w}{\partial t} + \dot{x}_j \frac{\partial w}{\partial x_j} = \frac{\partial w}{\partial t} + \dot{\boldsymbol{x}} \cdot \operatorname{grad} w. \tag{3.27}$$

The dot over a symbol denotes the material time derivative, whether the description is material or spatial. Thus, if the same field is denoted by w in the spatial description and W in the material description, so that

$$W(\boldsymbol{X}, t) = w[\boldsymbol{x}(\boldsymbol{X}, t), t], \tag{3.28}$$

then our notation is such that

$$\dot{w} = \dot{W} = \frac{\partial W}{\partial t} = \frac{\partial w}{\partial t} + \boldsymbol{v} \cdot \operatorname{grad} w \tag{3.29}$$

where $\boldsymbol{v} \equiv \dot{\boldsymbol{x}}$ is the material velocity.

3.3. Useful identities involving J. The identities (3.12) show that the quantities $\partial X_j/\partial x_k$ are the elements of F^{-1}, F being the matrix in (3.7). For each fixed k, (3.12) is a set of three linear algebraic equations in the three $\partial X_j/\partial x_k$, $j = 1, 2, 3$. The solution, by CRAMER's rule, is

$$\frac{\partial X_j}{\partial x_k} = \frac{1}{2J} \varepsilon_{jmp} \varepsilon_{kqr} \frac{\partial x_q}{\partial X_m} \frac{\partial x_r}{\partial X_p}. \tag{3.30}$$

Here ε_{ijk} is the alternating tensor, equal to $+1$ if the subscripts are cyclic permutations of 1, 2, 3 (i.e. 123, 231, 312), equal to -1 if the subscripts are 321, 213, 132), and otherwise equal to zero. The determinant J can be expressed as

$$J = \frac{1}{6} \varepsilon_{ijk} \varepsilon_{qrs} \frac{\partial x_q}{\partial X_i} \frac{\partial x_r}{\partial X_j} \frac{\partial x_s}{\partial X_k}. \tag{3.31}$$

By differentiation of (3.31) with respect to $(\partial x_p/\partial X_m)$ and comparison with (3.30), we obtain the useful result

$$\frac{\partial J}{\partial(\partial x_p/\partial X_m)} = J \frac{\partial X_m}{\partial x_p}. \tag{3.32}$$

Three identities involving J are conveniently demonstrated with the help of (3.32). The identity of EULER, PIOLA, and JACOBI is [TRUESDELL and TOUPIN, 1960, CFT, p. 246; THURSTON, 1964, p. 91]

$$\frac{\partial}{\partial x_j}\left(\frac{1}{J} \frac{\partial x_j}{\partial X_k}\right) = 0. \tag{3.33}$$

From (3.32) and the chain rule,

$$\frac{\partial J}{\partial x_j} = J \frac{\partial X_m}{\partial x_p} \frac{\partial}{\partial x_j}\left(\frac{\partial x_p}{\partial X_m}\right) \tag{3.34}$$

and hence

$$\begin{aligned} \frac{\partial x_j}{\partial X_k} \frac{\partial}{\partial x_j}\left(\frac{1}{J}\right) &= -\frac{1}{J} \frac{\partial X_m}{\partial x_p} \frac{\partial x_j}{\partial X_k} \frac{\partial}{\partial x_j}\left(\frac{\partial x_p}{\partial X_m}\right) = -\frac{1}{J} \frac{\partial X_m}{\partial x_p} \frac{\partial^2 x_p}{\partial X_k \partial X_m} \\ &= -\frac{1}{J} \frac{\partial}{\partial x_p}\left(\frac{\partial x_p}{\partial X_k}\right). \end{aligned} \tag{3.35}$$

We can now verify (3.33) in the steps

$$\frac{\partial}{\partial x_j}\left(\frac{1}{J}\frac{\partial x_j}{\partial X_k}\right)=\frac{1}{J}\frac{\partial}{\partial x_j}\left(\frac{\partial x_j}{\partial X_k}\right)+\frac{\partial x_j}{\partial X_k}\frac{\partial}{\partial x_j}\left(\frac{1}{J}\right)$$
$$=\frac{1}{J}\left[\frac{\partial}{\partial x_j}\left(\frac{\partial x_j}{\partial X_k}\right)-\frac{\partial}{\partial x_p}\left(\frac{\partial x_p}{\partial X_k}\right)\right]=0.$$

Another useful identity, called EULER's expansion formula [TRUESDELL and TOUPIN, 1960, CFT, p. 342], is the result that $\dot{J}/J$ equals the divergence of the velocity field in a spatial description:

$$\frac{\dot{J}}{J}=\frac{\partial \dot{x}_p}{\partial x_p}\equiv \operatorname{div} \boldsymbol{v}. \tag{3.36}$$

From (3.32) and the chain rule,

$$\dot{J}=J\frac{\partial X_m}{\partial x_p}\frac{\partial \dot{x}_p}{\partial X_m}=J\frac{\partial \dot{x}_p}{\partial x_p}, \tag{3.37}$$

which proves (3.36). In obtaining (3.37), it has been assumed that the reference configuration is fixed, so that $\dot{X}_k\equiv 0$ in (3.22).

The derivation of (3.32) can be used to show that [J. L. ERICKSEN, "Tensor Fields," pp. 794–858 in *Handbuch der Physik* (S. FLÜGGE, ed.), Vol. III/1, Eq. (38.16), p. 834. Berlin-Göttingen-Heidelberg: Springer 1960]

$$\frac{\partial \det \|A\|}{\partial A_{ij}}=A_{ji}^{-1}\det\|A\| \tag{3.38}$$

for any nonsingular 3×3 matrix A. Since F and its transpose have the same determinant,

$$\det\|C\|=\det(F^T F)=J^2. \tag{3.39}$$

Therefore, putting $A=C$ in (3.38) gives

$$2J\frac{\partial J}{\partial C_{ij}}=J^2 C_{ji}^{-1}, \tag{3.40}$$

or, in view of (3.2) and the symmetry of C_{ij}^{-1},

$$\frac{\partial J}{\partial V_{ij}}=J C_{ji}^{-1}=J C_{ij}^{-1}. \tag{3.41}$$

Eq. (3.41) is useful in relating strains to volume changes.

3.4. Transport theorem. Eq. (3.36) is helpful in establishing a formula for the time derivative of a volume integral, integrated over the moving region V occupied by a body at the variable time t. Let such a quantity be denoted by

$$P(t)\equiv\iiint_V w(x_1, x_2, x_3, t)\, dx_1\, dx_2\, dx_3. \tag{3.42}$$

From (3.28) and the rule for changing the variables of integration,

$$P(t)=\iiint_{\overline{V}} W J\, dX_1\, dX_2\, dX_3, \tag{3.43}$$

where $\overline{V}$ is the region occupied by the body in the reference configuration. Now the region of integration in (3.43) is independent of t, so that

$$\dot{P}=\iiint_{\overline{V}} (\dot{W}J+W\dot{J})\, dX_1\, dX_2\, dX_3. \tag{3.44}$$

When (3.44) is transformed back to spatial variables, the new integrand is $(1/J)$ times that in (3.44). Thus, using (3.36), we obtain

$$\dot{P} = \iiint_V (\dot{w} + w \operatorname{div} \boldsymbol{v})\, dx_1\, dx_2\, dx_3. \tag{3.45}$$

From (3.27) with $\dot{\boldsymbol{x}} = \boldsymbol{v}$ and the identity

$$\operatorname{div}(w\,\boldsymbol{v}) = w \operatorname{div} \boldsymbol{v} + \boldsymbol{v} \cdot \operatorname{grad} w, \tag{3.46}$$

(3.45) becomes

$$\dot{P} = \iiint_V \left[\frac{\partial w}{\partial t} + \operatorname{div}(w\,\dot{\boldsymbol{x}})\right] dx_1\, dx_2\, dx_3. \tag{3.47}$$

When the divergence theorem is used to transform the volume integral of the divergence of $w\,\boldsymbol{v}$ into the surface integral of the normal component of $w\,\boldsymbol{v}$ over the bounding surface, there results

$$\dot{P} = \iiint_V \frac{\partial w}{\partial t}\, dV + \oint\!\!\oint_A w\,\dot{\boldsymbol{x}} \cdot \boldsymbol{n}\, dA. \tag{3.48}$$

This result is called the transport theorem [TRUESDELL and TOUPIN, 1960, CFT, p. 347].

3.5. More identities. We give here some additional identities involving the deformation gradients. These identities are sometimes useful in revealing the structure of expressions that seem relatively opaque.

The inverse of (3.30), the result of solving (3.12) for $\partial x_j/\partial X_k$ instead of $\partial X_j/\partial x_k$, can be obtained from (3.30) by interchanging x with X and replacing J by $1/J$. Thus,

$$\frac{\partial x_j}{\partial X_k} = \frac{1}{2} J\, \varepsilon_{jmp}\, \varepsilon_{kqr} \frac{\partial X_q}{\partial x_m} \frac{\partial X_r}{\partial x_p}. \tag{3.49}$$

Multiplication by ε_{juv} and rearrangement of the right hand side after making use of

$$\varepsilon_{juv}\, \varepsilon_{jmp} \equiv \delta_{um}\, \delta_{vp} - \delta_{up}\, \delta_{vm} \tag{3.50}$$

gives

$$\varepsilon_{juv} \frac{\partial x_j}{\partial X_k} = J\, \varepsilon_{kqr} \frac{\partial X_q}{\partial x_u} \frac{\partial X_r}{\partial x_v}. \tag{3.51}$$

Similarly, multiplication of (3.30) by ε_{juv} yields

$$\varepsilon_{juv} \frac{\partial X_j}{\partial x_k} = \frac{1}{J}\, \varepsilon_{kqr} \frac{\partial x_q}{\partial X_u} \frac{\partial x_r}{\partial X_v}, \tag{3.52}$$

while multiplication of (3.30) and (3.49) respectively by ε_{kuv} gives

$$\varepsilon_{kuv} \frac{\partial X_j}{\partial x_k} = \frac{1}{J}\, \varepsilon_{jmp} \frac{\partial x_u}{\partial X_m} \frac{\partial x_v}{\partial X_p}, \tag{3.53}$$

$$\varepsilon_{kuv} \frac{\partial x_j}{\partial X_k} = J\, \varepsilon_{jmp} \frac{\partial X_u}{\partial x_m} \frac{\partial X_v}{\partial x_p}. \tag{3.54}$$

These identities are not limited to the deformation gradients, of course. From their derivation, it is clear that they connect the elements of any nonsingular 3×3 matrix with the elements of its inverse.

An interesting expression for the curl of the displacement field can be obtained from (3.30). From the definition $\boldsymbol{u} = \boldsymbol{x} - \boldsymbol{X}$, we see that

$$\frac{\partial u_j}{\partial x_k} = \delta_{jk} - \frac{\partial X_j}{\partial x_k}, \tag{3.55}$$

and therefore

$$(\text{curl}\,\boldsymbol{u})_i = \varepsilon_{ikj}\frac{\partial u_j}{\partial x_k} = -\varepsilon_{ikj}\frac{\partial X_j}{\partial x_k} = \varepsilon_{ijk}\frac{\partial X_j}{\partial x_k}$$
$$= \varepsilon_{ijk}\frac{1}{2J}\varepsilon_{jmp}\,\varepsilon_{kqr}\frac{\partial x_q}{\partial X_m}\frac{\partial x_r}{\partial X_p}. \tag{3.56}$$

Now $\varepsilon_{ijk} = \varepsilon_{jki} = \varepsilon_{kij}$. We can use (3.50) to simplify either the product $\varepsilon_{jki}\,\varepsilon_{jmp}$ or the product $\varepsilon_{kij}\,\varepsilon_{kqr}$ in (3.56). The two choices result in the following expressions which appear superficially different, but which are, in fact, identical:

$$(\text{curl}\,\boldsymbol{u})_i = \frac{1}{J}\varepsilon_{pqr}\frac{\partial x_q}{\partial X_p}\frac{\partial x_r}{\partial X_i} \equiv \frac{1}{J}\varepsilon_{jmp}\frac{\partial x_i}{\partial X_m}\frac{\partial x_j}{\partial X_p}. \tag{3.57}$$

Eq. (3.57) is a "material" representation of the "spatial" operation curl $\boldsymbol{u}$. The corresponding "material" operation is

$$\varepsilon_{ijk}\frac{\partial u_k}{\partial X_j} = \varepsilon_{ijk}\frac{\partial x_k}{\partial X_j}. \tag{3.58}$$

From (3.57), the results of the material and spatial operations are connected by the transformation

$$(\text{curl}\,\boldsymbol{u})_i = \frac{1}{J}\frac{\partial x_r}{\partial X_i}\varepsilon_{rpq}\frac{\partial x_q}{\partial X_p} = \frac{1}{J}\frac{\partial x_i}{\partial X_r}\varepsilon_{rpq}\frac{\partial x_q}{\partial X_p}, \tag{3.59}$$

this being merely a rewriting of Eq. (3.57). The inverse is

$$\varepsilon_{rpq}\frac{\partial x_q}{\partial X_p} = J\frac{\partial X_i}{\partial x_r}(\text{curl}\,\boldsymbol{u})_i = J\frac{\partial X_r}{\partial x_i}(\text{curl}\,\boldsymbol{u})_i. \tag{3.60}$$

Eq. (3.60) can also be obtained by contraction in either of the identities (3.51) or (3.54). Contraction in (3.52) or (3.53) yields (3.59).

4. Stress.

The stress at any point is described by a tensor $\boldsymbol{T}$ whose Cartesian component T_{ij} equals the j-component of the force per unit area acting across a surface whose normal is in the i-direction, exerted on the material situated on the $-i$ side of the surface, by the material on the other side. The i-component of force per unit area across a surface of unit normal $\boldsymbol{n}$ is $T_{ji}\,n_j$. From (3.9) and (3.19), this represents an i-component of force per unit area of

$$J f_{\boldsymbol{N}} T_{ji}\,n_j = J\,T_{ji}\,N_k\frac{\partial X_k}{\partial x_j} \equiv P_{ik}\,N_k$$

when referred to the area $\bar{A}$ of the surface in the reference configuration $\boldsymbol{X}$. The tensor of components

$$P_{ik} \equiv J\,T_{ji}\,\partial X_k/\partial x_j \tag{4.1}$$

is called the first Piola-Kirchhoff tensor or double-vector [TRUESDELL and TOUPIN, 1960, CFT, p. 553]. From (4.1),

$$T_{ij} = P_{jk}(\partial x_i/\partial X_k)/J. \tag{4.2}$$

The net i-component of force resulting from stress on a body is obtained by integrating $T_{ji}\,n_j$ over its bounding surface in the present configuration or $P_{ik}\,N_k$ over the surface in the reference configuration. By applying the divergence theorem to these surface integrals, it is found that the same net force is given by the volume integrals of $\partial T_{ji}/\partial x_j$ and $\partial P_{ik}/\partial X_k$ over the present and reference configurations, respectively. It follows that $\partial T_{ji}/\partial x_j$ and $\partial P_{ik}/\partial X_k$ may be interpreted as force components per unit volume and per unit reference volume, respectively.

The stress does work on the bounding surface of a body at the rate $\dot{x}_i T_{ji} n_j$ and $\dot{x}_i P_{ik} N_k$ per unit area referred to the present and reference configurations respectively. Application of the divergence theorem gives the power per unit volume in the present and reference configurations as

$$\frac{\partial}{\partial x_j}(\dot{x}_i T_{ji}) = \dot{x}_i \frac{\partial T_{ji}}{\partial x_j} + T_{ji} \frac{\partial \dot{x}_i}{\partial x_j}, \tag{4.3}$$

$$\frac{\partial}{\partial X_k}(\dot{x}_i P_{ik}) = \dot{x}_i \frac{\partial P_{ik}}{\partial X_k} + P_{ik} \frac{\partial \dot{x}_i}{\partial X_k}. \tag{4.4}$$

The first terms on the right hand sides of these expressions represent power of translation, being the scalar product of the velocity times the force per unit volume. The last terms represent power expended in stretching the material. With the help of (3.33), (4.1), and (4.2), it is readily verified that each term of (4.4) is J times the corresponding term of (4.3). When $T_{ji} = T_{ij}$, the last part of (4.3) can be written as

$$T_{ji} \frac{\partial \dot{x}_i}{\partial x_j} = T_{ij} d_{ij} \tag{4.5}$$

where d_{ij} is the stretching tensor defined in (3.25). The stretching power per unit of reference volume is $J T_{ij} d_{ij}$. The second Piola-Kirchhoff stress tensor t_{ij} is so defined as to make this expression equal to $t_{rs} \dot{V}_{rs}$. Thus, with d_{ij} from (3.26),

$$t_{rs} \dot{V}_{rs} \equiv J T_{ij} d_{ij} \equiv J T_{ij} \frac{\partial X_r}{\partial x_i} \frac{\partial X_s}{\partial x_j} \dot{V}_{rs} \tag{4.6}$$

whence

$$t_{rs} = J T_{ij} \frac{\partial X_r}{\partial x_i} \frac{\partial X_s}{\partial x_j}. \tag{4.7}$$

In view of (4.1), we can also write

$$t_{rs} = P_{jr} \frac{\partial X_s}{\partial x_j}, \quad P_{ir} = \frac{\partial x_i}{\partial X_s} t_{rs}. \tag{4.8}$$

In the nonpolar case, the tensors $\boldsymbol{T}$ and $\boldsymbol{t}$ are symmetric in their subscripts, but $\boldsymbol{P}$ is not. The result of solving (4.7) for $\boldsymbol{T}$ is

$$T_{pq} = \frac{1}{J} t_{rs} \frac{\partial x_p}{\partial X_r} \frac{\partial x_q}{\partial X_s}. \tag{4.9}$$

5. Energy flux and distributed sources.

Energy can be supplied to a body by doing mechanical work on it, or by nonmechanical means such as heating or applying an electric or magnetic field. The rate of doing mechanical work is by definition just the rate of working of the stress, discussed in Sect. 4. Any additional power added to a body is described by a nonmechanical energy flux vector integrated over its surface or by nonmechanical energy sources in the body or on its surface. The nonmechanical energy flux is specified by a vector $\boldsymbol{h}$, defined so as to render $\boldsymbol{h} \cdot \boldsymbol{n}$ the power per unit area flowing across a surface from the $-\boldsymbol{n}$ side to the $+\boldsymbol{n}$ side, $\boldsymbol{n}$ being a unit vector specifying the orientation of the surface. The power $\boldsymbol{h} \cdot \boldsymbol{n}\, dA$ crossing a material surface element $\boldsymbol{n}\, dA$ can be referred to the surface element in the reference configuration through Eqs. (3.9) and (3.19) and the equality

$$\boldsymbol{h} \cdot \boldsymbol{n}\, dA = \boldsymbol{H} \cdot \boldsymbol{N}\, d\bar{A}. \tag{5.1}$$

$\boldsymbol{H}$ is the nonmechanical energy flux vector, referred to material surface elements in the reference configuration. With dA and $d\bar{A}$ related by (3.19), and $\boldsymbol{n}$ and $\boldsymbol{N}$ by (3.9), it follows that the components of $\boldsymbol{H}$ are

$$H_j = J\, h_k\, \partial X_j/\partial x_k. \tag{5.2}$$

Power sources are sometimes introduced for convenience in representing the result of energy conversion processes such as electrical heating, chemical reactions, and the absorption of radiation. Let q denote the power supplied per unit of present volume from a source distributed throughout the volume. Then the power per unit of reference volume is

$$r \equiv q J. \tag{5.3}$$

An interesting remark can be made in connection with Eq. (5.2). Consider a flow of charge instead of energy. Because the current through a material surface element can be represented as $u\, \boldsymbol{v} \cdot \boldsymbol{n}\, dA$, where u is the charge density and $\boldsymbol{v}$ the relative velocity of the charge, we can associate $\boldsymbol{h}$ with the current density: $\boldsymbol{h} = u\, \boldsymbol{v}$. By the preceding derivation, the same current corresponds to the current density $\boldsymbol{H}$ when referred to material surface elements in the reference configuration. Now the charge density, referred to unit volume in the reference configuration, is $U = J u$. Therefore, from (5.2),

$$H_j = J\, u\, v_k\, \partial X_j/\partial x_k = U\, v_k\, \partial X_j/\partial x_k. \tag{5.4}$$

If $\boldsymbol{H}$ is represented as $\boldsymbol{H} = U\boldsymbol{V}$ where $\boldsymbol{V}$ is a fictitious velocity, the components of $\boldsymbol{V}$ are

$$V_j = v_k\, \partial X_j/\partial x_k. \tag{5.5}$$

We note that multiplication of the reference charge density U by the velocity $\boldsymbol{v}$ does not give the correct reference current density $\boldsymbol{H}$.

6. Equations of balance.

The basic equations of continuum mechanics are the equations of balance of mass, momentum, angular momentum, and energy.

The equation of balance of angular momentum can be used to prove the symmetry of the stress tensor T_{ij} in the absence of couple-stresses, body torques, and spin density in the medium, i.e. if all torque can be represented as the moment of force and all angular momentum as the moment of linear momentum. This is called the nonpolar case [Truesdell and Toupin, 1960, CFT, p. 546; Thurston, 1964, p. 12 and p. 18].

6.1 Mass balance. From (3.17) [and the definition (3.18)] it is seen that the balance of mass can be expressed by

$$J = \bar{\varrho}/\varrho \tag{6.1}$$

where ϱ and $\bar{\varrho}$ are the densities in the present and reference configurations, respectively. By taking the material derivative of (6.1) and using (3.36), we obtain the differential "continuity equation" commonly used in fluid dynamics:

$$\dot{\varrho} + \varrho \operatorname{div} \boldsymbol{v} = 0. \tag{6.2}$$

Thus, (6.1) can be viewed as an integrated form of (6.2), as noted by Truesdell and Noll [1965, NFT, p. 50]. An interesting derivation of (6.2) proceeds from (3.42) and (3.45) by letting w be the density ϱ. Then P in (3.42) is the constant mass of the body and the condition that $\dot{P}=0$ for every arbitrary body forces the integrand to vanish in (3.45), thus leading to (6.2).

6.2. A useful formula. If we let $w=\varrho g$ in (3.42), then (3.45) yields

$$\dot{P}=\iiint_V [g(\dot{\varrho}+\varrho \operatorname{div} \boldsymbol{v})+\varrho \dot{g}]\, dV$$

where dV has been written in place of $dx_1\, dx_2\, dx_3$. But since ϱ satisfies (6.2), this reduces to the simple and useful formula

$$\frac{d}{dt}\iiint_V \varrho g\, dV=\iiint_V \varrho \dot{g}\, dV. \tag{6.3}$$

Here g is to be interpreted as some additive property referred to unit mass, and V is the moving region occupied by a body, so that the two sides of (6.3) are equivalent expressions for the time derivative of a property of the body. The field g may be a vector in (6.3), as can be readily appreciated by noting that the equation must hold for each Cartesian component of the vector.

6.3. Equation of motion. The equation of motion expresses the principle that the resultant force on a body equals the rate of change of its momentum. The momentum per unit mass is just the velocity $\boldsymbol{v}\equiv\dot{\boldsymbol{x}}$. Thus, with a body force $\boldsymbol{g}$ per unit mass and the stress as discussed in Sect. 4, we have

$$\iiint_V \varrho \dot{v}_i\, dV=\iiint_V \varrho g_i\, dV+\oiint T_{ji}\, n_j\, dA=\iiint_V \left(\varrho g_i+\frac{\partial T_{ji}}{\partial x_j}\right) dV. \tag{6.4}$$

Since this equality must hold for every arbitrary body V, it implies the differential equation of motion

$$\varrho \dot{v}_i=\varrho g_i+\frac{\partial T_{ji}}{\partial x_j}. \tag{6.5}$$

Of course, suitable assumptions of differentiability and smoothness of T_{ji} are needed to convert the surface integral in (6.4) to a volume integral by means of the divergence theorem. When such assumptions are removed, appropriate arguments lead to jump conditions across the surfaces where smoothness breaks down [TRUESDELL and TOUPIN, 1960, CFT, pp. 525–530, 545–547, 549].

A formulation of the equation of motion in terms of integrals over a body and its surface in the reference configuration gives

$$\iiint_{\bar{V}} \bar{\varrho} \dot{v}_i\, d\bar{V}=\iiint_{\bar{V}} \bar{\varrho} g_i\, d\bar{V}+\iint P_{ik}\, N_k\, d\bar{A}=\iiint_{\bar{V}} \left(\bar{\varrho} g_i+\frac{\partial P_{ik}}{\partial X_k}\right) d\bar{V}, \tag{6.6}$$

from which, with sufficient smoothness assumptions,

$$\bar{\varrho} \dot{v}_i=\bar{\varrho} g_i+\frac{\partial P_{ik}}{\partial X_k}. \tag{6.7}$$

Eqs. (6.5) and (6.7) are equivalent.

6.4. Energy balance. An energy balance for a body states that the rate of increase of its kinetic and internal energy equals the power supplied by the working of body forces, stresses, and by nonmechanical energy flux and sources. Thus, with U denoting the internal energy per unit reference volume (so that $U/\bar{\varrho}$ = internal energy per unit mass),

$$\iiint_V \varrho (\dot{U}/\bar{\varrho} + \boldsymbol{v} \cdot \dot{\boldsymbol{v}})\, dV = \iiint_V (\varrho \boldsymbol{g} \cdot \boldsymbol{v} + q)\, dV + \oiint (v_i T_{ji} n_j - h_j n_j)\, dA$$
$$= \iiint_V \left[\varrho \boldsymbol{g} \cdot \boldsymbol{v} + q + \frac{\partial}{\partial x_j} (v_i T_{ji} - h_j)\right] dV. \tag{6.8}$$

Since this must hold for every arbitrary body,

$$\varrho (\dot{U}/\bar{\varrho} + v_i \dot{v}_i) = \varrho g_i v_i + q + \frac{\partial}{\partial x_j} (v_i T_{ji} - h_j). \tag{6.9}$$

By multiplying (6.5) by v_i and summing over i we obtain a separate energy equation connecting rate of kinetic energy change to translational power supplied by body forces and stress:

$$\varrho v_i \dot{v}_i = \varrho g_i v_i + v_i \frac{\partial T_{ji}}{\partial x_j}. \tag{6.10}$$

Subtraction of (6.10) from (6.9) and use of (4.5) gives the following energy equation connecting the stretching power of the stress to the nonmechanical terms:

$$\varrho \dot{U}/\bar{\varrho} = q + T_{ji} d_{ji} - \operatorname{div} \boldsymbol{h}. \tag{6.11}$$

An energy balance in terms of integrals over a body and its surfacc in the reference configuration reads

$$\iiint_{\bar{V}} (\dot{U} + \bar{\varrho} v_i \dot{v}_i)\, d\bar{V} = \iiint_{\bar{V}} (\bar{\varrho} g_i v_i + r)\, d\bar{V} + \oiint (v_i P_{ik} - H_k) N_k\, d\bar{A}$$
$$= \iiint_{\bar{V}} \left[\bar{\varrho} g_i v_i + r + \frac{\partial}{\partial X_k} (v_i P_{ik} - H_k)\right] d\bar{V}, \tag{6.12}$$

from which

$$(\dot{U} + \bar{\varrho} v_i \dot{v}_i) = \bar{\varrho} g_i v_i + r + \frac{\partial}{\partial X_k} (v_i P_{ik} - H_k). \tag{6.13}$$

(Here $r \equiv q J$, as noted in Sect. 5.)

The translational terms can now be separated out with the help of (6.7). The remaining result, equivalent to (6.11), is

$$\dot{U} = r + P_{ik} \frac{\partial v_i}{\partial X_k} - \frac{\partial H_k}{\partial X_k}. \tag{6.14}$$

By (4.1), (4.5), and (4.6), the sum $P_{ik}\, \partial v_i / \partial X_k$ can be written

$$P_{ik} \frac{\partial v_i}{\partial X_k} = P_{ik} \frac{\partial x_j}{\partial X_k} \frac{\partial v_i}{\partial x_j} = J T_{ij} d_{ij} = t_{ij} \dot{V}_{ij}. \tag{6.15}$$

Thus, (6.14) takes the form

$$\dot{U} = r + t_{ij} \dot{V}_{ij} - \frac{\partial H_k}{\partial X_k}. \tag{6.16}$$

Although some readers would find it physically more appealing to refer internal energy to unit mass, we are using the symbol U to denote internal energy per unit reference volume, in order to achieve a notational convenience in Sects. 8–10. The appearance of the formulas if U were to denote internal energy per unit mass can always be seen by replacing the present U by $\bar{\varrho} U$.

C. Equilibrium states.

7. Motivation. Mutative and nonmutative processes.

There is a wide range of experimental conditions in which it is useful to presume that a solid has a static equilibrium state that depends, for example, on the strain and internal energy density. This is the state to which the actual state tends if the strain and internal energy density are held constant for an indefinitely long time. Different static equilibrium states are related to each other in many cases by the relations of classical thermostatics.

There are different points of view that can be adopted concerning these relations: (1) They are relations that describe the instantaneous states of an idealized medium under any conditions. (2) They describe the states of an actual medium during an idealized process. (3) They pertain to the equilibrium states. The first two of these viewpoints are very close to definitions of what is meant by "idealized" in the respective statements. The third view can be justified as a theorem under rather general hypotheses. For example, M. E. GURTIN [1968], building on foundations laid primarily by B. D. COLEMAN, V. J. MIZEL, and W. NOLL, has obtained such relations when the free energy, stress, entropy, and heat flux depend on the histories of the temperature, temperature gradient, and first and second gradients of the deformation through functionals that obey a principle of fading memory.

Of course even the third view must be qualified. Obviously, the hypotheses are not satisfied if the memory does not fade. This rules out permanent deformations of the solid, for example. There may be a family of equilibrium states that are connected approximately if not exactly by the classical relations that apply before a permanent deformation takes place. After a permanent deformation, tempering, hardening, or annealing, process, there may be a new family of equilibrium states that are similarly connected, but no simple relations are necessarily expected to connect members of the new family with members of the old. A "permanently remembered process" (such as permanent deformation or annealing) creates new equilibrium functions—a "new material" as far as the equilibrium behavior is concerned.

In many experimental situations, the distinction between processes for which the memory is nonfading or fading is relatively sharp. A sample can be "permanently" changed (by neutron irradiation, magnetization to some remanent value, annealing, or permanent deformation, for example), or it can be subjected to processes from which it seems capable of recovering completely (such as stressing within its elastic limit or changes of temperature not too close to a phase transition). Admittedly, there are intermediate cases where some kind of combination of retentive and fading memory would seem to be appropriate. Nevertheless, it is unquestionably useful to have the concept of a family of equilibrium states, or set of equilibrium functions that applies as long as the material is not undergoing processes that produce permanent changes. In this view, "permanent" changes generate a new set of equilibrium functions. It is tempting to imagine that the different sets of equilibrium functions could all be derived from the same set of general response functionals by appropriate continuations of different histories.

To distinguish processes that produce permanent changes from those that don't, we call a process *nonmutative* if it leaves all the static equilibrium functions unchanged. A *mutative* process generates a new set of static equilibrium functions that applies to subsequent nonmutative processes.

8. Thermostatic assumptions. Basis of classical thermostatics.

Sect. 7 expresses a confidence that the classical thermostatic relations connecting equilibrium states have a firm position in the theory of solids. It seems most appropriate to state them as fundamental constitutive assumptions, because their validity may well reach beyond any other special set of assumptions we might use to derive them. Proceeding directly, then, we make the following three constitutive assumptions for homogeneous solids:

(1) If the strain V_{ij} and the internal energy density U are maintained constant in time and space for a long enough time, the solid approaches a static equilibrium state.

(2) With mutative processes excluded, the entropy per unit reference volume in the static equilibrium state is a function

$$S = S(U, V_{ij}) \tag{8.1}$$

that can be inverted to obtain U as a function

$$U = U(S, V_{ij}). \tag{8.2}$$

(3) With mutative processes excluded, the static equilibrium temperature T and the static equilibrium Piola-Kirchhoff stress τ_{ij} satisfy the classical thermostatic relations

$$T = (\partial U/\partial S)_V, \tag{8.3}$$

$$\tau_{ij} = (\partial U/\partial V_{ij})_S. \tag{8.4}$$

According to the terminology introduced in Sect. 7, the restriction to nonmutative processes is implicit in the use of an equilibrium function such as (8.1) that depends only on (U, V_{ij}) and not on the history also. The first assumption means that t_{ij}, for example, approaches some constant value τ_{ij}, while the third assumption asserts that this value satisfies (8.4). The restriction to homogeneous solids is made to avoid the appearance of $\boldsymbol{X}$ as an argument in (8.1) and (8.2) [cf. TRUESDELL and TOUPIN, 1960, CFT, pp. 619–620]. In (8.3) and similar expressions, the subscript V is a reminder that *all* strain components are constant in the differentiation. In (8.4), S and all but one of the V_{ij} are regarded as constant.

Now if U from (8.2) is substituted into (8.1), an identity results:

$$S = S[U(S, V_{ij}), V_{ij}]. \tag{8.5}$$

By setting $(\partial S/\partial S)_V \equiv 1$ and $(\partial S/\partial V_{ij})_S \equiv 0$ and applying the chain rule of partial differentiation, we find

$$\left(\frac{\partial S}{\partial S}\right)_V \equiv 1 \equiv \left(\frac{\partial S}{\partial U}\right)_V \left(\frac{\partial U}{\partial S}\right)_V, \tag{8.6}$$

$$\left(\frac{\partial S}{\partial V_{ij}}\right)_S \equiv 0 \equiv \left(\frac{\partial S}{\partial V_{ij}}\right)_U + \left(\frac{\partial S}{\partial U}\right)_V \left(\frac{\partial U}{\partial V_{ij}}\right)_S. \tag{8.7}$$

From the preceding equations,

$$\left(\frac{\partial S}{\partial U}\right)_V = \frac{1}{T}, \tag{8.8}$$

$$\left(\frac{\partial S}{\partial V_{ij}}\right)_U = -\frac{\tau_{ij}}{T}. \tag{8.9}$$

If energy, strain, temperature, and the static equilibrium stress are considered to be measurable, Eqs. (8.8) and (8.9) enable the static equilibrium entropy function

$S(U, V_{ij})$ to be found (to within an added constant of integration) from these measured quantities. On the other hand, if entropy were to be taken as primitive, (8.3) would define the temperature and (8.4) the static equilibrium stress.

We prefer to call S the entropy only when referring to the static equilibrium state. Obviously, however, substitution of the instantaneous local values of energy density and strain into (8.1) yields an instantaneous local value of S. This is certainly the value of the entropy density that would eventually be attained in a spatially homogeneous system if energy density and strain were to be frozen at their present values, but it may differ from an instantaneous local entropy density that one may wish to define through statistical physics or some other discipline. (An instantaneous local entropy density that increases during stress relaxation at constant strain and internal energy density would *necessarily* differ from S while such relaxation is taking place.)

Like S, T and τ_{ij} are functions of U and V_{ij} that have instantaneous local values determined by the instantaneous values of the arguments U and V_{ij}. Quantities τ_{ij} defined by (8.4) are called thermodynamic tensions [TRUESDELL and TOUPIN, 1960, CFT, p. 621]. Assumption 3 asserts that they equal the static equilibrium values of the Piola-Kirchhoff stress components, i.e., in static equilibrium, $t_{ij}=\tau_{ij}$.

From (8.2), (8.3), and (8.4), the material time derivatives of U, S, and V_{ij} are related by

$$\dot{U}=T\dot{S}+\tau_{ij}\dot{V}_{ij}. \tag{8.10}$$

Substitution of $\dot{U}$ from (8.10) into the energy balance (6.16) then yields

$$T\dot{S}=(t_{ij}-\tau_{ij})\dot{V}_{ij}-\frac{\partial H_k}{\partial X_k}+r. \tag{8.11}$$

An idealized thermoelastic medium can be defined as one in which there are no sources ($r=0$) and in which $T\dot{S}$ is always given simply by $-\partial H_k/\partial X_k$, for arbitrary $\dot{V}_{ij}$. Eq. (8.11) then requires $t_{ij}=\tau_{ij}$. In other words, the stress always has its static equilibrium value in an idealized thermoelastic medium.

By *classical thermostatic relations*, we mean all the mathematical formalism following from Eqs. (8.1)–(8.4) and additional assumptions of differentiability and invertibility on all the functions that arise or are defined. Included explicitly are the Maxwell relations that follow from equating the mixed partial derivatives of U and of the other thermodynamic potentials. Useful results of physical significance include the relation between isentropic and isothermal elastic coefficients, between the specific heat at constant equilibrium stress and at constant strain, the connection of the temperature derivative of the equilibrium stress at constant strain to expansion coefficients and elastic coefficients, and the connection of the isothermal derivatives of the specific heats with respect to strain or equilibrium stress to the temperature derivatives of the elastic and expansion coefficients. Derivations will be sketched and the relations summarized in Sect. 10.

The present choice of V_{ij} as thermodynamic substate variables is made in order to be specific. Other choices are possible. An excellent treatment of the mathematical formalism with the substate variables left unspecified has been given by TRUESDELL and TOUPIN [1960, CFT, Chap. E, pp. 607–660, especially pp. 621–633].

9. Tensor and abbreviated notation.

In the nonpolar case, to which we restrict our attention, the stress tensors T_{ij} and t_{ij} (and therefore also τ_{ij} by Assumption 3 of Sect. 8) are symmetric. V_{ij} is symmetric from its very definition (3.2). It is convenient to develop most of the remaining relations in an abbreviated notation in which the tensor subscripts are replaced in pairs according to the Voigt scheme [W. VOIGT, 1928, W. P. MASON, 1950].

$$\begin{aligned} &11\sim1, \quad 23=32\sim4,\\ &22\sim2, \quad 31=13\sim5,\\ &33\sim3, \quad 12=21\sim6. \end{aligned} \tag{9.1}$$

Each single-subscript thermodynamic tension and normal component of strain is defined to be equal to the corresponding two-subscript quantity, but each single-subscript shear strain is by convention twice the corresponding two-subscript shear strain. Thus,

$$\begin{aligned} &V_1=V_{11}, \quad V_2=V_{22}, \quad V_3=V_{33},\\ &V_4=2V_{23}, \quad V_2=2V_{13}, \quad V_6=2V_{12}. \end{aligned} \tag{9.2}$$

In differentiation with respect to the double-subscript strains or thermodynamic tensions, all nine components are regarded as formally independent, and the function is expressed symmetrically so that, for example, $\partial/\partial\tau_{12}=\partial/\partial\tau_{21}$. By this procedure, the constraints $V_{ij}=V_{ji}$, $\tau_{ij}=\tau_{ji}$ are preserved in all results. To transform to a function of the single-subscript thermodynamic tensions, τ_6 would be substituted for *both* t_{12} and t_{21}. Hence,

$$\partial/\partial\tau_6=2(\partial/\partial\tau_{12})=2(\partial/\partial\tau_{21}) \tag{9.3}$$

and similarly for derivatives with respect to τ_4 and τ_5. However, because of the factors 2 in (9.2),

$$\partial/\partial V_6=\partial/\partial V_{12}=\partial/\partial V_{21}. \tag{9.4}$$

With these conventions, the correct transcription of (8.4) into abbreviated notation is simply

$$\tau_\alpha=(\partial U/\partial V_\alpha)_S, \quad \alpha=1, 2, \dots, 6. \tag{9.5}$$

Further, the sum of the nine products in a sum like $\tau_{ij} V_{ij}$ or $\tau_{ij}\dot{V}_{ij}$ is correctly given by the sum of the six terms in the corresponding abbreviated representation. For example,

$$\tau_{ij} V_{ij}=\tau_\alpha V_\alpha. \tag{9.6}$$

We shall consistently use Greek letter subscripts when the range is from one to six.

The thermal expansion coefficients at constant thermodynamic tensions are defined by

$$\alpha_{ij}=\left(\frac{\partial V_{ij}}{\partial T}\right)_\tau, \quad \alpha_\gamma=\left(\frac{\partial V_\gamma}{\partial T}\right)_\tau \tag{9.7}$$

in tensor and abbreviated notation, respectively. Clearly, the double-subscript and single-subscript quantities are related in the same way as the strains in (9.2), and $\alpha_{ij}\,\tau_{ij}=\alpha_\gamma\,\tau_\gamma$.

For the temperature coefficients of the thermodynamic tensions at constant strain, we use the notation

$$\lambda_{ij}=-\left(\frac{\partial \tau_{ij}}{\partial T}\right)_V, \quad \lambda_\gamma=-\left(\frac{\partial \tau_\gamma}{\partial T}\right)_V. \tag{9.8}$$

The single-subscript and double-subscript lambdas are equal because of the equality of the taus. Further, a sum like $\alpha_{ij}\,\lambda_{ij}=\alpha_\gamma\,\lambda_\gamma$. This sum occurs in the expression for the difference between the specific heat at constant thermodynamic tensions and at constant strain.

Isentropic and isothermal second-order elastic stiffness coefficients are defined by

$$c^S_{ijkm} = \left(\frac{\partial \tau_{ij}}{\partial V_{km}}\right)_S, \quad c^S_{\alpha\beta} = \left(\frac{\partial \tau_\alpha}{\partial V_\beta}\right)_S,$$
$$c^T_{ijkm} = \left(\frac{\partial \tau_{ij}}{\partial V_{km}}\right)_T, \quad c^T_{\alpha\beta} = \left(\frac{\partial \tau_\alpha}{\partial V_\beta}\right)_T. \tag{9.9}$$

Third-order coefficients are defined by

$$c^S_{ijkmpq} = \left(\frac{\partial^2 \tau_{ij}}{\partial V_{pq}\,\partial V_{km}}\right)_S, \quad c^S_{\alpha\beta\gamma} = \left(\frac{\partial^2 \tau_\alpha}{\partial V_\gamma\,\partial V_\beta}\right)_S,$$
$$c^T_{ijkmpq} = \left(\frac{\partial^2 \tau_{ij}}{\partial V_{pq}\,\partial V_{km}}\right)_T, \quad c^T_{\alpha\beta\gamma} = \left(\frac{\partial^2 \tau_\alpha}{\partial V_\gamma\,\partial V_\beta}\right)_T. \tag{9.10}$$

By (9.4) and the equality of the taus, these definitions make the abbreviated cees equal to the corresponding tensor cees, and a sum like $\alpha_{km}\,c^T_{ijkm} = \alpha_\gamma\,c^T_{\beta\gamma}$, $(ij \sim \beta)$. These definitions of abbreviated third order cees are due to BRUGGER [1964]. They differ from some that occur in the older literature.

Isentropic second-order and third-order compliances are defined by

$$s^S_{ijkm} = \left(\frac{\partial V_{ij}}{\partial \tau_{km}}\right)_S, \quad s^S_{\alpha\beta} = \left(\frac{\partial V_\alpha}{\partial \tau_\beta}\right)_S,$$
$$s^S_{ijkmpq} = \left(\frac{\partial^2 V_{ij}}{\partial \tau_{pq}\,\partial \tau_{km}}\right)_S, \quad s^S_{\alpha\beta\gamma} = \left(\frac{\partial^2 V_\alpha}{\partial \tau_\gamma\,\partial \tau_\beta}\right)_S. \tag{9.11}$$

The isothermal quantities are defined analogously:

$$s^T_{ijkm} = \left(\frac{\partial V_{ij}}{\partial \tau_{km}}\right)_T, \quad s^T_{\alpha\beta} = \left(\frac{\partial V_\alpha}{\partial \tau_\beta}\right)_T,$$
$$s^T_{ijkmpq} = \left(\frac{\partial^2 V_{ij}}{\partial \tau_{pq}\,\partial \tau_{km}}\right)_T, \quad s^T_{\alpha\beta\gamma} = \left(\frac{\partial^2 V_\alpha}{\partial \tau_\beta\,\partial \tau_\gamma}\right)_T. \tag{9.12}$$

From (9.2), (9.3), and the above definitions, it can be seen that each abbreviated compliance coefficient is 2^n times the corresponding tensor coefficient, where n is the number of times any of 4, 5, or 6 appears as a subscript. For example, $s_{2323} = s_{44}/4$.

With either T or S constant,

$$c_{\alpha\beta}\,s_{\beta\gamma} = \frac{\partial \tau_\alpha}{\partial V_\beta}\,\frac{\partial V_\beta}{\partial \tau_\gamma} = \delta_{\alpha\gamma}. \tag{9.13}$$

Thus, the six by six matrices $\|c_{\alpha\beta}\|$ and $\|s_{\alpha\beta}\|$ are inverses of each other.

By differentiation of Eq. (9.13) with respect to V_λ with the help of the above definitions and the chain rule of partial differentiation, we obtain

$$c_{\alpha\beta\lambda}\,s_{\beta\gamma} + c_{\alpha\beta}\,c_{\mu\lambda}\,s_{\beta\gamma\mu} = 0. \tag{9.14}$$

Multiplication by $c_{\gamma\nu}$, summation over γ, and use of (9.13) gives

$$c_{\alpha\nu\lambda} = -c_{\alpha\beta}\,c_{\gamma\nu}\,c_{\mu\lambda}\,s_{\beta\gamma\mu}. \tag{9.15}$$

Analogous operations yield

$$s_{\nu\gamma\mu} = -s_{\nu\alpha}\,s_{\beta\gamma}\,s_{\lambda\mu}\,c_{\alpha\beta\lambda}. \tag{9.16}$$

The general relations connect either isentropic cees with isentropic esses or isothermal cees with isothermal esses.

10. Thermostatic relations and coefficients.

10.1. Thermodynamic potentials. By virtue of (9.6), the enthalpy per unit reference volume may be defined by either of the expressions

$$H = U - \tau_{ij} V_{ij} = U - \tau_\nu V_\nu. \tag{10.1}$$

The four classical thermodynamic potentials are the internal energy U, the enthalpy H, the free energy

$$A \equiv U - TS, \tag{10.2}$$

and the free enthalpy

$$G \equiv H - TS. \tag{10.3}$$

The entropy, specific heats, and these potentials are here referred to unit reference volume in order to free many formulas, such as (9.5) and (10.1), from the reference density $\bar{\varrho}$. Formulas with these quantities referred to unit mass have been published elsewhere [THURSTON, 1964].

Table 10.1 shows the appropriate independent variables for these potentials and certain relations involving them. With the indicated independent variables, the first partial derivatives are in each case one of the quantities T, S, τ_α, V_α, in justification of the name "potential." The second derivatives are expressed in terms of the specific heats defined below and the coefficients α_γ, λ_γ, $c^S_{\alpha\beta}$, $s^S_{\alpha\beta}$, $c^T_{\alpha\beta}$, $s^T_{\alpha\beta}$ defined in Sect. 9. The relations that follow from interchangeability of the order of differentiation in mixed partial derivatives like $\partial^2 U/\partial V_\alpha\, \partial S$ are called Maxwell relations, while interchanging the order in $\partial^2 U/\partial V_\alpha\, \partial V_\beta$ gives the symmetry condition $c^S_{\alpha\beta} = c^S_{\beta\alpha}$. In the abbreviated notation, the subscripts may be interchanged arbitrarily on each stiffness or compliance coefficient of any order, for such a coefficient is a derivative of one of the four thermodynamic potentials, and the order of differentiation is immaterial.

By the symmetry relation just noted, (9.15) and (9.16) can be put in the more appealing form

$$\begin{aligned} c_{\alpha\beta\gamma} &= -c_{\alpha\lambda}\, c_{\beta\mu}\, c_{\gamma\nu}\, s_{\lambda\mu\nu}, \\ s_{\lambda\mu\nu} &= -s_{\lambda\alpha}\, s_{\mu\beta}\, s_{\nu\gamma}\, c_{\alpha\beta\gamma}. \end{aligned} \tag{10.4}$$

10.2. Specific heats. The definitions of the specific heats can be motivated by (8.11). If $(t_{ij} - \tau_{ij})\, \dot{V}_{ij} = 0$, $T\dot{S}$ reduces to the net nonmechanical power input per unit reference volume, and $T\,dS$ is a differential amount of nonmechanical energy added. This is certainly true at constant strain, since $\dot{V}_{ij} = 0$. In a typical calorimetric experiment at constant stress, $\dot{V}_{ij} \neq 0$ because of thermal expansion, but t_{ij} and τ_{ij} are so nearly equal and $\dot{V}_{ij}$ so small that the time-integrated contribution of $(t_{ij} - \tau_{ij})\, \dot{V}_{ij}$ during any measured energy input is entirely negligible as compared with the energy input itself. In any case, the specific heat along any path in the space of thermostatic states, conceived physically as the limiting value (as $dT \to 0$) of the ratio of heat added to temperature change dT, is defined formally by the limiting value of the ratio $T\,dS/dT$, taken along the path [TRUESDELL and TOUPIN, 1960, CFT, p. 624]. Thus, the specific heats per unit reference volume, at constant strain and at constant thermodynamic tensions, are defined by

$$C_V = T\left(\frac{\partial S}{\partial T}\right)_V = \left(\frac{\partial U}{\partial T}\right)_V, \tag{10.5}$$

$$C_\tau = T\left(\frac{\partial S}{\partial T}\right)_\tau = \left(\frac{\partial H}{\partial T}\right), \tag{10.6}$$

Table 10.1. *Thermodynamic potentials and their derivatives.*

Potential	Independent variables	Differential	First Derivatives		Second derivatives		
			Mechanical	Thermal	Maxwell coefficients	Mechanical	Thermal
Internal Energy U	(V_α, S)	$dU = \tau_\alpha dV_\alpha + T dS$	$\tau_\alpha = \frac{\partial U}{\partial V_\alpha}$	$T = \frac{\partial U}{\partial S}$	$\frac{T}{C_V}\lambda_\alpha = -\frac{\partial \tau_\alpha}{\partial S} = -\frac{\partial T}{\partial V_\alpha}$	$c^S_{\alpha\beta} = \frac{\partial \tau_\alpha}{\partial V_\beta}$	$\frac{T}{C_V} = \frac{\partial T}{\partial S}$
Enthalpy $H = U - \tau_\alpha V_\alpha$	(τ_α, S)	$dH = -V_\alpha d\tau_\alpha + T dS$	$V_\alpha = -\frac{\partial H}{\partial \tau_\alpha}$	$T = \frac{\partial H}{\partial S}$	$\frac{T}{C_\tau}\alpha_\alpha = \frac{\partial V_\alpha}{\partial S} = -\frac{\partial T}{\partial \tau_\alpha}$	$s^S_{\alpha\beta} = \frac{\partial V_\alpha}{\partial \tau_\beta}$	$\frac{T}{C_\tau} = \frac{\partial T}{\partial S}$
Free energy $A = U - TS$	(V_α, T)	$dA = \tau_\alpha dV_\alpha - S dT$	$\tau_\alpha = \frac{\partial A}{\partial V_\alpha}$	$S = -\frac{\partial A}{\partial T}$	$\lambda_\alpha = -\frac{\partial \tau_\alpha}{\partial T} = \frac{\partial S}{\partial V_\alpha}$	$c^T_{\alpha\beta} = \frac{\partial \tau_\alpha}{\partial V_\beta}$	$\frac{C_V}{T} = \frac{\partial S}{\partial T}$
Free enthalpy $G = H - TS$	(τ_α, T)	$dG = -V_\alpha d\tau_\alpha - S dT$	$V_\alpha = -\frac{\partial G}{\partial \tau_\alpha}$	$S = -\frac{\partial G}{\partial T}$	$\alpha_\alpha = \frac{\partial V_\alpha}{\partial T} = \frac{\partial S}{\partial \tau_\alpha}$	$s^T_{\alpha\beta} = \frac{\partial V_\alpha}{\partial \tau_\beta}$	$\frac{C_\tau}{T} = \frac{\partial S}{\partial T}$

Note: $\gamma_\alpha = \lambda_\alpha / C_V$

where the second equalities follow from the expressions for dU and dH in Table 10.1, which show that $T=(\partial U/\partial S)_V=(\partial H/\partial S)_\tau$.

10.3. Maxwell coefficients. In compiling Table 10.1, use has been made of invertibility relations like (8.6), notably

$$1=\left(\frac{\partial S}{\partial T}\right)_V\left(\frac{\partial T}{\partial S}\right)_V=\left(\frac{\partial S}{\partial T}\right)_\tau\left(\frac{\partial T}{\partial S}\right)_\tau. \tag{10.7}$$

From its definition in (9.7), the expansion coefficient α_γ is the fourth Maxwell coefficient in Table 10.1. By considering $V_\gamma=V_\gamma[\tau_\alpha, T(\tau_\alpha, S)]$, one finds

$$\left(\frac{\partial V_\gamma}{\partial S}\right)_\tau=\left(\frac{\partial V_\gamma}{\partial T}\right)_\tau\left(\frac{\partial T}{\partial S}\right)_\tau. \tag{10.8}$$

Substitutions from (10.7) and (10.6) then enable the second Maxwell coefficient, $(\partial V_\gamma/\partial S)_\tau$, to be expressed as $T\alpha_\gamma/C_\tau$, as indicated. The connection of the first and third coefficients is entirely analogous: from $\tau_\gamma=\tau_\gamma[V_\alpha, T(V_\alpha, S)]$, one finds, using (10.7) and the definitions (9.8) and (10.5),

$$\left(\frac{\partial \tau_\gamma}{\partial S}\right)_V=\left(\frac{\partial \tau_\gamma}{\partial T}\right)_V\left(\frac{\partial T}{\partial S}\right)_V=-\lambda_\gamma T/C_V. \tag{10.9}$$

The third and fourth Maxwell coefficients, λ_γ and α_γ, are connected to each other through the elastic coefficients. From $S=S[\tau_\mu(V_\nu, T), T]$, the definitions (9.7), (9.8), (9.9), and the third and fourth Maxwell relations in Table 10.1,

$$\lambda_\nu=\left(\frac{\partial S}{\partial V_\nu}\right)_T=\left(\frac{\partial S}{\partial \tau_\mu}\right)_T\left(\frac{\partial \tau_\mu}{\partial V_\nu}\right)_T=\alpha_\mu c^T_{\mu\nu}. \tag{10.10}$$

Of course, also, by the inverse relation between stiffnesses and compliances,

$$\alpha_\mu=\lambda_\nu s^T_{\nu\mu}. \tag{10.11}$$

Similarly, the first and second Maxwell coefficients in Table 10.1 can be related to each other by considering $T=T[\tau_\mu(V_\nu, S), S]$, whence, by the first and second Maxwell relations,

$$-\frac{\lambda_\nu T}{C_V}=\left(\frac{\partial T}{\partial V_\nu}\right)_S=\left(\frac{\partial T}{\partial \tau_\mu}\right)_S\left(\frac{\partial \tau_\mu}{\partial V_\nu}\right)_S=-\frac{T\alpha_\mu}{C_\tau}c^S_{\mu\nu}. \tag{10.12}$$

This relation can be rewritten as

$$C_r\lambda_\nu=\alpha_\mu c^S_{\mu\nu}, \tag{10.13}$$

$$C_r\equiv C_\tau/C_V. \tag{10.14}$$

Comparison with (10.10) shows that for each ν,

$$C_r=\frac{\alpha_\mu c^S_{\mu\nu}}{\alpha_\mu c^T_{\mu\nu}}. \tag{10.15}$$

10.4. The differences $c^S_{\alpha\beta}-c^T_{\alpha\beta}$ and $s^T_{\alpha\beta}-s^S_{\alpha\beta}$. From $\tau_\alpha=\tau_\alpha[V_\beta, T(V_\beta, S)]$, the first and third Maxwell coefficients in Table 10.1, and the definitions (9.9),

$$c^S_{\alpha\beta}\equiv\left(\frac{\partial \tau_\alpha}{\partial V_\beta}\right)_S=\left(\frac{\partial \tau_\alpha}{\partial V_\beta}\right)_T+\left(\frac{\partial \tau_\alpha}{\partial T}\right)_V\left(\frac{\partial T}{\partial V_\beta}\right)_S=c^T_{\alpha\beta}+\frac{T\lambda_\alpha\lambda_\beta}{C_V}. \tag{10.16}$$

From $V_\mu=V_\mu[\tau_\nu, T(\tau_\nu, S)]$, the second and fourth Maxwell coefficients, and the definitions of the compliances,

$$s^S_{\mu\nu}\equiv\left(\frac{\partial V_\mu}{\partial \tau_\nu}\right)_S=\left(\frac{\partial V_\mu}{\partial \tau_\nu}\right)_T+\left(\frac{\partial V_\mu}{\partial T}\right)_\tau\left(\frac{\partial T}{\partial \tau_\nu}\right)_S=s^T_{\mu\nu}-\frac{T\alpha_\mu\alpha_\nu}{C_\tau}. \tag{10.17}$$

10.5. The difference $C_\tau - C_V$ and the ratio $C_r = C_\tau / C_V$. From (10.6), (10.1), and (9.7)

$$C_\tau = \left(\frac{\partial H}{\partial T}\right)_\tau = \left(\frac{\partial U}{\partial T}\right)_\tau - \tau_\nu \alpha_\nu . \tag{10.18}$$

From $U = U[V_\nu, S(V_\nu, T)]$, (8.3), (9.5), and the third Maxwell relation in Table 10.1

$$\left(\frac{\partial U}{\partial V_\nu}\right)_T = \left(\frac{\partial U}{\partial V_\nu}\right)_S + \left(\frac{\partial U}{\partial S}\right)_V \left(\frac{\partial S}{\partial V_\nu}\right)_T = \tau_\nu + T\lambda_\nu . \tag{10.19}$$

From $U = U[V_\nu(\tau_\mu, T), T]$, (10.5), and the above,

$$\left(\frac{\partial U}{\partial T}\right)_\tau = \left(\frac{\partial U}{\partial T}\right)_V + \left(\frac{\partial U}{\partial V_\nu}\right)_T \left(\frac{\partial V_\nu}{\partial T}\right)_\tau = C_V + (\tau_\nu + T\lambda_\nu)\,\alpha_\nu . \tag{10.20}$$

From (10.18) and (10.20),

$$C_\tau - C_V = T\lambda_\nu \alpha_\nu . \tag{10.21}$$

With the help of (10.10) and (10.13), this difference can be expressed as

$$C_\tau - C_V = T\alpha_\nu \alpha_\mu c^T_{\mu\nu} = T\alpha_\nu \alpha_\mu c^S_{\mu\nu} / C_r . \tag{10.22}$$

By multiplying both sides of (10.22) by $C_r / C_\tau = 1/C_V$, we find

$$C_r - 1 = T\alpha_\nu \alpha_\mu c^S_{\mu\nu} / C_\tau . \tag{10.23}$$

The coefficients $c^S_{\mu\nu}$ can be found to high precision from wave speeds. Thus, (10.23), relates C_r to the coefficients that are the most directly related to experimental measurements, α_ν, $c^S_{\mu\nu}$, and C_τ. The difference $C_\tau - C_V$ is then best found from

$$C_\tau - C_V = (C_r - 1)\, C_\tau / C_r . \tag{10.24}$$

With C_r from (10.23), (10.24) is equivalent to substituting C_r from (10.23) into (10.22). We remind the reader that the specific heat in (10.23) is referred to unit reference volume, not unit mass, and it is, of course, expressed in mechanical units.

10.6. Strain dependence of the specific heats and expansion coefficients. From (10.6) and the fourth Maxwell relation in Table 10.1,

$$\left(\frac{\partial C_\tau}{\partial \tau_\nu}\right)_T = T \frac{\partial^2 S}{\partial \tau_\nu \, \partial T} = T \left(\frac{\partial \alpha_\nu}{\partial T}\right)_\tau . \tag{10.25}$$

Similarly,

$$\left(\frac{\partial \alpha_\nu}{\partial \tau_\mu}\right)_T = \frac{\partial^2 V_\nu}{\partial \tau_\mu \, \partial T} = \left(\frac{\partial s^T_{\nu\mu}}{\partial T}\right)_\tau . \tag{10.26}$$

Hence, the derivatives of C_τ and α_ν with respect to the equilibrium stress can be found from the temperature derivatives of the expansion coefficients and compliance coefficients, respectively. The strain derivatives are given by

$$\left(\frac{\partial C_\tau}{\partial V_\mu}\right)_T = \left(\frac{\partial \tau_\nu}{\partial V_\mu}\right)_T \left(\frac{\partial C_\tau}{\partial \tau_\nu}\right)_T = T c^T_{\nu\mu} \left(\frac{\partial \alpha_\nu}{\partial T}\right)_T , \tag{10.27}$$

$$\left(\frac{\partial \alpha_\nu}{\partial V_\lambda}\right)_T = \left(\frac{\partial \alpha_\nu}{\partial \tau_\mu}\right)_T \left(\frac{\partial \tau_\mu}{\partial V_\lambda}\right)_T = \left(\frac{\partial s^T_{\nu\mu}}{\partial T}\right)_\tau c^T_{\mu\lambda} . \tag{10.28}$$

By differentiating the inverse relation between stiffnesses and compliances, $s^T_{\nu\mu} c^T_{\mu\lambda} = \delta_{\nu\lambda}$, with respect to T at constant τ, one sees that (10.28) can be put in the form

$$\left(\frac{\partial \alpha_\nu}{\partial V_\lambda}\right)_T = - s^T_{\nu\mu} \left(\frac{\partial c^T_{\mu\lambda}}{\partial T}\right)_\tau . \tag{10.29}$$

From (10.5) and the third Maxwell relation,

$$\left(\frac{\partial C_V}{\partial V_\mu}\right)_T = T\,\frac{\partial^2 S}{\partial V_\mu\,\partial T} = T\left(\frac{\partial \lambda_\mu}{\partial T}\right)_V. \tag{10.30}$$

By considering $\lambda_\mu = \lambda_\mu[T, V_\nu(T, \tau_\alpha)]$, we find

$$\left(\frac{\partial \lambda_\mu}{\partial T}\right)_\tau = \left(\frac{\partial \lambda_\mu}{\partial T}\right)_V + \left(\frac{\partial \lambda_\mu}{\partial V_\nu}\right)_T \left(\frac{\partial V_\nu}{\partial T}\right)_\tau$$

whence

$$\left(\frac{\partial \lambda_\mu}{\partial T}\right)_V - \left(\frac{\partial \lambda_\mu}{\partial T}\right)_\tau - \alpha_\nu \left(\frac{\partial \lambda_\mu}{\partial T_\nu}\right)_T. \tag{10.31}$$

From (10.30), (10.31), (10.10), and (10.29),

$$\left(\frac{\partial C_V}{\partial V_\mu}\right)_T = T\left[2\alpha_\nu \left(\frac{\partial c^T_{\nu\mu}}{\partial T}\right)_\tau + c^T_{\nu\mu}\left(\frac{\partial \alpha_\nu}{\partial T}\right)_\tau - \alpha_\nu\,\alpha_\gamma\,c^T_{\gamma\mu\nu}\right]. \tag{10.32}$$

From (10.27) and (10.32), the strain derivative of the difference $(C_\tau - C_V)$ is

$$\left(\frac{\partial (C_\tau - C_V)}{\partial V_\mu}\right)_T = -T\left[2\alpha_\nu \left(\frac{\partial c^T_{\nu\mu}}{\partial T}\right)_\tau - \alpha_\nu\,\alpha_\gamma\,c^T_{\gamma\mu\nu}\right]. \tag{10.33}$$

From (10.24), the isothermal strain derivative of $(C_r - 1)/C_r$ can be calculated as

$$\frac{\partial}{\partial V_\mu}\left(\frac{C_r - 1}{C_r}\right) = \frac{1}{C_r^2}\,\frac{\partial C_r}{\partial V_\mu} = -\frac{C_r - 1}{C_r\,C_\tau}\,\frac{\partial C_\tau}{\partial V_\mu} + \frac{1}{C_\tau}\,\frac{\partial}{\partial V_\mu}(C_\tau - C_V). \tag{10.34}$$

Substitution from (10.27) and (10.33) gives

$$\frac{1}{C_r^2}\left(\frac{\partial C_r}{\partial V_\mu}\right)_T = -\frac{C_r - 1}{C_r\,C_\tau}\,T c^T_{\nu\mu}\left(\frac{\partial \alpha_\nu}{\partial T}\right)_\tau - \frac{T}{C_\tau}\left[2\alpha_\nu \left(\frac{\partial c^T_{\nu\mu}}{\partial T}\right)_\tau - \alpha_\nu\,\alpha_\gamma\,c^T_{\gamma\mu\nu}\right]. \tag{10.35}$$

The formulas of this section are used to relate quantities that would be very difficult to measure directly to quantities that are more easily measured.

10.7. Internal energy as a function of other variables. With the independent variables (V_α, S), U has the derivatives shown in Table 10.1, but U may also be considered as a function of (V_α, T), (τ_α, S), or (τ_α, T). When so considered, U has the first partial derivatives shown in Table 10.2. Several of these have already been derived. From $U = U[V_\nu(\tau_\mu, T), T]$, (10.19), and (10.11),

$$\left(\frac{\partial U}{\partial \tau_\mu}\right)_T = \left(\frac{\partial U}{\partial V_\nu}\right)_T \left(\frac{\partial V_\nu}{\partial \tau_\mu}\right)_T = (\tau_\nu + T\lambda_\nu)\,s^T_{\nu\mu} = \tau_\nu\,s^T_{\nu\mu} + T\alpha_\mu. \tag{10.36}$$

From $U = U[V_\mu(\tau_\mu, S), S]$ and relations already established,

$$\left(\frac{\partial U}{\partial \tau_\nu}\right)_S = \left(\frac{\partial U}{\partial V_\mu}\right)_S \left(\frac{\partial V_\mu}{\partial \tau_\nu}\right)_S = \tau_\mu\,s^S_{\mu\nu}, \tag{10.37}$$

$$\left(\frac{\partial U}{\partial S}\right)_\tau = \left(\frac{\partial U}{\partial S}\right)_V + \left(\frac{\partial U}{\partial V_\mu}\right)_S \left(\frac{\partial V_\mu}{\partial S}\right)_\tau = T(1 + \tau_\mu\,\alpha_\mu/C_\tau). \tag{10.38}$$

In Table 10.2, we introduce as an abbreviation, the dimensionless variable

$$D \equiv 1 + \alpha_\nu\,\tau_\nu/C_\tau. \tag{10.39}$$

10.8. Internal energy as an independent variable. In the traditional formalism, U is a dependent variable, but it can be regarded as independent, consistent with our initial point of view in (8.1), (8.8), and (8.9). Table 10.3 shows the first partial

Table 10.2. *Internal energy as a function of* (V_α, T), (τ_α, T), (τ_α, S) $(D \equiv 1 + \alpha_\nu \tau_\nu / C_\tau)$.

Independent variables	First partial derivatives	
(V_α, T)	$\left(\frac{\partial U}{\partial V_\alpha}\right)_T = \tau_\alpha + T\lambda_\alpha,$	$\left(\frac{\partial U}{\partial T}\right)_V = C_V$
(τ_α, T)	$\left(\frac{\partial U}{\partial \tau_\alpha}\right)_T = T\alpha_\alpha + \tau_\beta s^T_{\beta\alpha},$	$\left(\frac{\partial U}{\partial T}\right)_\tau = DC_\tau$
(τ_α, S)	$\left(\frac{\partial U}{\partial \tau_\alpha}\right)_S = \tau_\beta s^S_{\beta\alpha},$	$\left(\frac{\partial U}{\partial S}\right)_\tau = DT$

derivatives of S, T, and τ_β with (V_ν, U) as independent variables. Table 10.4 shows the first partial derivatives of S, T, and V_β with (τ_ν, U) as independent.

Table 10.3. S, T, τ_β *as functions of* (V_α, U).

Function	First partial derivatives	
$S(V_\alpha, U)$	$\left(\frac{\partial S}{\partial V_\alpha}\right)_U = -\frac{\tau_\alpha}{T},$	$\left(\frac{\partial S}{\partial U}\right)_V = \frac{1}{T}$
$T(V_\alpha, U)$	$\left(\frac{\partial T}{\partial V_\alpha}\right)_U = -\frac{\tau_\alpha + T\lambda_\alpha}{C_V},$	$\left(\frac{\partial T}{\partial U}\right)_V = \frac{1}{C_V}$
$\tau_\beta(V_\alpha, U)$	$\left(\frac{\partial \tau_\beta}{\partial V_\alpha}\right)_U = c^S_{\beta\alpha} + \frac{\lambda_\beta \tau_\alpha}{C_V},$	$\left(\frac{\partial \tau_\beta}{\partial U}\right)_V = -\frac{\lambda_\beta}{C_V}$

Table 10.4. S, T, V_β *as functions of* (τ_α, U).

Function	First partial derivatives	
$S(\tau_\alpha, U)$	$\left(\frac{\partial S}{\partial \tau_\alpha}\right)_U = \frac{\tau_\beta s^S_{\beta\alpha}}{TD},$	$\left(\frac{\partial S}{\partial U}\right)_\tau = \frac{1}{TD}$
$T(\tau_\alpha, U)$	$\left(\frac{\partial T}{\partial \tau_\alpha}\right)_U = \frac{\tau_\beta s^T_{\beta\alpha} + T\alpha_\alpha}{DC_\tau},$	$\left(\frac{\partial T}{\partial U}\right)_V = \frac{1}{DC_\tau}$
$V_\beta(\tau_\alpha, U)$	$\left(\frac{\partial V_\beta}{\partial \tau_\alpha}\right)_U = s^S_{\beta\alpha} + \frac{\alpha_\beta \tau_\gamma s^S_{\gamma\alpha}}{DC_\tau},$	$\left(\frac{\partial V_\beta}{\partial U}\right)_\tau = \frac{\alpha_\beta}{C_\tau D}$

10.9. Grüneisen numbers. Dimensionless quantities $\gamma_\nu = \lambda_\nu / C_V$ are sometimes called tensor Grüneisen numbers. In the full tensor notation, they may be defined by

$$\gamma_{ij} \equiv -\frac{1}{T}\left(\frac{\partial T}{\partial V_{ij}}\right)_S \equiv \frac{\left(\frac{\partial S}{\partial V_{ij}}\right)_T}{T\left(\frac{\partial S}{\partial T}\right)_V} \equiv \frac{1}{C_V}\left(\frac{\partial S}{\partial V_{ij}}\right)_T. \tag{10.40}$$

Each two-subscript gamma is equal to the corresponding single-subscript gamma. From (10.10) and (10.12),

$$\gamma_\nu = \frac{\lambda_\nu}{C_V} = \frac{\alpha_\mu c^T_{\mu\nu}}{C_V} = \frac{\alpha_\mu c^S_{\mu\nu}}{C_\tau}. \tag{10.41}$$

From the Maxwell coefficients in the first and third rows of Table 10.1, we list

$$\gamma_v = -\frac{1}{T}\left(\frac{\partial \tau_v}{\partial S}\right)_V = -\frac{1}{T}\left(\frac{\partial T}{\partial V_v}\right)_S = \frac{1}{C_V}\left(\frac{\partial \tau_v}{\partial T}\right)_V = \frac{1}{C_V}\left(\frac{\partial S}{\partial V_v}\right)_T. \tag{10.42}$$

It may be noted that (10.41) and (10.23) enable the specific heat ratio C_r to expressed as

$$C_r = 1 + T\alpha_v\gamma_v. \tag{10.43}$$

11. Thermostatics under hydrostatic pressure.

The condition that the equilibrium stress be isotropic ("hydrostatic" or "spherical") is

$$\overline{T}_{ij} = -p\,\delta_{ij}, \tag{11.1}$$

where $\overline{T}_{ij}$ is the equilibrium stress and p is the hydrostatic pressure. By (4.7) and (3.11), the corresponding equilibrium value of the Piola-Kirchhoff stress $t_{rs} = \bar{t}_{rs} = \tau_{rs}$ is

$$\tau_{rs} = -pJ\frac{\partial X_r}{\partial x_i}\frac{\partial X_s}{\partial x_i} = -pJC^{-1}_{rs}. \tag{11.2}$$

When a solid is in equilibrium under hydrostatic pressure, its internal energy, entropy, and strain are expected to depend on only two variables, say (p, T). Some other possible choices of independent variables are (p, S), (ϱ, T), (ϱ, S), (p, U), or (ϱ, U). While there are in general infinitely many static equilibrium states of strain that correspond to a given (ϱ, T), there is presumably only one such state in which the stress is hydrostatic pressure. In cubic crystals or isotropic materials, hydrostatic pressure produces an isotropic strain. In general, however, the strain resulting from hydrostatic pressure is not isotropic, and when it is not, (11.2) shows that the Piola-Kirchhoff stress is not either.

11.1. Thermodynamic potentials. We refer the thermodynamic potentials to unit reference volume, as was done in Sect. 10. With the help of $(4.6)_1$, the contribution of $\tau_{ij}\dot{V}_{ij}$ to $\dot{U}$ in (8.10) may be transformed as

$$\tau_{ij}\dot{V}_{ij} = \bar{t}_{ij}\dot{V}_{ij} = J\,\overline{T}_{ij}\,d_{ij}. \tag{11.3}$$

In view of (11.1) and (3.36), this becomes

$$\tau_{ij}\dot{V}_{ij} = -pJd_{ss} = -p\dot{J}, \tag{11.4}$$

since the trace of d_{ij} is identically the same as div $\boldsymbol{v}$. Therefore, when the equilibrium stress $\overline{T}_{ij}$ is hydrostatic, we rewrite (8.10) as

$$\dot{U} = T\dot{S} - p\dot{J}. \tag{11.5}$$

If (8.2) is now replaced by

$$U = U(S, J), \tag{11.6}$$

it follows from (11.5) that

$$T = (\partial U/\partial S)_J, \tag{11.7}$$

$$p = -(\partial U/\partial J)_S. \tag{11.8}$$

J is still the Jacobian defined by (3.18). Its physical significance as a volume ratio (3.17) or as a "normalized dimensionless specific volume" $J = \bar{\varrho}/\varrho$ (6.1) is helpfully kept in mind while interpreting the formulas of this section. It should be remembered also that these are not general relations, but relations valid in the presence of the constraint (11.1).

Table 11.1. *Thermodynamic potentials and their derivatives under spherical stress.*

Potential	Independent variables	Differential	First derivatives		Second derivatives		
			Mechanical	Thermal	Maxwell coefficients	Mechanical	Thermal
Internal energy U	(J, S)	$dU = -p\,dJ + T\,dS$	$p = -\frac{\partial U}{\partial J}$	$T = \frac{\partial U}{\partial S}$	$\frac{TB\beta}{C_J} = \frac{\partial p}{\partial S} = -\frac{\partial T}{\partial J}$	$\frac{B^S}{J} = -\frac{\partial p}{\partial J}$	$\frac{T}{C_J} = \frac{\partial T}{\partial S}$
Enthalpy $H = U + pJ$	(p, S)	$dH = J\,dp + T\,dS$	$J = \frac{\partial H}{\partial p}$	$T = \frac{\partial H}{\partial S}$	$\frac{\beta TJ}{C_p} = \frac{\partial J}{\partial S} = \frac{\partial T}{\partial p}$	$\chi^S J = -\frac{\partial J}{\partial p}$	$\frac{T}{C_p} = \frac{\partial T}{\partial S}$
Free energy $A = U - TS$	(J, T)	$dA = -p\,dJ - S\,dT$	$p = -\frac{\partial A}{\partial J}$	$S = -\frac{\partial A}{\partial T}$	$\beta B = \frac{\partial p}{\partial T} = \frac{\partial S}{\partial J}$	$\frac{B}{J} = -\frac{\partial p}{\partial J}$	$\frac{C_J}{T} = \frac{\partial S}{\partial T}$
Free enthalpy $G = H - TS$	(p, T)	$dG = J\,dp - S\,dT$	$J = \frac{\partial G}{\partial p}$	$S = -\frac{\partial G}{\partial T}$	$\beta J = \frac{\partial J}{\partial T} = -\frac{\partial S}{\partial p}$	$\chi J = -\frac{\partial J}{\partial p}$	$\frac{C_p}{T} = \frac{\partial S}{\partial T}$
					$\gamma = -\frac{J}{T}\left(\frac{\partial T}{\partial J}\right)_S = \frac{J}{C_J}\left(\frac{\partial S}{\partial J}\right)_T = \frac{\beta BJ}{C_J} = \frac{\beta B^S J}{C_p}$		

Analogs to all the relations of Sect. 10 can be obtained by replacing τ_{ij} by $-p$ and V_{ij} by J. For example,

$$\left(\frac{\partial S}{\partial U}\right)_J = \frac{1}{T}, \tag{11.9}$$

$$\left(\frac{\partial S}{\partial J}\right)_U = \frac{p}{T}. \tag{11.10}$$

The free energy A is still defined by (10.2), namely

$$A \equiv U - TS, \tag{11.11}$$

but when the stress is hydrostatic, it is customary to define the enthalpy H by

$$H = U + pJ \tag{11.12}$$

instead of (10.1). Because $pJ \neq -\tau_{ij} V_{ij}$, even when the stress is hydrostatic, the enthalpy of Sect. 10 does not in general reduce to (11.12) under hydrostatic pressure. A similar remark applies to the free enthalpy G, still defined by (10.3):

$$G \equiv H - TS. \tag{11.13}$$

However, it may be remarked that a different choice of substate variables in Sect. 10 would make these functions agree [THURSTON, 1970]. The other functions, U, A, T, and S are of course the same as in Sect. 10.

Table 11.1 shows the appropriate independent variables for these potentials and certain relations involving them. The definitions and derivations are analogous to those of Table 10.1.

11.2. Compressibility, bulk modulus, and volumetric expansion coefficient. The isothermal bulk modulus B, isothermal compressibility $\chi \equiv (1/B)$, and constant-pressure volumetric expansion coefficient β are defined by the equations

$$\frac{1}{\chi} \equiv B \equiv -J\left(\frac{\partial p}{\partial J}\right)_T \equiv \varrho\left(\frac{\partial p}{\partial \varrho}\right)_T \equiv \frac{1}{(1/\varrho)\,(\partial \varrho/\partial p)_T} \equiv -\frac{1}{(1/J)\,(\partial J/\partial p)_T}. \tag{11.14}$$

$$\beta \equiv \frac{1}{J}\left(\frac{\partial J}{\partial T}\right)_p \equiv -\frac{1}{\varrho}\left(\frac{\partial \varrho}{\partial T}\right)_p \equiv -\left[\frac{\partial}{\partial T}(\ln \varrho)\right]_p. \tag{11.15}$$

Either the specific volume $V \equiv 1/\varrho$ or the actual volume of a system can be substituted for J without changing any of these definitions. From (11.14),

$$(\partial p/\partial \varrho)_T = B/\varrho. \tag{11.16}$$

By considering the identity $p \equiv p[\varrho(p, T), T]$, we find

$$0 = \left(\frac{\partial p}{\partial T}\right)_p = \left(\frac{\partial p}{\partial T}\right)_\varrho + \left(\frac{\partial p}{\partial \varrho}\right)_T \left(\frac{\partial \varrho}{\partial T}\right)_p,$$

whence

$$(\partial p/\partial T)_\varrho = \beta B. \tag{11.17}$$

The isentropic compressibility and bulk modulus are defined by expressions like (11.14), but with the constant-temperature derivatives replaced by derivatives at constant entropy:

$$\frac{1}{\chi^S} \equiv B^S \equiv \varrho\left(\frac{\partial p}{\partial \varrho}\right)_S \equiv -J\left(\frac{\partial p}{\partial J}\right)_S. \tag{11.18}$$

The above definitions and the relation (11.17) are incorporated in Table 11.1.

At zero pressure, β and χ can be expressed in terms of the tensor coefficients α_{ij} and s_{ijkm}, respectively. Since the zero of pressure coincides with the zero of thermodynamic tensions,

$$\left(\frac{\partial V_{ij}}{\partial T}\right)_{p=0} = \left(\frac{\partial V_{ij}}{\partial T}\right)_{\tau=0} = \alpha_{ij}(0). \tag{11.19}$$

With the reference configuration (where $C_{ij} = \delta_{ij}$ and $J = 1$) taken as the configuration at zero pressure at the temperature under consideration, differentiation of (11.2) gives

$$\left(\frac{\partial \tau_{ij}}{\partial p}\right)_{p=0} = -\delta_{ij}. \tag{11.20}$$

Also, from (3.41) and the above choice of reference configuration,

$$\left(\frac{\partial J}{\partial V_{ij}}\right)_{p=0} = (J C_{ij}^{-1}) = \delta_{ij}. \tag{11.21}$$

From (11.15), (11.19), (11.21), the chain rule, and the above choice of reference configuration,

$$\beta(0) = \left[\frac{1}{J}\left(\frac{\partial J}{\partial T}\right)_p\right]_{p=0} = \left(\frac{\partial J}{\partial V_{ij}}\,\frac{\partial V_{ij}}{\partial T}\right)_{p=0} = \delta_{ij}\,\alpha_{ij}(0) = \alpha_{ii}(0). \tag{11.22}$$

Similarly, from (11.14), 11.20), (11.21), and (9.12),

$$\chi(0) = \left[\frac{-1}{J}\left(\frac{\partial J}{\partial p}\right)_T\right]_{p=0} = \left(\frac{-\partial J}{\partial V_{ij}}\,\frac{\partial V_{ij}}{\partial \tau_{km}}\,\frac{\partial \tau_{km}}{\partial p}\right)_{p=0} = \delta_{ij}\,\delta_{km}\,s_{ijkm}^T(0) = s_{iikk}^T(0). \tag{11.23}$$

Obviously, a similar relation connects the isentropic compressibility with the isentropic compliances evaluated at zero pressure:

$$\chi^S(0) = s_{iikk}^S(0). \tag{11.24}$$

11.3. Specific heats at constant volume and constant pressure. The specific heat at constant volume is a well-defined quantity only when sufficient side conditions or constraints are specified that the thermodynamic functions are determined by the specific volume and the temperature. The condition that the stress be hydrostatic, (11.1), is such a set of constraints. When constant volume is preserved by applying hydrostatic pressure, the individual strain components do not remain zero, in general, and when they do not, it is reasonable to ask whether the specific heat at constant volume is the same as the specific heat at constant strain. This question does not arise in isotropic media or cubic crystals, where the application of hydrostatic pressure to keep the volume constant necessarily keeps the individual strain components constant also, as a result of symmetry considerations.

With the constraints (11.1) understood, we define the specific heats per unit reference volume at constant volume and constant pressure as

$$C_J \equiv T\left(\frac{\partial S}{\partial T}\right)_J = \left(\frac{\partial U}{\partial T}\right)_J, \quad C_p \equiv T\left(\frac{\partial S}{\partial T}\right)_p = \left(\frac{\partial H}{\partial T}\right)_p \tag{11.25}$$

respectively. With the definitions used in Sect. 10, zero pressure corresponds to zero thermodynamic tensions. Hence, *at zero pressure*, the specific heat at constant pressure equals the specific heat at constant thermodynamic tensions:

$$C_p(0) = C_\tau(0). \tag{11.26}$$

(By revising the definitions, as will be indicated in Sect. 11.4, the equality can be shifted to any other single pressure p_0.) By differentiating $S=S[T, J(p, T)]$, we find

$$\left(\frac{\partial S}{\partial T}\right)_p=\left(\frac{\partial S}{\partial T}\right)_J+\left(\frac{\partial S}{\partial J}\right)_T\left(\frac{\partial J}{\partial T}\right)_p. \tag{11.27}$$

Multiplying by T and making substitutions from (11.25) and Table 11.1 yields

$$C_p=C_J+T\beta^2 B J. \tag{11.28}$$

From (10.22) and (11.28),

$$(C_\tau-C_V)-(C_p-C_J)=T(\alpha_\nu\,\alpha_\mu\,c^T_{\mu\nu}-\beta^2 B J). \tag{11.29}$$

We shall be content to evaluate C_J-C_V at the reference configuration where $J=1$, taken as that corresponding to $p=0$. In view of (11.26) and (11.29),

$$C_J(0)-C_V(0)=-T(\beta^2 B-\alpha_\nu\,\alpha_\mu\,c^T_{\mu\nu}). \tag{11.30}$$

This difference need not vanish, in general.

Sufficient for the vanishing of this difference is the condition that at zero pressure,

$$\alpha_{ij}\equiv\left(\frac{\partial V_{ij}}{\partial T}\right)_p=-\beta\,B\left(\frac{\partial V_{ij}}{\partial p}\right)_T=\beta B s^T_{ijkk}, \tag{11.31}$$

which may be written as

$$\chi\,\alpha_{ij}=\beta\,s^T_{ijkk}, \tag{11.31a}$$

or, in abbreviated notation,

$$\chi\,\alpha_\mu=\beta\,(s^T_{\mu 1}+s^T_{\mu 2}+s^T_{\mu 3}). \tag{11.31b}$$

There is no thermodynamic requirement that (11.31) hold, in general. The contractions upon cooling are not necessarily proportional to those produced by pressure. However, it is true in general that at zero pressure

$$\chi\,\alpha_{ij}\,\delta_{ij}=\beta\,s^T_{ijkk}\,\delta_{ij} \tag{11.32}$$

because $\delta_{ij}\,\alpha_{ij}(0)=\beta(0)$ by (11.22) and $\delta_{ij}\,s^T_{ijkk}(0)=\chi(0)$ by (11.23). In isotropic media and cubic crystals, (11.31) is satisfied by virtue of symmetry, as can be seen from (11.32) and the fact that α_{ij} and s^T_{ijkk} are the same for $ij=11$, 22, and 33.

The difference, when not zero, between the specific heat at constant volume C_J and the specific heat at constant strain C_V, can be evaluated by experimental determination of the quantities on the right side of (11.30).

11.4. Arbitrariness of reference pressure. In the previous formalism, the zero of pressure plays an undeservedly unique role. For example, our definitions make the thermodynamic tensions equal to the second Piola-Kirchhoff stress, which is zero at zero pressure. The coincidence of the zero of thermodynamic tensions with the zero of stress is necessitated by our choice of the internal energy U as the thermodynamic potential for the independent variables (S, V_{ij}). Now changes in U correspond to the total work (per unit reference volume) needed to produce a deformation. If the sample is in a pressure vessel (or laboratory!) maintained at some pressure p_0, a part of this work may be supplied by the constraining pressure. The *additional* work per unit reference volume that must be supplied by another agent is not the change in U, but the change in

$$\varepsilon=U+J p_0. \tag{11.33}$$

If we keep V_{ij} as the substate variables, but use ε instead of U as a thermodynamic potential, then the new thermodynamic tensions $\bar{\tau}_{ij}$ are given by

$$\bar{\tau}_{ij}=\left(\frac{\partial\varepsilon}{\partial V_{ij}}\right)_S=\tau_{ij}+p_0\frac{\partial J}{\partial V_{ij}}. \tag{11.34}$$

Upon making use of (3.41) and (11.2), we find the tensions under hydrostatic pressure to be

$$\bar{\tau}_{ij}=(p_0-p)\,J\,C_{ij}^{-1}. \tag{11.35}$$

Thus, by adopting ε as the thermodynamic potential in place of U, any arbitrary constant pressure $p=p_0$ can be assigned the favored role previously played by $p=0$.

The advantage of this procedure is that coefficients defined "at constant thermodynamic tensions" can be made to coincide with coefficients defined at constant pressure for any constant pressure $p=p_0$, not just $p=0$.

$J p_0$ can be added to the other three classical thermodynamic potentials also, and the entire formalism of Sect. 10 can be reconstructed with the revised definitions.

11.5. Maxwell coefficients. By considering $J=J[p, S(p, T)]$, one finds

$$\left(\frac{\partial J}{\partial T}\right)_p=\left(\frac{\partial J}{\partial S}\right)_p\left(\frac{\partial S}{\partial T}\right)_p. \tag{11.36}$$

Hence, from the definitions of β and C_p in (11.15) and (11.25), the second Maxwell coefficient in Table 11.1, $(\partial J/\partial S)_p$, can be expressed as

$$(\partial J/\partial S)_p=\beta T J/C_p. \tag{11.37}$$

Similarly, from $p=p[J, S(J, T)]$, follows

$$\left(\frac{\partial p}{\partial T}\right)_J=\left(\frac{\partial p}{\partial S}\right)_J\left(\frac{\partial S}{\partial T}\right)_J. \tag{11.38}$$

Hence (11.17) and the definition of C_J in (11.25) enable the first Maxwell coefficient in Table 11.1 to be expressed as

$$\left(\frac{\partial p}{\partial S}\right)_J=\beta T B/C_J. \tag{11.39}$$

The relations (11.37) and (11.39) have been incorporated in Table 11.1, along with invertibility relations like (10.7).

It may be noted that the third Maxwell coefficient is $(\partial S/\partial T)_J$ times the first, and the fourth is $(\partial S/\partial T)_p$ times the second. This rule applies to Table 10.1 also, with appropriate revision of the variables held constant in taking $\partial S/\partial T$.

11.6. $B^S/B=C_p/C_J$. By considering $p=p[J, T(S, J)]$, we find

$$\left(\frac{\partial p}{\partial J}\right)_S=\left(\frac{\partial p}{\partial J}\right)_T+\left(\frac{\partial p}{\partial T}\right)_J\left(\frac{\partial T}{\partial J}\right)_S. \tag{11.40}$$

By means of substitutions from (11.18), (11.14), (11.17), Table 11.1, and (11.28), (11.40) shows that

$$B^S/B=C_p/C_J=\text{ratio of specific heats.} \tag{11.41}$$

11.7. Grüneisen number γ. The dimensionless Grüneisen number γ appropriate to a solid under hydrostatic pressure can be defined by

$$\gamma \equiv \frac{-J}{T}\left(\frac{\partial T}{\partial J}\right)_S = \frac{J\left(\frac{\partial S}{\partial J}\right)_T}{T\left(\frac{\partial S}{\partial T}\right)_J} = \frac{J}{C_J}\left(\frac{\partial S}{\partial J}\right)_T. \tag{11.42}$$

By (11.41) and the third Maxwell relation in Table 11.1, we have

$$\gamma = \frac{J}{C_J}\left(\frac{\partial p}{\partial T}\right)_J - \beta B J/C_J - \beta B^S J/C_p. \tag{11.43}$$

From (11.28) and (11.43), $(C_p - C_J)/C_J = \beta T \gamma$. Hence, in terms of the Grüneisen gamma,

$$\frac{B^S}{B} = \frac{\chi}{\chi^S} = \frac{C_p}{C_j} = 1 + \beta T \gamma. \tag{11.44}$$

From (11.44),

$$B^S - B = \frac{B^S \beta T \gamma}{1 + \beta T \gamma}. \tag{11.45}$$

The Grüneisen gamma can be expressed in terms of the tensorial Grüneisen numbers of (10.40) by considering $S = S[T, V_{ij}(T, J)]$, subject to the constraint (11.1). By the chain rule,

$$\left(\frac{\partial S}{\partial J}\right)_T = \left(\frac{\partial S}{\partial V_{ij}}\right)_T \left(\frac{\partial V_{ij}}{\partial J}\right)_T.$$

Upon inserting the definitions from (11.42) and (10.40), there results

$$\gamma = \frac{J C_V}{C_J}\left(\frac{\partial V_{ij}}{\partial J}\right)_T \gamma_{ij}, \tag{11.46}$$

in which $(\partial V_{ij}/\partial J)_T$ acquires a unique meaning through the constraint (11.1). From $V_{ij} = V_{ij}\{T, \tau_{km}[T, p(J, T)]\}$

$$\left(\frac{\partial V_{ij}}{\partial J}\right)_T = \left(\frac{\partial V_{ij}}{\partial \tau_{km}} \frac{\partial \tau_{km}}{\partial p} \frac{\partial p}{\partial J}\right)_T = -\frac{B}{J} s^T_{ijkm}\left(\frac{\partial \tau_{km}}{\partial p}\right)_T, \tag{11.47}$$

where the last step follows from the definitions (11.14) and (9.12). Finally, evaluation at the zero pressure reference configuration, where $(\partial \tau_{km}/\partial p)$ has the simple form (11.20), gives

$$\gamma(0) = \frac{B(0)\, C_V(0)}{C_J(0)} \gamma_{ij}(0)\, s^T_{ijkk}(0). \tag{11.48}$$

11.8. Pressure derivatives of the specific heats and expansion coefficient. By differentiating with respect to T at constant p in the fourth Maxwell relation of Table 11.1, multiplying the resulting equality by T, and using the definition of C_p (11.25), there results

$$\left(\frac{\partial C_p}{\partial p}\right)_T = -T\left(\frac{\partial^2 J}{\partial T^2}\right)_p = -JT\left[\left(\frac{\partial \beta}{\partial T}\right)_p + \beta^2\right]. \tag{11.49}$$

By differentiating with respect to p at constant T in $\beta J = (\partial J/\partial T)_p$, and introducing the isothermal compressibility χ,

$$\left[\frac{\partial}{\partial p}(\beta J)\right]_T = \frac{\partial^2 J}{\partial p\, \partial T} = -\left[\frac{\partial}{\partial T}(\chi J)\right]_p. \tag{11.50}$$

Since $\beta(\partial J/\partial p)_T = -\beta\chi J$ and $\chi(\partial J/\partial T)_p = \chi\beta J$, (11.50) reduces to

$$\left(\frac{\partial\beta}{\partial p}\right)_T = -\left(\frac{\partial\chi}{\partial T}\right)_p. \tag{11.51}$$

Thus, the pressure derivative of C_p can be found from the temperature dependence of β and the pressure derivative of β can be found from the temperature dependence of the compressibility χ. The corresponding derivatives with respect to J may be found by multiplying the pressure derivatives by $(\partial p/\partial J)_T = -B/J$:

$$\left(\frac{\partial C_p}{\partial J}\right)_T = \frac{BT}{J}\left(\frac{\partial^2 J}{\partial T^2}\right)_p = TB\left[\left(\frac{\partial\beta}{\partial T}\right)_p + \beta^2\right], \tag{11.52}$$

$$\left(\frac{\partial\beta}{\partial J}\right)_T = \frac{B}{J}\left(\frac{\partial\chi}{\partial T}\right)_p. \tag{11.53}$$

From (11.28), the pressure derivative of $(C_p - C_J)$ is

$$\begin{aligned}\left[\frac{\partial}{\partial p}(C_p - C_J)\right]_T &= 2\,T\beta BJ\left(\frac{\partial\beta}{\partial p}\right)_T + T\beta^2 J\left(\frac{\partial B}{\partial p}\right)_T + T\beta^2 B\left(\frac{\partial J}{\partial p}\right)_T \\ &= \beta^2 TJ\left[-\frac{2B}{\beta}\left(\frac{\partial\chi}{\partial T}\right)_p + \left(\frac{\partial B}{\partial p}\right)_T - 1\right].\end{aligned} \tag{11.54}$$

From (11.54) and (11.49),

$$\left(\frac{\partial C_J}{\partial p}\right)_T = \beta^2\,TJ\left[\frac{2B}{\beta}\left(\frac{\partial\chi}{\partial T}\right)_p - \left(\frac{\partial B}{\partial p}\right)_T - \frac{1}{\beta^2}\left(\frac{\partial\beta}{\partial T}\right)_p\right]. \tag{11.55}$$

The corresponding derivative with respect to J is

$$\left(\frac{\partial C_J}{\partial J}\right)_T = -\beta^2 BT\left[\frac{2B}{\beta}\left(\frac{\partial\chi}{\partial T}\right)_p - \left(\frac{\partial B}{\partial p}\right)_T - \frac{1}{\beta^2}\left(\frac{\partial\beta}{\partial T}\right)_p\right]. \tag{11.56}$$

From (11.49) and (11.54), the pressure derivative of the ratio (C_J/C_p) can be expressed as

$$\frac{\partial}{\partial p}(C_J/C_p) = -\frac{\beta^2 TJ}{C_p}\left[\left(\frac{\partial B}{\partial p}\right)_T - \frac{2B}{\beta}\left(\frac{\partial\chi}{\partial T}\right)_p + \frac{C_p - C_J}{C_p\beta^2}\frac{\partial\beta}{\partial T} - \frac{C_J}{C_p}\right]. \tag{11.57}$$

11.9. Internal energy as a function of other variables. With the independent variables (J, S), U has the derivatives shown in Table 11.1. When considered as a function of (J, T), (p, S), or (p, T), U has the first partial derivatives shown in Table 11.2. The derivations are straightforward:

$$\left(\frac{\partial U}{\partial J}\right)_T = \left(\frac{\partial U}{\partial J}\right)_S + \left(\frac{\partial U}{\partial S}\right)_J\left(\frac{\partial S}{\partial J}\right)_T = -p + \beta TB, \tag{11.58}$$

$$\left(\frac{\partial U}{\partial p}\right)_T = \left(\frac{\partial U}{\partial J}\right)_T\left(\frac{\partial J}{\partial p}\right)_T = \chi J(p - \beta TB) = J(\chi p - \beta T), \tag{11.59}$$

$$\left(\frac{\partial U}{\partial p}\right)_S = \left(\frac{\partial U}{\partial J}\right)_S\left(\frac{\partial J}{\partial p}\right)_S = pJ\chi^S = pJ/B^S, \tag{11.60}$$

$$\left(\frac{\partial U}{\partial S}\right)_p = \left(\frac{\partial U}{\partial S}\right)_J + \left(\frac{\partial U}{\partial J}\right)_S\left(\frac{\partial J}{\partial S}\right)_p = T - p\beta TJ/C_p, \tag{11.61}$$

$$\left(\frac{\partial U}{\partial T}\right)_p = \left(\frac{\partial U}{\partial S}\right)_p\left(\frac{\partial S}{\partial T}\right)_p = \frac{C_p}{T}\left(\frac{\partial U}{\partial S}\right)_p = C_p - p\beta J. \tag{11.62}$$

11.10. Internal energy as an independent variable. Table 11.3 shows the first partial derivatives of S, T, and p with (J, U) as independent variables. Table 11.4

Table 11.2. *Internal energy as a function of* (J, T), (p, T), (p, S).

Independent variables	First partial derivatives	
(J, T)	$\left(\frac{\partial U}{\partial J}\right)_T = -p + \beta T B$,	$\left(\frac{\partial U}{\partial T}\right)_J = C_J$
(p, T)	$\left(\frac{\partial U}{\partial p}\right)_T = J(\chi p - \beta T)$,	$\left(\frac{\partial U}{\partial T}\right)_p = C_p(1 - p\beta J/C_p)$
(p, S)	$\left(\frac{\partial U}{\partial p}\right)_S = \chi^S p J$,	$\left(\frac{\partial U}{\partial S}\right)_p = T(1 - p\beta J/C_p)$

shows the first partial derivatives of S, T, and J with (p, U) as independent variables.

Table 11.3. S, T, p *as functions of* (J, U).

Function	First partial derivatives	
$S(J, U)$	$\left(\frac{\partial S}{\partial J}\right)_U = \frac{p}{T}$,	$\left(\frac{\partial S}{\partial U}\right)_J = \frac{1}{T}$
$T(J, U)$	$\left(\frac{\partial T}{\partial J}\right)_U = \frac{p - \beta T B}{C_J}$,	$\left(\frac{\partial T}{\partial U}\right)_J = \frac{1}{C_J}$
$p(J, U)$	$\left(\frac{\partial p}{\partial J}\right)_U = \frac{B^S}{J} + \frac{p B \beta}{C_J}$,	$\left(\frac{\partial p}{\partial U}\right)_J = \frac{\beta B}{C_J}$

Table 11.4. S, T, J *as functions of* (p, U).

Function	First partial derivatives	
$S(p, U)$	$\left(\frac{\partial S}{\partial p}\right)_U = -\frac{\chi^S p J}{T(1 - p\beta J/C_p)}$,	$\left(\frac{\partial S}{\partial U}\right)_p = \frac{1}{T(1 - p\beta J/C_p)}$
$T(p, U)$	$\left(\frac{\partial T}{\partial p}\right)_U = -\frac{J(\chi p - \beta T)}{C_p(1 - p\beta J/C_p)}$,	$\left(\frac{\partial T}{\partial U}\right)_p = \frac{1}{C_p(1 - p\beta J/C_p)}$
$J(p, U)$	$\left(\frac{\partial J}{\partial p}\right)_U = -\chi^S J\left(1 - \frac{p\beta J/C_p}{1 - p\beta J/C_p}\right)$,	$\left(\frac{\partial J}{\partial U}\right)_p = \frac{1}{p + C_p/\beta J}$

11.11. Other derivatives with respect to pressure. To evaluate various quantities of interest, the first and second pressure derivatives of C_{ij}, $G_{ij} \equiv C_{ij}^{-1}$, and of the thermodynamic tensions are needed. For example, the derivatives of G_{ij} are related to the derivatives of the thickness shrink f_N [cf. (3.15)] through (3.10). We shall take the reference configuration where $J = 1$ and $G_{rs} = C_{rs} = \delta_{rs}$ as the configuration at $p = 0$. In the final formulas, the pressure derivatives are to be evaluated at $p = 0$. It should be noted, however, that formulas of the same form may be obtained at any other single pressure p_0 if the reference configuration is chosen as that corresponding to $p = p_0$ and if the thermodynamic definitions are revised as indicated in Sect. 11.4 above.

To make the strains depend on the single independent variable p, an additional constraint is needed besides (11.1). The simplest example of such a constraint is constant temperature. Another simple case is constant entropy. The formulas of this Sect. 11.11 that have elastic or compliance coefficients in them can be applied to either of these cases simply by using the isothermal or isentropic coefficients, respectively.

The needed derivatives of G_{ik} can be expressed in terms of the derivatives of its inverse C_{ik} by straightforward operations on the equation $C_{ik}(p)\, G_{kj}(p)=\delta_{ij}$. With derivatives denoted by primes, the results are

$$\begin{aligned} G'_{ij}(0) &= -C'_{ij}(0) = -2V'_{ij}(0) \\ G''_{ij}(0) &= -C''_{ij}(0) + 8A_{ij}, \\ A_{ij} &\equiv \tfrac{1}{4}\, C'_{ik}(0)\, C'_{kj}(0) = V'_{ik}(0)\, V'_{kj}(0). \end{aligned} \tag{11.63}$$

From (3.44) and the chain rule

$$J'(p) = \tfrac{1}{2} J(p)\, G_{ji}(p)\, C'_{ij}(p) \tag{11.64}$$

whence, by differentiation,

$$J'' = \tfrac{1}{2} J G_{ji}\, C''_{ij} + \tfrac{1}{2} (J G_{ji})'\, C'_{ij}. \tag{11.65}$$

Evaluated at $p=0$, these derivatives are

$$\begin{aligned} -\chi(0) &= J'(0) = \tfrac{1}{2} C'_{ii}(0) = V'_{ii}(0), \\ J''(0) &= \tfrac{1}{2} C''_{ii}(0) + [\chi(0)]^2 - 2A_{jj} \end{aligned} \tag{11.66}$$

where the equality $\chi(0) = -J'(0)$ follows from (11.14) and the condition $J=1$ at $p=0$. The first and second pressure derivatives of (11.2) are

$$\begin{aligned} \tau'_{rs} &= -J G_{rs} - p (J G_{rs})', \\ \tau''_{rs} &= -2 (J G_{rs})' - p (J G_{rs})''. \end{aligned} \tag{11.67}$$

Evaluation at $p=0$ gives

$$\begin{aligned} \tau'_{rs}(0) &= -\delta_{rs}, \\ \tau''_{rs}(0) &= 2C'_{rs}(0) + 2\,\delta_{rs}\,\chi(0). \end{aligned} \tag{11.68}$$

The derivatives of C_{ij} or V_{ij} are evaluated by "chain rule" formulas of the form $V'_{ij} = (\partial V_{ij}/\partial \tau_{km})\, \tau'_{km} = s_{ijkm}\, \tau'_{km}$. In abbreviated notation,

$$\begin{aligned} C'_{\alpha} &= 2V'_{\alpha} = 2(\partial V_{\alpha}/\partial \tau_{\beta})\, \tau'_{\beta} = 2 s_{\alpha\beta}\, \tau'_{\beta}, \\ C''_{\alpha} &= 2V''_{\alpha} = 2(s_{\alpha\beta\gamma}\, \tau'_{\beta}\, \tau'_{\gamma} + s_{\alpha\beta}\, \tau''_{\beta}). \end{aligned} \tag{11.69}$$

In order to present final formulas in the abbreviated notation, it is convenient to introduce the symbol δ_{α} as an "abbreviated Kronecker delta," equal to 1 for $\alpha=1, 2, 3$ and otherwise zero.

$$\begin{aligned} \delta_{\alpha} &= 1, \quad \alpha = 1, 2, 3 \\ &= 0, \quad \alpha = 4, 5, 6. \end{aligned} \tag{11.70}$$

We also introduce the abbreviations

$$s_{\alpha} = s_{\alpha\beta}\, \delta_{\beta} = s_{\alpha 1} + s_{\alpha 2} + s_{\alpha 3}, \tag{11.71}$$

$$g_{\alpha} = s_{\alpha\beta\gamma}\, \delta_{\beta}\, \delta_{\gamma} = s_{\alpha 11} + s_{\alpha 22} + s_{\alpha 33} + 2(s_{\alpha 12} + s_{\alpha 23} + s_{\alpha 13}), \tag{11.72}$$

$$\begin{aligned} h_{\alpha} &= \sum_{\beta=1}^{3} s_{\alpha\beta}\, s_{\beta} + \tfrac{1}{2} \sum_{\beta=4}^{6} s_{\alpha\beta}\, s_{\beta} \\ &= s_{\alpha 1}\, s_1 + s_{\alpha 2}\, s_2 + s_{\alpha 3}\, s_3 + \tfrac{1}{2}(s_{\alpha 4}\, s_4 + s_{\alpha 5}\, s_5 + s_{\alpha 6}\, s_6). \end{aligned} \tag{11.73}$$

For the rest of this section we shall understand the compressibility χ, the compliance coefficients $s_{\alpha\beta}$ and $s_{\alpha\beta\gamma}$, and the abbreviated quantities s_α, g_α, h_α to be evaluated at $p=0$. With this notation,

$$\tau'_\alpha(0) = -\delta_\alpha, \tag{11.74}$$

$$C'_\alpha(0) = 2V'_\alpha(0) = -2s_{\alpha\beta}\,\delta_\beta = -2s_\alpha, \tag{11.75}$$

$$\tau''_\alpha(0) = -2s_\alpha + (2\chi - 2s_\alpha)\,\delta_\alpha, \quad \textit{not} \text{ summed over } \alpha. \tag{11.76}$$

To obtain (11.76) from (11.68), it must be remembered that there are twos in (9.2) for $\alpha=4,5,6$ but that the single-subscript thermodynamic tension equal the double subscript values for all α. From (11.69), (11.74), (11.76), and the abbreviations (11.71)–(11.73),

$$C''_\alpha(0) = 2V''_\alpha(0) = 2g_\alpha + 4\chi\, s_\alpha - 8h_\alpha. \tag{11.77}$$

From (11.66), (11,75), and (11.77),

$$\begin{aligned}\chi(0) &= s_{\alpha\beta}\,\delta_\alpha\,\delta_\beta = s_\beta\,\delta_\beta = s_1 + s_2 + s_3\\ &= s_{11} + s_{22} + s_{33} + 2(s_{12} + s_{13} + s_{23}).\end{aligned} \tag{11.78}$$

The pressure derivative of the compressibility is

$$\chi' = -(J'/J)' = -J''/J + (J'/J)^2, \tag{11.79}$$

whence, from (11.66), (11.77), (11.63)$_3$, and (11.75),

$$\chi'(0) = -g_\alpha\,\delta_\alpha - 2\chi^2 + 6h_\alpha\,\delta_\alpha. \tag{11.80}$$

In writing (11.80), use has been made of the fact that the trace of A_{ij} equals $h_\alpha\,\delta_\alpha$:

$$A_{jj} = h_\alpha\,\delta_\alpha = s_1^2 + s_2^2 + s_3^2 + \tfrac{1}{2}(s_4^2 + s_5^2 + s_6^2). \tag{11.81}$$

The other type of sum in (11.80), $g_\alpha\,\delta_\alpha$, may be written out as

$$\begin{aligned}g_\alpha\,\delta_\alpha &= s_{111} + s_{222} + s_{333} + 6\,s_{123}\\ &\quad + 3(s_{112} + s_{113} + s_{122} + s_{133} + s_{223} + s_{233}).\end{aligned} \tag{11.82}$$

The pressure derivative of the bulk modulus can be expressed as

$$B' = -\chi'/\chi^2. \tag{11.83}$$

Of interest also are the pressure derivatives of the stretch $\lambda_{\boldsymbol{N}}$, determined from (3.16), and the pressure derivatives of the thickness shrink $f_{\boldsymbol{N}}$ and its reciprocal the thickness swell $(1/f_{\boldsymbol{N}})$, $f_{\boldsymbol{N}}$ being determined from (3.10). By differentiation of (3.10) and evaluation of the derivatives at $p=0$, making use of (11.63), we find

$$\begin{aligned}f'_{\boldsymbol{N}}(0) &= -\tfrac{1}{2}N_i\,N_k\,C'_{ik}(0),\\ f''_{\boldsymbol{N}}(0) &= -\tfrac{1}{2}N_i\,N_k[C''_{ik}(0) - 8A_{ik}] - [f'_{\boldsymbol{N}}(0)]^2.\end{aligned} \tag{11.84}$$

To express these relations in abbreviated notation, we introduce the following single-subscript notation for $N_i\,N_j$:

$$\begin{aligned} N_1^2 &= \nu_1, & N_2\,N_3 &= N_3\,N_2 = \nu_4,\\ N_2^2 &= \nu_2, & N_3\,N_1 &= N_1\,N_3 = \nu_5,\\ N_3^2 &= \nu_3, & N_1\,N_2 &= N_2\,N_1 = \nu_6.\end{aligned} \tag{11.85}$$

An abbreviated notation for A_{ij}, defined by $(11.63)_3$, is also needed. With the help of (11.75), the six distinct components may be written out as

$$\begin{aligned}
A_1 &= A_{11} = s_1^2 + \tfrac{1}{4}(s_6^2 + s_5^2),\\
A_2 &= A_{22} = s_2^2 + \tfrac{1}{4}(s_4^2 + s_6^2),\\
A_3 &= A_{33} = s_3^2 + \tfrac{1}{4}(s_5^2 + s_4^2),\\
A_4 &= 2A_{23} = \tfrac{1}{2}s_5 s_6 + s_4(s_2 + s_3),\\
A_5 &= 2A_{31} = \tfrac{1}{2}s_6 s_4 + s_5(s_3 + s_1),\\
A_6 &= 2A_{12} = \tfrac{1}{2}s_4 s_5 + s_6(s_1 + s_2).
\end{aligned} \tag{11.86}$$

With this abbreviated notation, (11.84) becomes

$$\begin{aligned}
f'_{\boldsymbol{N}}(0) &= -\tfrac{1}{2}\nu_\alpha C'_\alpha(0),\\
f''_{\boldsymbol{N}}(0) &= -\tfrac{1}{2}\nu_\alpha[C''_\alpha(0) - 8A_\alpha] - [f'_{\boldsymbol{N}}(0)]^2.
\end{aligned} \tag{11.87}$$

Finally, use of (11.75) and (11,77) yields

$$\begin{aligned}
f'_{\boldsymbol{N}}(0) &= \nu_\alpha s_{\alpha\beta}\delta_\beta = \nu_\alpha s_\alpha,\\
f''_{\boldsymbol{N}}(0) &= -\nu_\alpha g_\alpha - 2\chi\nu_\alpha s_\alpha + 4\nu_\alpha(h_\alpha + A_\alpha) - (\nu_\alpha s_\alpha)^2.
\end{aligned} \tag{11.88}$$

Similar manipulations yield the pressure derivatives of $(1/f_{\boldsymbol{N}})$ and $\lambda_{\boldsymbol{N}}$. The results are included in Table 11.5, which summarizes the results of this section on pressure derivatives.

Table 11.5. *Expressions for pressure derivatives evaluated at zero pressure.*

Quantity	First derivative	Second derivative
V_α	$-s_\alpha$	$g_\alpha + 2\chi s_\alpha - 4h_\alpha$
J	$-\chi = -s_\beta\delta_\beta$	$(g_\alpha - 6h_\alpha)\delta_\alpha + 3\chi^2$
χ	$-(g_\alpha - 6h_\alpha)\delta_\alpha - 2\chi^2$	—
ϱ	$\chi\varrho$	$\varrho(\chi' + \chi^2)$
$1/f_{\boldsymbol{N}}$	$-j \equiv -\nu_\alpha s_\alpha$	k
$f_{\boldsymbol{N}}$	$j \equiv \nu_\alpha s_\alpha$	$-k + 2j^2$
$\lambda_{\boldsymbol{N}}$	$-j \equiv -\nu_\alpha s_\alpha$	$k - 4j^2 + 4d$

$j \equiv \nu_\alpha s_\alpha$;
$d \equiv \nu_\alpha A_\alpha$;
$k \equiv 3j^2 + 2\chi j - 4d + \nu_\alpha(g_\alpha - 4h_\alpha)$.

Third-order elastic compliances enter the formulas only through the quantities $g_\alpha \equiv s_{\alpha\beta\gamma}\delta_\beta\delta_\gamma$. From (10.4),

$$g_\alpha = s_{\alpha\beta\gamma}\delta_\beta\delta_\gamma = -s_{\alpha\lambda}s_{\beta\mu}s_{\gamma\nu}c_{\lambda\mu\nu}\delta_\beta\delta_\gamma = -s_{\alpha\lambda}s_\mu s_\nu c_{\lambda\mu\nu} = s_{\alpha\lambda}s_\mu c'_{\lambda\mu} \tag{11.89}$$

where

$$c'_{\lambda\mu} = -s_\nu c_{\lambda\mu\nu}. \tag{11.90}$$

By differentiating $c_{\lambda\mu}(p)$ by the chain rule and using (11.75), we find

$$c'_{\lambda\mu} = \frac{\partial c_{\lambda\mu}}{\partial V_\nu}V'_\nu = -s_\nu c_{\lambda\mu\nu}, \tag{11.91}$$

thereby identifying $c'_{\lambda\mu}$ as the pressure derivative of the thermodynamic elastic coefficient $c_{\lambda\mu} \equiv \partial^2 U/\partial V_\lambda\,\partial V_\mu$. In these formulas, all quantities are evaluated at zero pressure. Eq. (11.89) provides the not-unexpected result that pressure derivatives of the strains can all be calculated from second-order elastic coefficients and their pressure derivatives. It is not necessary to know the third-order elastic coefficients individually.

It must be emphasized that the derivatives $c'_{\lambda\mu}$ may differ significantly from the derivatives of the *effective* elastic coefficients commonly found in the ultrasonics literature. This matter is discussed in Sect. 31.

11.12. Derivatives with respect to temperature. Formulas of the same general form as those derived in Sect. 11.11 may be obtained for derivatives with respect to temperature at constant pressure. From the definition of the expansion coefficients α_γ in Eq. (9.7), we have, at $p=0$,

$$\begin{aligned} V'_\gamma(T) &= \alpha_\gamma(T), \\ V''_\gamma(T) &= \alpha'_\gamma(T) \end{aligned} \tag{11.92}$$

where the prime now denotes the derivative with respect to temperature at constant pressure. The restriction to $p=0$ is needed to make constant pressure coincide with constant thermodynamic tensions according to the definitions in Sect. 8–10. By changing certain definitions, the reference can be shifted to any other constant pressure $p=p_0$, as noted in Sect. 11.4. Moreover, by a different choice of thermodynamic substate variables V_γ, constant pressure can be made to coincide with constant thermodynamic tensions for *all* pressures in the case of cubic crystals and isotropic media under hydrostatic pressure [THURSTON, 1970].

To obtain the temperature derivatives of J it is necessary only to reinterpret the independent variable as temperature in (11.64) and (11.65). To obtain simple formulas, the derivatives must be evaluated at the reference configuration, just as in the case of pressure derivatives. In other words, the reference configuration is taken as the one corresponding to zero pressure and the temperature at which the derivatives are evaluated, say $T=\bar{T}$. Then the first derivative of J with respect to temperature, evaluated at the reference configuration, is given by (11.22), which may be rewritten as

$$\beta(\bar{T}) = J'(\bar{T}) = \alpha_{jj}(\bar{T}) = \delta_\gamma\,\alpha_\gamma(\bar{T}) = \alpha_1(\bar{T}) + \alpha_2(\bar{T}) + \alpha_3(\bar{T}). \tag{11.93}$$

At the reference configuration, the volume expansion coefficient, defined in (11.15), *equals the trace of the expansion coefficient tensor*. Similarly,

$$J''(\bar{T}) = \alpha'_{jj}(\bar{T}) + [\beta(\bar{T})]^2 - 2\bar{A}_{jj} \tag{11.94}$$

where $\bar{A}_{jj}$, defined analogously to $(11.63)_3$, is given by a formula like (11.81) with s_γ replaced by $\alpha_\gamma(\bar{T})$:

$$\bar{A}_{jj} = \alpha_1^2 + \alpha_2^2 + \alpha_3^2 + \tfrac{1}{2}(\alpha_3^2 + \alpha_4^2 + \alpha_5^2). \tag{11.95}$$

In (11.95) and subsequent formulas, α_γ is used as an abbreviation for $\alpha_\gamma(\bar{T})$, the expansion coefficients evaluated at the reference configuration.

The temperature derivative of $f_{\boldsymbol{N}}$ at the reference configuration is

$$f'_{\boldsymbol{N}}(\bar{T}) = -\nu_\gamma\,\alpha_\gamma. \tag{11.96}$$

The coefficient of *linear* expansion in the direction $\boldsymbol{N}$ is

$$(1/f_{\boldsymbol{N}})' = N_i\,N_j\,\alpha_{ij} = \nu_\gamma\,\alpha_\gamma \tag{11.97}$$

when evaluated at the reference configuration. In practice, the tensor coefficients α_{ij} are of course determined from experimental measurements of the length changes in a sufficient number of directions.

11.13. Thermal expansion at constant pressure. In the formalism of Sects. 9 and 10, it was convenient to use thermal expansion coefficients at constant thermodynamic tensions, defined by (9.7). One may also define a thermal expansion tensor at constant *pressure* by the formula

$$\Pi_{ij} = \left(\frac{\partial V_{ij}}{\partial T}\right)_p, \qquad \Pi_\alpha = \left(\frac{\partial V_\alpha}{\partial T}\right)_p, \tag{11.98}$$

the constraint (11.1) being understood. With a common reference for strain, Π_{ij} and α_{ij} at $p=0$ are the same functions of temperature, but they are in principle different functions of pressure, because constant pressure does not correspond to constant thermodynamic tensions, in general. The situation is analogous to that which exists with respect to the specific heats C_p and C_τ.

By considering $V_{ij} = V_{ij}[\tau_{km}(p, T), T]$, we obtain

$$\left(\frac{\partial V_{ij}}{\partial T}\right)_p = \left(\frac{\partial V_{ij}}{\partial T}\right)_\tau + \left(\frac{\partial V_{ij}}{\partial \tau_{km}}\right)_T \left(\frac{\partial \tau_{km}}{\partial T}\right)_p. \tag{11.99}$$

From (11.2),

$$\left(\frac{\partial \tau_{km}}{\partial T}\right)_p = -p\left[\frac{\partial}{\partial T}(J C^{-1}_{km})\right]_p. \tag{11.100}$$

Hence, (11.99) can be written as

$$\Pi_{ij}(p, T) = \alpha_{ij}(p, T) - p\, s^T_{ijkm}(p, T)\left[\frac{\partial}{\partial T}(J C^{-1}_{km})\right]_p. \tag{11.101}$$

Straightforward operations on $C^{-1}_{ks} C_{st} = \delta_{kt}$ yield

$$\frac{\partial C^{-1}_{km}}{\partial T} = -C^{-1}_{ks} \frac{\partial C_{st}}{\partial T} C^{-1}_{tm}. \tag{11.102}$$

By invoking (11.102) and the definitions (3.2), (11.15), and (11.98), (11.101) becomes

$$\Pi_{ij} = \alpha_{ij} - p\, s^T_{ijkm}\, J(\beta\, C^{-1}_{km} - 2 C^{-1}_{ks}\, \Pi_{st}\, C^{-1}_{tm}). \tag{11.103}$$

With the reference configuration taken as that corresponding to zero pressure and the temperature under consideration, say $T = \overline{T}$, the zero-pressure values of the pressure derivatives of Π_{ij} and α_{ij} are related by

$$\Pi'_{ij}(0) = \alpha'_{ij}(0) - s^T_{ijkm}(0)\,[\delta_{km}\,\beta(0) - 2\Pi_{km}(0)], \tag{11.104}$$

in which, of course, $\Pi_{km}(0) = \alpha_{km}(0)$.

Now clearly

$$\left(\frac{\partial \alpha_{ij}}{\partial p}\right)_T = \frac{\partial^2 V_{ij}}{\partial \tau_{km}\,\partial T} \left(\frac{\partial \tau_{km}}{\partial p}\right)_T$$

whence

$$\alpha'_{ij}(0) = -\left(\frac{\partial s^T_{ijkk}}{\partial T}\right)_{\substack{T=\overline{T}\\ p=0}}. \tag{11.105}$$

In practice, the thermal expansivity is ordinarily measured only at essentially zero pressure (1 atmosphere or lower in a cryostat). Its dependence on stress, if needed, is inferred from the temperature dependence of the elastic properties through (10.26), (11.51), and (11.105).

In considering derivatives with respect to temperature at other than the zero-pressure reference configuration, the distinction between derivatives at constant pressure and at constant thermodynamic tensions, observed in (11.99), must be noted. However, the reference pressure can be shifted by changing the definitions as indicated in Sect. 11.4, and for cubic crystals and isotropic media under hydrostatic pressure, the coincidence of constant pressure and constant thermodynamic tensions can be preserved at all pressures by the choice of thermodynamic substate variables proposed by THURSTON (1970).

D. Electromechanical interactions.

12. Basic equations of electromagnetic theory in a material representation.

This section provides background for the extension of thermostatics to include electromechanical interactions. The nonlinear theories of the interaction of a material with an electromagnetic field typically introduce material representations of some or all of the fields in order to express rotational invariance of the stored energy conveniently and to refer the stored energy to unit reference volume. For example, $\mathcal{E}\cdot d\mathcal{D}$ is an energy increment per unit volume, and it proves convenient to define new vectors $\boldsymbol{E}$ and $\boldsymbol{D}$ such that $\boldsymbol{E}\cdot d\boldsymbol{D}$ is an energy increment per unit reference volume. It is desirable also for the material representations to have the property of remaining constant during a rigid rotation of the fields with a body, for then thermodynamic functions expressed in terms of such material representations will have built-in rotational invariance. This section will set forth the basic equations of electromagnetic theory, first in the familiar spatial representation, and then in a material representation. The purpose of the material representation is to facilitate the definition of properly invariant thermodynamic functions.

12.1. Maxwell's field equations. We shall consistently use script capital letters for the familiar fields that occur when MAXWELL's equations are written in the conventional spatial representation. These fields are so defined as to give MAXWELL's field equations the form

$$\begin{aligned} \operatorname{curl}\mathcal{E} &= -\partial\mathcal{B}/\partial t, \\ \operatorname{curl}\mathcal{H} &= \partial\mathcal{D}/\partial t + \mathcal{J}, \end{aligned} \tag{12.1}$$

$$\begin{aligned} \operatorname{div}\mathcal{B} &= 0, \\ \operatorname{div}\mathcal{D} &= \varrho_e, \end{aligned} \tag{12.2}$$

with

$$\begin{aligned} \mathcal{D} &= \varepsilon_0\,\mathcal{E} + \mathcal{P}, \\ \mathcal{H} &= \mathcal{B}/\mu_0 - \mathcal{M}. \end{aligned} \tag{12.3}$$

Here $\mathcal{P}$ and $\mathcal{M}$ are the electric and magnetic moment per unit volume, ϱ_e and $\mathcal{J}$ the densities of electric charge and current, and $\mathcal{E}$, $\mathcal{D}$, $\mathcal{B}$, and $\mathcal{H}$ are the usual vectors called electric field, electric displacement, magnetic flux density, and magnetic intensity. The divergence and curl operations are in spatial coordinates i.e., the independent variables are $(t, \boldsymbol{x})$. The above fields are all taken from the point of view of a fixed observer, with respect to which some particle of the material may have the velocity $\boldsymbol{v}$.

12.2. Units. In the rationalized *mks* system, μ_0 and ε_0 are given by

$$\begin{aligned} \mu_0 &= 4\pi \cdot 10^{-7} \text{ henry/meter}, \\ \varepsilon_0 &= (1/\mu_0 c^2) \text{ farad/meter} \doteq 8.854 \cdot 10^{-12} \text{ farad/meter}, \\ c &= \text{speed of light} \doteq 3 \cdot 10^8 \text{ m/sec}, \end{aligned}$$

while the field quantities have the following units:

$$\begin{aligned} \mathscr{E} &- \text{volt/meter} \\ \mathscr{P} \text{ and } \mathscr{D} &- \text{coulomb/(meter)}^2 \\ \mathscr{B} &- \text{webers/(meter)}^2 \\ \mathscr{M} \text{ and } \mathscr{H} &- \text{(ampere-turns)/meter}. \end{aligned}$$

All these units can be expressed in terms of the meter, kilogram, second, and coulomb by examining the equations of elementary electrical theory. For example

$$\begin{aligned} \text{ampere} &= \text{coul/sec}, \\ \text{volt} &= \frac{\text{watt}}{\text{ampere}} = \frac{\text{nt m}}{\text{coul}} = \frac{\text{kg m}^2}{\text{coul sec}^2}, \\ \text{henry} &= \frac{\text{volt}}{\text{amp/sec}} = \frac{\text{volt sec}^2}{\text{coul}} = \frac{\text{kg m}^2}{\text{coul}^2}, \\ \text{farad} &= \frac{\text{coul}}{\text{volt}} = \frac{\text{coul}^2 \text{ sec}^2}{\text{kg m}^2}, \\ \text{weber} &= \text{henry amp} = \text{volt sec} = \frac{\text{kg m}^2}{\text{coul sec}}. \end{aligned}$$

It is readily verified that $(\mu_0 \varepsilon_0)^{-1}$ is the square of a velocity:

$$\frac{(\text{meter})^2}{\text{henry} \cdot \text{farad}} = \frac{\text{m}^2 \text{ coul}^2 \text{ kg m}^2}{\text{kg m}^2 \text{ coul}^2 \text{ sec}^2} = \frac{\text{m}^2}{\text{sec}^2}.$$

In this system, COULOMB's law for the force between point charges of magnitudes q_1 and q_2 in a vacuum takes the form

$$F = \frac{q_1 q_2}{4\pi \varepsilon_0 r^2} \text{ newtons}$$

and hence the field of a point charge of q coulombs is

$$\mathscr{E} = \frac{q}{4\pi \varepsilon_0 r^2} \text{ volt/meter}.$$

The system is called rationalized because the factor 4π appears in COULOMB's law rather than MAXWELL's equations.

12.3. Effect of particle velocity. The fields at a moving particle are different, depending on whether they are taken from the point of view of an observer that moves with the particle or, as in (12.1), from the point of view of a fixed observer. In a wave situation, the ratio of the difference to the field itself is of the order of v/c where v is the particle velocity and c the speed of light. Because of the smallness of this ratio, the distinction is ordinarily of no practical importance. Nevertheless, a brief discussion is included here for completeness.

The fields taken from the point of view of an observer moving with the material will be denoted by adding primes. Although we do not intend to develop a Lorentz-invariant theory, there is no harm in indicating, for ready reference, the standard relations between the primed and unprimed fields that are taught in the special theory of relativity [e.g., J. A. STRATTON, Electromagnetic Theory, pp. 78–82. New York-London: McGraw-Hill 1941]

$$\begin{aligned}
&\mathscr{E}'_{\parallel}=\mathscr{E}_{\parallel}, && \mathscr{E}'_{\perp}=\gamma(\mathscr{E}+\boldsymbol{v}\times\mathscr{B})_{\perp}\\
&\mathscr{D}'_{\parallel}=\mathscr{D}_{\parallel}, && \mathscr{D}'_{\perp}=\gamma[\mathscr{D}+(\boldsymbol{v}\times\mathscr{H})/c^2]_{\perp}\\
&\mathscr{B}'_{\parallel}=\mathscr{B}_{\parallel}, && \mathscr{B}'_{\perp}=\gamma[\mathscr{B}-(\boldsymbol{v}\times\mathscr{E})/c^2]_{\perp}\\
&\mathscr{H}'_{\parallel}=\mathscr{H}_{\parallel}, && \mathscr{H}'_{\perp}=\gamma(\mathscr{H}-\boldsymbol{v}\times\mathscr{D})_{\perp}\\
&\mathscr{P}'_{\parallel}=\mathscr{P}_{\parallel}, && \mathscr{P}'_{\perp}=\gamma[\mathscr{P}_{\perp}-(\boldsymbol{v}\times\mathscr{M})/c^2]_{\perp}\\
&\mathscr{M}'_{\parallel}=\mathscr{M}_{\parallel}, && \mathscr{M}'_{\perp}=\gamma(\mathscr{M}+\boldsymbol{v}\times\mathscr{P})_{\perp}
\end{aligned}\tag{12.4}$$

$$\varrho'_e=\gamma(\varrho_e-\boldsymbol{v}\cdot\mathscr{J}/c^2),\quad \mathscr{J}'_{\parallel}=\gamma(\mathscr{J}-\boldsymbol{v}\,\varrho_e)_{\parallel},\quad \mathscr{J}'_{\perp}=\mathscr{J}_{\perp},$$

where

$$\gamma=(1-v^2/c^2)^{-\frac{1}{2}}\tag{12.5}$$

and the subscripts $\parallel$ and $\perp$ refer to the components parallel and perpendicular to $\boldsymbol{v}$.

The fact that the field values at a moving particle are different depending on whether the point of view is that of the local rest system or the laboratory introduces a complication that is needed only to first order in v/c, if at all.

To obtain orders of magnitude, consider a plane electromagnetic wave in a vacuum, for which the ratio of magnitude of $\mathscr{E}$ to $\mathscr{H}$ is readily shown to be

$$\mathscr{E}/\mathscr{H}=\eta_0=(\mu_0/\varepsilon_0)^{\frac{1}{2}}=\mu_0\,c\doteq 377\text{ ohms}.$$

Then $\mathscr{H}$, $\mathscr{B}$, and $\mathscr{D}$ have the following orders of magnitude

$$\begin{aligned}
&\mathscr{H}\sim\mathscr{E}/\mu_0\,c=c\,\mathscr{D},\\
&\mathscr{B}=\mu_0\,\mathscr{H}\sim\mathscr{E}/c,\\
&\mathscr{D}\sim\mathscr{H}/c.
\end{aligned}$$

It follows that the relative corrections to $\mathscr{E}$, $\mathscr{D}$, $\mathscr{B}$, and $\mathscr{H}$ given by (12.4) are all of order v/c. Of course, if the fields are applied statically instead of being associated with each other in a wave, these smallness relations do not hold. For example, $\boldsymbol{v}\times\mathscr{B}$ could be tremendous, even though $\mathscr{E}=0$. The relative importance of the corrections to the polarization, magnetization, and source densities depend strongly on the properties of the medium. However, if it is assumed that $\mathscr{M}$ and $\mathscr{P}$ are of the same order as $\mathscr{H}$ and $\mathscr{D}$ respectively, then $\mathscr{M}$ is of the order of $c\,\mathscr{P}$, and the relative corrections to $\mathscr{P}$ and $\mathscr{M}$ can also be of order v/c.

All corrections of order v/c can be accounted for by setting $\gamma=1$. To first order in v/c, the transformations of $\mathscr{P}$ and $\mathscr{M}$ can be written

$$\begin{aligned}
&\mathscr{P}\doteq\mathscr{P}'+(\boldsymbol{v}\times\mathscr{M}')/c^2,\\
&\mathscr{M}\doteq\mathscr{M}'-\boldsymbol{v}\times\mathscr{P}'.
\end{aligned}\tag{12.6}$$

Substitution of (7) into (3) gives

$$\begin{aligned}
&\mathscr{D}=\varepsilon_0\,\mathscr{E}+\mathscr{P}'+(\boldsymbol{v}\times\mathscr{M}')/c^2,\\
&\mathscr{H}=\mathscr{B}/\mu_0-(\mathscr{M}'-\boldsymbol{v}\times\mathscr{P}').
\end{aligned}\tag{12.7}$$

Now in a mechanical wave, the ratio of v to the sound speed is of the same order as the dynamic strain, which in a solid rarely exceeds 10^{-3}, while the sound speed is typically of the order of $10^{-5}\,c$. Hence, when the particle motion is associated with a mechanical wave, the ratio v/c will ordinarily be 10^{-7} or less. Unless relatively large static fields are involved, the distinction between the primed and unprimed fields is of no practical importance.

12.4. Material representations of fields and their properties. We shall consistently use script capitals for the familiar fields that occur in (12.1)–(12.3), and ordinary capitals for their material representations. Let us call $\mathcal{E}$, $\mathcal{H}$, and $\mathcal{M}$ force-like, and $\mathcal{B}$, $\mathcal{D}$, $\mathcal{P}$, and $\mathcal{J}$ flux-like. With the scalar J denoting the Jacobian of the deformation gradient matrix, let the force-like fields be connected to their material representations by

$$\begin{aligned} E_j &= \mathcal{E}_i\, \partial x_i/\partial X_j, \\ \mathcal{E}_i &= E_j\, \partial X_j/\partial x_i, \end{aligned} \tag{12.8}$$

with similar equations connecting $\boldsymbol{H}$ with $\mathcal{H}$ and $\boldsymbol{M}$ with $\mathcal{M}$. Let the flux-like fields be connected to their material representations by

$$\begin{aligned} B_j &= J\mathcal{B}_i\, \partial X_j/\partial x_i, \\ \mathcal{B}_i &= B_j(\partial x_i/\partial X_j)/J, \end{aligned} \tag{12.9}$$

with similar equations connecting $\boldsymbol{D}$ with $\mathcal{D}$, $\mathcal{P}$ with $\boldsymbol{P}$, and $\boldsymbol{J}$ with $\mathcal{J}$. [We trust that no confusion will result from the use of $\boldsymbol{H}$ to denote the material representation of both $\mathcal{H}$ and the "nonmechanical energy flux vector" h, as in (5.2).]

Logically, it is the transformation formulas (12.8) and (12.9) that should be distinguished as force-like and flux-like, not the fields themselves. As discussed more fully below, the force-like formula makes $\boldsymbol{E}\cdot d\boldsymbol{X}=\mathcal{E}\cdot d\boldsymbol{x}$, and the flux-like formula makes $\boldsymbol{B}\cdot\boldsymbol{N}\,d\bar{A}=\mathcal{B}\cdot\boldsymbol{n}\,dA$. The classification of the fields serves only as a help in remembering that we shall always use (12.8) for $\mathcal{E}$, $\mathcal{H}$, $\mathcal{M}$, and (12.9) for $\mathcal{B}$, $\mathcal{D}$, $\mathcal{P}$, $\mathcal{J}$.

Other material representations are possible. These are not unique. The relation connecting the flux-like variables is motivated by (5.2). That is, it makes $\boldsymbol{B}\cdot\boldsymbol{N}\,d\bar{A}=\mathcal{B}\cdot\boldsymbol{n}\,dA$ where $\boldsymbol{n}\,dA$ is a vector material element of area in the present configuration, and $\boldsymbol{N}\,d\bar{A}$ is the same element in the reference configuration. Related to this is the fact that for any flux-like variable subject to (12.9), say $\mathcal{D}$,

$$\operatorname{div}\mathcal{D}=\frac{\partial\mathcal{D}_i}{\partial x_i}=\frac{\partial}{\partial x_i}\left(\frac{1}{J}\frac{\partial x_i}{\partial X_j}D_j\right)=\frac{1}{J}\frac{\partial D_j}{\partial X_j} \tag{12.10}$$

since

$$\frac{\partial}{\partial x_i}\left(\frac{1}{J}\frac{\partial x_i}{\partial X_j}\right)\equiv 0$$

by the identity of EULER, PIOLA, and JACOBI (3.33).

By comparison of (12.9) with (3.14), it can be seen that the directions of flux-like variables and their material representations are connected by the rotation of material line elements, while the magnitudes satisfy relations like

$$|\mathcal{P}|=|\boldsymbol{P}|\,\lambda_{\boldsymbol{N}}/J. \tag{12.11}$$

Eq. (12.8) makes $\mathcal{E}\cdot d\boldsymbol{x}=\boldsymbol{E}\cdot d\boldsymbol{X}$ where $d\boldsymbol{x}$ and $d\boldsymbol{X}$ represent the same material line element in the present and reference configurations, and similarly for the other force-like fields. As a corollary to this property, it can be said that if there is a scalar $\varphi(\boldsymbol{x})$ such that $\mathcal{E}=-\operatorname{grad}\varphi$, and if $\varphi(\boldsymbol{x})=\Phi(\boldsymbol{X})$, then (12.8) makes $E_j=-\partial\Phi/\partial X_j$. When the force-like field is irrotational, we can picture its equipotential surfaces, $\varphi(\boldsymbol{x})=$constant, as material surfaces. The corresponding equipotential surfaces $\Phi(\boldsymbol{X})=$constant are these same material surfaces in the reference configuration. In a homogeneous deformation, the normal to a surface rotates in accordance with (3.9), and the perpendicular distance between a pair of surfaces changes in accordance with (3.15). These geometric relationships are reflected in (12.8) by putting $E_j = \partial\Phi/\partial X_j$ and $\mathscr{E}_i = \partial\varphi/\partial x_i$. The directions of $\mathscr{E}$ and $\boldsymbol{E}$ are connected by the rotation of material surface elements. Their magnitudes are connected by the thickness shrink $f_{\boldsymbol{N}}$ in the direction of $\boldsymbol{E}$.

Closely related to the property $\mathscr{E}\cdot d\boldsymbol{x} = \boldsymbol{E}\cdot d\boldsymbol{X}$ is the fact that curl $\mathscr{E}$ is related to the corresponding material operation $\varepsilon_{jkm}\,\partial E_m/\partial X_k$ by the relation (12.9) between flux-like variables and their material representations. In this sense, the curl of a force-like field is flux-like. This is demonstrated as follows:

$$\varepsilon_{jkm}\frac{\partial E_m}{\partial X_k} = \varepsilon_{jkm}\frac{\partial}{\partial X_k}\left(\mathscr{E}_s\frac{\partial x_s}{\partial X_m}\right) = \varepsilon_{jkm}\frac{\partial x_s}{\partial X_m}\frac{\partial x_n}{\partial X_k}\frac{\partial \mathscr{E}_s}{\partial x_n} \tag{12.12}$$

since the sums $\varepsilon_{jkm}(\partial^2 x_s/\partial X_k\,\partial X_m)$ vanish. By employing the identity (3.53), (12.12) becomes

$$\varepsilon_{jkm}\frac{\partial E_m}{\partial X_k} = J\,\varepsilon_{ins}\frac{\partial X_j}{\partial x_i}\frac{\partial \mathscr{E}_s}{\partial x_n} = J\frac{\partial X_j}{\partial x_i}(\text{curl}\,\mathscr{E})_i \tag{12.13}$$

in agreement with (12.9), as was asserted.

Another important consequence of the definitions (12.8) and (12.9) is that the material representations remain constant in a rigid rotation of the fields with a body. Hence, a thermodynamic function expressed in terms of the material representations has built-in rotational invariance. The constancy of the material representations in a rigid rotation of the fields and body follows from statements already made. For suppose $\mathscr{E}$ lies normal to some material surface element; then $\boldsymbol{E}$ lies normal to the reference direction of the same material element, which is a constant direction. Moreover, $|\boldsymbol{E}| = |\mathscr{E}|$ in a rigid rotation, since $f_{\boldsymbol{N}} = 1$. Hence $\dot{\boldsymbol{E}} = 0$ in this case, and similarly for any force-like field. Analogous arguments apply to any flux-like field, such as $\mathscr{B}$. In a rigid rotation of $\mathscr{B}$ with a body, the direction of $\boldsymbol{B}$ is constant along the reference direction of the material line element that has the direction of $\mathscr{B}$, while $|B| = |\mathscr{B}| =$ constant because of (12.11).

To evaluate the time derivative of the material representation of a flux-like variable, we need a formula for the time derivative of $(\partial X_j/\partial x_i)$. We begin by differentiating

$$\frac{\partial X_j}{\partial x_i}\frac{\partial x_i}{\partial X_k} = \delta_{jk}$$

to obtain

$$\frac{\partial X_j}{\partial x_i}\frac{\partial v_i}{\partial X_k} + \frac{\partial x_i}{\partial X_k}\frac{d}{dt}\left(\frac{\partial X_j}{\partial x_i}\right) = 0.$$

Upon multiplying by $\partial X_k/\partial x_m$, we obtain the formula

$$\frac{d}{dt}\left(\frac{\partial X_j}{\partial x_m}\right) = -\frac{\partial X_j}{\partial x_i}\frac{\partial v_i}{\partial x_m}. \tag{12.14}$$

Recalling $\dot{J} = J\,\text{div}\,\boldsymbol{v}$ from (3.36), we now differentiate (12.9) to obtain the general formula

$$\dot{B}_j = J\frac{\partial X_j}{\partial x_i}\left(\mathscr{B}_i\,\text{div}\,\boldsymbol{v} + \dot{\mathscr{B}}_i - \mathscr{B}_m\frac{\partial v_i}{\partial x_m}\right) = J\frac{\partial X_j}{\partial x_i}\overset{*}{\mathscr{B}}_i, \tag{12.15}$$

where

$$\begin{aligned}\overset{*}{\mathscr{B}} &= \dot{\mathscr{B}} + \mathscr{B}\,\text{div}\,\boldsymbol{v} - (\mathscr{B}\cdot\nabla)\,\boldsymbol{v}\\ &= \partial\mathscr{B}/\partial t + \text{curl}\,(\mathscr{B}\times\boldsymbol{v}) + \boldsymbol{v}\,\text{div}\,\mathscr{B}.\end{aligned} \tag{12.16}$$

$\overset{*}{\mathscr{B}}$ is called the convected time derivative (or convected time flux) of $\mathscr{B}$ [Truesdell and Toupin, 1960, CFT, p. 448; Toupin, 1963, p. 106].

The definition is motivated by the fact that it makes (12.15) fit into the pattern of (12.9), and therefore results in (12.41) below. The two forms of (12.16) are equivalent in view of the vector analog of (3.29), namely,

$$\dot{\mathscr{B}} = \frac{\partial \mathscr{B}}{\partial t} + (\boldsymbol{v} \cdot \nabla)\, \mathscr{B}, \tag{12.17}$$

and the identity

$$\operatorname{curl} (\mathscr{B} \times \boldsymbol{v}) = (\boldsymbol{v} \cdot \nabla)\, \mathscr{B} - (\mathscr{B} \cdot \nabla)\, \boldsymbol{v} + \mathscr{B} \operatorname{div} \boldsymbol{v} - \boldsymbol{v} \operatorname{div} \mathscr{B}. \tag{12.18}$$

Clearly, the same relation (12.15) holds between the convected time derivatives of all the flux-like fields and the time derivatives of their material representations:

$$\dot{B}_j = J \frac{\partial X_j}{\partial x_i} \overset{*}{\mathscr{B}}_i, \quad \overset{*}{\mathscr{B}}_i = \frac{1}{J} \frac{\partial x_i}{\partial X_j} \dot{B}_j, \tag{12.19}$$

and similarly for $\mathscr{D}$, $\mathscr{P}$, and $\mathscr{J}$.

To show formally that $\dot{\boldsymbol{B}} = 0$ in a rigid rotation of $\mathscr{B}$ with the body, it suffices to show that $\overset{*}{\mathscr{B}} = 0$. In a rigid rotation of the body and $\mathscr{B}$ with angular velocity $\boldsymbol{\omega}(t)$, we have $\boldsymbol{v} = \boldsymbol{\omega} \times \boldsymbol{x}$ and $\dot{\mathscr{B}} = \boldsymbol{\omega} \times \mathscr{B}$, whence

$$v_i = \varepsilon_{ikm}\, \omega_k\, x_m, \qquad \dot{\mathscr{B}}_i = \varepsilon_{ikm}\, \omega_k\, \mathscr{B}_m.$$

Then $\operatorname{div} \boldsymbol{v} = 0$ and

$$\mathscr{B}_m \frac{\partial v_i}{\partial x_m} = \mathscr{B}_m\, \varepsilon_{ikm}\, \omega_k = (\boldsymbol{\omega} \times \mathscr{B})_i.$$

Hence $\dot{\mathscr{B}} \equiv (\mathscr{B} \cdot \nabla)\, \boldsymbol{v}$ and $\overset{*}{\mathscr{B}} = 0$. The same proof applies to $\boldsymbol{D}$, $\boldsymbol{P}$, and $\boldsymbol{J}$.

For a force-like field, in general,

$$\dot{E}_j = \dot{\mathscr{E}}_i \frac{\partial x_i}{\partial X_j} + \mathscr{E}_i \frac{\partial v_i}{\partial x_m} \frac{\partial x_m}{\partial X_j}. \tag{12.20}$$

In a rigid rotation of $\mathscr{E}$ and the body with angular velocity $\boldsymbol{\omega}$, this reduces to

$$\dot{E}_j = \varepsilon_{ikm} \left(\omega_k\, \mathscr{E}_m \frac{\partial x_i}{\partial X_j} + \omega_k\, \mathscr{E}_i \frac{\partial x_m}{\partial X_j} \right) = 0. \tag{12.21}$$

The sum is zero in (12.21) because interchanging i and m leaves the factor in parentheses unchanged while reversing the sign of ε_{ikm}.

12.5. Material representation of the field equations. By expressing $\partial \mathscr{B}/\partial t$ in terms of $\overset{*}{\mathscr{B}}$ through (12.16), we find that (12.1), becomes

$$\operatorname{curl} \mathscr{E}' = -\overset{*}{\mathscr{B}}, \tag{12.22}$$

$$\mathscr{E}' = \mathscr{E} + \boldsymbol{v} \times \mathscr{B}, \tag{12.23}$$

since $\operatorname{div} \mathscr{B} = 0$. To terms of order v/c, $\mathscr{E}'$ is the field from the point of view of an observer moving with the material, as set forth in Sect. 12.3. Similarly, by using (12.16) to express $\partial \mathscr{D}/\partial t$ in terms of $\overset{*}{\mathscr{D}}$, we find that $(12.1)_2$ gives

$$\operatorname{curl} \mathscr{H}' = \overset{*}{\mathscr{D}} + \mathscr{J}', \tag{12.24}$$

$$\mathscr{H}' = \mathscr{H} - \boldsymbol{v} \times \mathscr{D}, \tag{12.25}$$

$$\mathscr{J}' = \mathscr{J} - \boldsymbol{v} \operatorname{div} \mathscr{D}. \tag{12.26}$$

From (12.4), we see that the primed fields $\mathcal{H}'$ and $\mathcal{J}'$ are, to terms of order v/c, the fields from the point of view of an observer moving with the material. By defining $\boldsymbol{E}'$, $\boldsymbol{H}'$, and $\boldsymbol{J}'$ as the material representations of $\mathcal{E}'$, $\mathcal{H}'$, and $\mathcal{J}'$, and defining also a vector $\boldsymbol{V}$ as in (5.5), such that

$$V_j = v_i\, \partial X_j / \partial x_i, \tag{12.27}$$

(12.22) and (12.24) become

$$\varepsilon_{jkm} \frac{\partial E'_m}{\partial X_k} = -\dot{B}_j, \tag{12.28}$$

$$\varepsilon_{jkm} \frac{\partial H'_m}{\partial X_k} = \dot{D}_j + J'_j, \tag{12.29}$$

where

$$\begin{aligned} \boldsymbol{E}' &= \boldsymbol{E} + \boldsymbol{V} \times \boldsymbol{B}, \\ \boldsymbol{H}' &= \boldsymbol{H} - \boldsymbol{V} \times \boldsymbol{D}, \\ \boldsymbol{J}' &= \boldsymbol{J} - \boldsymbol{V}\, \partial D_k / \partial X_k. \end{aligned} \tag{12.30}$$

From (12.10) and (12.2),

$$\partial B_k / \partial X_k = 0, \tag{12.31}$$

$$\partial D_k / \partial X_k = J\, \varrho_e. \tag{12.32}$$

Eqs. (12.30) imply that $(\boldsymbol{V} \times \boldsymbol{B})$ should be the material representation of $(\boldsymbol{v} \times \mathcal{B})$, considering the vector product as *force-like* (and similarly of course with $\mathcal{B}$ replaced by any other flux-like vector) and also that $\boldsymbol{V}\, \partial D_k / \partial X_k$ should be the material representation of the flux-like vector $\boldsymbol{v}$ div $\mathcal{D}$. The latter relation is verified in the steps

$$V_j \frac{\partial D_k}{\partial X_k} = V_j\, J\, \varrho_e = J \frac{\partial X_j}{\partial x_s} v_s \operatorname{div} \mathcal{D}, \tag{12.33}$$

and the former as follows:

$$(\boldsymbol{V} \times \boldsymbol{B})_j = \varepsilon_{jkm} V_k B_m = \varepsilon_{jkm} v_s \frac{\partial X_k}{\partial x_s} J \frac{\partial X_m}{\partial x_n} \mathcal{B}_n.$$

Calling on the identity (3.51), we continue this chain of equalities as

$$(\boldsymbol{V} \times \boldsymbol{B})_j = \varepsilon_{isn} \frac{\partial x_i}{\partial X_j} v_s \mathcal{B}_n = \frac{\partial x_i}{\partial X_j} (\boldsymbol{v} \times \mathcal{B})_i, \tag{12.34}$$

which was to be shown.

From the definitions (12.8) and (12.9), Eqs. (12.3) become

$$\begin{aligned} D_j &= \varepsilon_0\, J G_{jk} E_k + P_j, \\ B_j &= \mu_0\, J G_{jk} (H_k + M_k), \end{aligned} \tag{12.35}$$

where G_{jk} is the tensor defined in (3.11).

The material representations of (12.6) are

$$\begin{aligned} \boldsymbol{P} &\doteq \boldsymbol{P}' + (\overline{\boldsymbol{V}} \times \boldsymbol{M}')/c^2, \\ \boldsymbol{M} &\doteq \boldsymbol{M}' - \boldsymbol{V} \times \boldsymbol{P}', \end{aligned} \tag{12.36}$$

where

$$\overline{V}_j = v_i\, \partial x_i / \partial X_j. \tag{12.37}$$

The transformation of $(\boldsymbol{v} \times \mathcal{M}')$ to a material representation, as in $(12.36)_1$, motivates the introduction of $\overline{\boldsymbol{V}}$. By direct calculation, invoking (3.53),

$$
\begin{aligned}
(\boldsymbol{v}\times\mathcal{M})_i &= \varepsilon_{ijk}\, v_j \mathcal{M}_k = \varepsilon_{ijk}\, v_j\, M_s \frac{\partial X_s}{\partial x_k} = v_j\, M_s\, \varepsilon_{kij} \frac{\partial X_s}{\partial x_k} \\
&= v_j\, M_s \frac{1}{J}\, \varepsilon_{smp} \frac{\partial x_i}{\partial X_m} \frac{\partial x_j}{\partial X_p} = \frac{1}{J} \frac{\partial x_i}{\partial X_m}\, \varepsilon_{mps} \frac{\partial x_j}{\partial X_p}\, v_j\, M_s \\
&= \frac{1}{J} \frac{\partial x_i}{\partial X_m}\, \varepsilon_{mps}\, \overline{V}_p\, M_s = \frac{1}{J} \frac{\partial x_i}{\partial X_m} (\overline{\boldsymbol{V}}\times \boldsymbol{M})_m .
\end{aligned}
\tag{12.38}
$$

Thus, $(\overline{\boldsymbol{V}}\times\boldsymbol{M})$ is the flux-like material representation of $(\boldsymbol{v}\times\mathcal{M})$.

12.6. Integral forms of the equations. The transformation between the differential and integral forms of the field equations are based on the transformations of GREEN and KELVIN, more commonly called the divergence theorem and STOKES's theorem [J. L. ERICKSEN, Tensor Fields, Appendix to The Classical Field Theories, in Handbuch der Physik, Vol. III/1, pp. 794–858, pp. 815–817. Berlin-Göttingen-Heidelberg: Springer 1960]. Integration of the tangential components of $\mathcal{E}$ and $\mathcal{H}$ around closed circuits gives, in view of (12.1),

$$
\begin{aligned}
\oint \mathcal{E}\cdot d\boldsymbol{x} &= \iint (\operatorname{curl}\mathcal{E})\cdot \boldsymbol{n}\, dA = -\iint \frac{\partial \mathcal{B}}{\partial t}\cdot \boldsymbol{n}\, dA ,\\
\oint \mathcal{H}\cdot d\boldsymbol{x} &= \iint (\operatorname{curl}\mathcal{H})\cdot \boldsymbol{n}\, dA = \iint \left(\frac{\partial \mathcal{D}}{\partial t} + \mathcal{J}\right)\cdot \boldsymbol{n}\, dA .
\end{aligned}
\tag{12.39}
$$

Integration of (12.2) over a volume gives

$$
\begin{aligned}
\iiint \operatorname{div}\mathcal{B} &= \oiint \mathcal{B}\cdot \boldsymbol{n}\, dA = 0 ,\\
\iiint \operatorname{div}\mathcal{D} &= \oiint \mathcal{D}\cdot \boldsymbol{n}\, dA = \iiint \varrho_e\, dV .
\end{aligned}
\tag{12.40}
$$

Let us consider the rate of change of flux of a vector field, say $\mathcal{D}$, through a material surface bounded by a material circuit. This quantity is given by

$$
\frac{d}{dt}\iint \mathcal{D}\cdot \boldsymbol{n}\, dA
$$

where the integral is taken over the moving material surface. From (12.9), (3.9), and (3.19), $\mathcal{D}\cdot\boldsymbol{n}\, dA = \boldsymbol{D}\cdot\boldsymbol{N}\, d\bar{A}$, and similarly for $\overset{*}{\mathcal{D}}$ and $\dot{\boldsymbol{D}}$. Hence

$$
\frac{d}{dt}\iint \mathcal{D}\cdot \boldsymbol{n}\, dA = \frac{d}{dt}\iint \boldsymbol{D}\cdot\boldsymbol{N}\, d\bar{A} = \iint \dot{\boldsymbol{D}}\cdot\boldsymbol{N}\, d\bar{A} = \iint \overset{*}{\mathcal{D}}\cdot \boldsymbol{n}\, dA . \tag{12.41}
$$

This relation motivates the definition of $\overset{*}{\mathcal{D}}$. When (12.16) is substituted (with $\mathcal{B}$ replaced by $\mathcal{D}$, of course), (12.41) becomes

$$
\begin{aligned}
\frac{d}{dt}\iint \mathcal{D}\cdot \boldsymbol{n}\, dA &= \iint \left[\frac{\partial \mathcal{D}}{\partial t} + \operatorname{curl}(\mathcal{D}\times\boldsymbol{v}) + \boldsymbol{v}\operatorname{div}\mathcal{D}\right]\cdot \boldsymbol{n}\, dA \\
&= \iint \left(\frac{d\mathcal{D}}{dt} + \boldsymbol{v}\operatorname{div}\mathcal{D}\right)\cdot \boldsymbol{n}\, dA + \oint (\mathcal{D}\times\boldsymbol{v})\cdot d\boldsymbol{x}
\end{aligned}
\tag{12.42}
$$

where the line integral is taken around the circuit that bounds the material surface. Since

$$
(\mathcal{D}\times\boldsymbol{v})\cdot d\boldsymbol{x} = \mathcal{D}\cdot(\boldsymbol{v}\times d\boldsymbol{x}) ,
$$

the line integral has an easily pictured interpretation. The quantity $(\boldsymbol{v}\times d\boldsymbol{x})$ is the rate at which the element $d\boldsymbol{x}$ sweeps out area, and hence the line integral is the rate of change of flux attributable to motion of the boundary. Eq. (12.42) is a strictly kinematical result for moving surfaces analogous to the transport theorem (3.48) for moving volumes.

A given material circuit can serve as a boundary for arbitrarily many different material surfaces, and the rate of change of flux in (12.42) will in general be different for the different surfaces. However, if the vector field is solenoidal ($\operatorname{div} \mathscr{D} \equiv 0$), then the same value is obtained for any surface bounded by the same circuit.

From (12.42), since $\operatorname{div} \mathscr{B}=0$, the rate of change of flux of $\mathscr{B}$ through a material circuit is

$$\frac{d}{dt}\iint \mathscr{B}\cdot \boldsymbol{n}\, dA=\iint \frac{\partial \mathscr{B}}{\partial t}\cdot \boldsymbol{n}\, dA+\oint \mathscr{B}\cdot(\boldsymbol{v}\times d\boldsymbol{x}). \tag{12.43}$$

The surface integral on the right expresses what the rate of change of flux would be if the circuit were fixed, and the line integral adds the effect of the motion ("cutting" of flux lines).

From (12.41) and (12.28),

$$\frac{d}{dt}\iint \mathscr{B}\cdot \boldsymbol{n}\, dA=\iint \dot{\boldsymbol{B}}\cdot \boldsymbol{N}\, d\bar{A}=-\oint \boldsymbol{E}'\cdot d\boldsymbol{X}. \tag{12.44}$$

The last line integral is taken over the circuit in its reference configuration. This line integral can be transformed to the present configuration. Since $\boldsymbol{E}'\cdot d\boldsymbol{X}=\mathscr{E}'\cdot d\boldsymbol{x}$, the result is simply

$$\frac{d}{dt}\iint \mathscr{B}\cdot \boldsymbol{n}\, dA=-\oint \boldsymbol{E}'\cdot d\boldsymbol{X}=-\oint (\mathscr{E}+\boldsymbol{v}\times\mathscr{B})\cdot d\boldsymbol{x}, \tag{12.45}$$

a formula which can also be obtained by substituting from $(12.39)_1$ into (12.43), and noting that $\mathscr{B}\cdot(\boldsymbol{v}\times d\boldsymbol{x})=-(\boldsymbol{v}\times\mathscr{B})\cdot d\boldsymbol{x}$.

Similarly, from (12.29), (12.16), and (12.26),

$$\begin{aligned}\oint \mathscr{H}'\cdot d\boldsymbol{x}=\oint \boldsymbol{H}'\cdot d\boldsymbol{X}&=\iint (\dot{\boldsymbol{D}}+\boldsymbol{J}')\cdot \boldsymbol{N}\, d\bar{A}=\iint (\overset{*}{\mathscr{D}}+\mathscr{J}')\cdot \boldsymbol{n}\, dA\\ &=\iint\left(\frac{\partial \mathscr{D}}{\partial t}+\mathscr{J}\right)\cdot \boldsymbol{n}\, dA-\oint(\boldsymbol{v}\times\mathscr{D})\cdot d\boldsymbol{x}.\end{aligned} \tag{12.46}$$

12.7. Electrodynamic potentials. In a spatial description, take

$$\mathscr{B}=\operatorname{curl}\mathscr{A}. \tag{12.47}$$

Then $(12.1)_1$ can be written as

$$\operatorname{curl}\left(\mathscr{E}+\frac{\partial \mathscr{A}}{\partial t}\right)=0. \tag{12.48}$$

Since $(\mathscr{E}+\partial\mathscr{A}/\partial t)$ has zero curl, it can be expressed as the spatial gradient of a potential φ. Thus

$$\mathscr{E}+\frac{\partial \mathscr{A}}{\partial t}=-\operatorname{grad}\varphi \tag{12.49}$$

whence

$$\mathscr{E}=-\operatorname{grad}\varphi-\frac{\partial \mathscr{A}}{\partial t}. \tag{12.50}$$

In a material description, take

$$\boldsymbol{B}=\operatorname{Curl}\boldsymbol{A} \quad \text{or} \quad B_i=\varepsilon_{ijk}\,\partial A_k/\partial X_j. \tag{12.51}$$

[The curl, gradient, and divergence operations in a material description (differentiation with respect to X_j) will be denoted by Curl, Grad, Div. Note the initial capital letter.] Then (12.28) can be written as

$$\operatorname{Curl}(\boldsymbol{E}'+\dot{\boldsymbol{A}})=0. \tag{12.52}$$

Since $(\boldsymbol{E}'+\dot{\boldsymbol{A}})$ has zero Curl, it can be expressed as the material gradient of a potential Φ. Thus,

$$\boldsymbol{E}'+\dot{\boldsymbol{A}}=-\operatorname{Grad}\Phi \tag{12.53}$$

whence

$$\boldsymbol{E}'=-\dot{\boldsymbol{A}}-\operatorname{Grad}\Phi. \tag{12.54}$$

Let us now explore the relations between the material and spatial electrodynamic potentials. From (12.13) with $\boldsymbol{E}$ and $\mathscr{E}$ replaced by $\boldsymbol{A}$ and $\boldsymbol{\mathscr{A}}$, we see that the correct relation (12.9) between $\boldsymbol{B}$ and $\boldsymbol{\mathscr{B}}$ will hold if $\boldsymbol{A}$ and $\boldsymbol{\mathscr{A}}$ are related by

$$A_j=\mathscr{A}_i\frac{\partial x_i}{\partial X_j}+\frac{\partial\psi}{\partial X_j}. \tag{12.55}$$

The term Grad ψ in $\boldsymbol{A}$ does not affect $\boldsymbol{B}$ because Curl Grad $\psi\equiv 0$ for all functions ψ. From (12.55),

$$\dot{A}_j=\dot{\mathscr{A}}_i\frac{\partial x_i}{\partial X_j}+\mathscr{A}_i\frac{\partial v_i}{\partial X_j}+\frac{\partial\dot{\psi}}{\partial X_j}. \tag{12.56}$$

From (12.16) and (12.47), since div $\boldsymbol{\mathscr{B}}=0$,

$$\overset{*}{\mathscr{B}}=\operatorname{curl}\left(\frac{\partial\boldsymbol{\mathscr{A}}}{\partial t}-\boldsymbol{v}\times\operatorname{curl}\boldsymbol{\mathscr{A}}\right). \tag{12.57}$$

Hence, the relation (12.19) between $\dot{\boldsymbol{B}}$ and $\overset{*}{\mathscr{B}}$ implies

$$(\operatorname{Curl}\dot{\boldsymbol{A}})_j=\boldsymbol{J}\frac{\partial X_j}{\partial x_i}\left[\operatorname{curl}\left(\frac{\partial\boldsymbol{\mathscr{A}}}{\partial t}-\boldsymbol{v}\times\operatorname{curl}\boldsymbol{\mathscr{A}}\right)\right]_i. \tag{12.58}$$

By comparison with (12.13), we know that (12.58) will be satisfied if

$$\dot{A}_j=\frac{\partial x_i}{\partial X_j}\left(\frac{\partial\boldsymbol{\mathscr{A}}}{\partial t}-\boldsymbol{v}\times\operatorname{curl}\boldsymbol{\mathscr{A}}\right)_i. \tag{12.59}$$

By a standard identity,

$$\dot{\boldsymbol{\mathscr{A}}}=\frac{\partial\boldsymbol{\mathscr{A}}}{\partial t}+\boldsymbol{v}\cdot\nabla\boldsymbol{\mathscr{A}}=\frac{\partial\boldsymbol{\mathscr{A}}}{\partial t}-\boldsymbol{v}\times\operatorname{curl}\boldsymbol{\mathscr{A}}+(\nabla\boldsymbol{\mathscr{A}})\cdot\boldsymbol{v}. \tag{12.60}$$

Hence (12.59) becomes

$$\dot{A}_j=\frac{\partial x_i}{\partial X_j}\left(\dot{\mathscr{A}}_i-v_s\frac{\partial\mathscr{A}_s}{\partial x_i}\right)=\dot{\mathscr{A}}_i\frac{\partial x_i}{\partial X_j}-v_i\frac{\partial\mathscr{A}_i}{\partial X_j}. \tag{12.61}$$

Reconciliation of (12.61) and (12.56) is possible if Grad $\dot{\psi}=-\operatorname{Grad}(\boldsymbol{\mathscr{A}}\cdot\boldsymbol{v})$, or, in particular, if

$$\dot{\psi}=-\boldsymbol{\mathscr{A}}\cdot\boldsymbol{v}. \tag{12.62}$$

From (12.59), (12.49), and the definitions of $\mathscr{E}'$ and $\boldsymbol{E}'$, (12.53) becomes

$$-\frac{\partial\Phi}{\partial X_j}=\frac{\partial x_i}{\partial X_j}\left(\frac{\partial\boldsymbol{\mathscr{A}}}{\partial t}-\boldsymbol{v}\times\boldsymbol{\mathscr{B}}+\mathscr{E}+\boldsymbol{v}\times\boldsymbol{\mathscr{B}}\right)_i=\frac{\partial x_i}{\partial X_j}\left(-\frac{\partial\varphi}{\partial x_i}\right)=-\frac{\partial\varphi}{\partial X_j}. \tag{12.63}$$

Hence, we can take $\Phi(\boldsymbol{X},t)=\varphi(\boldsymbol{x},t)$.

12.8. Poynting's theorem in spatial and material representations. "POYNTING'S theorem" is based on MAXWELL's equations and the vector identity

$$\mathscr{E}\cdot\operatorname{curl}\mathscr{H}-\mathscr{H}\cdot\operatorname{curl}\mathscr{E}=-\operatorname{div}(\mathscr{E}\times\mathscr{H}). \tag{12.64}$$

Now the left side of (12.64) can be obtained by dotting $\mathscr{E}$ into $(12.1)_2$ and subtracting the result of dotting $\mathscr{H}$ into $(12.1)_1$. Hence, if the field vectors satisfy MAXWELL's equations, we have

$$\mathscr{E}\cdot\left(\frac{\partial\mathscr{D}}{\partial t}+\mathscr{J}\right)+\mathscr{H}\cdot\frac{\partial\boldsymbol{\mathscr{B}}}{\partial t}=-\operatorname{div}(\mathscr{E}\times\mathscr{H}). \tag{12.65}$$

By the divergence theorem,

$$\iiint \left[\mathscr{E}\cdot\left(\frac{\partial\mathscr{D}}{\partial t}+\mathscr{J}\right)+\mathscr{H}\cdot\frac{\partial\mathscr{B}}{\partial t}\right]dV=-\oiint(\mathscr{E}\times\mathscr{H})\cdot\boldsymbol{n}\,dA\,. \tag{12.66}$$

Similar operations in the material description lead to

$$\boldsymbol{E}'\cdot(\dot{\boldsymbol{D}}+\boldsymbol{J}')+\boldsymbol{H}'\cdot\dot{\boldsymbol{B}}=-\mathrm{Div}\,(\boldsymbol{E}'\times\boldsymbol{H}')\,, \tag{12.67}$$

$$\iiint[\boldsymbol{E}'\cdot(\dot{\boldsymbol{D}}+\boldsymbol{J}')+\boldsymbol{H}'\cdot\dot{\boldsymbol{B}}]\,d\bar{V}=-\oiint(\boldsymbol{E}'\times\boldsymbol{H}')\cdot N\,d\bar{A}\,. \tag{12.68}$$

By direct calculation, using (12.8) and the identity (3.54),

$$(\mathscr{E}\times\mathscr{H})_i=\varepsilon_{ijk}\,\mathscr{E}_j\,\mathscr{H}_k=\varepsilon_{ijk}\frac{\partial X_p}{\partial x_j}\frac{\partial X_q}{\partial x_k}E_p\,H_k=\frac{1}{J}\frac{\partial x_i}{\partial X_r}\varepsilon_{rpq}\,E_p\,H_q\,. \tag{12.69}$$

Thus, a flux-like transformation of $\mathscr{E}\times\mathscr{H}$ gives $\boldsymbol{E}\times\boldsymbol{H}$. The quantity $(\boldsymbol{E}'\times\boldsymbol{H}')$ that appears naturally in (12.67) and (12.68) is a different vector. It is the result of a flux-like transformation of $(\mathscr{E}'\times\mathscr{H}')$.

13. Results based on the electrodynamical theory of Tiersten and Tsai (1972).

13.1. Introduction. The electrodynamics of deformable bodies is a subject that appears to be still maturing. The classical linear theories as taught by VOIGT (1928), MASON (1950, 1966), and others have been adequate for many practical purposes. Substantial progress on a nonlinear theory was initiated by TOUPIN (1963). LAX and NELSON (1971) have constructed a nonlinear theory from first principles, based on a microscopic model. TIERSTEIN and TSAI (1972) have also constructed a nonlinear theory from first principles based on a *macroscopic* model that includes two interpenetrating ionic continua that together comprise a lattice continuum, two electronic charge continua (one for each ionic continuum), and an electronic spin continuum to account for the magnetization.

The fundamental fact that made the recent theories necessary is that assumptions beyond those of classical mechanics and electromagnetism have to be made in order to treat electromechanical interactions in deformable bodies. For these assumptions, the present treatment leans on the work of TIERSTEN and TSAI (1972). We use their expression for the electromagnetic body force per unit volume, $\boldsymbol{f}$, and for the power per unit volume Σ supplied to the material from the electromagnetic field.

In both the theory of LAX and NELSON and that of TIERSTEN and TSAI, the densities of free charge and current are assumed to be zero, i.e.,

$$\mathrm{div}\,\mathscr{B}=0\,, \tag{13.1}$$

$$\mathscr{J}=0\,, \tag{13.2}$$

and the electromagnetic contribution to the body force per unit volume is found to be

$$\boldsymbol{f}=\mathscr{P}\cdot\nabla\mathscr{E}'+\overset{*}{\mathscr{P}}\times\mathscr{B}+(\nabla\mathscr{B})\cdot\mathscr{M}' \tag{13.3}$$

where the notation is as in the previous section. Eq. (13.3) is the complete expression obtained by TIERSTEN and TSAI, but LAX and NELSON include additional terms involving the electric quadrupole moment.

In view of the form of the equation of motion (6.5), it would be naive to expect a uniquely "correct" expression for the body force. The acceleration is determined by the sum $\varrho\,g_i+\partial T_{ji}/\partial x_j$, and different theories may have equivalent equations of motion with different expressions for the body force compensated by different expressions for the stresses. In both of the aforementioned theories, Eq. (13.3) is a result that is *derived* by applying fundamental principles to the respective models.

In addition to their more general theory, TIERSTEN and TSAI indicate a simpler theory that is applicable when material resonances are suppressed. The next section will sketch the results of this simpler theory.

13.2. The theory of Tiersten and Tsai. In their simpler theory, the equation of balance of energy is expressed as

$$\frac{d}{dt}\iiint \varrho\left(\frac{1}{2}\boldsymbol{v}\cdot\boldsymbol{v}+\varepsilon\right)dV=\iiint \Sigma\, dV+\oint\!\!\oint (v_i T_{ji}-q_j)\, n_j\, dA \tag{13.4}$$

where $\boldsymbol{q}$ is the heat flux vector, ε is the internal stored energy per unit mass, and Σ is the power per unit volume supplied to the material from the electromagnetic field. From their model, TIERSTEN and TSAI derive the result

$$\Sigma=\mathscr{E}\cdot\varrho\,\dot{\boldsymbol{\pi}}+\mathscr{P}\cdot\nabla\mathscr{E}\cdot\boldsymbol{v}-\mathscr{M}'\cdot\partial\mathscr{B}/\partial t \tag{13.5}$$

where

$$\pi=\mathscr{P}/\varrho. \tag{13.6}$$

From (13.6) and the continuity equation (6.2),

$$\varrho\,\dot{\boldsymbol{\pi}}=\dot{\mathscr{P}}+\mathscr{P}\,\mathrm{div}\,\boldsymbol{v}=\overset{*}{\mathscr{P}}+\mathscr{P}\cdot\nabla\boldsymbol{v}. \tag{13.7}$$

Their equation of motion is

$$\varrho\,\dot{v}_i=f_i+\partial T_{ji}/\partial x_j \tag{13.8}$$

with the body force given by (13.3) above. A thermodynamic theory of electromechanical interactions can be based on these equations.

From (13.4),

$$\varrho\, v_i\,\dot{v}_i+\varrho\,\dot{\varepsilon}=\Sigma+v_i\frac{\partial T_{ji}}{\partial x_j}+T_{ji}\frac{\partial v_i}{\partial x_j}-\frac{\partial q_j}{\partial x_j}. \tag{13.9}$$

Multiplication of (13.8) by v_i, summing over i, and subtraction of the result from (13.9) leaves

$$\varrho\,\dot{\varepsilon}=\Sigma-\boldsymbol{v}\cdot\boldsymbol{f}+T_{ji}\,\partial v_i/\partial x_j-\partial q_j/\partial x_j. \tag{13.10}$$

From (13.5) and (13.3), after some rearrangement,

$$\Sigma-\boldsymbol{v}\cdot\boldsymbol{f}=\mathscr{E}'\cdot\varrho\,\dot{\boldsymbol{\pi}}-\mathscr{M}'\cdot\dot{\mathscr{B}}=\mathscr{E}'\cdot(\dot{\mathscr{P}}+\mathscr{P}\,\mathrm{div}\,\boldsymbol{v})-\mathscr{M}'\cdot\dot{\mathscr{B}}. \tag{13.11}$$

Subject to (13.11) and the idealization

$$\varrho T\dot{S}/\bar{\varrho}=-\partial q_j/\partial x_j, \tag{13.12}$$

(13.10) becomes

$$\varrho\,\dot{\varepsilon}=\mathscr{E}'\cdot\dot{\mathscr{P}}-\mathscr{M}'\cdot\dot{\mathscr{B}}+\varrho T\dot{S}/\bar{\varrho}+(T_{ji}+\mathscr{E}'\cdot\mathscr{P}\,\delta_{ij})\,\partial v_i/\partial x_j. \tag{13.13}$$

Here S is the entropy per unit reference volume as in Sect. 8. In view of the relation

$$\frac{\partial v_i}{\partial x_j}=\frac{\partial X_k}{\partial x_j}\frac{\partial v_i}{\partial X_k}=\frac{\partial X_k}{\partial x_j}\frac{d}{dt}\left(\frac{\partial x_i}{\partial X_k}\right), \tag{13.14}$$

(13.13) motivates the assumption that ε is a function of the independent variables $\mathscr{P}$, $\mathscr{B}$, S, and $\partial x_i/\partial X_k$ with first partial derivatives that can be read off from (13.13) and (13.14). With

$$\varepsilon=\hat{\varepsilon}(\mathscr{P},\mathscr{B},S,\partial x_i/\partial X_k), \tag{13.15}$$

we have

$$\begin{aligned}
\varrho\,\partial\hat{\varepsilon}/\partial\mathscr{P}_i &= \mathscr{E}'_i,\\
\varrho\,\partial\hat{\varepsilon}/\partial\mathscr{B}_i &= -\mathscr{M}'_i,\\
\bar{\varrho}\,\partial\hat{\varepsilon}/\partial S &= T,\\
\varrho\,\partial\hat{\varepsilon}/\partial(\partial x_i/\partial X_k) &= (T_{ji}+\boldsymbol{\mathscr{E}}'\cdot\boldsymbol{\mathscr{P}}\,\delta_{ij})\,\partial X_k/\partial x_j.
\end{aligned}\tag{13.16}$$

The rotational invariance of ε can be guaranteed by changing the independent variables to $\boldsymbol{P}$, $\boldsymbol{B}$, S, and the strain components, all of which are constant in a rigid rotation of the fields with the body. We shall do this in two steps, first introducing $\boldsymbol{P}$ and $\boldsymbol{B}$, and finally the strain components V_{ij}. From (12.9) and (3.36),

$$\begin{aligned}
\dot{\mathscr{B}}_i &= \frac{1}{J}\left(\frac{\partial x_i}{\partial X_j}\,\dot{B}_j + B_j\,\frac{\partial v_i}{\partial X_j}\right) - \mathscr{B}_i\,\operatorname{div}\boldsymbol{v}\\
&= \frac{1}{J}\,\frac{\partial x_i}{\partial X_j}\,\dot{B}_j + \mathscr{B}_k\,\frac{\partial v_i}{\partial x_k} - \mathscr{B}_i\,\operatorname{div}\boldsymbol{v}
\end{aligned}\tag{13.17}$$

and similarly for $\dot{\mathscr{P}}_i$. Hence (13.13) becomes

$$\bar{\varrho}\,\dot{\varepsilon} = \boldsymbol{E}'\cdot\dot{\boldsymbol{P}} - \boldsymbol{M}'\cdot\dot{\boldsymbol{B}} + T\dot{S} + J\,(T_{ji} + \mathscr{E}'_i\mathscr{P}_j - \mathscr{M}'_i\mathscr{B}_j + \boldsymbol{\mathscr{M}}'\cdot\boldsymbol{\mathscr{B}}\,\delta_{ij})\,\frac{\partial v_i}{\partial x_j}. \tag{13.18}$$

Recalling (13.14), we see from (13.18) that with

$$\varepsilon = \bar{\varepsilon}(\boldsymbol{P}, \boldsymbol{B}, S, \partial x_i/\partial X_k),$$

we have

$$\begin{aligned}
\bar{\varrho}\,\partial\bar{\varepsilon}/\partial P_i &= E'_i,\\
\bar{\varrho}\,\partial\bar{\varepsilon}/\partial B_i &= -M'_i,\\
\bar{\varrho}\,\partial\bar{\varepsilon}/\partial S &= T,\\
\bar{\varrho}\,\frac{\partial\bar{\varepsilon}}{\partial(\partial x_i/\partial X_k)} &= J\,\frac{\partial X_k}{\partial x_j}\,(T_{ji} + \mathscr{E}'_i\mathscr{P}_j - \mathscr{M}'_i\mathscr{B}_j + \boldsymbol{\mathscr{M}}'\cdot\boldsymbol{\mathscr{B}}\,\delta_{ij}).
\end{aligned}\tag{13.19}$$

To introduce the strain components as independent variables, we differentiate (3.2) to obtain

$$\frac{\partial V_{pq}}{\partial(\partial x_i/\partial X_k)} = \frac{1}{2}\left(\frac{\partial x_i}{\partial X_p}\,\delta_{qk} + \frac{\partial x_i}{\partial X_p}\,\delta_{pk}\right). \tag{13.20}$$

Then, with

$$\varepsilon = \varepsilon(\boldsymbol{P}, \boldsymbol{B}, S, V_{pq}) = \bar{\varepsilon}(\boldsymbol{P}, \boldsymbol{B}, S, \partial x_i/\partial X_k), \tag{13.21}$$

we have

$$\begin{aligned}
T_{ji} + \mathscr{E}'_i\mathscr{P}_j - \mathscr{M}'_i\mathscr{B}_j + \boldsymbol{\mathscr{M}}'\cdot\boldsymbol{\mathscr{B}}\,\delta_{ij} &= \varrho\,\frac{\partial x_j}{\partial X_k}\,\frac{\partial\bar{\varepsilon}}{\partial(\partial x_i/\partial X_k)}\\
&= \varrho\,\frac{\partial x_j}{\partial X_k}\,\frac{\partial V_{pq}}{\partial(\partial x_i/\partial X_k)}\,\frac{\partial\varepsilon}{\partial V_{pq}}\\
&= \varrho\,\frac{\partial x_i}{\partial X_p}\,\frac{\partial x_j}{\partial X_q}\,\frac{\partial\varepsilon}{\partial V_{pq}}.
\end{aligned}\tag{13.22}$$

The partial derivatives of $\varepsilon\,(\boldsymbol{P}, \boldsymbol{B}, S, V_{ij})$ with respect to P_i and B_i are the same as those of $\bar{\varepsilon}$ in (13.19) since constant deformation gradients imply constant strain. Thus, with $\varepsilon = \varepsilon(\boldsymbol{P}, \boldsymbol{B}, S, V_{ij})$,

$$\begin{aligned}
\bar{\varrho}\,\partial\varepsilon/\partial P_i &= E'_i,\\
\bar{\varrho}\,\partial\varepsilon/\partial B_i &= -M'_i,\\
\bar{\varrho}\,\partial\varepsilon/\partial S &= T.
\end{aligned}\tag{13.23}$$

Eqs. (13.22) and (13.23) are equivalent to

$$\begin{aligned} T_{ji} &= \mathcal{M}_i' \mathcal{B}_j - \mathcal{E}_i' \mathcal{P}_j - \boldsymbol{\mathcal{M}}' \cdot \boldsymbol{\mathcal{B}}\, \delta_{ij} + \varrho \frac{\partial x_i}{\partial X_p} \frac{\partial x_j}{\partial X_q} \frac{\partial \varepsilon}{\partial V_{pq}}, \\ \mathcal{E}_j' &= \mathcal{E}_j + (\boldsymbol{v} \times \boldsymbol{\mathcal{B}})_j = \bar{\varrho} \frac{\partial X_i}{\partial x_j} \frac{\partial \varepsilon}{\partial P_i}, \\ \mathcal{M}_j' &= -\bar{\varrho} \frac{\partial X_i}{\partial x_j} \frac{\partial \varepsilon}{\partial B_i}. \end{aligned} \tag{13.24}$$

It may be noted that

$$T_{ij} - T_{ji} = \mathcal{M}_j' \mathcal{B}_i - \mathcal{M}_i' \mathcal{B}_j + \mathcal{P}_j \mathcal{E}_i' - \mathcal{P}_i \mathcal{E}_j' \tag{13.25}$$

and hence, from (6.21), the implied body torque per unit volume is

$$\boldsymbol{G} = \boldsymbol{\mathcal{M}}' \times \boldsymbol{\mathcal{B}} + \boldsymbol{\mathcal{P}} \times \boldsymbol{\mathcal{E}}', \tag{13.26}$$

in agreement with intuitive concepts.

TIERSTEN and TSAI find it convenient to introduce a function F, defined by

$$F = \varepsilon - \boldsymbol{\mathcal{E}}' \cdot \boldsymbol{\mathcal{P}}/\varrho - TS/\bar{\varrho} = \varepsilon - (\boldsymbol{E}' \cdot \boldsymbol{P} + TS)/\varrho. \tag{13.27}$$

Then (13.13) becomes

$$\varrho \dot{F} = -\boldsymbol{\mathcal{P}} \cdot \dot{\boldsymbol{\mathcal{E}}}' - \boldsymbol{\mathcal{M}}' \cdot \dot{\boldsymbol{\mathcal{B}}} - \varrho S \dot{T}/\bar{\varrho} + T_{ji}\, \partial v_i/\partial x_j, \tag{13.28}$$

or

$$\begin{aligned} \bar{\varrho} \dot{F} = &-\boldsymbol{P} \cdot \dot{\boldsymbol{E}}' - \boldsymbol{M}' \cdot \dot{\boldsymbol{B}} - S \dot{T} \\ &+ J \frac{\partial X_k}{\partial x_i} \frac{\partial X_m}{\partial x_j} (T_{ji} - \mathcal{M}_i' \mathcal{B}_j + \mathcal{E}_i' \mathcal{P}_j + \boldsymbol{\mathcal{M}}' \cdot \boldsymbol{\mathcal{B}}\, \delta_{ij}) \dot{V}_{km}. \end{aligned} \tag{13.29}$$

With

$$F = F(\boldsymbol{E}', \boldsymbol{B}, T, V_{km}),$$

we obtain immediately from (13.29),

$$\begin{aligned} \bar{\varrho}\, \partial F/\partial E_i' &= -P_i, \\ \bar{\varrho}\, \partial F/\partial B_i &= -M_i', \\ \bar{\varrho}\, \partial F/\partial T &= -S, \\ \bar{\varrho}\, \partial F/\partial V_{km} &= J \frac{\partial X_k}{\partial x_i} \frac{\partial X_m}{\partial x_j} (T_{ji} - \mathcal{M}_i' \mathcal{B}_j + \mathcal{E}_i' \mathcal{P}_j + \boldsymbol{\mathcal{M}}' \cdot \boldsymbol{\mathcal{B}}\, \delta_{ij}). \end{aligned} \tag{13.30}$$

The explicit expressions for $\boldsymbol{\mathcal{P}}$, $\boldsymbol{\mathcal{M}}'$, and T_{ji} are, from (12.8), (12.9), and (13.30),

$$\begin{aligned} \mathcal{P}_j &= \frac{\varrho}{\bar{\varrho}} \frac{\partial x_j}{\partial X_i} P_i = -\varrho \frac{\partial x_j}{\partial X_i} \frac{\partial F}{\partial E_i'}, \\ \mathcal{M}_j' &= M_i' (\partial X_i/\partial x_j) = -\bar{\varrho} \frac{\partial X_i}{\partial x_j} \frac{\partial F}{\partial B_i} \\ T_{ji} &= \varrho \frac{\partial x_i}{\partial X_k} \frac{\partial x_j}{\partial X_m} \frac{\partial F}{\partial V_{km}} + \mathcal{M}_j' \mathcal{B}_j - \mathcal{P}_j \mathcal{E}_i' - \boldsymbol{\mathcal{M}}' \cdot \boldsymbol{\mathcal{B}}\, \delta_{ij}. \end{aligned} \tag{13.31}$$

In place of $\boldsymbol{B}$, TIERSTEN and TSAI use an independent variable $\boldsymbol{Z}$ defined analogously to $\boldsymbol{E}$, i.e.,

$$Z_j = \mathcal{B}_i (\partial x_i/\partial X_j) = C_{jk} B_k/J. \tag{13.32}$$

With

$$F = \overline{F}(\boldsymbol{E}', Z, T, V_{km}), \tag{13.33}$$

they find

$$\begin{aligned}\mathscr{P}_j &= -\varrho\,\frac{\partial x_j}{\partial X_i}\,\frac{\partial \bar F}{\partial E_i'},\\ \mathscr{M}_j' &= -\varrho\,\frac{\partial x_j}{\partial X_i}\,\frac{\partial \bar F}{\partial Z_i},\\ T_{ji} &= \varrho\,\frac{\partial x_i}{\partial X_k}\,\frac{\partial x_j}{\partial X_m}\,\frac{\partial \bar F}{\partial V_{km}} - \mathscr{M}_j'\mathscr{B}_i - \mathscr{P}_j\mathscr{E}_i'.\end{aligned} \tag{13.34}$$

The formula for the stress in (13.34) appears different from that in (13.31) because $\bar F$ has $\boldsymbol{Z}$ as an independent variable whereas F in (13.31) has $\boldsymbol{B}$. This illustrates how a change of independent variable can cause equivalent expressions to look different.

13.3. The total stored energy. For their simpler model, TIERSTEN and TSAI write the balance of total energy as

$$\begin{aligned}&\frac{d}{dt}\iiint\left[\varrho\left(\frac{1}{2}\boldsymbol{v}\cdot\boldsymbol{v}+\varepsilon\right)+\mathscr{U}^{\mathrm F}\right]dV\\ &\qquad=\oint\!\!\!\oint\left[T_{jk}\,v_k+v_j(\mathscr{U}^{\mathrm F}+\mathscr{P}\cdot\mathscr{E})-(\mathscr{E}\times\mathscr{H})_j-q_j\right]n_j\,dA,\end{aligned} \tag{13.35}$$

in which $\mathscr{U}^{\mathrm F}$ is the field energy, defined by

$$\mathscr{U}^{\mathrm F}=\tfrac12(\varepsilon_0\,\mathscr{E}\cdot\mathscr{E}+\mathscr{B}\cdot\mathscr{B}/\mu_0)\,. \tag{13.36}$$

The left member differs from (13.4) by the addition of the time derivative of the volume integral of $\mathscr{U}^{\mathrm F}$. By the transport theorem (3.48), this is

$$\frac{d}{dt}\iiint\mathscr{U}^{\mathrm F}\,dV=\iiint\frac{\partial\mathscr{U}^{\mathrm F}}{\partial t}\,dV+\oint\!\!\!\oint\mathscr{U}^{\mathrm F}\,\boldsymbol{v}\cdot\boldsymbol{n}\,dA\,. \tag{13.37}$$

The result of subtracting (13.4) from (13.35) can therefore be put in the form

$$\begin{aligned}\iiint\left(\frac{\partial\mathscr{U}^{\mathrm F}}{\partial t}+\Sigma\right)dV&=\oint\!\!\!\oint\left[(\mathscr{P}\cdot\mathscr{E})\,\boldsymbol{v}-(\mathscr{E}\times\mathscr{H})\right]\cdot\boldsymbol{n}\,dA\,.\\ &=\iiint\operatorname{div}\left[(\mathscr{P}\cdot\mathscr{E})\,\boldsymbol{v}-(\mathscr{E}\times\mathscr{H})\right]dV\,.\end{aligned} \tag{13.38}$$

Since (13.38) must hold for arbitrary volumes if (13.35) and (13.4) are to be consistent, this requires

$$\Sigma=-\frac{\partial\mathscr{U}^{\mathrm F}}{\partial t}+\operatorname{div}\left[(\mathscr{P}\cdot\mathscr{E})\,\boldsymbol{v}-(\mathscr{E}\times\mathscr{H})\right]. \tag{13.39}$$

Taking div $(\mathscr{E}\times\mathscr{H})$ from (12.65) with $\mathscr{J}=0$, and making use of (12.3) and (13.36), we reduce (13.39) to

$$\Sigma=\mathscr{E}\cdot\frac{\partial\mathscr{P}}{\partial t}-\mathscr{M}\cdot\frac{\partial\mathscr{B}}{\partial t}+\operatorname{div}\left[(\mathscr{P}\cdot\mathscr{E})\,\boldsymbol{v}\right]. \tag{13.40}$$

It can be verified that (13.40) is equivalent to (13.5). Note that (13.5) contains $\mathscr{M}'$ and (13.40) contains $\mathscr{M}$. Working only to first order in v/c, we write

$$-\mathscr{M}'\cdot\frac{\partial\mathscr{B}}{\partial t}=-(\mathscr{M}+\boldsymbol{v}\times\mathscr{P})\cdot\frac{\partial\mathscr{B}}{\partial t}=-\mathscr{M}\cdot\frac{\partial\mathscr{B}}{\partial t}+(\boldsymbol{v}\times\mathscr{P})\cdot\operatorname{curl}\mathscr{E} \tag{13.41}$$

where $(12.1)_1$ has been used to replace $-\partial\mathscr{B}/\partial t$ by curl $\mathscr{E}$ in the term containing $(\boldsymbol{v}\times\mathscr{P})$. By direct expansion, it is readily verified that

$$(\boldsymbol{v}\times\mathscr{P})\cdot\operatorname{curl}\mathscr{E}=v_k\,\frac{\partial\mathscr{E}_i}{\partial x_k}\,\mathscr{P}_i-\mathscr{P}_k\,\frac{\partial\mathscr{E}_i}{\partial x_k}\,v_i=\boldsymbol{v}\cdot\nabla\mathscr{E}\cdot\mathscr{P}-\mathscr{P}\cdot\nabla\mathscr{E}\cdot\boldsymbol{v}\,. \tag{13.42}$$

In view of (13.41) and (13.42), (13.5) becomes

$$\Sigma = \mathcal{E} \cdot \varrho \dot{\pi} + \boldsymbol{v} \cdot \nabla \mathcal{E} \cdot \mathcal{P} - \mathcal{M} \cdot \partial \mathcal{B} / \partial t. \tag{13.43}$$

With the help of (13.7) and the relation $\dot{\mathcal{P}} = \partial \mathcal{P}/\partial t + \boldsymbol{v} \cdot \nabla \mathcal{P}$, the reader may now readily verify that (13.40) and (13.43) are equivalent.

Let us tentatively regard $(\varrho \varepsilon + \mathcal{U}^{\mathrm{F}})$ in (13.35) as the *total* stored energy per unit volume:

$$\mathcal{U} \equiv \varrho \varepsilon + \mathcal{U}^{\mathrm{F}}. \tag{13.44}$$

Then the stored energy per unit reference volume would be

$$U = J\mathcal{U} = \bar{\varrho} \varepsilon + J \mathcal{U}^{\mathrm{F}} \tag{13.45}$$

whence

$$\dot{U} = J(\varrho \dot{\varepsilon} + \dot{\mathcal{U}}^{\mathrm{F}} + \mathcal{U}^{\mathrm{F}} \operatorname{div} \boldsymbol{v}). \tag{13.46}$$

In the conventional linearized theories, the differential of electromagnetic energy per unit volume is ordinarily expressed as $\mathcal{E} \cdot d\mathcal{D} + \mathcal{H} \cdot d\mathcal{B}$. It is therefore of interest to see whether U can be expressed as a function of the independent variables $\mathcal{D}$, $\mathcal{B}$, S, $\partial x_i/\partial X_k$. With $\varrho \dot{\varepsilon}$ from (13.13), $\mathcal{U}^{\mathrm{F}}$ from (13.36), and $\mathcal{D}$ and $\mathcal{B}$ from (12.3), Eq. (13.46) can be expressed as

$$\dot{U}/J = \psi + \mathcal{E} \cdot \dot{\mathcal{D}} + \mathcal{H} \cdot \dot{\mathcal{B}} + T\dot{S}/J + [T_{ji} + (\mathcal{E}' \cdot \mathcal{P} + \mathcal{U}^{\mathrm{F}})\,\delta_{ij}]\,\partial v_i/\partial x_j \tag{13.47}$$

where

$$\begin{aligned} \psi &\equiv (\boldsymbol{v} \times \mathcal{B}) \cdot \dot{\mathcal{P}} + (\mathcal{P} \times \boldsymbol{v}) \cdot \dot{\mathcal{B}} = -\boldsymbol{v} \cdot \frac{d}{dt}(\mathcal{P} \times \mathcal{B}) \\ &= (\boldsymbol{v} \times \mathcal{B}) \cdot \dot{\mathcal{D}} + (\mathcal{D} \times \boldsymbol{v}) \cdot \dot{\mathcal{B}} + \boldsymbol{v} \cdot \frac{d}{dt}(\varepsilon_0\, \mathcal{E} \times \mathcal{B}). \end{aligned} \tag{13.48}$$

Without further assumptions, the term in ψ containing $\dot{\mathcal{P}}$ (or $\dot{\mathcal{E}}$) prevents U from being expressed as a function of $\mathcal{D}$, $\mathcal{B}$, S, and $\partial x_i/\partial X_k$ alone. However, by using the last expression for ψ, we obtain the interesting form

$$\begin{aligned} \dot{U}/J = {} & \mathcal{E}' \cdot \dot{\mathcal{D}} + \mathcal{H}' \cdot \dot{\mathcal{B}} + T\dot{S}/J + \boldsymbol{v} \cdot \frac{d}{dt}(\varepsilon_0\, \mathcal{E} \times \mathcal{B}) \\ & + [T_{ji} + (\mathcal{E}' \cdot \mathcal{P} + \mathcal{U}^{\mathrm{F}})\,\delta_{ij}]\,\delta v_i/\partial x_j. \end{aligned} \tag{13.49}$$

13.4. Thermostatics. The theory of TIERSTEN and TSAI is a *dynamical* theory that takes account of the "Lorentz force" $(\mathcal{E}' - \mathcal{E}) = \boldsymbol{v} \times \mathcal{B}$ and the apparent magnetic moment per unit volume of a polarized medium in motion, $(\mathcal{M} - \mathcal{M}') = \mathcal{P} \times \boldsymbol{v}$. To outline a strictly thermo*static* theory, we may ignore ψ in (13.47) and drop the distinction between $\mathcal{E}$ and $\mathcal{E}'$. The resulting thermostatic approximation to (13.47) is then

$$\begin{aligned} \dot{U}/J = {} & \mathcal{E} \cdot \dot{\mathcal{D}} + \mathcal{H} \cdot \dot{\mathcal{B}} + T\dot{S}/J \\ & + [T_{ji} + (\mathcal{E} \cdot \mathcal{P} + \mathcal{U}^{\mathrm{F}})\,\delta_{ij}]\,\frac{\partial X_k}{\partial x_j}\,\frac{d}{dt}\left(\frac{\partial x_i}{\partial X_k}\right) \end{aligned} \tag{13.50}$$

in which (13.14) has been used. Eq. (13.50) motivates the assumption that *as far as thermostatics is concerned*, U may be taken to depend on $\mathcal{D}$, $\mathcal{B}$, S, and $\partial x_i/\partial X_k$.

Because of relations like (13.17) for $\dot{\mathcal{D}}$ and $\dot{\mathcal{B}}$, we have

$$\begin{aligned} J\mathcal{E} \cdot \dot{\mathcal{D}} &= \boldsymbol{E} \cdot \dot{\boldsymbol{D}} + J(\mathcal{E}_i\, \mathcal{D}_j - \mathcal{E} \cdot \mathcal{D}\, \partial_{ij})\,\partial v_i/\partial x_j, \\ J\mathcal{H} \cdot \dot{\mathcal{B}} &= \boldsymbol{H} \cdot \dot{\boldsymbol{B}} + J(\mathcal{H}_i\, \mathcal{B}_j - \mathcal{H} \cdot \mathcal{B}\, \delta_{ij})\,\partial v_i/\partial x_j. \end{aligned} \tag{13.51}$$

Then (13.50) becomes

$$\begin{aligned}\dot{U} &= \boldsymbol{E}\cdot\dot{\boldsymbol{D}} + \boldsymbol{H}\cdot\dot{\boldsymbol{B}} + T\dot{S}\\ &\quad + J\,[T_{ji} + \mathscr{E}_i\,\mathscr{D}_j + \mathscr{H}_i\,\mathscr{B}_j + (\mathscr{M}\cdot\mathscr{B} - \mathscr{U}^{\mathrm{F}})\,\delta_{ij}]\,\frac{\partial X_k}{\partial x_j}\,\frac{d}{dt}\left(\frac{\partial x_i}{\partial X_k}\right).\end{aligned} \tag{13.52}$$

The definitions (13.36) and (12.3) have been used to reduce the quantity in parentheses multiplying δ_{ij}:

$$(\mathscr{P}\cdot\mathscr{E} + \mathscr{U}^{\mathrm{F}} - \mathscr{E}\cdot\mathscr{D} - \mathscr{H}\cdot\mathscr{B}) = (\mathscr{M}\cdot\mathscr{B} - \mathscr{U}^{\mathrm{F}}).$$

Let us now assume that *in the thermostatic approximation,*

$$U = \bar{U}\,(\boldsymbol{D}, \boldsymbol{B}, S, \partial x_i/\partial X_k). \tag{13.53}$$

We then have

$$\begin{gathered}\partial\bar{U}/\partial D_i = E_i, \qquad \partial\bar{U}/\partial B_i = H_i, \qquad \partial\bar{U}/\partial S = T,\\ \frac{\partial\bar{U}}{\partial(\partial x_i/\partial X_k)} = J\,\frac{\partial X_k}{\partial x_j}\,[T_{ji} + \mathscr{E}_i\,\mathscr{D}_j + \mathscr{H}_i\,\mathscr{B}_j + (\mathscr{M}\cdot\mathscr{B} - \mathscr{U}^{\mathrm{F}})\,\delta_{ij}].\end{gathered} \tag{13.54}$$

As in (13.22), the strain components can be introduced by means of (13.20). With

$$U = \bar{U} = U(\boldsymbol{D}, \boldsymbol{B}, S, V_{ij}), \tag{13.55}$$

we find

$$\begin{gathered}\partial U/\partial D_i = E_i, \qquad \partial U/\partial B_i = H_i, \qquad \partial U/\partial T = S,\\ [T_{ji} + \mathscr{E}_i\,\mathscr{D}_j + \mathscr{H}_i\,\mathscr{B}_j + (\mathscr{M}\cdot\mathscr{B} - \mathscr{U}^{\mathrm{F}})\,\delta_{ij}] = \frac{1}{J}\,\frac{\partial x_i}{\partial X_p}\,\frac{\partial x_j}{\partial X_q}\,\frac{\partial U}{\partial V_{pq}}.\end{gathered} \tag{13.56}$$

The symmetry of V_{pq} has been used. In view of this symmetry, $(13.56)_4$ and (6.21) give the body torque as

$$G = \mathscr{D}\times\mathscr{E} - \mathscr{H}\times\mathscr{B} = \mathscr{P}\times\mathscr{E} + \mathscr{M}\times\mathscr{B}. \tag{13.57}$$

Comparison with (13.26) shows that, as expected, the effect of the particle velocity has been lost in the thermostatic approximation.

13.5. Equation of motion. For use in the equation of motion (13.8), it is the sum $(f_i + \partial T_{ji}/\partial x_j)$ that is of interest. TIERSTEN and TSAI note that the f_i of (13.3) can be expressed as

$$f_i = \partial T_{ji}^{\mathrm{EM}}/\partial x_j - \frac{\partial}{\partial t}\,(\varepsilon_0\,\mathscr{E}\times\mathscr{B})_i \tag{13.58}$$

where

$$T_{ji}^{\mathrm{EM}} = \mathscr{P}_j\,\mathscr{E}'_i - \mathscr{M}'_i\,\mathscr{B}_j + \varepsilon_0\,\mathscr{E}_j\,\mathscr{E}_i + \frac{1}{\mu_0}\,\mathscr{B}_j\,\mathscr{B}_i - (\mathscr{U}^{\mathrm{F}} - \mathscr{M}'\cdot\mathscr{B})\,\delta_{ij}. \tag{13.59}$$

It follows that

$$f_i + \frac{\partial T_{ji}}{\partial x_j} = \frac{\partial}{\partial x_j}\,(T_{ji} + T_{ji}^{\mathrm{EM}}) - \frac{\partial}{\partial t}\,(\varepsilon_0\,\mathscr{E}\times\mathscr{B}). \tag{13.60}$$

From (13.22) and (13.59),

$$T_{ji} + T_{ji}^{\mathrm{EM}} = \varrho\,\frac{\partial x_i}{\partial X_p}\,\frac{\partial x_j}{\partial X_q}\,\frac{\partial\varepsilon}{\partial V_{pq}} + \varepsilon_0\,\mathscr{E}_i\,\mathscr{E}_j + \frac{1}{\mu_0}\,\mathscr{B}_i\,\mathscr{B}_j - \mathscr{U}^{\mathrm{F}}\,\delta_{ij}. \tag{13.61}$$

With the help of (12.1)–(12.3) and the assumptions $\mathscr{J} = 0$ and $\operatorname{div}\mathscr{D} = 0$, we find

$$\frac{\partial}{\partial t}\,(\varepsilon_0\,\mathscr{E}\times\mathscr{B}) = \varepsilon_0\,\mathscr{E}\cdot\nabla\mathscr{E} + \frac{1}{\mu_0}\,\mathscr{B}\cdot\nabla\mathscr{B} - \nabla\mathscr{U}^{\mathrm{F}} + \mathscr{B}\times\left(\frac{\partial\mathscr{P}}{\partial t} + \operatorname{curl}\mathscr{M}\right) \tag{13.62}$$

and

$$\frac{\partial}{\partial x_j}\left(\varepsilon_0\,\mathscr{E}_i\,\mathscr{E}_j+\frac{1}{\mu_0}\,\mathscr{B}_i\mathscr{B}_j-\mathscr{U}^{\mathrm{F}}\delta_{ij}\right)$$
$$=\left(\varepsilon_0\,\mathscr{E}\cdot\nabla\mathscr{E}+\frac{1}{\mu_0}\,\mathscr{B}\cdot\nabla\mathscr{B}-\nabla\mathscr{U}^{\mathrm{F}}-\mathscr{E}\,\nabla\cdot\mathscr{P}\right)_i. \tag{13.63}$$

Also, since $\varrho=\bar{\varrho}/J$ with $\bar{\varrho}$ constant, the identity of EULER, PIOLA and JACOBI (3.33) can be used to show that

$$\frac{\partial}{\partial x_j}\left(\varrho\,\frac{\partial x_i}{\partial X_p}\,\frac{\partial x_j}{\partial X_q}\,\frac{\partial\varepsilon}{\partial V_{pq}}\right)=\frac{\bar{\varrho}}{J}\,\frac{\partial}{\partial X_q}\left(\frac{\partial x_i}{\partial X_p}\,\frac{\partial\varepsilon}{\partial V_{pq}}\right). \tag{13.64}$$

Finally, (13.60) becomes

$$f_i+\frac{\partial T_{ji}}{\partial x_j}=\frac{\bar{\varrho}}{J}\,\frac{\partial}{\partial X_q}\left(\frac{\partial x_i}{\partial X_p}\,\frac{\partial\varepsilon}{\partial V_{pq}}\right)-\mathscr{E}_i\,\mathrm{div}\,\mathscr{P}-\left[\mathscr{B}\times\left(\frac{\partial\mathscr{P}}{\partial t}+\mathrm{curl}\,\mathscr{M}\right)\right]_i. \tag{13.65}$$

It may be recalled from (13.21) that ε has the independent variables $\boldsymbol{P}$, $\boldsymbol{B}$, S, V_{pq}.

In terms of the function $F(\boldsymbol{E}', \boldsymbol{B}, T, V_{km})$ defined in (13.27), we get precisely the same equation (13.65) with ε replaced by F, since comparison of $(13.24)_1$ and $(13.30)_4$ shows that

$$\bar{\varrho}\;\partial\varepsilon/\partial V_{km}=\bar{\varrho}\;\partial F/\partial V_{km}. \tag{13.66}$$

Similar calculations show that in the approximation (13.56), with terms dependent on the particle velocity ignored,

$$f_i+\frac{\partial T_{ji}}{\partial x_j}=\frac{1}{J}\,\frac{\partial}{\partial X_q}\left(\frac{\partial x_i}{\partial X_p}\,\frac{\partial U}{\partial V_{pq}}\right)-\frac{\partial}{\partial t}\left(\varepsilon_0\,\mathscr{E}\times\mathscr{B}\right). \tag{13.67}$$

For use in the equation of motion (6.7), one needs $(F_i+\partial P_{ik}/\partial X_k)$ where $F_i=J f_i$, which denotes the body force per unit reference volume, replaces $\bar{\varrho}\,g_i$ in (6.7). Since

$$F_i+\partial P_{ik}/\partial X_k=J(f_i+\partial T_{ji}/\partial x_j), \tag{13.68}$$

the dynamical theory of TIERSTEN and TSAI (13.65) yields

$$\bar{\varrho}\,\ddot{u}_i=F_i+\frac{\partial P_{ik}}{\partial X_k}=\bar{\varrho}\,\frac{\partial}{\partial X_q}\left(\frac{\partial x_i}{\partial X_p}\,\frac{\partial\varepsilon}{\partial V_{pq}}\right)$$
$$-J\left[\mathscr{E}\,\mathrm{div}\,\mathscr{P}+\mathscr{B}\times\left(\frac{\partial\mathscr{P}}{\partial t}+\mathrm{curl}\,\mathscr{M}\right)\right]_i, \tag{13.69}$$

while the approximation (13.67) gives

$$\bar{\varrho}\,\ddot{u}_i=F_i+\frac{\partial P_{ik}}{\partial X_k}=\frac{\partial}{\partial X_q}\left(\frac{\partial x_i}{\partial X_p}\,\frac{\partial U}{\partial V_{pq}}\right)-J\,\frac{\partial}{\partial t}\left(\varepsilon_0\,\mathscr{E}\times\mathscr{B}\right). \tag{13.70}$$

It was remarked earlier that equivalent theories can have the same equation of motion with different expressions for the body force compensated by different expressions for the stress. From (13.64) and (13.65), and no other information, one would be tempted to put the body force as

$$\boldsymbol{f}^*=\left(\frac{\partial\mathscr{P}}{\partial t}+\mathrm{curl}\,\mathscr{M}\right)\times\mathscr{B}-\mathscr{E}\;\mathrm{div}\;\mathscr{P}, \tag{13.71}$$

and the stress as

$$T_{ji}^*=\varrho\,\frac{\partial x_i}{\partial X_p}\,\frac{\partial x_j}{\partial X_q}\,\frac{\partial\varepsilon}{\partial V_{pq}}. \tag{13.72}$$

The logical derivation in the paper of TIERSTEN and TSAI gives different expressions. From (13.72) and $(13.24)_1$,

$$T_{ji} = T_{ji}^* + \mathcal{M}'_i \mathcal{B}_j - \mathcal{E}'_i \mathcal{P}_j - \mathcal{M}' \cdot \mathcal{B}\, \delta_{ij}, \tag{13.73}$$

while analysis shows that

$$\begin{aligned} \boldsymbol{f}^* &= (\overset{*}{\mathcal{P}} + \operatorname{curl} \mathcal{M}') \times \mathcal{B} - \mathcal{E}' \operatorname{div} \mathcal{P} \\ &= \boldsymbol{f} - \mathcal{P} \cdot \nabla \mathcal{E}' - \mathcal{E}' \operatorname{div} \mathcal{P} - \nabla \mathcal{B} \cdot \mathcal{M}' + (\operatorname{curl} \mathcal{M}') \times \mathcal{B} \\ &= \boldsymbol{f} - \nabla \cdot (\mathcal{P} \mathcal{E}') - \nabla \mathcal{B} \cdot \mathcal{M}' + \mathcal{B} \cdot \nabla \mathcal{M}' - \nabla \mathcal{M}' \cdot \mathcal{B} \end{aligned} \tag{13.74}$$

since $(\operatorname{curl} \mathcal{M}') \times \mathcal{B} = \mathcal{B} \cdot \nabla \mathcal{M}' - \nabla \mathcal{M}' \cdot \mathcal{B}$.

14. Extension of thermostatics to include electromechanical interactions.

14.1. Material coefficients. Explicit constitutive relations in terms of material coefficients can be obtained by expressing the thermodynamic functions as power series in their independent variables. For example, the stored energy ε can be expanded in powers and products of P_i, B_i, $(S - S_0)$, and V_{ij} with coefficients that are properties of the material and of the initial state. To indicate the coefficients for electromechanical interactions, it suits our purpose to expand U of (13.55) as a power series in D_i and V_{ij}, assuming $\boldsymbol{B} = 0$ and $S = S_0 =$ constant. This procedure parallels that of MASON (1950, 1966). The differencc is that the present choice of variables makes the theory appropriate for finite deformations. The reader is reminded that the V_{ij} are the finite strain components defined in (3.2), and that Eqs. (12.8) and (12.9) relate the present $\boldsymbol{E}$ and $\boldsymbol{D}$ to the conventional fields $\mathcal{E}$ and $\mathcal{D}$. The fact that MASON's formulation can be put into a rotationally invariant setting by this system of definitions was pointed out by TOUPIN (1963). A similar remark applies to publications on nonlinear effects by MCMAHON (1968) and LJAMOV (1972).

For the derivatives of U at constant entropy and with $\boldsymbol{B} = 0$ we adopt the notation and names in Table 14.1. We make free use of the symmetries allowed by interchange of the order of differentiation. As presented in Table 14.1, the partial derivatives of U are functions of the same independent variables as U itself, i.e., $\boldsymbol{D}$, $\boldsymbol{B}$, S, and V_{ij}. On the other hand, when the derivatives are used as coefficients in power series expansions for the thermodynamic tensions τ_{pq}^* and E_i, they must be evaluated at a particular state, say $S = S_0, \boldsymbol{B} = 0, \boldsymbol{D} = 0, V_{ij} = 0$. To account for the effect of entropy changes in such an expansion, the derivatives of all the coefficients with respect to entropy would be needed also. Of these, we note only the following:

$$\begin{aligned} \frac{\partial^2 U}{\partial D_i \partial S} &= \frac{\partial E_i}{\partial S} = \frac{\partial T}{\partial D_i} = -q_i^{V,S}, \\ \frac{\partial^2 U}{\partial V_{ij}\, \partial S} &= \frac{\partial T}{\partial V_{ij}} = \frac{\partial \tau_{ij}^*}{\partial S} = -T \gamma_{ij}^D, \\ \frac{\partial^2 U}{\partial S^2} &= \frac{\partial T}{\partial S} = \frac{T}{C_V^D}. \end{aligned} \tag{14.1}$$

Here $q_i^{V,S}$ is a measure of the pyroelectric effect, and the other relations have already been encountered in Sect. 10, the only difference being the specification that the electrical variable D is held constant.

With the coefficients interpreted as *constants*, the derivatives being evaluated at $(S, D, V_{ij}) = (S, 0, 0)$, we have

Table 14.1. Derivatives of $U(\boldsymbol{D}, \boldsymbol{B}, S, V_{ij})$.

Quantity	Definition
Thermodynamic tension	$\tau^*_{ij} = \partial U / \partial V_{ij}$
Material electric field	$E_i = \partial U / \partial D_i$
Impermittivity	$\beta_{ij} = \partial^2 U / \partial D_i \, \partial D_j = \partial E_j / \partial D_i$
Piezoelectric h-coefficient	$-h_{ikm} = \dfrac{\partial^2 U}{\partial D_i \, \partial V_{km}} = \dfrac{\partial E_i}{\partial V_{km}} = \dfrac{\partial \tau^*_{km}}{\partial D_i}$
Elastic coefficient	$c^{D,S}_{ijkm} = \dfrac{\partial^2 U}{\partial V_{ij} \, \partial V_{km}} = \dfrac{\partial \tau^*_{km}}{\partial V_{ij}}$
Electro-optic coefficient	$\beta_0 r_{ijk} = \dfrac{\partial^3 U}{\partial D_i \, \partial D_j \, \partial D_k} = \dfrac{\partial^2 E_k}{\partial D_i \, \partial D_j} = \dfrac{\partial \beta_{jk}}{\partial D_i} \quad (\beta_0 \equiv 1/\varepsilon_0)$
Piezo-optic coefficient	$\beta_0 m_{ijpq} = \dfrac{\partial^3 U}{\partial V_{ij} \, \partial D_p \, \partial D_q} = \dfrac{\partial^2 E_q}{\partial V_{ij} \, \partial D_p} = \dfrac{\partial^2 \tau^*_{ij}}{\partial D_p \, \partial D_q} = \dfrac{\partial \beta_{pq}}{\partial V_{ij}} = -\dfrac{\partial h_{qij}}{\partial D_p}$
	$\beta_0 b_{kmpqi} = \dfrac{\partial^3 U}{\partial V_{km} \, \partial V_{pq} \, \partial D_i} = \dfrac{\partial^2 E_i}{\partial V_{km} \, \partial V_{pq}} = \dfrac{\partial^2 \tau^*_{pq}}{\partial V_{km} \, \partial D_i} = \dfrac{-\partial h_{ikm}}{\partial V_{pq}} = \dfrac{\partial c^{D,S}_{kmpq}}{\partial D_i}$

Third-order elastic coefficient	$c^{D,S}_{ijkmpq}=\dfrac{\partial^3 U}{\partial V_{ij}\,\partial V_{km}\,\partial V_{pq}}=\dfrac{\partial^2 \tau^*_{pq}}{\partial V_{ij}\,\partial V_{km}}=\dfrac{\partial c^{D,S}_{kmpq}}{\partial V_{ij}}$
Quadratic electro-optic coefficients	$\beta_0 f_{ijkm}=\dfrac{\partial^4 U}{\partial D_i\,\partial D_j\,\partial D_k\,\partial D_m}=\dfrac{\partial^3 E_i}{\partial D_j\,\partial D_k\,\partial D_m}=\dfrac{\partial^2 \beta_{ij}}{\partial D_k\,\partial D_m}=\beta_0\dfrac{\partial r_{ijk}}{\partial D_m}$
Fourth-order elastic coefficient	$c^{D,S}_{ijkmpquv}=\dfrac{\partial^4 U}{\partial V_{ij}\,\partial V_{km}\,\partial V_{pq}\,\partial V_{uv}}=\dfrac{\partial^3 \tau^*_{uv}}{\partial V_{ij}\,\partial V_{km}\,\partial V_{pq}}=\dfrac{\partial^2 c^{D,S}_{pquv}}{\partial V_{ij}\,\partial V_{km}}=\dfrac{\partial c^{D,S}_{kmpquv}}{\partial V_{ij}}$
	$h_{ijkpq}=\dfrac{\partial^4 U}{\partial D_i\,\partial D_j\,\partial D_k\,\partial V_{pq}}=\dfrac{\partial^3 E_i}{\partial D_j\,\partial D_k\,\partial V_{pq}}=\dfrac{\partial^3 \tau^*_{pq}}{\partial D_i\,\partial D_j\,\partial D_k}=-\dfrac{\partial^2 h_{kpq}}{\partial D_i\,\partial D_j}=\dfrac{\partial^2 \beta_{ij}}{\partial D_k\,\partial V_{pq}}=\beta_0\dfrac{\partial m_{pqjk}}{\partial D_i}=\beta_0\dfrac{\partial r_{ijk}}{\partial V_{pq}}$
	$w_{ijkmpq}=\dfrac{\partial^4 U}{\partial D_i\,\partial D_j\,\partial V_{km}\,\partial V_{pq}}=\dfrac{\partial^3 E_i}{\partial D_j\,\partial V_{km}\,\partial V_{pq}}=\dfrac{\partial^3 \tau^*_{pq}}{\partial D_i\,\partial D_j\,\partial V_{km}}=\dfrac{-\partial^2 h_{ipq}}{\partial D_j\,\partial V_{km}}=\dfrac{\partial^2 \beta_{ij}}{\partial V_{km}\,\partial V_{pq}}=\dfrac{\partial^2 c^{D,S}_{kmpq}}{\partial D_i\,\partial D_j}=\beta_0\dfrac{\partial b_{kmpqi}}{\partial D_j}=\beta_0\dfrac{\partial m_{kmij}}{\partial V_{pq}}$
	$b_{uijkmpq}=\dfrac{\partial^4 U}{\partial D_u\,\partial V_{ij}\,\partial V_{km}\,\partial V_{pq}}=\dfrac{\partial^3 E_u}{\partial V_{ij}\,\partial V_{km}\,\partial V_{pq}}=\dfrac{\partial^3 \tau^*_{pq}}{\partial D_u\,\partial V_{ij}\,\partial V_{km}}=-\dfrac{\partial^2 h_{uij}}{\partial V_{km}\,\partial V_{pq}}=\dfrac{\partial^2 c^{D,S}_{kmpq}}{\partial D_u\,\partial V_{ij}}=\beta_0\dfrac{\partial b_{kmpqu}}{\partial V_{ij}}=\dfrac{\partial c^{D,S}_{ijkmpq}}{\partial D_u}$

$$\begin{aligned} E_i(S, D_j, V_{pq}) - E_i(S, 0, 0) = {} & \beta_{ij} D_j - h_{ipq} V_{pq} \\ & + \tfrac{1}{2}\beta_0 r_{ijk} D_j D_k + \beta_0 m_{pqji} V_{pq} D_j \\ & + \tfrac{1}{2}\beta_0 b_{kmpqi} V_{km} V_{pq} + \tfrac{1}{6}\beta_0 f_{ijkm} D_j D_k D_m \\ & + \tfrac{1}{2} h_{ijkpq} D_j D_k V_{pq} + \tfrac{1}{2} w_{ijkmpq} D_j V_{km} V_{pq} \\ & + \tfrac{1}{6} b_{ikmpqrs} V_{km} V_{pq} V_{rs} + \cdots, \end{aligned} \tag{14.2}$$

$$\begin{aligned} \tau^*_{km}(S, D_j, V_{pq}) - \tau^*_{km}(S, 0, 0) = {} & -h_{ikm} D_i + c^{D,S}_{kmpq} V_{pq} + \tfrac{1}{2}\beta_0 m_{kmij} D_i D_j \\ & + \beta_0 b_{kmpqi} V_{pq} D_i + \tfrac{1}{2} c^{D,S}_{kmpqrs} V_{pq} V_{rs} \\ & + \tfrac{1}{6} h_{ijukm} D_i D_j D_u + \tfrac{1}{2} w_{ijpqkm} D_i D_j V_{pq} \\ & + \tfrac{1}{2} b_{ipqrskm} D_i V_{pq} V_{rs} \\ & + \tfrac{1}{6} c^{D,S}_{ijpqrskm} V_{ij} V_{pq} V_{rs} + \cdots. \end{aligned} \tag{14.3}$$

These expressions are appropriate when changes take place at constant entropy. The effect of small changes of entropy on E_i and τ^*_{km} can be expressed by

$$\begin{aligned} E_i(S, 0, 0) - E_i(S_0, 0, 0) &= -(S - S_0)\, q_i^{V,D} + \cdots \\ \tau^*_{km}(S, 0, 0) - \tau^*_{km}(S_0, 0, 0) &= -T(S - S_0)\, \gamma^D_{km} + \cdots. \end{aligned} \tag{14.4}$$

The choices of variables S or T, $\boldsymbol{D}$ or $\boldsymbol{E}$, V_{km} or τ^*_{km} give rise to a total of eight different thermodynamic potentials, two corresponding to each of the four in Table 10.1. Table 14.2 lists these eight functions and the names and symbols assigned by MASON (1966). The strains V_{ij} and thermodynamic tensions τ^*_{ij} are reduced to a single-subscript notation as described in Sect. 9. The mechanical-electrical Maxwell coefficients, consisting of the mixed partial derivatives of a thermodynamic potential with respect to an electrical variable ($\boldsymbol{D}$ or $\boldsymbol{E}$) and a mechanical variable (strain V_α or thermodynamic tension τ^*_α) serve to define various forms of piezoelectric coefficients. The electrical-thermal Maxwell coefficients, which consist of the mixed partial derivatives of a thermodynamic potential with respect to an electrical variable and a thermal variable (S or T), are various forms of pyroelectric coefficients. The entire mechanical-thermal formalism of Sect. 10 can be carried over in two ways, with $\boldsymbol{E}$ constant, and with $\boldsymbol{D}$ constant. The mechanical-thermal Maxwell coefficients, not shown in Table 14.2, remain essentially as given in Table 10.1, except that there are now two forms of each coefficient, one which applies at constant $\boldsymbol{E}$, and another at constant $\boldsymbol{D}$. Similarly, there are now four forms of elastic and compliance coefficients corresponding to the constancy of the pairs $(S, \boldsymbol{D})$, $(S, \boldsymbol{E})$, $(T, \boldsymbol{D})$, $(T, \boldsymbol{E})$. Likewise, the second derivatives of the eight thermodynamic potentials with respect to the electrical variable yield four kinds of permittivities ε_{ij} and impermittivities β_{ij}.

With the same mechanical and thermal variables held constant,

$$\frac{\partial D_i}{\partial E_j} \frac{\partial E_j}{\partial D_k} = \delta_{ik},$$

and hence the permittivities $\varepsilon_{ij} = \partial D_i / \partial E_j$ form a 3×3 matrix that is the inverse of the matrix of impermittivities $\beta_{jk} = \partial E_j / \partial D_k$:

$$\varepsilon_{ij} \beta_{jk} = \delta_{jk} \qquad \text{(same variables held constant)}. \tag{14.5}$$

At either constant strain or at constant thermodynamic tensions, we can write, in view of $D_i = D_i[T, E_j(T, D_i)]$,

$$\left(\frac{\partial D_i}{\partial T}\right)_{\boldsymbol{D}} = 0 = \left(\frac{\partial D_i}{\partial T}\right)_{\boldsymbol{E}} + \left(\frac{\partial D_i}{\partial E_j}\right)_T \left(\frac{\partial E_j}{\partial T}\right)_{\boldsymbol{D}}. \tag{14.6}$$

Table 14.2. *Thermodynamic potentials including electrical variables.* (After MASON, 1966.)

Potential	Independent variables	Differential	Maxwell coefficients (Mechanical-electrical)	Maxwell coefficients (Electrical-thermal)
Internal energy U	(V_α, D_i, S)	$\tau^*_\alpha\, dV_\alpha + E_i\, dD_i + T\, dS$	$-h_{i\alpha} = \frac{\partial \tau^*_\alpha}{\partial D_i} = \frac{\partial E_i}{\partial V_\alpha}$	$-\frac{T}{C^D_V} q^V_i = -q^{V,S}_i = \frac{\partial E_i}{\partial S} = \frac{\partial T}{\partial D_i} = -\beta^{V,S}_{ij} p^{V,S}_j$
Free energy $A = U - TS$	(V_α, D_i, T)	$\tau^*_\alpha\, dV_\alpha + E_i\, dD_i - S\, dT$	$-h^T_{i\alpha} = \frac{\partial \tau^*_\alpha}{\partial D_i} = \frac{\partial E_i}{\partial V_\alpha}$	$-q^V_i = \frac{\partial E_i}{\partial T} = -\frac{\partial S}{\partial D_i} = -\beta^{V,T}_{ij} p^V_j$
Enthalpy $H = U - \tau^*_\alpha V_\alpha - E_i D_i$	(τ^*_α, E_i, S)	$-V_\alpha\, d\tau^*_\alpha - D_i\, dE_i + T\, dS$	$d_{i\alpha} = \frac{\partial V_\alpha}{\partial E_i} = \frac{\partial D_i}{\partial \tau^*_\alpha}$	$-\frac{T}{C^E_\tau} p^\tau_i = -p^{\tau,S}_i = -\frac{\partial D_i}{\partial S} = \frac{\partial T}{\partial E_i} = -\varepsilon^{\tau,S}_{ij} q^{\tau,S}_j$
Elastic enthalpy $H_1 = U - \tau^*_\alpha V_\alpha$	(τ^*_α, D_i, S)	$-V_\alpha\, d\tau^*_\alpha + E_i\, dD_i + T\, dS$	$g_{i\alpha} = \frac{\partial V_\alpha}{\partial D_i} = -\frac{\partial E_i}{\partial \tau^*_\alpha}$	$-\frac{T}{C^D_\tau} q^\tau_i = -q^{\tau,S}_i = \frac{\partial E_i}{\partial S} = \frac{\partial T}{\partial D_i} = -\beta^{\tau,S}_{ij} p^{\tau,S}_j$
Electrical enthalpy $H_2 = U - E_i D_i$	(V_α, E_i, S)	$\tau^*_\alpha\, dV_\alpha - D_i\, dE_i + T\, dS$	$-e_{i\alpha} = \frac{\partial \tau^*_\alpha}{\partial E_i} = -\frac{\partial D_i}{\partial V_\alpha}$	$-\frac{T}{C^E_V} p^V_i = -p^{V,S}_i = -\frac{\partial D_i}{\partial S} = \frac{\partial T}{\partial E_i} = -\varepsilon^{V,S}_{ij} q^{V,S}_j$
Gibbs function $G = U - \tau^*_\alpha V_\alpha - E_i D_i - TS$	(τ^*_α, E_i, T)	$-V_\alpha\, d\tau^*_\alpha - D_i\, dE_i - S\, dT$	$d^T_{i\alpha} = \frac{\partial V_\alpha}{\partial E_i} = \frac{\partial D_i}{\partial \tau^*_\alpha}$	$p^\tau_i = \frac{\partial D_i}{\partial T} = \frac{\partial S}{\partial E_i} = \varepsilon^{\tau,T}_{ij} q^\tau_j$
Elastic Gibbs function $G_1 = U - \tau^*_\alpha V_\alpha - TS$	(τ^*_α, D_i, T)	$-V_\alpha\, d\tau^*_\alpha + E_i\, dD_i - S\, dT$	$g^T_{i\alpha} = \frac{\partial V_\alpha}{\partial D_i} = -\frac{\partial E_i}{\partial \tau^*_\alpha}$	$-q^\tau_i = \frac{\partial E_i}{\partial T} = -\frac{\partial S}{\partial D_i} = -\beta^{\tau,T}_{ij} p^\tau_j$
Electric gibbs function $G_2 = U - E_i D_i - TS$	(V_α, E_i, T)	$\tau^*_\alpha\, dV_\alpha - D_i\, dE_i - S\, dT$	$-e^T_{i\alpha} = \frac{\partial \tau^*_\alpha}{\partial E_i} = -\frac{\partial D_i}{\partial V_\alpha}$	$p^V_i = \frac{\partial D_i}{\partial T} = \frac{\partial S}{\partial E_i} = \varepsilon^{V,T}_{ij} q^V_j$

Hence

$$p_i^V = \varepsilon_{ij}^{V,T} q_i^V, \qquad p_i^\tau = \varepsilon_{ij}^{\tau,T} q_i^\tau. \tag{14.7}$$

The pyroelectric coefficient is called p or q when the *dependent* electrical variable in the corresponding potential is $\boldsymbol{D}$ or $\boldsymbol{E}$ respectively. The superscripts on the pyroelectric coefficients indicate the independent mechanical and thermal variables of the corresponding potential, except that the thermal superscript is omitted when temperature is independent. Since a relation similar to (14.6) applies with T replaced by S, there are relations similar to (14.7) when the independent variable is S. The present q_i is the $(-q_i)$ of MASON (1966).

The four specific heats are defined analogously to (10.5) and (10.6):

$$C_V^D = \left(\frac{\partial U}{\partial T}\right)_{V,D} = \frac{\partial U}{\partial S}\left(\frac{\partial S}{\partial T}\right)_{V,D} = T\left(\frac{\partial S}{\partial T}\right)_{V,D}; \qquad C_V^E = T\left(\frac{\partial S}{\partial T}\right)_{V,E}; \quad C_\tau^E = T\left(\frac{\partial S}{\partial T}\right)_{\tau,E}; \quad C_\tau^D = T\left(\frac{\partial S}{\partial T}\right)_{\tau,D}. \tag{14.8}$$

With the same mechanical and electrical independent variables held constant, there are relations like (10.7) and

$$\frac{\partial (D_i \text{ or } E_i)}{\partial S} = \frac{\partial T}{\partial S}\,\frac{\partial (D_i \text{ or } E_i)}{\partial T} = \frac{T}{C}\,\frac{\partial (D_i \text{ or } E_i)}{\partial T} \tag{14.9}$$

in which C is either C_V^E or C_τ^E for the differentiation of D_i and C_V^D or C_τ^D for the differentiation of E_i. The relations (14.9) make four connections between the eight electrical-thermal Maxwell coefficients. These connections are indicated in Table 14.2. For example,

$$q_i^{V,S} \equiv \frac{-\partial^2 U}{\partial D_i\,\partial S} = -\frac{T}{C_V^D}\,\frac{\partial^2 A}{\partial D_i\,\partial T} = \frac{T}{C_V^D}\,q_i^V. \tag{14.10}$$

The relations following from (14.6) and (14.9) connect pyroelectric coefficients at constant strain with other pyroelectric coefficients at constant strain, and pyroelectric coefficients at constant thermodynamic tensions with other pyroelectric coefficients at constant thermodynamic tensions. A connection between p_i^V and p_i^τ can be obtained by considering that at constant $\boldsymbol{E}$, the variable $\boldsymbol{D}$ can be expressed as a function $D_i[T, V_\beta(T, \tau_\alpha^*)]$. Then

$$\left(\frac{\partial D_i}{\partial T}\right)_{\tau^*} = \left(\frac{\partial D_i}{\partial T}\right)_V + \left(\frac{\partial D_i}{\partial V_\beta}\right)_T \left(\frac{\partial V_\beta}{\partial T}\right)_{\tau^*}. \tag{14.11}$$

By a simple generalization of Table 10.1, the derivative $\partial V_\beta / \partial T$ is the expansion coefficient at constant τ^* and $\boldsymbol{E}$, which we may denote by α_β^E. Hence, referring to Table 14.2 for the symbols, we find that (14.11) can be expressed as

$$p_i^\tau = p_i^V + e_{i\beta}^T \alpha_\beta^E. \tag{14.12}$$

By interchanging $\boldsymbol{D}$ and $\boldsymbol{E}$ in (14.11), we find

$$q_i^\tau = q_i^V + h_{i\beta}^T \alpha_\beta^D. \tag{14.13}$$

The piezoelectric d- and g-constants can be related to each other and the permittivies by considering that at constant τ_α^* and constant S or constant T, the strain can be considered as a function only of the electrical variable. Thus, $V_\alpha = V_\alpha[\boldsymbol{D}(\boldsymbol{E})]$ and

$$\left(\frac{\partial V_\alpha}{\partial E_i}\right)_{\tau^*} = \left(\frac{\partial D_j}{\partial E_i}\right)_{\tau^*} \left(\frac{\partial V_\alpha}{\partial D_j}\right)_{\tau^*}. \tag{14.14}$$

Hence

$$\begin{aligned} d_{i\alpha} &= \varepsilon_{ij}^{\tau,S} g_{j\alpha}, \\ d_{i\alpha}^T &= \varepsilon_{ij}^{\tau,T} g_{j\alpha}^T. \end{aligned} \tag{14.15}$$

Similarly, at constant strain V_α and constant S or T, we have a relation like (14.14) with V_α and τ_α^* interchanged. This leads to

$$\begin{aligned} e_{i\alpha} &= \varepsilon_{ij}^{\tau,S} h_{j\alpha}, \\ e_{i\alpha}^T &= \varepsilon_{ij}^{\tau,T} h_{j\alpha}^T. \end{aligned} \tag{14.16}$$

Table 14.3. *Connections among the various coefficients.*

Inverse relations that apply when the same variables are fixed. (The superscripts have been suppressed)

$$\beta_{ik}\,\varepsilon_{kj} = \delta_{ij}; \qquad s_{\alpha\gamma}\,c_{\gamma\beta} = \delta_{\alpha\beta}$$

Relations that apply when the same thermal variable (T or S) is fixed. (The thermal superscript has been suppressed)

$$\begin{aligned} d_{i\beta} &= \varepsilon_{ij}^{\tau} g_{j\beta} = e_{i\alpha}\, s_{\alpha\beta}^E \\ e_{i\beta} &= \varepsilon_{ij}^{V} h_{j\beta} = d_{i\alpha}\, c_{\alpha\beta}^E \\ g_{i\beta} &= \beta_{ij}^{\tau} d_{j\beta} = h_{i\alpha}\, s_{\alpha\beta}^D \\ h_{i\beta} &= \beta_{ij}^{V} e_{j\beta} = g_{i\alpha}\, c_{\alpha\beta}^D \\ \varepsilon_{ij}^{\tau} &= \varepsilon_{ij}^{V} + d_{i\alpha} e_{j\alpha} \\ \beta_{ij}^{\tau} &= \beta_{ij}^{V} - g_{i\alpha} h_{j\alpha} \\ c_{\alpha\beta}^D &= c_{\alpha\beta}^E + e_{i\alpha} h_{i\beta} \\ s_{\alpha\beta}^D &= s_{\alpha\beta}^E - d_{i\alpha} g_{i\beta} \end{aligned}$$

Relations involving pyroelectric coefficients

$$\begin{aligned} d_{i\beta}^T &= d_{j\beta} + \alpha_\beta^E p_j^{\tau,S} = d_{j\beta} + \frac{T\alpha_\beta^E p_j^\tau}{C_\tau^E} \\ e_{j\beta}^T &= e_{j\beta} + \lambda_\beta^E p_j^{V,S} = e_{j\beta} + \frac{T\lambda_\beta^E p_j^V}{C_V^E} \\ g_{j\beta}^T &= g_{j\beta} + \alpha_\beta^D q_j^{\tau,S} = g_{j\beta} + \frac{T\alpha_\beta^D q_j^\tau}{C_\tau^D} \\ h_{j\beta}^T &= h_{j\beta} + \lambda_\beta^D q_j^{V,S} = h_{j\beta} + \frac{T\lambda_\beta^D q_j^V}{C_V^D} \end{aligned}$$

$$\begin{aligned} \alpha_\beta^E &= \alpha_\beta^D + p_j^\tau g_{j\beta}^T \\ \lambda_\beta^E &= \lambda_\beta^D + q_j^V e_{j\beta}^T \\ C_V^E - C_V^D &= T\beta_{ij}^{V,T} p_i^V p_j^V \\ C_\tau^E - C_\tau^D &= T\beta_{ij}^{\tau} p_i^\tau p_j^\tau \end{aligned}$$

$$\varepsilon_{ij}^{V,T} - \varepsilon_{ij}^{V,S} = \frac{T p_i^V p_j^V}{C_V^E}, \qquad \varepsilon_{ij}^{\tau,T} - \varepsilon_{ij}^{\tau,S} = \frac{T p_i^\tau p_j^\tau}{C_\tau^E}$$

$$\beta_{ij}^{V} - \beta^{S,}{}_{ij}^{V,T} = \frac{T q_i^V q_j^V}{C_V^D}, \qquad \beta_{ij}^{\tau,S} - \beta_{ij}^{\tau,T} = \frac{T q_i^\tau q_j^\tau}{C_\tau^D}$$

The other pairs of piezoelectric coefficients, (d, e) and (g, h), are connected through the elasticities. At constant entropy S or constant temperature, $\tau_\beta^* = \tau_\beta^* [V_\alpha(\tau_\alpha^*, E_i), E_i]$ so that

$$\left(\frac{\partial \tau_\beta^*}{\partial E_i}\right)_{\tau^*} = 0 = \left(\frac{\partial \tau_\beta^*}{\partial E_i}\right)_V + \left(\frac{\partial \tau_\beta^*}{\partial V_\alpha}\right)_E \left(\frac{\partial V_\alpha}{\partial E_i}\right)_{\tau^*}. \tag{14.17}$$

Hence,

$$\begin{aligned} e_{i\beta} &= d_{i\alpha}\, c_{\alpha\beta}^{S,E}, \\ e_{i\beta}^T &= d_{i\alpha}^T\, c_{\alpha\beta}^{T,E}. \end{aligned} \tag{14.18}$$

Consideration of the relation like (14.17) with E_i replaced by D_i leads to

$$\begin{aligned} h_{i\beta} &= g_{i\alpha}\, c_{\alpha\beta}^{S,D}, \\ h_{i\beta}^T &= g_{i\alpha}^T\, c_{\alpha\beta}^{T,D}. \end{aligned} \tag{14.19}$$

Table 14.3 lists these and other relations that can be verified by straightforward derivations. It is worth noting that purely pyroelectric relations, with either V_α constant of τ_α^* constant, follow from the formalism of Sect. 10 upon making suitable replacements of the variables. Suitable replacements are indicated in Table 14.4.

Table 14.4. *Replacements for obtaining pyroelectric relations from thermoelastic relations.*

Strains V_α constant	Thermodynamic tensions τ_α^* constant
$V_\alpha \to D_i$	$V_\alpha \to D_i$
$\tau_\beta \to E_j$	$\tau_\beta \to E_j$
$c_{\alpha\beta}^S \to \beta_{ij}^{V,S}$	$c_{\alpha\beta}^S \to \beta_{ij}^{\tau,S}$
$s_{\alpha\beta}^S \to \varepsilon_{ij}^{V,S}$	$s_{\alpha\beta}^S \to \varepsilon_{ij}^{\tau,S}$
$c_{\alpha\beta}^T \to \beta_{ij}^{V,T}$	$c_{\alpha\beta}^T \to \beta_{ij}^{\tau,T}$
$s_{\alpha\beta}^T \to \varepsilon_{ij}^{V,T}$	$s_{\alpha\beta}^T \to \varepsilon_{ij}^{\tau,T}$
$\alpha_\beta \to p_j^V$	$\alpha_\beta \to p_j^\tau$
$\lambda_\beta \to q_j^V$	$\lambda_\beta \to q_j^\tau$
$C_V \to C_V^D$	$C_V \to C_\tau^D$
$C_\tau \to C_V^E$	$C_\tau \to C_\tau^E$

As an example it may be verified that (14.7) corresponds to (10.11). Corresponding to (10.21) we have the two equations

$$\begin{aligned} C_V^E - C_V^D &= T q_j^V p_j^V = T \beta_{ij}^{V,T} p_i^V p_j^V, \\ C_\tau^E - C_\tau^D &= T q_j^\tau p_j^\tau = T \beta_{ij}^\tau p_i^\tau p_j^\tau. \end{aligned} \tag{14.20}$$

As a final example, it follows from (10.15) that for each j, $j = 1, 2, 3$,

$$\begin{aligned} \frac{C_V^E}{C_V^D} &= \frac{p_i^V \beta_{ij}^{V,S}}{p_i^V \beta_{ij}^{V,T}}, \\ \frac{C_\tau^E}{C_\tau^D} &= \frac{p_i^\tau \beta_{ij}^{\tau,S}}{p_i^\tau \beta_{ij}^{\tau,T}}. \end{aligned} \tag{14.21}$$

14.2. Linear piezoelectric equations. The linear piezoelectric equations connect V_α, τ^*_α, $\boldsymbol{E}$, and $\boldsymbol{D}$ by linear equations. In these equations, we can retain the view that the V_α are the finite strain components, τ^*_α the thermodynamic tensions, and $\boldsymbol{E}$ and $\boldsymbol{D}$ are the material representations of the $\mathscr{E}$ and $\mathscr{D}$ fields. From this point of view, the equations represent the linear terms in power series expansions of the dependent variables in terms of the independent variables, at constant entropy or at constant temperature. Additional approximations are needed in the applications in order to obtain a linear system of differential equations. These consist of the usual small displacement and small displacement-gradient approximations, such as approximating the finite strain components by the "infinitesimal" ones, etc.

There are four equivalent sets of equations, corresponding to the four choices of independent variables (V_α, D_i), (τ^*_α, E_i), (τ^*_α, D_i), and (V_α, E_i). From the notation in Table 14.2 and a simple generalization of Table 10.1 (to incorporate constant $\boldsymbol{E}$ or constant $\boldsymbol{D}$), the equations may be listed as

$$\begin{aligned} \tau^*_\gamma &= c^D_{\gamma\alpha} V_\alpha - h_{j\gamma} D_j, \\ E_i &= -h_{i\alpha} V_\alpha + \beta^V_{ij} D_j, \end{aligned} \tag{14.22}$$

$$\begin{aligned} V_\gamma &= s^E_{\gamma\alpha} \tau^*_\alpha + d_{j\gamma} E_j, \\ D_i &= d_{i\alpha} \tau^*_\alpha + \varepsilon^\tau_{ij} E_j, \end{aligned} \tag{14.23}$$

$$\begin{aligned} V_\gamma &= s^D_{\gamma\alpha} \tau^*_\alpha + g_{j\gamma} D_j, \\ E_i &= -g_{i\alpha} \tau^*_\alpha + \beta^\tau_{ij} D_j, \end{aligned} \tag{14.24}$$

$$\begin{aligned} \tau^*_\gamma &= c^E_{\gamma\alpha} V_\alpha - e_{j\gamma} E_j, \\ D_i &= e_{i\alpha} V_\alpha + \varepsilon^V_{ij} E_j. \end{aligned} \tag{14.25}$$

The equations apply either at constant entropy or constant temperature, with isentropic or isothermal coefficients, respectively. The thermal superscripts on the coefficients have been suppressed. To each equation there could be added a term linear in the temperature difference or entropy difference from an initial state, and in each set there could be included a thermal equation. For example, corresponding to (14.25) are

$$\begin{aligned} \tau^*_\gamma &= c^{S,E}_{\gamma\alpha} V_\alpha - e_{j\gamma} E_j - \frac{T}{C^E_V} \lambda^E_\gamma (S - S_0), \\ D_i &= e_{i\alpha} V_\alpha + \varepsilon^{V,S}_{ij} E_j + \frac{T}{C^E_V} p^V_i (S - S_0), \\ T - T_0 &= -\frac{T}{C^E_V} [\lambda^E_\alpha V_\alpha + p^V_j E_j - (S - S_0)], \end{aligned} \tag{14.26}$$

and

$$\begin{aligned} \tau^*_\gamma &= c^{T,E}_{\gamma\alpha} V_\alpha - e^T_{j\gamma} E_j - \lambda^E_\gamma (T - T_0), \\ D_i &= e^T_{i\alpha} V_\alpha + \varepsilon^{V,T}_{ij} E_j + p^V_i (T - T_0), \\ S - S_0 &= \lambda^E_\alpha V_\alpha + p^V_j E_j + \frac{C^E_V}{T} (T - T_0). \end{aligned} \tag{14.27}$$

The result of substituting $(T - T_0)$ from (14.26) into the first two equations of (14.27) is simply to reproduce the first two equations of (14.26), when account is taken of the connections between the isentropic and isothermal coefficients.

For many purposes, the thermal variations need not be included in the equations. It is customary to use the forms in (14.22) to (14.25), taking account of thermal effects only to the extent of using isothermal coefficients for static problems and isentropic coefficients for problems in wave propagation.

E. Material symmetry.

15. Isotropy groups, Laue groups, and crystal point groups.

The symmetry of a crystal or other material imposes constraints on the coefficients that appear in constitutive relations such as (14.2) and (14.3). The fundamental ideas have been generalized by NOLL and summarized in this encyclopedia, Vol. III/3 by TRUESDELL and NOLL (1965). The discussion is in the context of the mechanics of simple materials. A material is called simple (or grade 1) if the local response to deformations at a particle X is determined by deformations that are homogeneous in the neighborhood of X, i.e., the response is independent of the deformation gradients higher than the first. Assuming the response of a material to be given by a response functional, we can consider its *isotropy group* to be the group of transformations of the reference configuration for which the response functional remains unchanged. In this sense, transformations belonging to the isotropy group connect configurations of the material that are indistinguishable, in as much as they have indistinguishable responses to any given deformation history. According to NOLL's definition, the isotropy group depends on the reference configuration that is chosen. *Undistorted states* are local reference configurations for which the corresponding isotropy group is a subgroup of the orthogonal group. As far as mechanical response is concerned, NOLL shows that *every* isotropy group contains the inversion. Twelve *proper* subgroups (rotation groups) of the orthogonal groups then seem adequate to distinguish the kinds of symmetries that occur in the pure mechanics of anisotropic materials. A thirteenth case is the fully isotropic material, in which there are no preferred directions at all.

When additional interactions are involved, such as piezoelectricity, the concepts of isotropy groups and undistorted states, made precise for pure mechanics by NOLL, need to be extended. The essence of the matter is simply that certain different reference configurations have indistinguishable responses to whatever excitation is applied. With electromechanical interactions included, the number of types of isotropy groups needed to characterize the kinds of symmetry is increased beyond the thirteen needed for pure mechanics. The inclusion of magnetic effects produces a further increase.

Crystals are classified into thirty-two point groups, each characterized by a different set of symmetry elements. These thirty-two point groups have been partitioned into eleven "Laue groups," which correspond to eleven of the rotation groups referred to above. The twelfth rotation group is transverse isotropy, corresponding to an arbitrary rotation about a specified axis.

Table 15.1 lists the eleven Laue groups, the generators of the associated rotation group (called g_0) as given by TRUESDELL and NOLL (1965), and the point groups that are included in each Laue group. The symbol for a generator is $R_{\boldsymbol{N}}^n$, which signifies a right-handed rotation through the angle $2\pi/n$ about an axis in the direction of the unit vector $\boldsymbol{N}$. In the table, $\boldsymbol{i}, \boldsymbol{j}, \boldsymbol{k}$ denote a right-handed orthonormal basis, and $\boldsymbol{p} = (\boldsymbol{i}+\boldsymbol{j}+\boldsymbol{k})/3^{\frac{1}{2}}$. The identity is denoted by 1.

Table 15.1. *Laue groups, generators of associated rotation groups, and point groups included in each Laue group.*

Crystal system	Laue group	Generators of g_0	Point groups included
Triclinic	N	1	$1, \bar{1}$
Monoclinic	M	$\mathcal{R}_k^2$	$2, m, \frac{2}{m}$
Orthorhombic	O	$\mathcal{R}_i^2, \mathcal{R}_j^2$	$222, mm2, \frac{2}{m}\frac{2}{m}\frac{2}{m}$
Tetragonal	TII	$\mathcal{R}_k^4$	$4, \bar{4}, \frac{4}{m}$
Tetragonal	TI	$\mathcal{R}_k^4, \mathcal{R}_i^2$	$422, 4mm, \bar{4}2m, \frac{4}{m}\frac{2}{m}\frac{2}{m}$
Cubic	CII	$\mathcal{R}_i^2, \mathcal{R}_j^2, \mathcal{R}_p^3$	$23, \frac{2}{m}\bar{3}$
Cubic	CI	$\mathcal{R}_i^4, \mathcal{R}_j^4, \mathcal{R}_k^4$	$432, \bar{4}3m, \frac{4}{m}\bar{3}\frac{2}{m}$
Rhombohedral (trigonal)	RII	$\mathcal{R}_k^3$	$3, \bar{3}$
Rhombohedral (trigonal)	RI	$\mathcal{R}_i^2, \mathcal{R}_k^3$	$32, 3m, \bar{3}\frac{2}{m}$
Hexagonal	HII	$\mathcal{R}_k^6$	$6, \bar{6}, \frac{6}{m}$
Hexagonal	HI	$\mathcal{R}_i^2, \mathcal{R}_k^6$	$622, 6mm, \bar{6}m2, \frac{6}{m}\frac{2}{m}\frac{2}{m}$

The symmetry elements associated with crystal point groups are the following:

- n-fold rotation axes, corresponding to rotations of $2\pi/n$, where n can be 1, 2, 3, 4, or 6. Such axes are denoted by the number n, i.e., 1, 2, 3, 4, or 6. The 2-, 3-, 4-, and 6-fold rotation axes are sometimes called dyad, triad, tetrad, and hexad axes, respectively;
- reflection planes—A reflection plane is denoted by the letter m, suggestive of *mirror*;
- inversion center—This corresponds to the operation $\boldsymbol{X} \rightarrow -\boldsymbol{X}$. An inversion center is denoted by $\bar{1}$.
- n-fold rotation-inversion axes, where n can be 1, 2, 3, 4, or 6. The corresponding operation is rotation followed by an inversion. The rotation-inversion axes are denoted by the symbols $\bar{1}$, $\bar{2}$, $\bar{3}$, $\bar{4}$, $\bar{6}$.

It may be noted that $\bar{2}$ is equivalent to mirror reflection in a plane normal to the 2-fold rotation-inversion axis. Therefore, one writes $\bar{2} = m$. Also, $\bar{3}$ implies a 3-fold rotation axis and an inversion center, while $\bar{6}$ is equivalent to a 3-fold rotation axis with a mirror plane normal to it.

The symbols for point groups express some rudimentary information about the associated symmetry. There are two forms of the symbols, full and short. The full symbol always starts with a number $\bar{n}$ or n that gives the highest order rotation or rotation-inversion axis that is present, except in the two classes of the cubic system that comprise the CII Laue group. In ten of the classes, this completes the symbol. When a mirror plane is normal to the first-indicated axis, the m is written as the denominator of a fraction with the order n of the axis in the numerator. An

m simply following the number indicates a mirror plane or planes parallel to the indicated axis. Thus,

$\frac{n}{m}$ or n/m indicates an n-fold rotation axis with a mirror plane normal to it.

nm indicates an n-fold rotation axis with a mirror plane or planes parallel to it.

In the cubic system, the second *number* in the full symbol is always a 3, suggestive of the four 3-fold axes that characterize this system. In all other cases, the second number, if present, is a 2, signifying a 2-fold axis normal to the first-indicated axis. Thus,

$n2$ denotes an n-fold rotation axis with a 2-fold rotation (dyad) axis or axes normal to it.

$\bar{n}2$ denotes an n-fold rotation-inversion axis with a 2-fold rotation axis or axes normal to it.

The third number, if present, is always a 2, and this signifies a dyad axis perpendicular to the other two indicated axes, except in the cubic system. Other symbols that occur are explained as follows: $mm2$ indicates two mirror planes intersecting in a dyad axis. $4mm$ denotes a tetrad axis with two mirror planes parallel to it.

Table 15.2 lists the 32 crystal point groups, the short and full "international symbols", developed by HERMANN (1928, 1929, 1930) and MAUGUIN (1931) as well as other symbols due to NIGGLI (1919) and SCHOENFLIES (1923) that have been widely used. The 32 classes were assigned names and numbers by GROTH. These are also indicated in Table 15.2.

The Niggli-Schoenflies symbols for the point groups may be explained as follows:

C_n describes a group having a single n-fold rotation axis.
D_n describes a group having an n-fold axis and n two-fold axes at right angles to it.
S_n describes a group having an n-fold rotation-inversion axis.
V describes a group with three mutually perpendicular two-fold axes.
T describes a group with four three-fold axes, placed to correspond to the symmetry of a regular tetrahedron (tetrahedral group).
O describes a group with three mutually perpendicular four-fold axes (cubic group).

- Subscript v (vertical) signifies the presence of a reflection plane containing the symmetry axis.
- Subscript h (horizontal) signifies the presence of a reflection plane perpendicular to a symmetry axis.
- Subscript d (dihedral) signifies the presence of a reflection plane bisecting the angle between two two-fold axes.
- Subscript i (inversion) means that the group contains the inversion.
- Subscript s means that the group contains a reflection plane.

Additional properties of the 32 point groups are shown in Table 15.3, which lists the order of each group and the number of associated symmetry operations of each kind. In each case, the total number of symmetry operations is equal to the order of the group. In groups having inversion operations (an inversion center or rotation-inversion axes are present), the number of inversion operations is equal to the number of proper rotation operations. For $n \geq 3$, there are two operations

Table 15.2. *The thirty-two crystal point groups.*

Laue group	International symbol short	International symbol full	Schoenflies symbol	Crystal class name*	No.*
N	1	1	C_1	Triclinic asymmetric	1
	$\bar{1}$	$\bar{1}$	$C_i = S_2$	Triclinic pinacoidal	2
M	2	2	C_2	Monoclinic sphenoidal	3
	m	$\bar{2}$	$C_s = C_{1h}$	Monoclinic domatic	4
	$2/m$	$\frac{2}{m}$	C_{2h}	Monoclinic prismatic	5
O	222	222	$D_2 = V$	Orthorhombic disphenoidal	6
	$mm2$	$mm2$	C_{2v}	Orthorhombic pyramidal	7
	mmm	$\frac{2}{m}\frac{2}{m}\frac{2}{m}$	$D_{2h} = V_h$	Orthorhombic dipyramidal	8
TII	4	4	C_4	Tetragonal pyramidal	10
	$\bar{4}$	$\bar{4}$	S_4	Tetragonal disphenoidal	9
	$4/m$	$\frac{4}{m}$	C_{4h}	Tetragonal dypyramidal	13
TI	422	422	D_4	Tetragonal trapezohedral	12
	$4mm$	$4mm$	C_{4v}	Ditetragonal-pyramidal	14
	$\bar{4}2m$	$\bar{4}2m$	$D_{2d} = V_d$	Tetragonal scalenohedral	11
	$4/mmm$	$\frac{4}{m}\frac{2}{m}\frac{2}{m}$	D_{4h}	Ditetragonal-dipyramidal	15
RII	3	3	C_3	Trigonal pyramidal	16
	$\bar{3}$	$\bar{3}$	$C_{3i} = S_6$	Trigonal rhombohedral	17
RI	32	32	D_3	Trigonal trapezohedral	18
	$3m$	$3m$	C_{3v}	Ditrigonal-pyramidal	20
	$\bar{3}m$	$\bar{3}\frac{2}{m}$	D_{3d}	Ditrigonal-scalenohedral	21
HII	6	6	C_6	Hexagonal pyramidal	23
	$\bar{6}$	$\bar{6}$	C_{3h}	Trigonal dipyramidal	19
	$6/m$	$\frac{6}{m}$	C_{6h}	Hexagonal dipyramidal	25
HI	622	622	D_6	Hexagonal trapezohedral	24
	$6mm$	$6mm$	C_{6v}	Dihexagonal-pyramidal	26
	$6m2$	$6m2$	D_{3h}	Ditrigonal-dipyramidal	22
	$6/mm$	$\frac{6}{m}\frac{2}{m}\frac{2}{m}$	D_{6h}	Dihexagonal-dipyramidal	27
CII	23	23	T	Cubic tetrahedral-pentagonal-dodecahedral (tetartoidal)	28
	$m3$	$\frac{2}{m}\bar{3}$	T_h	cubic dyakis-dodecahedral (diploidal)	30
CI	432	432	O	Cubic pentagonal icositetrahedral (gyroidal)	29
	$\bar{4}3m$	$\bar{4}3m$	T_d	Cubic hexakis tetrahedral (hextetrahedral)	31
	$m3m$	$\frac{4}{m}\bar{3}\frac{2}{m}$	O_h	Cubic hexakis-octahedral (hexoctahedral)	32

* After GROTH.

Table 15.3. *Order of the point groups and number of symmetry operations of each kind.*

Group	Order	1	2	3	4	6	$\bar{1}$	$m=\bar{2}$	$\bar{3}$	$\bar{4}$	$\bar{6}$
1	1	1									
$\bar{1}$	2	1					1				
2	2	1	1								
m	2	1						1			
$2/m$	4	1	1				1	1			
222	4	1	3								
$mm2$	4	1	1					2			
mmm	8	1	3				1	3			
3	3	1		2							
$\bar{3}$	6	1		2			1		2		
32	6	1	3	2							
$3m$	6	1		2				3			
$\bar{3}m$	12	1	3	2			1	3	2		
$\bar{4}$	4	1	1							2	
4	4	1	1		2						
$4/m$	8	1	1		2		1	1		2	
422	8	1	5		2						
$4mm$	8	1	1		2			4			
$4/mmm$	16	1	5		2		1	5		2	
$\bar{4}2m$	8	1	3					2		2	
$\bar{6}$	6	1		2				1			2
6	6	1	1	2		2					
$6/m$	12	1	1	2		2	1	1	2		2
622	12	1	7	2		2					
$6mm$	12	1	1	2		2		6			
$6/mmm$	24	1	7	2		2	1	7	2		2
$\bar{6}m2$	12	1	3	2				4			2
23	12	1	3	8							
$m3$	24	1	3	8			1	3	8		
432	24	1	9	8	6						
$\bar{4}3m$	24	1	3	8				6		6	
$m3m$	48	1	9	8	6		1	9	8	6	

corresponding to each n-fold axis, a right-handed rotation of $2\pi/n$ about the axis, and its inverse, a left-handed rotation of the same amount. Thus, the number of n-fold *axes* is half the tabulated number of operations for $n=3$, 4, 6. For $n=2$, each axis corresponds to only a single operation, since the left-handed and right-handed rotations produce the same final configuration. (The operation is its own inverse.)

The traditional partitioning of the crystal classes into seven crystal *systems* has also been indicated in Table 15.1. The symmetry essential to each system is indicated in Table 15.4. The trigonal (rhombohedral) and hexagonal systems are not always separated.

Table 15.4. *Essential symmetry of the crystal systems.*

System	Essential symmetry
Triclinic	None
Monoclinic	One 2-fold axis or one plane
Orthorhombic	Three mutually perpendicular 2-fold axes or two planes intersecting in a 2-fold axis
Trigonal (rhombohedral)	One 3-fold axis, but no 6 or $\bar{6}$ axes
Hexagonal	Either a 6-fold rotation axis or a 6-fold rotation-inversion axis (6 or $\bar{6}$)
Tetragonal	Either a 4-fold rotation axis or a 4-fold rotation-inversion axis (4 or $\bar{4}$)
Cubic	Four 3-fold rotation axes (which imply three two-fold axes)

For additional discussion of crystal classification and symmetry operations, the reader may refer to texts on crystallography and to the article by H. JAGODZINSKI (1955) in this encyclopedia, Vol. VII/1.

In discussing piezoelectric interactions, it is important to know whether a crystal class has an inversion center. This is because inversion ($\boldsymbol{X} \to -\boldsymbol{X}$) can be pictured as reversing the material representation of the electric field $E_j = \mathscr{E}_i \partial x_i / \partial X_j$ while leaving the strain components and thermodynamic tensions unchanged. Consequently, in an expansion such as (14.2) or (14.3), taken with respect to an appropriately generalized "undeformed state" ($\boldsymbol{E}$ zero, V_{pq} zero, stress spherical), one can argue that any given coefficient h_{ipq} must be equal to its negative, and therefore zero. Thus, the so-called "linear piezoelectric effect" is prohibited in the crystal classes that have an inversion center. A similar argument applies to all coefficients that correspond to polar tensors of odd rank. The 11 point groups that have inversion symmetry can be seen by inspecting the column headed by $\bar{1}$ in Table 15.3. (The linear piezoelectric effect vanishes also in a twelfth crystal class, 432, not because the point group includes the inversion operation, but because it is ruled out by the other symmetry operations that are present.)

The restriction to an appropriately generalized "undeformed state" (here "zero field state") is crucial in arguing the odd rank material coefficients to zero. For example, the partial derivative $\partial^2 U / \partial D_i \partial V_{km}$ will generally not be zero when evaluated at other states, because of its dependence on $\boldsymbol{D}$ as expressed through the piezo-optic coefficients m_{ijpq} (cf. Table 14.1).

16. Effect of symmetry on material coefficients.

By material coefficients, we here mean derivatives of a thermodynamic potential, evaluated at some generalized undistorted, zero-field state. In the simplest case, the only nonzero *first* derivative of any thermodynamic potential is its derivative with respect to the thermal variable, either temperature or entropy. A hydrostatic pressure at the undistorted state is permissible [cf. Sect. 11.4, especially Eq. (11.35)].

The elastic, dielectric, piezoelectric, and other coefficients of crystals are ordinarily specified with reference to a set of rectangular Cartesian axes that are

Table 16.1. *Forms of a first-rank polar tensor referred to the conventional cartesian system.*

Point groups	Tensor components
1	$\|q_1, q_2, q_3\|$
2	$\|0, q_2, 0\|$
$\bar{2} = m$	$\|q_1, 0, q_3\|$
$mm2$, 4, $4mm$, 3, $3m$, 6, $6mm$	$\|0, 0, q_3\|$
All others	$\|0, 0, 0\|$

Table 16.2. *Second-rank polar tensors.*

Crystal systems	Tensor terms	Reduction for symmetric form
Triclinic 1, $\bar{1}$ 9 constants	$\begin{Vmatrix} \pi_{11} & \pi_{12} & \pi_{13} \\ \pi_{21} & \pi_{22} & \pi_{23} \\ \pi_{31} & \pi_{32} & \pi_{33} \end{Vmatrix}$	If symmetric $\pi_{12} = \pi_{21}$; $\pi_{13} = \pi_{31}$; $\pi_{23} = \pi_{32}$; 6 constants
Monoclinic 2, m, $2/m$ 5 constants	$\begin{Vmatrix} \pi_{11} & 0 & \pi_{13} \\ 0 & \pi_{22} & 0 \\ \pi_{31} & 0 & \pi_{33} \end{Vmatrix}$	If symmetric $\pi_{13} = \pi_{31}$ 4 constants
Orthorhombic 222, $mm2$, mmm 3 constants	$\begin{Vmatrix} \pi_{11} & 0 & 0 \\ 0 & \pi_{22} & 0 \\ 0 & 0 & \pi_{11} \end{Vmatrix}$	Same for symmetric tensor
Trigonal, tetragonal, hexagonal 3, $\bar{3}$, 4, $\bar{4}$, $4/m$, 6, $\bar{6}$, $6/m$ 3 constants	$\begin{Vmatrix} \pi_{11} & \pi_{12} & 0 \\ -\pi_{12} & \pi_{11} & 0 \\ 0 & 0 & \pi_{33} \end{Vmatrix}$	If symmetric $\pi_{12} = 0$ 2 constants
Trigonal, tetragonal, hexagonal 32, $3m$, $\bar{3}m$ 422, $4mm$, $\bar{4}2m$, $4/mmm$ 622, $6mm$, $\bar{6}m2$, $6/mmm$ Transverse isotropy 2 constants	$\begin{Vmatrix} \pi_{11} & 0 & 0 \\ 0 & \pi_{11} & 0 \\ 0 & 0 & \pi_{33} \end{Vmatrix}$	If symmetric the same number of constants
Cubic or isotropic 23, $m3$, $\bar{6}3m$ 432, $m3m$ 1 constant	$\begin{Vmatrix} \pi_{11} & 0 & 0 \\ 0 & \pi_{11} & 0 \\ 0 & 0 & \pi_{11} \end{Vmatrix}$	Same for symmetric tensor

chosen by convention. The conventional method of relating a set of rectangular axes to the crystallographic axes has been standardized through the work of BOND (1949) and others, and is described in publications of MASON (1950, 1966).

The material coefficients considered here are polar tensors of rank 1, 2, 3, 4, 5, and 6.

The possible forms of these tensors, referred to the conventional Cartesian axes, are indicated in Tables 16.1 to 16.6. Tables 16.1 to 16.4 are adapted from MASON

Table 16.3. *Third-rank polar tensors*

Group	Components	Group	Components
1	$\begin{Vmatrix} e_{11} & e_{12} & e_{13} & e_{14} & e_{15} & e_{16} \\ e_{21} & e_{22} & e_{23} & e_{24} & e_{25} & e_{26} \\ e_{31} & e_{32} & e_{33} & e_{34} & e_{35} & e_{36} \end{Vmatrix}$	2	$\begin{Vmatrix} 0 & 0 & 0 & e_{14} & 0 & e_{16} \\ e_{21} & e_{22} & e_{23} & 0 & e_{25} & 0 \\ 0 & 0 & 0 & e_{34} & 0 & e_{36} \end{Vmatrix}$
m	$\begin{Vmatrix} e_{11} & e_{12} & e_{13} & 0 & e_{15} & 0 \\ 0 & 0 & 0 & e_{24} & 0 & e_{26} \\ e_{31} & e_{32} & e_{33} & 0 & e_{35} & 0 \end{Vmatrix}$	222	$\begin{Vmatrix} 0 & 0 & 0 & e_{14} & 0 & 0 \\ 0 & 0 & 0 & 0 & e_{25} & 0 \\ 0 & 0 & 0 & 0 & 0 & e_{36} \end{Vmatrix}$
$mm2$	$\begin{Vmatrix} 0 & 0 & 0 & 0 & e_{15} & 0 \\ 0 & 0 & 0 & e_{24} & 0 & 0 \\ e_{31} & e_{32} & e_{33} & 0 & 0 & 0 \end{Vmatrix}$	$\overline{4}$	$\begin{Vmatrix} 0 & 0 & 0 & e_{14} & e_{15} & 0 \\ 0 & 0 & 0 & -e_{15} & e_{14} & 0 \\ e_{31} & -e_{31} & 0 & 0 & 0 & e_{36} \end{Vmatrix}$
4, 6	$\begin{Vmatrix} 0 & 0 & 0 & e_{14} & e_{15} & 0 \\ 0 & 0 & 0 & e_{15} & -e_{14} & 0 \\ e_{31} & e_{31} & e_{31} & 0 & 0 & 0 \end{Vmatrix}$	$\overline{4}2m$	$\begin{Vmatrix} 0 & 0 & 0 & e_{14} & 0 & 0 \\ 0 & 0 & 0 & 0 & e_{14} & 0 \\ 0 & 0 & 0 & 0 & 0 & e_{36} \end{Vmatrix}$
422 622	$\begin{Vmatrix} 0 & 0 & 0 & e_{14} & 0 & 0 \\ 0 & 0 & 0 & 0 & -e_{14} & 0 \\ 0 & 0 & 0 & 0 & 0 & 0 \end{Vmatrix}$	$4mm$ $6mm$ Transverse isotropy	$\begin{Vmatrix} 0 & 0 & 0 & 0 & e_{15} & 0 \\ 0 & 0 & 0 & e_{15} & 0 & 0 \\ e_{31} & e_{31} & e_{33} & 0 & 0 & 0 \end{Vmatrix}$
3	$\begin{Vmatrix} e_{11} & -e_{11} & 0 & e_{14} & e_{15} & -e_{22} \\ -e_{22} & e_{22} & 0 & e_{15} & -e_{14} & -e_{11} \\ e_{31} & e_{31} & e_{33} & 0 & 0 & 0 \end{Vmatrix}$	32	$\begin{Vmatrix} e_{11} & -e_{11} & 0 & e_{14} & 0 & 0 \\ 0 & 0 & 0 & 0 & -e_{14} & -e_{11} \\ 0 & 0 & 0 & 0 & 0 & 0 \end{Vmatrix}$
$\overline{6}$	$\begin{Vmatrix} e_{11} & -e_{11} & 0 & 0 & 0 & -e_{22} \\ -e_{22} & e_{22} & 0 & 0 & 0 & -e_{11} \\ 0 & 0 & 0 & 0 & 0 & 0 \end{Vmatrix}$	$3m$	$\begin{Vmatrix} 0 & 0 & 0 & 0 & e_{15} & -e_{22} \\ -e_{22} & e_{22} & 0 & e_{15} & 0 & 0 \\ e_{31} & e_{31} & e_{33} & 0 & 0 & 0 \end{Vmatrix}$
$\overline{6}m2$	$\begin{Vmatrix} e_{11} & -e_{11} & 0 & 0 & 0 & 0 \\ 0 & 0 & 0 & 0 & 0 & -e_{11} \\ 0 & 0 & 0 & 0 & 0 & 0 \end{Vmatrix}$	23 $\overline{4}3m$	$\begin{Vmatrix} 0 & 0 & 0 & e_{14} & 0 & 0 \\ 0 & 0 & 0 & 0 & e_{14} & 0 \\ 0 & 0 & 0 & 0 & 0 & e_{14} \end{Vmatrix}$

All components are required to be zero in the following twelve groups: $\overline{1}$, $2/m$, mmm, $4/m$, $4/mmm$, $\overline{3}$, $\overline{3}m$, $6/mmm$, $6/m$, $m3$, $m3m$, 432.

(1966), Table 16.5 is from NELSON and LAX (1971), and Table 16.6 is from BRUGGER (1965).

The pyroelectric coefficients p_i or q_i are an example of a first rank polar tensor. Pyroelectricity is possible in only ten of the 32 crystal classes.

Examples of second rank polar tensors are the dielectric permittivity ε_{ij} and the expansion coefficients α_{ij}.

Examples of third rank polar tensors are the piezoelectric coefficients and the electro-optic coefficients. One pair of indices is reduced to a single-subscript notation by the scheme described in Sect. 9. For the piezoelectric coefficients, this is the pair that refers to the mechanical variable, the last two indices of the three-subscript symbol, consistent with the six columns in the matrices of Table 16.3. In the case of the electro-optic coefficients, either the first two indices or the last two may be paired.

Table 16.4. *Fourth-rank polar tensors.*

(Type M_{ijkm}, $i\to j$, $k\to m$; Type c_{ijkm}, $i\to j$, $k\to m$, $ij\to km$; Type K_{ijkm}, $i\to j\to k\to m$)

Group							Remarks
Group I	M_{11}	M_{12}	M_{13}	M_{14}	M_{15}	M_{16}	c constants the same except $c_{ab}=c_{ba}$, resulting in 21 constants.
Triclinic	M_{21}	M_{22}	M_{23}	M_{24}	M_{25}	M_{26}	K constants the same as c, except $K_{44}=K_{23}$, $K_{55}=K_{13}$, $K_{66}=K_{12}$, $K_{46}=K_{25}$, $K_{56}=K_{14}$, $K_{45}=K_{36}$, resulting in 15 constants.
$1, \bar{1}$	M_{31}	M_{32}	M_{33}	M_{34}	M_{35}	M_{36}	
36 constants	M_{41}	M_{42}	M_{43}	M_{44}	M_{45}	M_{46}	
	M_{51}	M_{52}	M_{53}	M_{54}	M_{55}	M_{56}	
	M_{61}	M_{62}	M_{63}	M_{64}	M_{65}	M_{66}	
Group II	M_{11}	M_{12}	M_{13}	0	M_{15}	0	c constants the same except $c_{ab}=c_{ba}$, resulting in 13 constants.
Monoclinic	M_{21}	M_{22}	M_{23}	0	M_{25}	0	K constants the same as c, except $K_{44}=K_{23}$, $K_{55}=K_{13}$, $K_{66}=K_{12}$, $K_{46}=K_{25}$, resulting in 9 constants.
$2, m, 2/m$,	M_{31}	M_{32}	M_{33}	0	M_{35}	0	
20 constants	0	0	0	M_{44}	0	M_{46}	
$y=$ unique axis	M_{51}	M_{52}	M_{53}	0	M_{55}	0	
	0	0	0	M_{64}	0	M_{66}	
Group III	M_{11}	M_{12}	M_{13}	0	0	0	c constants the same except $c_{ab}=c_{ba}$, resulting in 9 constants.
Orthorhombic	M_{21}	M_{22}	M_{23}	0	0	0	K constants the same as c, except $K_{44}=K_{23}$, $K_{55}=K_{13}$, $K_{66}=K_{12}$, resulting in 6 constants.
$mm2$, 222, mmm	M_{31}	M_{32}	M_{33}	0	0	0	
12 constants	0	0	0	M_{44}	0	0	
	0	0	0	0	M_{55}	0	
	0	0	0	0	0	M_{66}	
Group IV	M_{11}	M_{12}	M_{13}	M_{14}	$-M_{25}$	$2M_{62}$	c constants the same except $c_{ab}=c_{ba}$ and $c_{62}=c_{45}=0$. $c_{46}=c_{25}$; $c_{56}=c_{14}$, resulting in 7 constants.
Trigonal	M_{12}	M_{11}	M_{13}	$-M_{14}$	M_{25}	$-2M_{62}$	K constants the same as c constants, except $K_{44}=K_{23}$, resulting in 6 constants.
$3, \bar{3}$	M_{31}	M_{31}	M_{33}	0	0	0	
12 constants	M_{41}	$-M_{41}$	0	M_{44}	M_{45}	$2M_{52}$	
	$-M_{52}$	M_{52}	0	$-M_{45}$	M_{44}	$2M_{41}$	
	$-M_{62}$	M_{62}	0	M_{25}	M_{14}	$M_{11}-M_{12}$	

Table 16.4 (continued).

(Type M_{ijkm}, $i \to j$, $k \to m$;	Type c_{ijkm}, $i \to j$, $k \to m$, $ij \to km$;	Type K_{ijkm}, $i \to j \to k \to m$)
Group V Trigonal $3m$, 32, $\bar{3}m$ 8 constants	$\begin{Vmatrix} M_{11} & M_{12} & M_{13} & M_{14} & 0 & 0 \\ M_{12} & M_{11} & M_{13} & -M_{14} & 0 & 0 \\ M_{31} & M_{31} & M_{33} & 0 & 0 & 0 \\ M_{41} & -M_{41} & 0 & M_{44} & 0 & 0 \\ 0 & 0 & 0 & 0 & M_{44} & 2M_{41} \\ 0 & 0 & 0 & 0 & M_{14} & M_{11}-M_{12} \end{Vmatrix}$	c constants the same except $c_{ab}=c_{ba}$, $2c_{41}=c_{14}=c_{56}$, 6 constants. K constants the same as the c constants, except $K_{44}=K_{23}$, 5 constants.
Group VI Tetragonal 4, $\bar{4}$, $4/m$ 10 constants	$\begin{Vmatrix} M_{11} & M_{12} & M_{13} & 0 & 0 & M_{16} \\ M_{12} & M_{11} & M_{13} & 0 & 0 & -M_{16} \\ M_{31} & M_{31} & M_{33} & 0 & 0 & 0 \\ 0 & 0 & 0 & M_{44} & M_{45} & 0 \\ 0 & 0 & 0 & -M_{45} & M_{44} & 0 \\ M_{61} & -M_{61} & 0 & 0 & 0 & M_{66} \end{Vmatrix}$	c constants the same except $c_{13}=c_{31}$, $c_{16}=c_{61}$, $c_{45}=0$, 7 constants. K constants the same except $K_{44}=K_{23}$, $K_{66}=K_{12}$. 5 constants.
Group VII Tetragonal $4mm$, $\bar{4}2m$, $4/mmm$ 7 constants	$\begin{Vmatrix} M_{11} & M_{12} & M_{13} & 0 & 0 & 0 \\ M_{12} & M_{11} & M_{13} & 0 & 0 & 0 \\ M_{31} & M_{31} & M_{33} & 0 & 0 & 0 \\ 0 & 0 & 0 & M_{44} & 0 & 0 \\ 0 & 0 & 0 & 0 & M_{44} & 0 \\ 0 & 0 & 0 & 0 & 0 & M_{66} \end{Vmatrix}$	c constants the same except $c_{13}=c_{31}$, 6 constants. K constants the same except $K_{44}=K_{23}$, $K_{66}=K_{12}$, 4 constants.
Group VIII Hexagonal $\bar{6}$, 6, $6/m$ 8 constants	$\begin{Vmatrix} M_{11} & M_{12} & M_{13} & 0 & 0 & 2M_{61} \\ M_{12} & M_{11} & M_{13} & 0 & 0 & -2M_{61} \\ M_{31} & M_{31} & M_{33} & 0 & 0 & 0 \\ 0 & 0 & 0 & M_{44} & M_{45} & 0 \\ 0 & 0 & 0 & -M_{45} & M_{44} & 0 \\ M_{61} & -M_{61} & 0 & 0 & 0 & M_{11}-M_{12} \end{Vmatrix}$	c constants the same except $c_{13}=c_{31}$, $c_{61}=0$, $c_{45}=0$, 5 constants. K constants the same as c constants, except $K_{44}=K_{23}$, 4 constants.

Table 16.4 (continued).

(Type M_{ijkm}, $i \to j$, $k \to m$; Type c_{ijkm}, $i \to j$, $k \to m$, $ij \to km$; Type K_{ijkm}, $i \to j \to k \to m$)							
Group IX	M_{11}	M_{12}	M_{13}	0	0	0	c constants the same except $c_{13} = c_{31}$,
Hexagonal	M_{12}	M_{11}	M_{13}	0	0	0	5 constants.
$\bar{6}m2$, 622, $6mm$, $6/mmm$	M_{31}	M_{31}	M_{33}	0	0	0	K constants the same as c constants,
6 constants	0	0	0	M_{44}	0	0	except $K_{44} = K_{23}$,
	0	0	0	0	M_{44}	0	4 constants.
	0	0	0	0	0	$M_{11} - M_{12}$	
Group X	M_{11}	M_{12}	M_{13}	0	0	0	c constants the same except $c_{12} = c_{13}$,
Cubic	M_{13}	M_{11}	M_{12}	0	0	0	3 constants.
23, $m3$	M_{12}	M_{13}	M_{11}	0	0	0	K constants the same as c constants,
4 constants	0	0	0	M_{44}	0	0	except $K_{44} = K_{12}$,
	0	0	0	0	M_{44}	0	2 constants.
	0	0	0	0	0	M_{44}	
Group XI	M_{11}	M_{12}	M_{12}	0	0	0	c constants the same,
Cubic	M_{12}	M_{11}	M_{12}	0	0	0	3 constants.
$\bar{4}3m$, 432, $m3m$	M_{12}	M_{12}	M_{12}	0	0	0	K constants the same as c constants,
3 constants	0	0	0	M_{44}	0	0	except $K_{44} = K_{12}$,
	0	0	0	0	M_{44}	0	2 constants.
	0	0	0	0	0	M_{44}	
Group XII	M_{11}	M_{12}	M_{12}	0	0	0	c and K constants the same,
Isotropic	M_{12}	M_{11}	M_{12}	0	0	0	2 constants.
2 constants	M_{12}	M_{12}	M_{11}	0	0	0	
	0	0	0	$M_{11} - M_{12}$	0	0	
	0	0	0	0	$M_{11} - M_{12}$	0	
	0	0	0	0	0	$M_{11} - M_{12}$	

There are three types of fourth rank tensor indicated in Table 6.4: (1) a type M, typified by the piezo-optic coefficients

$$\beta_0 m_{ijkm} = \frac{\partial^3 U}{\partial V_{ij}\, \partial D_k\, \partial D_m}, \tag{16.1}$$

in which the first two subscripts can be interchanged with each other, and the last two can be interchanged with each other, but the first two cannot necessarily be interchanged with the last two; (2) a type c, typified by the elastic coefficients

$$c_{ijkm} = \frac{\partial^2 U}{\partial V_{ij}\, \partial V_{km}}, \tag{16.2}$$

which has the symmetry of M plus a permitted interchange of the first two with the last two; (3) a type K, typified by the "quadratic electro-optic coefficients"

$$\beta_0 f_{ijkm} = \frac{\partial^4 U}{\partial D_i\, \partial D_j\, \partial D_k\, \partial D_m}, \tag{16.3}$$

in which the ordering of the indices is completely irrelevant.

Examples of fifth rank tensors are

$$\beta_0 b_{kmpqi} = \frac{\partial^3 U}{\partial V_{km}\, \partial V_{pq}\, \partial D_i} \tag{16.4}$$

and

$$h_{ijkpq} = \frac{\partial^4 U}{\partial D_i\, \partial D_j\, \partial D_k\, \partial V_{pq}}. \tag{16.5}$$

The tensor considered by NELSON and LAX (1971), for which results are given in Table 16.5, is assumed to possess the symmetry corresponding to interchange of indices within each of two pairs, but not of the pairs with each other. This covers both (16.4) and (16.5) as special cases. The pairs of interchangeable indices are abbreviated to a single subscript running from one to six in the usual manner, and the tensor is presented in the form of three 6×6 arrays, one array for each value of the remaining index. (The tensor can be represented as $H_{\alpha\beta i} = H_{i\alpha\beta}$, with $i=1$, 2, 3, and α and β each running from one to six. For each i, there is a 6×6 array.) All the elements are zero for the eleven point groups that have inversion symmetry. It should be noted that in the case of groups 3, $3m$, 32, and $\bar{6}m$ 2, the axes used in Table 16.5 should be rotated $-30°$ about Z in order to agree with the IEEE Standard No. 176 (1949) [M. LAX, private communication, 1972]. This can be accomplished by the scheme $(-x \to Y, y \to X)$, where (x, y) are the axes in Table 16.5 and (X, Y) the IEEE axes.

Finally, as an example of coefficients forming a sixth rank tensor, Table 16.6 shows the possible third-order elastic constants for the eleven Laue groups and for the fully isotropic case, after BRUGGER (1965). This table differs from others in the literature for the trigonal groups. The difference presumably arises from a different choice of axes. BRUGGER has followed the IEEE Standard No. 176 [BOND, 1949]. The Laue groups are designated as shown in Table 15.1.

Unlike the case of electrical and mechanical properties, the symmetry of macroscopic magnetic properties is not determined by the point group alone. The reason is that an additional symmetry operation, the reversal of all microscopic currents, must be taken into account [see, e.g., LANDAU and LIFSCHITZ (1950), BIRSS (1964), and the references they give]. The additional operation results in 90 *magnetic crystal classes*.

Table 16.5. *Fifth-rank polar tensors.*

$$H_{i\alpha\beta}=H_{\alpha\beta i};\quad i=1,2,3;\quad \alpha=1,2,\ldots,6;\quad \beta=1,2,\ldots,6.$$

Related elements are expressed in terms of the same letters except when only X appears, for which cases (triclinic, monoclinic, and orthorhombic) no relation exists between the nonzero elements. The triclinic case 1 is not listed, as all 108 components are nonzero and distinct. In groups 3, 32, $3m$, and $\bar{6}m2$, the axes should be rotated $-30°$ about Z $(-x\to Y,\ y\to X)$ in order to agree with the IEEE standard.

$i=1$						$i=2$						$i=3$					
Monoclinic 2																	
0	0	0	X	0	X	X	X	X	0	X	0	0	0	0	X	0	X
0	0	0	X	0	X	X	X	X	0	X	0	0	0	0	X	0	X
0	0	0	X	0	X	X	X	X	0	X	0	0	0	0	X	0	X
X	X	X	0	X	0	0	0	0	X	0	X	X	X	X	0	X	0
0	0	0	X	0	X	X	X	X	0	X	0	0	0	0	X	0	X
X	X	X	0	X	0	0	0	0	X	0	X	X	X	X	0	X	0
Monoclinic m																	
X	X	X	0	X	0	0	0	0	X	0	X	X	X	X	0	X	0
X	X	X	0	X	0	0	0	0	X	0	X	X	X	X	0	X	0
X	X	X	0	X	0	0	0	0	X	0	X	X	X	X	0	X	0
0	0	0	X	0	X	X	X	X	0	X	0	0	0	0	X	0	X
X	X	X	0	X	0	0	0	0	X	0	X	X	X	X	0	X	0
0	0	0	X	0	X	X	X	X	0	X	0	0	0	0	X	0	X
Orthorhombic 222																	
0	0	0	X	0	0	0	0	0	0	X	0	0	0	0	0	0	X
0	0	0	X	0	0	0	0	0	0	X	0	0	0	0	0	0	X
0	0	0	X	0	0	0	0	0	0	X	0	0	0	0	0	0	X
X	X	X	0	0	0	0	0	0	0	0	X	0	0	0	0	X	0
0	0	0	0	0	X	X	X	X	0	0	0	0	0	0	X	0	0
0	0	0	0	X	0	0	0	0	X	0	0	X	X	X	0	0	0

Orthorhombic $mm2$																	
0	0	0	0	X	0	0	0	0	X	0	0	X	X	X	0	0	0
0	0	0	0	X	0	0	0	0	X	0	0	X	X	X	0	0	0
0	0	0	0	X	0	0	0	0	X	0	0	X	X	X	0	0	0
0	0	0	0	0	X	X	X	X	0	0	0	0	0	c	X	0	0
X	X	X	0	0	0	0	0	0	0	0	X	0	0	c	0	X	0
0	0	0	X	0	0	0	0	0	0	X	0	0	0	0	0	0	X

Tetragonal 4																	
0	0	0	I	M	0	0	0	0	N	$-J$	0	Q	R	S	0	0	Z
0	0	0	J	N	0	0	0	0	M	$-I$	0	R	Q	S	0	0	$-Z$
0	0	0	K	z	0	0	0	0	z	$-K$	0	T	T	U	0	0	0
A	B	C	0	0	D	F	E	G	0	0	$-H$	0	0	0	V	W	0
E	F	G	0	0	H	B	$-A$	$-C$	0	0	D	0	0	0	$-W$	V	0
0	0	0	L	P	0	0	0	0	$-P$	L	0	Y	$-Y$	0	0	0	X

Tetragonal $\bar{4}$																	
0	0	0	I	M	0	0	0	0	$-N$	J	0	Q	R	S	0	0	Z
0	0	0	J	N	0	0	0	0	$-M$	I	0	$-R$	$-Q$	$-S$	0	0	Z
0	0	0	K	W	0	0	0	0	$-W$	K	0	T	$-T$	0	0	0	X
A	B	C	0	0	D	$-F$	$-E$	$-G$	0	0	H	0	0	0	V	0	0
E	F	G	0	0	H	$-B$	A	C	0	0	$-D$	0	0	0	0	$-V$	0
0	0	0	L	P	0	0	0	0	P	$-L$	0	Y	Y	U	0	0	0

Tetragonal 422																	
0	0	0	I	0	0	0	0	0	0	$-J$	0	0	0	0	0	0	Z
0	0	0	J	0	0	0	0	0	0	$-I$	0	0	0	0	0	0	$-Z$
0	0	0	K	0	0	0	0	0	0	$-K$	0	0	0	0	0	0	0
A	B	C	0	0	0	0	0	0	0	0	$-H$	0	0	0	0	W	0
0	0	0	0	0	H	$-B$	$-A$	$-C$	0	0	0	0	0	0	$-W$	0	0
0	0	0	0	P	0	0	0	0	$-P$	0	0	Y	$-Y$	0	0	0	0

Table 16.5 (continued).

$i=1$						$i=2$						$i=3$					
Hexagonal 6 2 2																	

Hexagonal 6 2 2 is the same as Tetragonal 4 2 2 except that

$$H=\tfrac{1}{2}(B-A), \quad P=\tfrac{1}{2}(J-I).$$

Tetragonal $4mm$

0	0	0	0	M	0	0	0	0	N	0	0	Q	R	S	0	0	0
0	0	0	0	N	0	0	0	0	M	0	0	R	Q	S	0	0	0
0	0	0	0	W	0	0	0	0	W	0	0	T	T	U	0	0	0
0	0	0	0	0	D	F	E	G	0	0	0	0	0	0	V	0	0
E	F	G	0	0	0	0	0	0	0	0	D	0	0	0	0	V	0
0	0	0	L	0	0	0	0	0	0	L	0	0	0	0	0	0	X

Hexagonal $6mm$

Hexagonal $6mm$ is the same as Tetragonal $4mm$ except that

$$L=\tfrac{1}{2}(M-N), \quad D=\tfrac{1}{2}(E-F), \quad X=\tfrac{1}{2}(Q-R).$$

Tetragonal $\bar{4}2m$

0	0	0	I	0	0	0	0	0	0	J	0	0	0	0	0	0	Z
0	0	0	J	0	0	0	0	0	0	I	0	0	0	0	0	0	Z
0	0	0	K	0	0	0	0	0	0	K	0	0	0	0	0	0	X
A	B	C	0	0	0	0	0	0	0	0	H	0	0	0	0	W	0
0	0	0	0	0	H	B	A	C	0	0	0	0	0	0	W	0	0
0	0	0	0	P	0	0	0	0	P	0	0	Y	Y	U	0	0	0

Hexagonal 6

0	0	0	$2D$	$2J$	0	0	0	0	$2K$	$-2E$	0	$2M$	$2N$	Z	0	0	T
0	0	0	$2E$	$2K$	0	0	0	0	$2J$	$-2D$	0	$2N$	$2M$	Z	0	0	$-T$
0	0	0	F	L	0	0	0	0	L	$-F$	0	P	P	Q	0	0	0
$2A$	$2B$	C	0	0	$G-H$	$2H$	$2G$	T	0	0	$A-B$	0	0	0	R	S	0
$2G$	$2H$	I	0	0	$B-A$	$-2B$	$-2A$	$-C$	0	0	$G-H$	0	0	0	$-S$	R	0
0	0	0	$J-K$	$E-D$	0	0	0	0	$D-E$	$J-K$	0	$-T$	T	0	0	0	$M-N$

Hexagonal $\bar{6}$ $[D \equiv -(A+B+C);\quad M \equiv -(J+K+L)]$

$2A$	$2B$	E	0	0	$L-M$	$2J$	$2K$	N	0	0	$D-C$	0	0	0	R	$-I$	0
$2C$	$2D$	$-E$	0	0	$J-K$	$2L$	$2M$	$-N$	0	0	$B-A$	0	0	0	$-R$	I	0
F	$-F$	0	0	0	Z	Z	$-Z$	0	0	0	$-F$	0	0	0	0	0	0
0	0	0	G	P	0	0	0	0	$-P$	G	0	Q	$-Q$	0	0	0	H
0	0	0	P	$-G$	0	0	0	0	G	P	0	$-H$	H	0	0	0	Q
$K-M$	$J-L$	N	0	0	$B+C$	$D-B$	$C-A$	$-E$	0	0	$K+L$	0	0	0	I	R	0

Hexagonal 622 (see above)

Hexagonal $6\,mm$ (see above)

Hexagonal $\bar{6}m2$ $[M \equiv -(J+K+L)]$*

0	0	0	0	0	$L-M$	$2J$	$2K$	N	0	0	0	0	0	0	R	0	0
0	0	0	0	0	$J-K$	$2L$	$2M$	$-N$	0	0	0	0	0	0	$-R$	0	0
0	0	0	0	0	Z	Z	$-Z$	0	0	0	0	0	0	0	0	0	0
0	0	0	0	P	0	0	0	0	$-P$	0	0	Q	$-Q$	0	0	0	0
0	0	0	P	0	0	0	0	0	0	P	0	0	0	0	0	0	Q
$K-M$	$J-L$	N	0	0	0	0	0	0	0	0	$K+L$	0	0	0	0	R	0

Table 16.5 (continued).

$i=1$						$i=2$						$i=3$					
Cubic 2 3																	
0	0	0	I	0	0	0	0	0	0	K	0	0	0	0	0	0	J
0	0	0	J	0	0	0	0	0	0	I	0	0	0	0	0	0	K
0	0	0	K	0	0	0	0	0	0	J	0	0	0	0	0	0	I
A	B	C	0	0	0	0	0	0	0	0	P	0	0	0	0	H	0
0	0	0	0	0	H	C	A	B	0	0	0	0	0	0	P	0	0
0	0	0	0	P	0	0	0	0	H	0	0	B	C	A	0	0	0
Cubic 4 3 2																	
0	0	0	0	0	0	0	0	0	0	$-J$	0	0	0	0	0	0	J
0	0	0	J	0	0	0	0	0	0	0	0	0	0	0	0	0	$-J$
0	0	0	$-J$	0	0	0	0	0	0	J	0	0	0	0	0	0	0
0	B	$-B$	0	0	0	0	0	0	0	0	$-H$	0	0	0	0	H	0
0	0	0	0	0	H	$-B$	0	B	0	0	0	0	0	0	$-H$	0	0
0	0	0	0	$-H$	0	0	0	0	H	0	0	B	$-B$	0	0	0	0
Cubic $\bar{4}3m$																	
0	0	0	I	0	0	0	0	0	0	J	0	0	0	0	0	0	J
0	0	0	J	0	0	0	0	0	0	I	0	0	0	0	0	0	J
0	0	0	J	0	0	0	0	0	0	J	0	0	0	0	0	0	I
A	B	B	0	0	0	0	0	0	0	0	H	0	0	0	0	H	0
0	0	0	0	0	H	B	A	B	0	0	0	0	0	0	H	0	0
0	0	0	0	H	0	0	0	0	H	0	0	B	B	A	0	0	0

Trigonal 3 $[D \equiv -(A+B+C);\ m = -(j+k+l)]$ *

$2A$	$2B$	$-E$	$2d$	$2I$	$l-m$	$2j$	$2k$	n	$2J$	$-2e$	$D-C$	$2N$	$2P$	Q	r	$-U$	t
$2C$	$2D$	E	$2e$	$2J$	$j-k$	$2l$	$2m$	$-n$	$2I$	$-2d$	$B-A$	$2P$	$2N$	Q	$-r$	U	$-t$
$-F$	F	0	f	K	z	z	$-z$	0	K	$-f$	F	R	R	S	0	0	0
$2a$	$2b$	c	M	p	$G-H$	$2H$	$2G$	L	$-p$	M	$a-b$	q	$-q$	0	T	s	Z
$2G$	$2H$	L	p	$-M$	$b-a$	$-2b$	$-2a$	$-c$	M	p	$G-H$	$-Z$	Z	0	$-s$	T	q
$k-m$	$j-l$	n	$I-J$	$e-d$	$B+C$	$D-B$	$C-A$	E	$d-e$	$I-J$	$k+l$	$-t$	t	0	U	r	$N-P$

Trigonal $3m$ $[D \equiv -(A+B+C)]$ *

$2A$	$2B$	$-E$	0	$2I$	0	0	0	0	$2J$	0	$D-C$	$2N$	$2P$	Q	0	$-U$	0
$2C$	$2D$	E	0	$2J$	0	0	0	0	$2I$	0	$B-A$	$2P$	$2N$	Q	0	U	0
$-F$	F	0	0	K	0	0	0	0	K	0	F	R	R	S	0	0	0
0	0	0	M	0	$G-H$	$2H$	$2G$	L	0	M	0	0	0	0	T	0	Z
$2G$	$2H$	L	0	$-M$	0	0	0	0	M	0	$G-H$	$-Z$	Z	0	0	T	0
0	0	0	$I-J$	0	$B+C$	$D-B$	$C-A$	E	0	$I-J$	0	0	0	0	U	0	$N-P$

Trigonal 3 2 $[m \equiv -(j+k+l)]$ *

0	0	0	$2d$	0	$l-m$	$2j$	$2k$	n	0	$-2e$	0	0	0	0	r	0	t
0	0	0	$2e$	0	$j-k$	$2l$	$2m$	$-n$	0	$-2d$	0	0	0	0	$-r$	0	$-t$
0	0	0	f	0	z	z	$-z$	0	0	$-f$	0	0	0	0	0	0	0
$2a$	$2b$	c	0	p	0	0	0	0	$-p$	0	$a-b$	q	$-q$	0	0	s	0
0	0	0	p	0	$b-a$	$-2b$	$-2a$	$-c$	0	p	0	0	0	0	$-s$	0	q
$k-m$	$j-l$	n	0	$e-d$	0	0	0	0	$d-e$	0	$k+l$	$-t$	t	0	0	r	0

* In groups 3, 3 2, $3m$, and $\bar{6}m2$, the axes should be rotated $-30°$ about $Z(-x \to Y,\ y \to X)$ in order to agree with the IEEE standard.

Table 16.6. *Third-order elastic constants for the eleven Laue groups and for isotropic media (I).*

N	M	O	TII	TI	CII	CI	RII	RI	HII	HI	I
111	111	111	111	111	111	111	111	111	111	111	111
112	112	112	112	112	112	112	112	112	112	112	112
113	113	113	113	113	113	112	113	113	113	113	112
114	0	0	0	0	0	0	114	114	0	0	0
115	115	0	0	0	0	0	115	0	0	0	0
116	0	0	116	0	0	0	116	0	116	0	0
122	122	122	112	112	113	112	122[a]	122[a]	122[a]	122[a]	112
123	123	123	123	123	123	123	123	123	123	123	123
124	0	0	0	0	0	0	124	124	0	0	0
125	125	0	0	0	0	0	125	0	0	0	0
126	0	0	0	0	0	0	−116	0	−116	0	0
133	133	133	133	133	112	112	133	133	133	133	112
134	0	0	0	0	0	0	134	134	0	0	0
135	135	0	0	0	0	0	135	0	0	0	0
136	0	0	136	0	0	0	0	0	0	0	0
144	144	144	144	144	144	144	144	144	144	144	144[l]
145	0	0	145	0	0	0	145	0	145	0	0
146	146	0	0	0	0	0	146[b]	0	0	0	0
155	155	155	155	155	155	155	155	155	155	155	155[m]
156	0	0	0	0	0	0	156[c]	156[e]	0	0	0
166	166	166	166	166	166	155	166[d]	166[d]	166[d]	166[d]	155[m]
222	222	222	111	111	111	111	222	222	222	222	111
223	223	223	113	113	112	112	113	113	113	113	112
224	0	0	0	0	0	0	224[e]	224[e]	0	0	0
225	225	0	0	0	0	0	225[f]	0	0	0	0
226	0	0	−116	0	0	0	116	0	116	0	0
233	233	233	133	133	113	112	133	133	133	133	112
234	0	0	0	0	0	0	−134	−134	0	0	0
235	235	0	0	0	0	0	−135	0	0	0	0
236	0	0	−136	0	0	0	0	0	0	0	0
244	244	244	155	155	166	155	155	155	155	155	155[m]
245	0	0	−145	0	0	0	−145	0	−145	0	0
246	246	0	0	0	0	0	246[g]	0	0	0	0
255	255	255	144	144	144	144	144	144	144	144	144[l]
256	0	0	0	0	0	0	256[h]	256[h]	0	0	0
266	266	266	166	166	155	155	266[i]	266[i]	266[i]	266[i]	155[m]
333	333	333	333	333	111	111	333	333	333	333	111
334	0	0	0	0	0	0	0	0	0	0	0
335	335	0	0	0	0	0	0	0	0	0	0
336	0	0	0	0	0	0	0	0	0	0	0
344	344	344	344	344	155	155	344	344	344	344	155[m]
345	0	0	0	0	0	0	0	0	0	0	0
346	346	0	0	0	0	0	−135	0	0	0	0
355	355	355	344	344	166	155	344	344	344	344	155[m]
356	0	0	0	0	0	0	134	134	0	0	0
366	366	366	366	366	144	144	366[j]	366[j]	366[j]	366[j]	144[l]
444	0	0	0	0	0	0	444	444	0	0	0
445	445	0	0	0	0	0	445	0	0	0	0
446	0	0	446	0	0	0	145	0	145	0	0
455	0	0	0	0	0	0	−444	−444	0	0	0
456	456	456	456	456	456	456	456[k]	456[k]	456[k]	456[k]	456[n]
466	0	0	0	0	0	0	124	124	0	0	0
555	555	0	0	0	0	0	−445	0	0	0	0
556	0	0	−446	0	0	0	−145	0	−145	0	0
566	566	0	0	0	0	0	125	0	0	0	0
666	0	0	0	0	0	0	−116	0	−116	0	0

[a] $c_{122} = c_{111} + c_{112} - c_{222}$.
[b] $c_{146} = \frac{1}{2}(-c_{115} - 3\,c_{125})$.
[c] $c_{156} = \frac{1}{2}(c_{114} + 3\,c_{124})$.
[d] $c_{166} = \frac{1}{4}(-2\,c_{111} - c_{112} + 3\,c_{222})$.
[e] $c_{224} = -c_{114} - 2\,c_{124}$.
[f] $c_{225} = -c_{114} - 2\,c_{125}$.
[g] $c_{246} = \frac{1}{2}(-c_{115} + c_{125})$.
[h] $c_{256} = \frac{1}{2}(c_{114} - c_{124})$.
[i] $c_{266} = \frac{1}{4}(2\,c_{111} - c_{112} - c_{222})$.
[j] $c_{366} = \frac{1}{2}(c_{113} - c_{123})$.
[k] $c_{456} = \frac{1}{2}(-c_{144} + c_{155})$.
[l] $c_{144} = \frac{1}{2}(c_{112} - c_{123})$.
[m] $c_{155} = \frac{1}{4}(c_{111} - c_{112})$.
[n] $c_{456} = \frac{1}{8}(c_{111} - 3\,c_{112} + 2\,c_{123})$.

F. Exponentially damped plane waves.

17. Complex representation of waves.

17.1. Waves sinusoidal in time, attenuated in space. In many experimental situations, one attempts to excite trains of plane waves that are essentially sinusoidal in time. These appear to be exponentially attenuated in space. Let us therefore consider the formal representation

$$u = u_0\, \mathrm{e}^{-\alpha Z} \cos \omega (t - Z/c) \tag{17.1}$$

for a wave traveling in the Z direction with plase velocity c and attenuation α per unit length. Eq. (17.1) is equivalent to

$$u = \mathrm{Re}(u^*), \quad u^* = u_0\, \mathrm{e}^{j\omega t - (\alpha + j\omega/c) Z}. \tag{17.2}$$

Eq. $(17.2)_2$ can be expressed in the forms

$$u^* = u_0\, \mathrm{e}^{j\omega t - \Gamma Z} = u_0\, \mathrm{e}^{j(\omega t - k^* Z)} = u_0\, \mathrm{e}^{j\omega(t - Z/c^*)}, \tag{17.3}$$

in which the complex quantities Γ, k^*, and c^* may be expressed in terms of the real quantities α and c by comparing (17.2) and (17.3). Clearly,

$$\Gamma = \alpha + j\,\omega/c, \qquad k^* = \frac{\omega}{c} - j\alpha = \frac{\omega}{c^*}. \tag{17.4}$$

The quantity c^* is called the complex phase velocity. In terms of the real quantities, α, c, and ω, we have

$$\frac{1}{c^*} = \frac{1 - j\alpha\, c/\omega}{c} = \frac{1}{c} - \frac{j\alpha}{\omega}. \tag{17.5}$$

The "quality" Q associated with the disturbance (17.1) is defined as

$$Q = \omega/2\alpha\, c. \tag{17.6}$$

Q is the ratio of the energy carried by the wave to the energy dissipated per radian of phase shift [see, e.g., THURSTON, 1964, p. 56]. Now from (17.5),

$$\tan \underline{/c^*} = \frac{\alpha c}{\omega} \tag{17.7}$$

where $\underline{/c^*}$ denotes the angle of the complex number c^*. From (17.6) and (17.7),

$$Q^{-1} = 2 \tan \underline{/c^*}. \tag{17.8}$$

Subject to the validity of (17.1), a medium can be characterized by $\alpha(\omega)$ and $c(\omega)$ for the waves that can be propagated.

The complex phase velocity c^* can be considered as the negative of the ratio of complex particle velocity $\partial u^*/\partial t$ to the complex displacement gradient $\partial u^*/\partial Z$:

$$c^* = -\frac{\partial u^*/\partial t}{\partial u^*/\partial Z}. \tag{17.9}$$

The relations of this section are all based only on the assumption (17.1).

17.2. Waves sinusoidal in space decaying in time. One can also consider solutions of the form

$$u_i = \mathrm{Re}(u_i^*), \qquad u_i^* = u_0\, \mathrm{e}^{st - jkZ} = u_0\, \mathrm{e}^{j(\omega^* t - kZ)} \tag{17.10}$$

where k is real but s and ω^* are complex with

$$s = j\,\omega^* = j\,2\pi\nu - \sigma, \tag{17.11}$$

$$\omega^* = 2\pi\nu + j\,\sigma \tag{17.12}$$

so that

$$u_i^* = u_0\,\mathrm{e}^{-\sigma t}\,\mathrm{e}^{j(2\pi\nu t - kZ)}. \tag{17.13}$$

The Q for waves decaying in time can be defined analogously to that for waves damped in space. The Q is the ratio of the energy in the wave to the energy loss per radian. If energy is assumed proportional to the square of the amplitude, the energy at time t is proportional to $\exp(-2\sigma t)$ and the rate of loss is therefore proportional to

$$-\frac{\partial}{\partial t}(\mathrm{e}^{-2\sigma t}) = 2\sigma\,\mathrm{e}^{-2\sigma t}$$

and hence

$$\frac{\text{rate of loss}}{\text{energy}} = 2\sigma.$$

This is referred to one radian by multiplying by $1/(2\nu\pi)$ and hence

$$Q_t = \pi\,\nu/\sigma. \tag{17.14}$$

In terms of Q_t.

$$\omega^* = 2\pi\nu\left(1 + \frac{j\sigma}{2\pi\nu}\right) = 2\pi\nu\left(1 + \frac{j}{2Q}\right). \tag{17.15}$$

17.3. Inhomogeneous plane waves. In the preceding sections, we have considered waves in which the planes of constant phase are also planes of constant amplitude. A surface wave on an elastic half-space is an example of a case in which there are planes of constant phase perpendicular to the surface, but the planes of constant amplitude are parallel to the surface. The wave propagates along the surface, and may be said to be "guided" by it. A good discussion of surface waves may be found in the book by VIKTOROV (1967). Guided waves in general provide numerous examples of cases in which there are planes (or near-planes) of constant phase that do not coincide with the surfaces of constant amplitude. A plane wave is called homogeneous if the planes of constant phase are planes of constant amplitude and inhomogeneous if they are not. Inhomogeneous waves have to be considered in discussing the reflection and refraction of an incident homogeneous wave at the interface between two viscoelastic media [COOPER, 1967; BUCHEN, 1971].

To describe inhomogeneous plane waves in which the surfaces of constant amplitude are planes, (17.1) can be generalized to

$$u_i(t, \boldsymbol{X}) = u_i^0 \exp(-\alpha\,\boldsymbol{A}\cdot\boldsymbol{X}) \cos\omega(t + t_i^0 - \boldsymbol{N}\cdot\boldsymbol{X}/c) \tag{17.16}$$

in which $\boldsymbol{A}$ and $\boldsymbol{N}$ are constant unit vectors that specify the directions of planes of constant amplitude and phase, respectively. The most general particle orbits allowed by (17.16) are ellipses resulting from the superposition of simple harmonic motions of the same frequency along the three rectangular axes. Without additional information, however, the plane of the ellipse may have an arbitrary orientation.

18. Stress and deformation in exponentially damped plane sinusoidal waves.

What can be said concerning the stress-deformation relation that permits waves of the form (17.1) and (17.10) to propagate? From (6.7), disregarding the body force, we have the equation of motion

$$\bar{\varrho}\,\ddot{u}_i = \frac{\partial P_{ik}}{\partial X_k}\,. \tag{18.1}$$

Suppose there are variations only with respect to $X_3 = Z$. Then, taking

$$\begin{aligned} u_i &= \mathrm{Re}\,(u_i^*), \qquad u_i^* = u_{0\,i}\,\mathrm{e}^{j(\omega t - k^* Z)}, \\ P_{i3} &= \mathrm{Re}\,(P_{i3}^*), \end{aligned} \tag{18.2}$$

we find that (18.1) implies

$$P_{i3}^* = \frac{\bar{\varrho}\,\omega^2\,u_i^*}{j\,k^*} \tag{18.3}$$

while

$$\partial u_i^*/\partial Z = -j\,k^*\,u_i^*\,. \tag{18.4}$$

Here k^* is the complex phase factor of Eqs. (17.3) and (17.4).

A complex elastic modulus $E^*(\omega)$ may be defined for this wave as the ratio

$$E^*(\omega) \equiv \frac{P_{i3}^*}{\partial u_i^*/\partial Z} = \bar{\varrho}\,\omega^2/k^{*\,2} = \bar{\varrho}\,c^{*\,2}\,. \tag{18.5}$$

Z^*, the complex characteristic impedance of the medium for this wave is defined as the negative of the ratio of P_{i3}^* to the complex particle velocity $j\,\omega\,u_i^*$. Hence, from (18.3) and (17.4)

$$Z^*(\omega) \equiv -\frac{P_{i3}^*}{j\,\omega\,u_i^*} = \bar{\varrho}\,\omega/k^* = \bar{\varrho}\,c^*\,. \tag{18.6}$$

In the context of small deformations, there is no distinction between the Piola-Kirchhoff stress tensors P_{ik} and t_{rs} and the ordinary stress T_{ij}, nor between $\partial u_i/\partial Z$ and the strain. For small deformations, $E^*(\omega)$ is the ratio of complex stress to complex strain. Its magnitude is the ratio of stress amplitude to strain amplitude and its angle is the angle by which the stress leads the strain. The complex wave impedance is a similar ratio relating stress to particle velocity.

Unless $\alpha = 0$, α and c must have the same sign in (17.1) to indicate attenuation in the direction of travel. Hence, referring to (17.7), we see that

$$0 \leq \underline{/c^*} < \pi/2, \tag{18.7}$$

while from (18.5),

$$\underline{/E^*} = 2\,\underline{/c^*}\,. \tag{18.8}$$

When $\alpha = 0$ ($c^* = c$, a real wave speed), stress and strain are in phase. Otherwise, the stress has to lead the strain. From measurements of $\alpha(\omega)$ and $c(\omega)$, the phase difference can be calculated as

$$\underline{/E^*(\omega)} = 2\,\underline{/c^*} = 2\tan^{-1}(\alpha\,c/\omega)\,. \tag{18.9}$$

From (17.8), the relation of Q to $\underline{/E^*}$ is

$$Q^{-1} = 2\tan\left(\tfrac{1}{2}\,\underline{/E^*}\right). \tag{18.10}$$

Thus, for very small damping, $Q^{-1} = \underline{/E^*}$.

The stress associated with the solution (17.10) is also obtained from the equation of motion. From (18.1) we find

$$P_{i3}^* = \bar{\varrho}\, \omega^{*2}\, u_i^* / j k \tag{18.11}$$

while direct differentiation of (17.10) gives

$$\partial u_i^* / \partial Z = -j k u_i^* . \tag{18.12}$$

For this solution, the complex elastic modulus is therefore

$$E_t^* (k) = \frac{P_{i3}^*}{\partial u_i^* / \partial Z} = \bar{\varrho}\, \omega^{*2} / k^2 . \tag{18.13}$$

Similarly, the complex characteristic impedance is

$$Z_t^* (k) = \frac{P_{i3}^*}{\dot{u}_i^*} = \bar{\varrho}\, \omega^* / k \tag{18.14}$$

and the ratio that corresponded to c^* in the time-periodic case is

$$c_t^* (k) = -\frac{\dot{u}_i^*}{\partial u_i^* / \partial Z} = \omega^* / k . \tag{18.15}$$

It should be emphasized that the loci of $E^* (\omega)$ and $E_t^* (k)$ are curves in the complex plane that cannot in general be expected to intersect or coincide. There is no reason to expect a complex elastic modulus $E_t^* (k)$ for waves decaying in time to be found among the possible values of $E^* (\omega)$ for time-sinusoidal waves in the same medium. The same remark applies to $Z_t^* (k)$ and $Z^* (\omega)$ and $c_t^* (k)$ and $c^* (\omega)$.

G. Linear viscoelastic interactions.

19. The linear viscoelastic model.

Linear viscoelasticity provides a means of describing a combination of elastic and viscous properties. Aside from the assumptions of linearity, an important shortcoming of the model is its inability to take proper account of thermal effects. Nevertheless, the model is capable of describing a wide variety of material behavior. A thorough treatment of the classical theory has been given by GURTIN and STERNBERG (1962). Additional results are given by FISHER and LEITMAN (1972) in this encyclopedia, Vol. VI a/3. In this section, we review some of the simpler basic relations and indicate the parameters that appear in the solutions for sinusoidal waves.

If the strain $V_\beta(t)$ is in "Heaviside class H^1," i.e., zero for $t<0$, and in C^1 on $[0, \infty)$, we express the relaxation integral law of linear viscoelasticity as [cf. GURTIN and STERNBERG, 1962, p. 315]

$$t_\alpha(t) = \mathring{V}_\beta\, G_{\alpha\beta}(t) + \int_0^t G_{\alpha\beta}(t-\tau)\, \dot{V}_\beta(\tau)\, d\tau . \tag{19.1}$$

Here $\mathring{V}_\beta$ is the magnitude of the step of strain at $t=0$; i.e., at $t=0$, the strain $V_\beta(t)$ jumps from 0 to $\mathring{V}_\beta = V_\beta(0+)$. $G_{\alpha\beta}(t)$, called the relaxation function, is the local stress $t_\alpha(t)$ that would result from a hypothetical *unit* step of local strain V_β applied at $t=0$. The relaxation function is assumed to be in H^1.

Subject to analogous assumptions, linear viscoelastic properties may also be described in terms of a creep function $J_{\alpha\beta}(t)$ by

$$V_\alpha(t) = \overset{\circ}{t}_\beta\, J_{\alpha\beta}(t) + \int_0^t J_{\alpha\beta}(t-\tau)\, \dot{t}_\beta(\tau)\, d\tau \tag{19.2}$$

or in terms of differential operators $\mathscr{P}(D)$ and $\mathscr{Q}(D)$ by the differential equation

$$\begin{gathered} \mathscr{P}(D)\, t_\alpha = \mathscr{Q}_{\alpha\beta}(D)\, V_\beta, \\ \mathscr{P}(D) = \sum_{n=0}^{N} p_n D^n, \quad \mathscr{Q}_n = \sum_{n=0}^{N} q_n^{\alpha\beta} D^n. \end{gathered} \tag{19.3}$$

$J_{\alpha\beta}(t)$, called the creep function, is the local strain $V_\alpha(t)$ that would result from a hypothetical unit step of local stress $\overset{\circ}{t}_\beta\, h(t)$ applied at $t=0$. In (19.3), D is the time derivative operator, and $\mathscr{P}$ and $\mathscr{Q}_{\alpha\beta}$ are understood to be polynominals in D, the coefficients p_n and $q_n^{\alpha\beta}$ being constants. In all these equations, the $\boldsymbol{X}$-dependence has been suppressed.

By comparing the Laplace transforms of (19.1), (19.2), and (19.3), one sees that

$$s\,\overline{G}_{\alpha\beta}(s) = \frac{\mathscr{Q}_{\alpha\beta}(s)}{\mathscr{P}(s)}, \tag{19.4}$$

$$s^2\,\overline{G}_{\alpha\beta}(s)\,\overline{J}_{\beta\gamma}(s) = \delta_{\alpha\gamma}, \tag{19.5}$$

where the bars denote Laplace transforms.

Precise conditions for the transition between a relaxation integral law (19.1), a creep integral law (19.2), and a differential operator formulation (19.3) have been discussed by GURTIN and STERNBERG (1962).

For adiabatic deformations of a hypothetical purely *elastic solid*

$$G_{\alpha\beta}(t) = c_{\alpha\beta}^{S}\, h(t), \tag{19.6}$$

where $h(t)$ denotes the Heaviside unit step function. While it should not be presumed that the theory is entirely appropriate in the case of finite deformations, it is attractive to interpret t_α as the second Piola-Kirchhoff stress (4.7), in order to maintain a closer parallel with (9.9) in this special case. Another interesting special case is the "standard linear solid," for which

$$G_{\alpha\beta}(t) = (c_{\alpha\beta} + K_{\alpha\beta}\, e^{-t/\tau_0})\, h(t). \tag{19.7}$$

Some other special cases will be indicated in the next section.

20. One-dimensional linear viscoelastic models.

20.1. Introduction. If only a single component of stress and strain are of interest, the subscripts can be dropped, and (19.4) and (19.5) may be replaced by

$$s\,\overline{G}(s) = \frac{1}{s\,\overline{J}(s)} = \frac{\mathscr{Q}(s)}{\mathscr{P}(s)}. \tag{20.1}$$

In such "one-dimensional" cases, linear viscoelastic properties may also be described in terms of a "complex elastic modulus" $E^*(\omega)$ that relates local stress and strain in the same way that a complex driving point impedance relates voltage and current in a circuit. For steady-state sinusoidal excitations, its magnitude is the ratio of stress amplitude to strain amplitude, and its angle is the angle by which the stress leads the strain.

20.2. Sinusoidal time variations. To relate $E^*(\omega)$ to the preceding functions, let as denote "real part of" by Re, and write

$$\begin{aligned} &\text{stress } t_\alpha = \mathrm{Re}(t^*), \qquad t^* = A_t\, \mathrm{e}^{j\omega t}, \\ &\text{strain } V_\alpha = \mathrm{Re}(V^*), \qquad V^* = A_V\, \mathrm{e}^{j\omega t}. \end{aligned} \tag{20.2}$$

Because of its linearity, (19.3) can be taken to apply to the complex stress t^* and the complex strain V^*. Substitution of t^* and V^* into (19.3) then gives

$$\mathscr{P}(j\,\omega)\; t^* = \mathscr{Q}(j\,\omega)\; V^*. \tag{20.3}$$

The complex elastic modulus $E^*(\omega)$ is then by definition, $E^* = t^*/V^*$. From (20.3) and (20.1),

$$E^*(\omega) = \frac{t^*}{V^*} = \frac{\mathscr{Q}(j\,\omega)}{\mathscr{P}(j\,\omega)} = j\,\omega\,\bar{G}(j\,\omega) = \frac{1}{j\,\omega\,\bar{J}\,(j\,\omega)}\,. \tag{20.4}$$

We have used the same symbol $E^*(\omega)$ in (20.4) and in (18.5) because there is no significant distinction between them in the context of small deformations. From (20.4) and (18.5),

$$\bar{\varrho}\, c^{*2} = j\,\omega\,\bar{G}(j\,\omega) = \frac{1}{j\,\omega\,\bar{J}\,(j\,\omega)} = E^*(\omega)\,. \tag{20.5}$$

Conceptually, wave propagation measurements of α and c, made as a function of frequency on waves that are sinusoidal in time, could be used to determine the relaxation function or creep function, subject to the validity of the linear visco-elastic model. In such a program, c^* could be determined from (17.5), and $E^*(\omega) = j\,\omega\,\bar{G}(j\,\omega)$ from (20.5). The substitution $s = j\,\omega$ would then yield $s\,\bar{G}(s)$, and inversion of the transform $\bar{G}(s)$ would yield $G(t)$. Such a program would in principle require $\alpha(\omega)$ and $c(\omega)$ on the entire frequency range from zero to infinity. Besides the necessity of making assumptions concerning the behavior outside the measured range, there are the difficulties associated with obtaining suitable representations of the functions and inverting the transforms.

In a nonhomogeneous medium, the relaxation and creep functions (and the operators $\mathscr{P}$ and $\mathscr{Q}$) may depend on position. In any application to a real material, they would be expected to depend on some thermal variable, such as the equilibrium temperature or entropy.

20.3. Decaying time variations. With (17.10) in mind, we may substitute

$$t^* = A_t\, \mathrm{e}^{st}, \qquad V^* = A_V\, \mathrm{e}^{st}$$

into (19.3) to obtain

$$s\,\bar{G}(s) = \frac{t^*}{V^*} = \frac{\mathscr{Q}(s)}{\mathscr{P}(s)}\,. \tag{20.6}$$

In the solution (17.10), to each value of k there corresponds a value of

$$s = j\,\omega^* = j\,2\pi\,\nu(k) - \sigma(k)\,. \tag{20.7}$$

We can now write

$$\bar{\varrho}\,\omega^{*2}/k^2 = E_t^*(k) = s(k)\,\bar{G}[s(k)]\,. \tag{20.8}$$

It is clear from (20.4) and (20.8) that the complex elastic moduli $E^*(\omega)$ and $E_t^*(k)$ correspond to different assignments of s in the function $s\,\bar{G}(s)$. This fact is related to the remark at the end of Sect. 18.

Table 20.1. *Functions associated with five special models of viscoelasticity.*

Function	1 Elastic medium	2 Hypothetical distortionless viscoelastic medium	3 "Standard linear solid"	4 Maxwell model	5 Kelvin-Voigt model
$\frac{G(t)}{\bar{\varrho} v^2}$	$h(t)$	$e^{-\alpha v t}(1-\alpha v t)\, h(t)$	$(1+K^2 e^{-t/\tau_0})\, h(t)$	$e^{-t/\tau_0} h(t)$	$h(t)+\tau_0\, \delta(t)$
$\frac{\bar{G}(s)}{\bar{\varrho} v^2}$	$\frac{1}{s}$	$\frac{s}{(\alpha v+s)^2}$	$\frac{1}{s}+\frac{K^2}{s+1/\tau_0}$	$\frac{1}{s+1/\tau_0}$	$\frac{1}{s}+\tau_0$
$\bar{\varrho} v^2 \bar{J}(s)$	$\frac{1}{s}$	$\frac{1}{s}+\frac{2\alpha v}{s^2}+\frac{\alpha^2 v^2}{s^3}$	$\frac{1}{s}-\frac{K^2/(1+K^2)}{s+\frac{1}{\tau_0(1+K^2)}}$	$\frac{1}{s}+\frac{1}{\tau_0 s^2}$	$\frac{1}{s}-\frac{1}{s+1/\tau_0}$
$\bar{\varrho} v^2 J(t)$	$h(t)$	$\left(1+2\alpha v+\frac{1}{2}\alpha^2 v^2 t^2\right) h(t)$	$\left(1-\frac{K^2}{1+K^2}\exp\frac{-t}{\tau_0(1+K^2)}\right) h(t)$	$\left(1+\frac{t}{\tau_0}\right) h(t)$	$(1-e^{-t/\tau_0})\, h(t)$
$\frac{\mathcal{Q}(s)}{\bar{\varrho} v^2}$	1	s^2	$\frac{1}{\tau_0}+(1+K^2)\, s$	s	$1+\tau_0 s$
$\mathcal{P}(s)$	1	$(\alpha v+s)^2$	$\frac{1}{\tau_0}+s$	$\frac{1}{\tau_0}+s$	1
$\frac{v}{c^*}$	1	$1-j\frac{\alpha v}{\omega}$	$\left(1+\frac{\omega^2\tau_0^2 K^2+j\omega\tau_0 K^2}{1+\omega^2\tau_0^2}\right)^{-\frac{1}{2}}$	$\left(1+\frac{1}{j\omega\tau_0}\right)^{\frac{1}{2}}$	$(1+j\omega\tau_0)^{-\frac{1}{2}}$
$\alpha(\omega)$	0	$\alpha=$ constant	$\frac{\omega}{v C(\omega)}\sin\frac{1}{2}\varphi(\omega)$	$\frac{\omega}{v B(\omega)}\sin\frac{1}{2}\Theta(\omega)$	$\frac{\omega}{v A(\omega)}\sin\frac{1}{2}\psi(\omega)$
$c(\omega)$	v	v	$v C(\omega)\sec\frac{1}{2}\varphi(\omega)$	$v B(\omega)\sec\frac{1}{2}\Theta(\omega)$	$v A(\omega)\sec\frac{1}{2}\psi(\omega)$
$Q^{-1}(\omega)$	0	$\frac{2\alpha v}{\omega}$	$2\tan\frac{1}{2}\varphi(\omega)$	$2\tan\frac{1}{2}\Theta(\omega)$	$2\tan\frac{1}{2}\psi(\omega)$

Abbreviations: $A=(1+\omega^2\tau_0^2)^{\frac{1}{4}}$, $B=\left(\frac{\omega^2\tau_0^2}{1+\omega^2\tau_0^2}\right)^{\frac{1}{4}}$, $\psi=\tan^{-1}\omega\tau_0$, $\Theta=\cot^{-1}\omega\tau_0=\tan^{-1}\left(\frac{1}{\omega\tau_0}\right)$,

$$C=\frac{\{[1+\omega^2\tau_0^2(1+K^2)]^2+K^4\omega^2\tau_0^2\}^{\frac{1}{4}}}{(1+\omega^2\tau_0^2)^{\frac{1}{2}}},\quad \varphi=\tan^{-1}\left(\frac{\omega\tau_0 K^2}{1+\omega^2\tau_0^2(1+K^2)}\right).$$

20.4. Special cases of linear viscoelasticity. Special formulas for several of the functions associated with a viscoelastic medium are given in Table 20.1 for five special cases of linear viscoelasticity. The constant v is the limiting (real) value of c^* as $\omega\to\infty$ in cases 2 and 4, and as $\omega\to 0$ in cases 3 and 5. It is also the constant value of c in case 2, and of course it is the constant value of both c and c^* in the purely elastic medium (case 1). In Table 20.1, $h(t)$ denotes the Heaviside step function, and $\delta(t)$ the delta function. Included in the table are the functions $G(t)$, $\bar{G}(s)$, $\bar{J}(s)$, $J(t)$, $\mathcal{Q}(s)$, $\mathcal{P}(s)$, v/c^*, $\alpha(\omega)$, $c(\omega)$, $Q^{-1}(\omega)$.

$G(t)$ for the Kelvin-Voigt model contains $\delta(t)$. This may be viewed as preventing the infinite strain rate needed to apply a step of strain. Since $s\bar{G}(s)=\bar{\varrho}v^2(1+s\tau_0)$, the locus in the complex plane of the complex elastic modulus $j\omega\bar{G}(j\omega)$ is the half-line $\bar{\varrho}v^2(1+j\omega\tau_0)$. The material may be considered to be elastic in the low-frequency limit and viscous in the high-frequency limit. From $J(t)$, we see that a step of stress has no instantaneous effect (a manifestation of the viscosity) but eventually the expected elastic response is attained. The Maxwell model, on the other hand, permits a steady creep in response to a step of stress, while its $G(t)$ shows the response to a step of strain to be an initial elastic response followed by a relaxation of the stress all the way to zero. The "standard linear solid" exhibits both creep and stress relaxation, but the stress does not relax all the way to zero, and the creep is stabilized, dropping off to zero as $t\to\infty$ when the stress is held constant. The hypothetical distortionless viscoelastic medium is so named because under this constitutive law, the equation of motion has solutions of the form $e^{-\alpha Z} f(t-Z/v)$ where f is an arbitrary function. A disturbance $f(t)$ at $Z=0$ is reproduced at $Z=v\tau$, except for the delay τ and attenuation $e^{-\alpha Z}$. The ω-independence of the attenuation α and phase velocity c are responsible for this property.

To demonstrate formally the attenuating but distortionless property in case 2, we suppose u_i to be given by

$$u_i=u_0\, e^{-\alpha Z} f(t-Z/v) \tag{20.9}$$

where α and v are constants, and

$$f(x)\equiv 0 \quad \text{on} \quad x<0.$$

For simplicity, we also assume that f has continuous second derivatives everywhere and that $f'(0)=f''(0)=0$. We shall consider the solution of the equation of motion in the halfspace $Z>0$. With these assumptions, the Laplace transform of the equation of motion (18.1) reads

$$\bar{\varrho}\, s^2\, \bar{u}_i(s,Z)=\partial\bar{P}_{i3}(s,Z)/\partial Z. \tag{20.10}$$

Taking $\partial u_i/\partial Z$ as the significant strain component and P_{i3} as the stress (assuming small deformations), we write the relaxation integral law (19.1) as

$$P_{i3}=\int_0^t G(t-\tau)\,\frac{\partial\dot{u}_i(\tau,Z)}{\partial Z}\,d\tau=\frac{\partial}{\partial Z}\int_0^\infty G(t-\tau)\,\dot{u}_i(\tau,Z)\,d\tau. \tag{20.11}$$

The upper limit can be replaced by ∞ because the relaxation function G vanishes for negative values of its argument. Since the integral in (20.11) represents the convolution of G and $\dot{u}_i$, its Laplace transform is the product $\bar{G}(s)\,\bar{\dot{u}}_i(s,Z)$, and we have

$$\bar{P}_{i3}=\frac{\partial}{\partial Z}\bar{G}(s)\,\bar{\dot{u}}_i(s,Z). \tag{20.12}$$

From the assumed form (20.9),

$$\begin{aligned}\bar{u}_i(s,Z) &= u_0\,\mathrm{e}^{-\alpha Z}\int_0^\infty \mathrm{e}^{-st} f(t-Z/v)\,dt\\ &= u_0\,\mathrm{e}^{-\left(\alpha+\frac{s}{v}\right)Z}\int_0^\infty \mathrm{e}^{-sx} f(x)\,dx\\ &= u_0\exp\left[-(\alpha+s/v)\,Z\right]\bar{f}(s)\,.\end{aligned} \tag{20.13}$$

Similarly,

$$\bar{\dot{u}}_i(s,Z)=u_0\exp\left[-(\alpha+s/v)\,Z\right]s\bar{f}(s)=s\,\bar{u}_i(s,Z). \tag{20.14}$$

From (20.14) and (20.12)

$$\frac{\partial \bar{P}_{i3}(s,Z)}{\partial Z}=(\alpha+s/v)^2\,s\,\bar{G}(s)\,\bar{u}_i(s,Z)\,. \tag{20.15}$$

Substitution from (20.15) and (20.13) into (20.10) gives

$$\bar{\varrho}\,s^2=(\alpha+s/v)^2\,s\,\bar{G}(s)\,. \tag{20.16}$$

Hence, the assumed form (20.9) is a solution if

$$\bar{G}(s)=\bar{\varrho}\,\frac{s}{(\alpha+s/v)^2}=\bar{\varrho}\,v^2\,\frac{s}{(s+\alpha v)^2}\,. \tag{20.17}$$

Since (20.17) is just the $\bar{G}(s)$ given in Table 20.1, this completes the demonstration. The assumption concerning continuity of f' and f'' can be removed by considering the propagation of singular surfaces.

21. A difficulty: Thermal effects.

Linear viscoelasticity is sometimes called isothermal linear viscoelasticity because there is no provision to account for thermal effects once the relaxation function is set. This difficulty has been overcome in a general thermodynamics of materials with memory developed by Coleman $(1964)_1$, Gurtin (1968), Gurtin and Pipkin (1968), and others. The general theory is nonlinear, of course.

A difficulty in applying linear viscoelasticity (or any other linear theory) to adiabatic changes is that the effect of energy dissipation gets linearized away. Dissipation results in a change of the "equilibrium" functions with respect to which a sensible linearization can be made.

To illustrate the difficulty, let us consider the effect of different strain rates in the "standard linear solid" by considering the spatially homogeneous strain

$$\begin{aligned}&V_\beta(t)=0 \quad \text{for} \quad t<0,\\ &V_\beta(t)=\varepsilon_\beta\,t/t_0 \quad \text{on} \quad 0\le t\le t_0,\\ &V_\beta(t)=\varepsilon_\beta=\text{constant} \quad \text{for} \quad t>t_0.\end{aligned} \tag{21.1}$$

For fixed final strain ε_β, the strain rate varies inversely with the duration of the time interval t_0 on $0\le t\le t_0$, being zero before time zero and after time t_0.

Substitution from (21.1) into (19.1), using the relaxation function for the "standard linear solid" (19.7), yields

$$\begin{aligned}&t_\alpha=0, && t<0,\\ &t_\alpha=c_{\alpha\beta}\,V_\beta(t)+K_{\alpha\beta}\,\varepsilon_\beta\,\frac{\tau_0}{t_0}\left(1-\mathrm{e}^{-t/\tau_0}\right), && 0\le t\le t_0,\\ &t_\alpha=c_{\alpha\beta}\,\varepsilon_\beta+K_{\alpha\beta}\,\varepsilon_\beta\,\frac{\tau_0}{t_0}\left(\mathrm{e}^{-(t-t_0)/\tau_0}-\mathrm{e}^{-t/\tau_0}\right), && t>t_0.\end{aligned} \tag{21.2}$$

For a given final strain, ε_β, the final stress $t_\alpha(\infty)$ is seen to be the same, independently of t_0:

$$t_\alpha(\infty) = c_{\alpha\beta}\,\varepsilon_\beta. \tag{21.3}$$

This is a shortcoming of the model, for different deformation rates will produce different amounts of dissipated energy, and this should result in different final stresses. If ΔU_{D} is the difference in dissipated energy between two processes, the final stress difference after equilibrium is again attained should be approximately

$$\Delta\tau_\alpha(\infty) \doteq \left(\frac{\partial \tau_\alpha}{\partial U}\right)_V \Delta U_{\mathrm{D}} = -\frac{\lambda_\alpha}{C_V}\,\Delta U_{\mathrm{D}}. \tag{21.4}$$

(We refer to Table 10.3 for the second equality.)

That the dissipated energy does indeed depend on the strain rate in this model is easily verified. For an adiabatic deformation, we may define the energy dissipation rate as the excess of $\dot{U}$ over what $\dot{U}$ would be if the deforming were to take place at constant entropy. From (8.10) and (8.11), this is

$$\text{dissipation rate} = \dot{U} - \tau_\alpha\,\dot{V}_\alpha = T\dot{S} = (t_\alpha - \tau_\alpha)\,\dot{V}_\alpha. \tag{21.5}$$

(We assume adiabatic conditions, i.e., no heat flux and no heat sources.) From (21.1) and (21.2), taking $\tau_\alpha = c_{\alpha\beta}\,V_\beta(t)$, we have

$$T\dot{S} = (t_\alpha - \tau_\alpha)\,\dot{V}_\alpha = K_{\alpha\beta}\,\varepsilon_\alpha\,\varepsilon_\beta\,\frac{\tau_0}{t_0^2}\,(1 - \mathrm{e}^{-t/\tau_0}), \qquad 0 \leq t \leq t_0. \tag{21.6}$$

Then the energy dissipated is

$$\begin{aligned} U_D(t_0) &= \int_0^{t_0} T\dot{S}\,dt = K_{\alpha\beta}\,\varepsilon_\alpha\,\varepsilon_\beta\,\frac{\tau_0}{t_0^2}\int_0^{t_0}(1 - \mathrm{e}^{-t/\tau_0})\,dt \\ &= K_{\alpha\beta}\,\varepsilon_\alpha\,\varepsilon_\beta\,\frac{\tau_0}{t_0}\left[1 + \frac{\tau_0}{t_0}\,(\mathrm{e}^{-t_0/\tau_0} - 1)\right]. \end{aligned} \tag{21.7}$$

If the deformation is performed arbitrarily slowly ($t_0 \to \infty$), the dissipated energy U_{D} goes to zero. In the opposite extreme of an arbitrarily rapid deformation ($t_0 \to 0$), we can expand the exponential in powers of (t_0/τ_0) to obtain

$$\operatorname*{Limit}_{t_0 \to 0} U_{\mathrm{D}}(t_0) = K_{\alpha\beta}\,\varepsilon_\beta\,\varepsilon_\alpha\left(\frac{1}{2} - \frac{1}{3!}\,\frac{t_0}{\tau_0} + \cdots\right). \tag{21.8}$$

There are many situations in which dissipative heating is negligible, and linearization with respect to a fixed equilibrium state gives useful results. It should be recognized, however, that dissipation produces a change with time in the base equilibrium state that is not accounted for in the linear theory.

H. Thermoviscoelastic media.

22. General relations.

The goal of this chapter is to indicate constitutive relations (or assumptions) from which the instantaneous stress and heat flux during a nonmutative process can be found in principle when the strain and heat source distribution are prescribed functions of time and position in the material. This is an ambitious goal, and it cannot be achieved with the same degree of confidence and explicitness with which static equilibrium behavior can be described.

Fundamental to the present discussion are the static equilibrium relations of Chap. C, and Eq. (8.11), which we rewrite as

$$\begin{aligned} T(t,\boldsymbol{X})\,\dot{S}(t,\boldsymbol{X}) &= \Phi(t,\boldsymbol{X}) - \frac{\partial H_k}{\partial X_k}(t,\boldsymbol{X}) + r(t,\boldsymbol{X}), \\ \Phi(t,\boldsymbol{X}) &= [t_\alpha(t,\boldsymbol{X}) - \tau_\alpha(t,\boldsymbol{X})]\,\dot{V}_\alpha(t,\boldsymbol{X}). \end{aligned} \tag{22.1}$$

Summation over k from 1 to 3 and α from 1 to 6 is understood. It is worth recalling that (22.1) is in essence the equation of balance of energy, obtained merely by substituting the *thermostatic* relation (8.10) into the equation of balance, (6.16). For every $(t,\boldsymbol{X})$, values of $V_\alpha(t,\boldsymbol{X})$ and $U(t,\boldsymbol{X})$ generate values of all the static equilibrium functions, which we call *present values*. For example

$$S(t,\boldsymbol{X}) = S[U(t,\boldsymbol{X}), V_\alpha(t,\boldsymbol{X})] \tag{22.2}$$

is the present value of the static equilibrium entropy. Similarly

$$\tau_\alpha(t,\boldsymbol{X}) = \tau_\alpha[S(t,\boldsymbol{X}), V_\alpha(t,\boldsymbol{X})], \tag{22.3}$$

is the present static equilibrium stress, while

$$T(t,\boldsymbol{X}) = T[S(t,\boldsymbol{X}), V_\alpha(t,X)] \tag{22.4}$$

is the present static equilibrium temperature. This terminology allows instantaneous values of the *physical* variables entropy and temperature either to be defined differently from S and T, or to remain undefined. The present static *equilibrium* values are simply the values that pertain to the static equilibrium state that would be approached if strain and internal energy density were held constant (in *both* time and space) at their present values.

It is essential to keep track of the present static equilibrium state, determined, say, by (U, V_α), or (S, V_α). With $V_\alpha(t,\boldsymbol{X})$ and $r(t,\boldsymbol{X})$ given, we could get U from (6.16) if we knew the actual stress and the heat flux. Alternatively, we could extract S from (22.1) if we knew the temperature, the heat flux and $(t_\alpha - \tau_\alpha)$, the difference of the stress from its present static equilibrium value.

We assume the heat flux $\boldsymbol{H}$ and stress difference $t_\alpha - \tau_\alpha$ to be initially zero, and given by constitutive relations of the general form

$$t_\alpha(t) - \tau_\alpha(t) = \int_0^t [g_{\alpha\beta}(t,\tau)\,\dot{V}_\beta(\tau) + f_{\alpha j}(t,\tau)\,g_j(\tau)]\,d\tau, \tag{22.5}$$

$$H_i(t) = -\int_0^t N_{i\beta}(t,\tau)\,\dot{V}_\beta(\tau) + L_{ij}(t,\tau)\,g_j(\tau)]\,d\tau, \tag{22.6}$$

where $\boldsymbol{g}$ is the temperature gradient, of Cartesian components $g_j = \partial T/\partial X_j$. In (22.5) and (22.6), only the time dependence is shown explicitly. $\boldsymbol{X}$ is understood to be an argument of every function that occurs.

With the nonequilibrium constitutive behavior specified by (22.5) and (22.6) in terms of functions $g_{\alpha\beta}$, $f_{\alpha j}$, $N_{i\beta}$, L_{ij} that are presumed known, one can in principle keep track of the present static equilibrium properties like $\tau_\alpha(t)$ and $T(t)$ by time integration of

$$\dot{\tau}_\alpha = \left(\frac{\partial \tau_\alpha}{\partial V_\beta}\right)_S \dot{V}_\beta + \left(\frac{\partial \tau_\alpha}{\partial S}\right)_V \dot{S} = c^S_{\alpha\beta}\,\dot{V}_\beta - \gamma_\alpha\,T\dot{S}, \tag{22.7}$$

$$\dot{T} = \left(\frac{\partial T}{\partial V_\beta}\right)_S \dot{V}_\beta + \left(\frac{\partial T}{\partial S}\right)_V \dot{S} = -\gamma_\beta\,T\dot{V}_\beta + \frac{1}{C_V}\,T\dot{S}, \tag{22.8}$$

where $T\dot{S}$ is to be obtained through (22.1), (22.5), and (22.6). The second forms of (22.7) and (22.8) follow from Table 10.1 with $\gamma_\alpha = \lambda_\alpha / C_V$ as in (10.41). Of course, the coefficients of $\dot{V}_\beta$ and $T\dot{S}$ in (22.7) and (22.8) are themselves static equilibrium functions that need to be evaluated by a similar scheme when their variation is important. All the needed equilibrium functions are presumed known in terms of (V_β, T), say, and these functions can be converted to any other set of independent variables that specify the thermostatic state.

With the strain $V_\alpha(t, \boldsymbol{X})$ and the heat source distributed $r(t, \boldsymbol{X})$ specified, Eqs. (22.1), (22.5), (22.6), (22.7), and (22.8) may be regarded as 17 simultaneous equations that in principle determine the 17 unknowns consisting of $T\dot{S}$, T, the six τ_α, the six present stresses t_α, and the three components of the heat flux vector $\boldsymbol{H}$, all as functions of time and position $(t, \boldsymbol{X})$.

It should be remarked that relations more general than (22.5) and (22.6) have been studied. For example, GURTIN (1972) allows the stress, internal energy, entropy, and heat flux at time t to depend on the histories of the deformation gradient, the temperature, and the temperature gradient. The present discussion, on the other hand, retains the internal energy and entropy as thermodynamic *functions* in the sense discussed earlier, imposes the restriction of equilibrium at $t=0$, and does not include the temperature *history* as an influence that affects the present stress or heat flux.

23. A special model for thermoviscoelasticity.

A connection of the stress difference $(t_\alpha - \tau_\alpha)$ to the strain rate $\dot{V}_\beta$ and of the heat flux to the temperature gradient is expected. On the other hand, the responses implied by $f_{\alpha j}$ and $N_{i\beta}$ in (22.5) and (22.6) are not ordinarily considered. While it would be rash to assert that $f_{\alpha j}$ and $N_{i\beta}$ are always zero or that their effects are never important, we do not know of any experiments that have detected them. It is safe to say that the heat flux is ordinarily determined principally by the temperature gradient and the stress difference by the straining. We therefore drop the terms in $\boldsymbol{f}_\alpha$ and $\boldsymbol{N}_\beta$.

In general, the functions $g_{\alpha\beta}$ and $\boldsymbol{L}$ are expected to depend explicitly on the strain gradients ΔV_γ and the strain rates $\dot{V}_\gamma$. As an additional simplifying assumption (or approximation), such dependence will be ignored.

Further elucidation of the nature of $g_{\alpha\beta}$ and $\boldsymbol{L}$ seems to require assumptions that are far more special than those already made. To illustrate the kind of behavior that is a qualitatively plausible approximation, although admittedly very special, let us make the outright assumption that the stress differences $t_\alpha - \tau_\alpha$ satisfy the system of first-order differential equations

$$\dot{t}_\alpha - \dot{\tau}_\alpha + p_{\alpha\beta}(t_\beta - \tau_\beta) = A_{\alpha\gamma} \dot{V}_\gamma. \tag{23.1}$$

The coefficients $p_{\alpha\beta}$ and $A_{\alpha\gamma}$ are assumed to depend on the present static equilibrium state. For example, they can be regarded as functions of (S, V_α) or (T, V_α), in which case their time dependence, needed for the integration of (23.1), is derived from the time dependence of (S, V_α) or (T, V_α). Let p and A denote the six-by-six square matrices of elements $p_{\alpha\beta}$ and $A_{\alpha\beta}$, and let y and V denote the single-column matrices of elements $t_\alpha - \tau_\alpha$ and V_α respectively. Then, with the $\boldsymbol{X}$ dependence suppressed, (23.1) is

$$\dot{y}(t) + p(t)\, y(t) = A(t)\, \dot{V}(t). \tag{23.2}$$

If $p(t)$ commutes with $\int p(s)\,ds$, then $\exp[\int p(s)\,ds]$ is an integrating factor, and the solution that satisfies $y(0)=0$ can be expressed as

$$y(t)=\int_0^t g(t,\tau)\,\dot{V}(\tau)\,d\tau, \qquad g(t,\tau)=\exp\left[-\int_\tau^t p(s)\,ds\right]A(\tau). \tag{23.3}$$

Eq. (23.3) can be written as

$$\begin{aligned} t_\alpha-\tau_\alpha &= \int_0^t g_{\alpha\beta}(t,\tau)\,\dot{V}_\beta(\tau)\,d\tau, \\ g_{\alpha\beta}(t,\tau) &- E_{\alpha\gamma}(t,\tau)\,A_{\gamma\beta}(\tau), \\ E_{\alpha\gamma}(t,\tau) &\equiv \left(\exp\left[-\int_\tau^t p(s)\,ds\right]\right)_{\alpha\gamma}. \end{aligned} \tag{23.4}$$

Thus, in this special case, $(23.4)_2$ and $(23.4)_3$ express the relaxation functions $g_{\alpha\beta}(t,\tau)$ in terms of coefficients $p_{\alpha\beta}$ and $A_{\alpha\beta}$ that are presumed to be static equilibrium properties.

To obtain a similar formula for the heat flux, we make a similar *ad hoc* assumption, namely that the components of $\boldsymbol{H}$ satisfy

$$\dot{H}_i+q_{ik}(H_k+\varkappa_{kj}\,g_j)=0 \tag{23.5}$$

where the coefficients q_{ik} and $\varkappa_{ij}$ are assumed to depend on the present static equilibrium state. It is easy to see that $\varkappa_{ij}$ is the steady-state thermal conductivity, for in the time-independent case, (23.5) is satisfied by FOURIER's law of heat conduction,

$$H_k=-\varkappa_{kj}\,g_j. \tag{23.6}$$

Let $\varkappa$ and q denote the 3×3 square matrices of elements $\varkappa_{ij}$ and q_{ij}. Let H and g denote the single-column matrices of elements H_i and g_i. Then, with the $\boldsymbol{X}$-dependence suppressed, (23.5) is

$$\dot{H}+q(t)\,H(t)=-q(t)\,\varkappa(t)\,g(t). \tag{23.7}$$

If $q(t)$ commutes with $\int q(s)\,ds$, $\exp[\int q(s)\,ds]$ is an integrating factor, and the solution that has $H(0)=0$ can be expressed as

$$\begin{aligned} H(t) &= -\int_0^t \exp\left[-\int_\tau^t q(s)\,ds\right] q(\tau)\,\varkappa(\tau)\,g(\tau)\,d\tau \\ &= -\int_0^t \frac{d}{d\tau}\left[\exp\left(-\int_\tau^t q(s)\,ds\right)\right]\varkappa(\tau)\,g(\tau)\,d\tau. \end{aligned} \tag{23.8}$$

Eq. (23.8) can be written as

$$\begin{aligned} H_i(t) &= -\int_0^t L_{ij}(t,\tau)\,g_j(\tau)\,d\tau, \\ L_{ij}(t,\tau) &= F_{ik}(t,\tau)\,q_{km}(\tau)\,\varkappa_{mj}(\tau), \\ F_{ik}(t,\tau) &= \left(\exp\left[-\int_\tau^t q(s)\,ds\right]\right)_{ik}. \end{aligned} \tag{23.9}$$

Through $(23.9)_2$ and $(23.9)_3$, the heat flux relaxation functions $L_{ij}(t,\tau)$ for this special case are expressed in terms of coefficients q_{km} and $\varkappa_{km}$ that are presumed to be static equilibrium properties.

Writing (23.8) as

$$H(t) = -\int_0^t L(t,\tau)\, g(\tau)\, d\tau \tag{23.10}$$

and integrating by parts with

$$L(t,\tau) = \frac{d}{d\tau} K(t,\tau), \tag{23.11}$$

we obtain

$$H(t) = -K(t,t)\, g(t) + K(t,0)\, g(0) + \int_0^t K(t,\tau)\, \dot{g}(\tau)\, d\tau. \tag{23.12}$$

It is easy to see that an arbitrary function of time t may be added to $K(t,\tau)$ with no change in $H(t)$; if such a function is added, the additional contribution of the integral in (23.12) is just compensated by the additional contributions of the other two terms. It follows that either $K(t,t)$ or $K(t,0)$ may be chosen arbitrarily. However, from (23.11),

$$K(t,t) = K(t,\tau) + \int_\tau^t L(t,r)\, dr. \tag{23.13}$$

It is useful to choose $K(t,t)$ to be equal to the steady-state conductivity matrix $\varkappa(t)$:

$$K(t,t) = \varkappa(t). \tag{23.14}$$

Then, from (23.13) and (23.14),

$$K(t,\tau) = \varkappa(t) - \int_\tau^t L(t,r)\, dr. \tag{23.15}$$

The special form of (23.15) that results from using the L in (23.8) is

$$K(t,\tau) = \varkappa(t) - \int_\tau^t \frac{d}{dr}\left[\exp\left(-\int_r^t q(s)\, ds\right)\right] \varkappa(r)\, dr. \tag{23.16}$$

An integration by parts puts (23.16) in the form

$$K(t,\tau) = \varkappa(\tau)\left[\exp\left(-\int_\tau^t q(s)\, ds\right)\right] + \int_\tau^t \dot{\varkappa}(r)\left[\exp\left(-\int_r^t q(s)\, ds\right)\right] dr. \tag{23.17}$$

If the matrices $p(t)$ in (23.2) and $q(t)$ in (23.7) have time-dependent eigenvectors, $\int p(s)\, ds$ will not in general commute with its derivative $p(t)$, nor will $q(t)$ commute with $\int q(s)\, ds$, in general. In this case, the solutions cannot be expressed in the concise explicit forms shown here.

We emphasize again the very special nature of the equations of this section. They are only partially motivated by physical considerations, and while they are presumably an improvement over the equations for an elastic medium with Newtonian viscosity and heat conduction according to FOURIER's law, their range of validity is expected to be limited.

24. Linearized steady-state response.

Let the matrices p and A be constant in (23.2), and consider a complex exponential "excitation" of the form

$$V(t) = V_0\, e^{st}, \qquad s = j\omega + \sigma, \tag{24.1}$$

where σ and ω are real. Then Eq. (23.2) has a complex exponential solution, which may be expressed formally as

$$\begin{aligned} y(t) &= y_0\, e^{st} = Y(s)\, V_0\, e^{st}, \\ Y(s) &= s\,(sI+p)^{-1} A, \end{aligned} \tag{24.2}$$

I being the unit matrix. With $s=j\omega$, (24.1) is a "steady-state" sinusoidal excitation. With y as the response, $Y(j\omega)$ is the steady-state response matrix (function) in the language of linear systems. $Y(s)$ can be called the generalized steady-state response function.

The Laplace transform and its inverse provide a means of expressing the response to any Laplace-transformable excitation in terms of $Y(s)$. Such an excitation $V(t)$ can be expressed as

$$V(t) = \mathcal{L}^{-1}[\overline{V}(s)] = \frac{1}{2\pi j}\int_{\Gamma} \overline{V}(s)\, e^{st}\, ds \tag{24.3}$$

where $\overline{V}(s)$ is the Laplace transform of $V(t)$, given by

$$\overline{V}(s) = \mathcal{L}[V(t)] = \int_0^{\infty} V(t)\, e^{-st}\, dt \tag{24.4}$$

and Γ is any contour in the s-plane extenting from $-j\infty$ to $+j\infty$ and passing to the right of all poles of $\overline{V}(s)$. Eq. (24.3) enables us to picture the excitation $V(t)$ as the limit of the sum of complex exponential excitations of the form e^{st} and amplitude $\overline{V}(s)\, ds/2\pi j$. From (24.2), each such complex exponential excitation produces the response $dy = Y(s)\, \overline{V}(s)\, e^{st}\, ds/2\pi j$. Superposition (allowed by the assumed linearity) then lets us picture the complete response as the integral

$$y(t) = \frac{1}{2\pi j}\int_{\Gamma} Y(s)\, \overline{V}(s)\, e^{st}\, ds = \mathcal{L}^{-1}[Y(s)\, \overline{V}(s)]. \tag{24.5}$$

The right side of (24.5) is just the inverse Laplace transform of the product $Y(s)\, \overline{V}(s)$. Hence, the Laplace transform of the response $y(t)$ is

$$\mathcal{L}[y(t)] \equiv \bar{y}(s) = Y(s)\, \overline{V}(s). \tag{24.6}$$

Analogous formulas can be proved rigorously for linear systems under very general conditions: if the Laplace transforms exist, the Laplace transform of the response is the generalized steady-state response function times the Laplace transform of the excitation. (This simple statement requires initial conditions to be construed as part of the excitation.)

With the matrices p and A constant in (23.2), exp pt is an integrating factor and $g(t, \tau)$ in (23.3) reduces to a function of $(t-\tau)$, namely $\exp[-(t-\tau)p]A$. To simplify the treatment of this case, define

$$\begin{aligned} g(t,\tau) \equiv g(t-\tau) &= \exp[-(t-\tau)p]\, A, \qquad t \geqq \tau \\ &= 0, \qquad\qquad\qquad\qquad\quad t < \tau, \end{aligned} \tag{24.7}$$

that is,

$$g(t) = \exp(-pt)\, A\, h(t) \tag{24.7a}$$

where $h(t)$ is the Heaviside unit step function. Then, because $g(t, \tau)$ is by definition zero for $\tau > t$, the integral in $(23.3)_1$ can run from 0 to ∞:

$$y(t) = \int_0^{\infty} g(t-\tau)\, \dot{V}(\tau)\, d\tau. \tag{24.8}$$

By the theorems on Laplace transforms of convolutions and derivatives, the Laplace transform of (24.8) is

$$\bar{y}(s) = s\,\bar{g}(s)\,\bar{V}(s). \tag{24.9}$$

By comparison with (24.6), the generalized steady-state response function (matrix) is

$$Y(s) = s\,\bar{g}(s). \tag{24.10}$$

For the present model, comparison with $(24.2)_2$ gives

$$\bar{g}(s) = (sI + p)^{-1} A, \tag{24.11}$$

a formula which can be verified directly by applying the definition of the Laplace transform, (24.4), to the $g(t)$ in (24.7a).

The matrix p is real. If its effect in (23.2) is to tend to cause a return toward equilibrium, its eigenvalues λ_i must be real and positive. Therfore we expect the determinant of $(sI+p)$ to vanish only at the points in the s-plane on the negative real axis, $s = -\lambda_i$, $i = 1, 2, \ldots, 6$. The inverse matrix $(sI+p)^{-1}$ will then be defined everywhere else.

Interesting variants of the above results can be obtained by taking p as constant but keeping $A(\tau)$ as a variable that depends on the static equilibrium state at the time τ.

In the foregoing, the column-matrix y, of components $t_\alpha - \tau_\alpha$ expresses the difference of the Piola-Kirchhoff stress t_α from its static equilibrium value τ_α. Analogous considerations applied to a linearized version of (23.7) relate the heat flux to the temperature gradient. With q and $\varkappa$ constant and with the temperature gradient column-matrix specified by $g(t) = g_0\, e^{st}$, (23.7) has the formal solution

$$H(t) = -(sI+q)^{-1} q \varkappa g_0\, e^{st}, \tag{24.12}$$

from which we conclude that the Laplace transforms of the heat flux and temperature gradient in the present linearized model are related by

$$\bar{H}(s) = -(sI+q)^{-1} q \varkappa \bar{g}(s). \tag{24.13}$$

Eq. (23.9) takes the form

$$H(t) = -\int_0^t L(t-\tau)\, g(\tau)\, d\tau \tag{24.14}$$

with

$$L(t-\tau) = \exp[-(t-\tau)\, q]\, q \varkappa h(t-\tau). \tag{24.15}$$

The Laplace transform of (24.14) is

$$\bar{H}(s) = -\bar{L}(s)\, \bar{g}(s). \tag{24.16}$$

Comparison with (24.13) shows that for the present model

$$\bar{L}(s) = (sI+q)^{-1} q \varkappa, \tag{24.17}$$

a formula which can be verified directly from (24.15).

From (23.15) and (24.15), the special form of $K(t,\tau)$ in the present linearization is

$$K(t,\tau) = \exp[-(t-\tau)\, q]\, \varkappa h(t-\tau). \tag{24.18}$$

The equilibrium conductivity matrix $\varkappa = K(t,t)$ satisfies

$$\varkappa = K(t,t) = \int_0^\infty L(s)\, ds \tag{24.19}$$

with $L(s)$ from (24.15) [cf. GURTIN and PIPKIN, 1968; GURTIN, 1972].

Interesting variants of (24.12)–(24.18) can be obtained by taking q as constant, but allowing $\varkappa(\tau)$ to remain as a variable that depends on the static equilibrium state at the time τ.

In a more general model, $\bar{g}(s)$ and $\bar{L}(s)$ would be given by more general expressions instead of the special relations (24.11) and (24.17). Thus, a frequency dependence of the linearized steady-state response incompatible with that predicted by putting $s=j\omega$ in (24.2) and (24.12) would indicate a failure of the special model of Sect. 23, but not necessarily of the more general forms (22.5) and (22.6).

I. Small-amplitude waves that are sinusoidal in time.

25. Thermoviscoelastic medium.

Many experiments on waves in solids are done by exciting a transducer with a signal that is essentially a pulse-modulated sine wave. Each resulting pulse, or wave train is characterized by the frequency of the sine waves that comprise it and a pulse duration that is long compared with their period. The pulse repetition rate is sufficiently low that the medium is presumed to be in static equilibrium when the next pulse arrives. Thus, each wave train is assumed to be propagating into an initially undisturbed medium.

The system of equations that describes wave propagation can be attacked by expanding the variables, where necessary, in powers of the strain and difference in entropy ΔS (or temperature ΔT) from that of the undisturbed medium ahead of the wave. To obtain theoretical expressions for the velocity and attenuation of small-amplitude waves, it is sufficient to have the stress correct to first order in the strain and ΔS (or ΔT). Through solutions of the resulting linear equations, material parameters are related to the velocity and attenuation of waves in the small-amplitude limit. (More exact analysis is required to reveal amplitude-dependent effects, such as amplitude-dependent damping and dissipative heating of the medium arising from internal friction.)

We shall illustrate the procedure by considering small-amplitude sinusoidal wave propagation in the special thermoviscoelastic medium of Sects. 23 and 24.

Consider plane waves of particle displacement $\boldsymbol{u}$ of components

$$u_i(t, \boldsymbol{X}) = u_i^0 \exp(-\alpha \boldsymbol{N}\cdot\boldsymbol{X}) \cos\omega(t + t_i^0 - \boldsymbol{N}\cdot\boldsymbol{X}/c) \tag{25.1}$$

where $\boldsymbol{N}$ is a unit vector in the direction of propagation (and where the range of $\boldsymbol{N}\cdot\boldsymbol{X}$ is appropriately limited, e.g., $\boldsymbol{N}\cdot\boldsymbol{X}>0$). The quantities u_i^0, α, ω, t_i^0, and c are all real; c is the phase velocity, α the attenuation per unit length, ω the angular frequency, u_i^0 the amplitude of the i component of particle displacement, and the angles ωt_i^0 $(i=1, 2, 3)$ allow for adjustment of the relative phase of the harmonic oscillations of u_i. The most general particle orbits allowed by (15.1) are ellipses. In special cases, each particle moves along a straight line, and the wave is said to be linearly polarized. This is the case if the phase differences are all either zero or π.

The three functions $u_i(t, \boldsymbol{X})$ in (25.1) can be represented as

$$\begin{aligned} u_i(t, \boldsymbol{X}) &= \varepsilon\lambda_0 \operatorname{Re}[u_i^*(t, \boldsymbol{X})], \\ u_i^*(t, \boldsymbol{X}) &= U_i \exp(j\omega t - \Gamma \boldsymbol{N}\cdot\boldsymbol{X}), \\ \Gamma &= \alpha + j\omega/c \end{aligned} \tag{25.2}$$

where ε is a small dimensionless parameter and λ_0 is a reference wavelength. The symbol Re stands for "real part of". By comparison with (25.1),

$$u_i^0 = \varepsilon \lambda_0 |U_i|, \qquad \omega t_i^0 = \underline{/U_i}, \tag{25.3}$$

where $|U_i|$ and $\underline{/U_i}$ represent the magnitude and angle of the complex number $U_i = |U_i| \exp j \underline{/U_i}$. Without loss of generality, we take the complex vector $\boldsymbol{U}$ to be normalized such that the inner product of $\boldsymbol{U}$ with its complex conjugate, $\tilde{\boldsymbol{U}}$, is 1, i.e.,

$$\tilde{\boldsymbol{U}} \cdot \boldsymbol{U} = \tilde{U}_i U_i = 1. \tag{25.4}$$

If the wave is linearly polarized, $\boldsymbol{U}$ can be taken as a real unit vector, in which case the real components U_i are simply the direction cosines of the polarization direction. This is the case with which we shall be concerned, provided the equations have solutions for real $\boldsymbol{U}$. In the contrary case, if any such exists, a complex $\boldsymbol{U}$ can be interpreted in terms of *elliptically* polarized waves, as above.

To relate the phase velocity c and attenuation α to material properties, we need the stress to first order in ε, for use in the equation of motion. To first order in ε, the displacement field (25.2) results in the strain components

$$V_{km}(t, \boldsymbol{X}) = \mathrm{Re}\,[V_{km}^*(t, \boldsymbol{X})], \qquad V_{km}^*(t, \boldsymbol{X}) = (\Delta V)_{km}^* \exp(j\omega t - \Gamma \boldsymbol{N} \cdot \boldsymbol{X}), \tag{25.5}$$

in which

$$(\Delta V)_{km}^* = -\tfrac{1}{2}\varepsilon (N_m U_k + N_k U_m) \Gamma \lambda_0. \tag{25.6}$$

The difference of the stress from the present static equilibrium stress is obtained, to order ε, as the real part of (24.2) with $s = j\omega$. Thus, in the matrix notation of (23.2) and (24.2),

$$y(t, \boldsymbol{X}) = \mathrm{Re}\, y^*(t, \boldsymbol{X}), \qquad y^*(t, \boldsymbol{X}) = Y(j\omega)\, V^*(t, \boldsymbol{X}). \tag{25.7}$$

To get the static equilibrium stress, we need to know how the present static equilibrium entropy S or temperature T varies in the wave. From (22.1), we see that changes of S may arise from Φ, $\nabla \cdot \boldsymbol{H}$, or r. We assume $r=0$ (no heat source distribution). Because neither y (the stress differences $t_\alpha - \tau_\alpha$) nor $\dot{V}_\alpha$ has a term of order zero in ε, Φ gives rise to changes of S that are of order two in ε, but not to any first-order change. The only possible first-order change in S arises from the divergence of the heat flux. There will be a first-order contribution from this term if there is a temperature variation that is of first-order in ε, for such a temperature variation will have a spatial dependence like that of $V(t, \boldsymbol{X})$. We see from Table 10.1 that if $\lambda_\alpha = 0$ for all the strain components V_α that are present in the wave, then strain variations at constant entropy do not produce any first-order temperature change (nor do strain variations at constant temperature produce any first-order entropy change). In this case, constant temperature and constant entropy are mutually compatible. *A strain wave having only strain components V_α for which $\lambda_\alpha = 0$ propagates without producing any first-order changes in temperature or entropy.* The temperature and entropy variations accompanying such a wave are of order two and higher in the strain amplitude. This is often the case with the purely transverse modes. On the other hand, if the wave has strain components V_α for which $\lambda_\alpha \neq 0$, then *both* the temperature and entropy undergo first-order variations. To calculate them, we call on the energy equation (22.1), Table 10.1, the constitutive relation for the heat flux here taken as (24.16), and the equation of motion (6.7).

Let T_0, S_0, and $\tau_\alpha^0 = t_\alpha^0$ be the constant static equilibrium values of T, S, and t_α in the absence of the wave. Having in mind an expansion of $T - T_0$, $S - S_0$, and $t_\alpha - \tau_\alpha^0$ in powers of ε, with coefficients that are functions of $(t, \boldsymbol{X})$, but seeking only the terms linear in ε, we write

$$\begin{aligned} T - T_0 &= \mathrm{Re}\, T^*(t, \boldsymbol{X}) + \cdots, \\ T^*(t, \boldsymbol{X}) &= (\Delta T)^* \exp(j\omega t - \Gamma \boldsymbol{N} \cdot \boldsymbol{X}), \end{aligned} \tag{25.8}$$

$$\begin{aligned} S - S_0 &= \mathrm{Re}\, S^*(t, \boldsymbol{X}) + \cdots, \\ S^*(t, \boldsymbol{X}) &= (\Delta S)^* \exp(j\omega t - \Gamma \boldsymbol{N} \cdot \boldsymbol{X}), \end{aligned} \tag{25.9}$$

$$\tau_\alpha - \tau_\alpha^0 = \mathrm{Re}\, \tau_\alpha^*(t, \boldsymbol{X}) + \cdots, \tag{25.10}$$

$$t_\alpha - \tau_\alpha^0 = \mathrm{Re}\, t_\alpha^*(t, \boldsymbol{X}) + \cdots. \tag{25.11}$$

The terms not written explicitly are understood to be of order two and higher in ε.

The next step is to relate $(\Delta T)^*$ and $(\Delta S)^*$ to each other through the energy equation (22.1) and the constitutive relation for the heat flux (24.14). From (25.8), the temperature gradient column-matrix $g(t, \boldsymbol{X})$, of elements $\partial T/\partial X_i$, $i = 1, 2, 3$, is

$$\begin{aligned} g(t, \boldsymbol{X}) &= \mathrm{Re}\, g^*(t, \boldsymbol{X}) + \cdots \\ g^*(t, \boldsymbol{X}) &= -\Gamma N T^*(t, X) \end{aligned} \tag{25.12}$$

where N is the column-matrix of elements N_i, N_i being the components of the unit vector $\boldsymbol{N}$. From (24.16), the column-matrix H (whose elements are the components of the heat flux vector) can be expressed as

$$\begin{aligned} H(t, \boldsymbol{X}) &= \mathrm{Re}\, H^*(t, \boldsymbol{X}) + \cdots \\ H^*(t, \boldsymbol{X}) &= -\bar{L}(j\omega) g^*(t, \boldsymbol{X}) = \bar{L}(j\omega) N \Gamma T^*(t, \boldsymbol{X}). \end{aligned} \tag{25.13}$$

It should be recalled that $\bar{L}$ is a square 3×3 matrix. From (22.1), (25.13), and (25.8),

$$T\dot{S} = -\frac{\partial H_k}{\partial X_k} + \cdots = \mathrm{Re}\left[L_{\boldsymbol{N}}(\omega)\Gamma^2 T^*(t, \boldsymbol{X})\right] + \cdots \tag{25.14}$$

where

$$L_{\boldsymbol{N}}(\omega) = \bar{L}_{ki}(j\omega) N_k N_i. \tag{25.15}$$

But from (25.9),

$$T\dot{S} = T_0\, \mathrm{Re}\left[j\omega S^*(t, \boldsymbol{X})\right] + \cdots. \tag{25.16}$$

Thus, the "energy equation" (22.1) is satisfied to first order in ε by taking

$$(\Delta S)^* = \frac{1}{j\omega T_0} L_{\boldsymbol{N}}(\omega) \Gamma^2 (\Delta T)^*. \tag{25.17}$$

To relate $(\Delta S)^*$ and $(\Delta T)^*$ to the complex strain amplitudes $(\Delta V)^*_{mn}$, we can expand T as a Taylor series in the strain and S, giving

$$(\Delta T)^* = \left(\frac{\partial T}{\partial V_{mn}}\right)_S (\Delta V)^*_{mn} + \left(\frac{\partial T}{\partial S}\right)_V (\Delta S)^* = -T_0 \gamma_{mn} (\Delta V)^*_{mn} + \frac{T_0}{C_V} (\Delta S)^*. \tag{25.18}$$

In (25.18), relations from Table 10.1 and Eq. (10.40) have been used. The coefficients γ_{mn} and C_V are evaluated at the static equilibrium state in the absence of the wave. Use of (25.18) in (25.17) yields

$$(\Delta S)^* = -\frac{L_{\boldsymbol{N}} \Gamma^2 \gamma_{mn} (\Delta V)^*_{mn}/j\omega}{1 - L_{\boldsymbol{N}} \Gamma^2/j\omega C_V}. \tag{25.19}$$

For future reference, we record also

$$(\Delta T)^* = -\frac{T_0 \gamma_{mn} (\Delta V)^*_{mn}}{1 - L_{\boldsymbol{N}} \Gamma^2 / j\omega C_V}. \tag{25.20}$$

To determine the unknown Γ through the equation of motion, we need an expression for the stress. The Taylor series expansion of $\tau_\alpha(S, V_\beta)$ is, from Table 10.1,

$$\tau_\alpha - \tau_\alpha^0 = c^S_{\alpha\beta} V_\beta - T_0 \gamma_\alpha (S - S_0) + \cdots, \tag{25.21}$$

where $\gamma_\alpha = \lambda_\alpha / C_V$ as in (10.41), and where $c^S_{\alpha\beta}$ and γ_α are understood to be evaluated at the static equilibrium state that pertains in the absence of the wave. In view of (25.5), (25.10), and (25.9), the leading terms of (25.21) may be expressed as

$$\tau^*_\alpha(t, \boldsymbol{X}) = c^S_{\alpha\beta} V^*_\beta(t, \boldsymbol{X}) - T_0 \gamma_\alpha S^*(t, \boldsymbol{X}). \tag{25.22}$$

Similarly, by considering $\tau_\alpha = \tau_\alpha(T, V_\beta)$, we obtain

$$\tau^*_\alpha(t, \boldsymbol{X}) = c^T_{\alpha\beta} V^*_\beta(t, \boldsymbol{X}) - \lambda_\alpha T^*(t, \boldsymbol{X}). \tag{25.23}$$

Rewriting $(25.7)_2$ as

$$t^*_\alpha - \tau^*_\alpha = Y_{\alpha\beta}(j\omega) V^*_\beta(t, \boldsymbol{X}), \tag{25.24}$$

we combine (25.22) and (25.24) to obtain

$$t^*_\alpha(t, \boldsymbol{X}) = [c^S_{\alpha\beta} + Y_{\alpha\beta}(j\omega)] V^*_\beta(t, \boldsymbol{X}) - T_0 \gamma_\alpha S^*(t, \boldsymbol{X}). \tag{25.25}$$

Similarly, from (25.23) and (25.24),

$$t^*_\alpha(t, \boldsymbol{X}) = [c^T_{\alpha\beta} + Y_{\alpha\beta}(j\omega)] V^*_\beta(t, \boldsymbol{X}) - \lambda_\alpha T^*(t, \boldsymbol{X}). \tag{25.26}$$

In view of (25.8), (25.9), (25.19), and (25.20), Eqs. (25.25) and (25.26) each take the form

$$t^*_\alpha(t, \boldsymbol{X}) = c^*_{\alpha\beta} V^*_\beta(t, \boldsymbol{X}), \tag{25.27}$$

with the complex elastic constant $c^*_{\alpha\beta}(\omega)$ given by

$$\begin{aligned} c^*_{\alpha\beta} &= c^S_{\alpha\beta} + Y_{\alpha\beta}(j\omega) + \frac{T_0 \lambda_\alpha \gamma_\beta L_{\boldsymbol{N}} \Gamma^2 / j\omega C_V}{1 - L_{\boldsymbol{N}} \Gamma^2 / j\omega C_V} \\ &= c^T_{\alpha\beta} + Y_{\alpha\beta}(j\omega) + \frac{T_0 \lambda_\alpha \gamma_\beta}{1 - L_{\boldsymbol{N}} \Gamma^2 / j\omega C_V}. \end{aligned} \tag{25.28}$$

The equality of these two expressions for $c^*_{\alpha\beta}$ follows easily from (10.16), which may be put in the form

$$c^S_{\alpha\beta} - c^T_{\alpha\beta} = T \lambda_\alpha \gamma_\beta. \tag{25.29}$$

We are finally ready to see whether a wave of the type under consideration can satisfy the equation of motion. In view of (4.8) and the equality $\dot{v}_i = \ddot{u}_i$, the equation of motion (6.7) becomes, after disposing of the body force g_i,

$$\bar{\varrho} \ddot{u}_i = \frac{\partial}{\partial X_k} \left(\frac{\partial x_i}{\partial X_s} t_{ks} \right). \tag{25.30}$$

From (3.4) and (25.2),

$$\frac{\partial x_i}{\partial X_s} = \delta_{is} - \varepsilon \lambda_0 \operatorname{Re}[\Gamma N_s u^*_i(t, \boldsymbol{X})]. \tag{25.31}$$

From (25.11), (25.27), and (25.5)–(25.6), in full tensor notation,

$$\begin{aligned} t_{ks} &= \tau^0_{ks} + \operatorname{Re} t^*_{ks} + \cdots, \\ t^*_{ks} &= \Delta t^*_{ks} \exp(j\omega t - \Gamma \boldsymbol{N} \cdot \boldsymbol{X}), \\ \Delta t^*_{ks} &= -\tfrac{1}{2} \varepsilon \Gamma \lambda_0 (N_m U_n + N_n U_m) c^*_{ksmn}. \end{aligned} \tag{25.32}$$

Because $c^*_{ksmn}=c^*_{ksnm}$, the complex stress amplitude Δt^*_{ks} can be expressed as

$$\Delta t^*_{ks} = -\varepsilon\Gamma\lambda_0 c^*_{ksmn} N_m U_n. \tag{25.33}$$

When the equation of motion (25.30) is expanded in powers of ε, the ε-independent part is obviously satisfied: $\partial\tau^0_{ki}/\partial X_k=0$. This is true even without assuming τ^0_{ki} to be constant in space, provided only that the state in the absence of the wave is a static equilibrium state. The terms linear in ε will satisfy the equation of motion if the following complex equation holds.

$$-\bar{\rho}\omega^2 u^*_i = -\frac{\partial}{\partial X_k}(\Gamma\tau^0_{ks} N_s u^*_i + \Gamma c^*_{kimn} N_m u^*_n)$$

or

$$-\bar{\varrho}\omega^2 U_i = (\tau^0_{km}\delta_{in} + c^*_{kimn}) N_k N_m U_n \Gamma^2. \tag{25.34}$$

Upon introducing the complex velocity c^*, defined by

$$\frac{1}{c^*} = \frac{\Gamma}{j\omega} = \frac{1}{c} + \frac{\alpha}{j\omega}, \tag{25.35}$$

we see that (25.34) has the form

$$[Q_{in}(\omega, \bar{\varrho}c^{*2}) - \bar{\varrho}c^{*2}\delta_{in}] U_n = 0 \tag{25.36}$$

with the complex matrix Q_{in} given by

$$Q_{in} = (\tau^0_{km}\delta_{in} + c^*_{kimn}) N_k N_m. \tag{25.37}$$

By making the definition

$$B^S_{kimn} = \tau^0_{km}\delta_{in} + c^S_{kimn}, \tag{25.38}$$

and using (25.35) and (25.28), we may put Q_{in} in the form

$$Q_{in} = \left(B^S_{kimn} + Y_{kimn} + \frac{C_V T_0 (j\omega\bar{\varrho}L_{\boldsymbol{N}}/C_V)}{\bar{\varrho}c^{*2} - (j\omega\bar{\varrho}L_{\boldsymbol{N}}/C_V)}\gamma_{ki}\gamma_{mn}\right) N_k N_m. \tag{25.39}$$

Q_{in} acquires its complex nature not only through the j that occurs explicitly, but also through the complex functions of frequency Y_{kimn}, $L_{\boldsymbol{N}}$, and $\bar{\varrho}c^{*2}$ itself. The quantities B^S_{kimn} and the Grüneisen numbers γ_{mn} are of course frequency-independent real coefficients.

With all quantities in (25.39) except $\bar{\varrho}c^{*2}$ considered as known, the discussion of (25.36) can proceed along classical lines: Nontrivial solutions for U_n in (25.36) can occur only for values of $\bar{\varrho}c^{*2}$ that make the determinant of the coefficient matrix vanish. The real wave speeds c and attenuations α of the possible waves are determined by these values of $\bar{\varrho}c^{*2}$. Corresponding to each such value of $\bar{\varrho}c^{*2}$ is a set of three (possibly complex) numbers, U_1, U_2, U_3 that indicate the type of motion in accordance with (25.2). The angles of these complex numbers indicate relative phases and their magnitudes indicate the maximum excursions along the respective coordinate axes. From the form of Q_{in} in (25.39), it can be seen that the vanishing of the determinant of the coefficient matrix in (25.36) may lead, in general, to an equation of sixth degree in $\bar{\varrho}c^{*2}$. We expect three of the roots to correspond (at least in the case of small damping) to the three modes for the purely elastic case, in which Q_{in} is given simply by $B^S_{kimn} N_k N_m$. As Y_{kimn} and $L_{\boldsymbol{N}}$ are varied from zero to their actual values, these roots move away from their positions on the positive real axis to the positions in the complex plane that give the values of $\bar{\varrho}c^{*2}$ for the corresponding damped waves. The other three roots are introduced by the coupling of mechanical strain to thermal effects through the Grüneisen numbers γ_{ki} and the effect of heat flux on S through (22.1). In the case of fluids, heat conduction introduces one such additional root. A careful discussion of waves in fluids, governed by the Navier-Stokes equations and FOURIER's law of heat conduction, has been given by TRUESDELL ["Precise Theory of the Absorption and Dispersion of Forced Plane Infinitesimal Waves According to the Navier-Stokes Equations", J. Rat. Mech. Anal. 2, 643–741, 1953].

Our present interest is in the use of wave propagation measurements to determine the material coefficients occurring in (25.36)–(25.39), rather than in an analysis of the possible wave motions when the coefficients are assumed to be known. The determination of elastic properties in materials of low loss is ordinarily based on an analysis that neglects damping entirely. This is a completely acceptable procedure, for small damping contributes only a very slight shift in velocity from the purely elastic value.

26. Elastic medium with Newtonian viscosity and heat conduction according to Fourier's law.

If (23.1) is multiplied by $p_{\zeta\alpha}^{-1}$, there results

$$t_\zeta - \tau_\zeta = p_{\zeta\alpha}^{-1} A_{\alpha\gamma} \dot{V}_\gamma - p_{\zeta\alpha}^{-1} (\dot{t}_\alpha - \dot{\tau}_\alpha). \tag{26.1}$$

By neglecting the terms in $\dot{t}_\alpha - \dot{\tau}_\alpha$, and defining viscosity coefficients

$$\mu_{\zeta\gamma} = p_{\zeta\alpha}^{-1} A_{\alpha\gamma}, \tag{26.2}$$

one obtains

$$t_\zeta - \tau_\zeta = \mu_{\zeta\gamma} \dot{V}_\gamma. \tag{26.3}$$

Eq. (26.3) gives the difference of the Piola-Kirchhoff stress from its static equilibrium value in a medium with Newtonian viscosity. By considering the response to a complex exponential excitation of the form (24.1) when the viscosity coefficients are constant, it is seen that the steady-state response matrix $Y(j\omega)$ is simply

$$Y_{\zeta\gamma}(j\omega) = j\omega \mu_{\zeta\gamma}. \tag{26.4}$$

This can be viewed as a special case of (24.10) in which the matrix $\bar{g}_{\zeta\gamma}(s)$ is $\mu_{\zeta\gamma}$, independent of s. Consistent with this view, $g(t-\tau)$ in (24.8) is then the viscosity matrix times the delta function, $\delta(t-\tau)$. (In a theory that does not admit the delta function as a relaxation function (because of smoothness restrictions) this case must be viewed as a *limiting* case, rather than a *special* one.)

Similarly, if (23.5) is multiplied by q_{mi}^{-1} and the terms in $\dot{H}_i$ are dropped, there results FOURIER's law of heat conduction (23.6). In the case of constant conductivity $\varkappa_{kj}$, the resulting matrix $\bar{L}(s)$ in an expression of the form (24.16) is simply the conductivity matrix $\varkappa$:

$$\bar{L}_{km}(s) = \varkappa_{km}, \qquad \text{independent of } s. \tag{26.5}$$

In this case, the matrix $L(t-\tau)$ in (24.14) is the conductivity matrix times the delta function $\delta(t-\tau)$.

The analysis of small-amplitude wave propagation in an elastic medium with Newtonian viscosity and heat conduction according to FOURIER's law proceeds formally just as in Sect. 25. In particular, (25.28) and (25.36)–(25.39) hold. In this case, $Y_{\alpha\beta}(j\omega)$ is given simply by (26.4) and, in view of (26.5) and (25.15), L_N is the real, frequency-independent sum

$$L_N = \varkappa_{ki} N_k N_i. \tag{26.6}$$

The thermodynamics of elastic materials with heat conduction and viscosity has been discussed by COLEMAN and NOLL (1963).

27. Idealized thermoelastic medium. Elastic medium.

In an idealized thermoelastic medium, the stress t_α is always equal to its present static equilibrium value τ_α. The linearized equations are (25.28) and (25.36)–(25.39) with $Y_{\alpha\beta}=0$. Heat conduction can be presumed to take place according to either FOURIER's law or a more general model.

When heat conduction is ignored as well as the difference $t_\alpha - \tau_\alpha$, one obtains the purely elastic case with $c^*_{\alpha\beta}=c^S_{\alpha\beta}$. In this case Q_{in} is the real matrix

$$Q_{in}=(\tau^0_{km}\,\delta_{in}+c^S_{kimn})\,N_k\,N_m. \tag{27.1}$$

This model with $\tau^0_{km}=0$ is practically always used for the determination of elastic constants from wave speeds. In this case, ignoring the heat flux allows temperature changes to take place without any changes in entropy.

Another perfectly elastic case is the limiting case in which heat conduction is so perfect that temperature equalization takes place with negligible $\nabla \cdot \boldsymbol{H}$. This case is obtained formally by setting $Y_{\alpha\beta}=0$, and letting $L_{\boldsymbol{N}}\,\Gamma^2/j\,\omega\,C_V\to\infty$ in (25.28) or $j\,\omega\,L_{\boldsymbol{N}}/C_V\to\infty$ in (25.39). This case has $c^*_{\alpha\beta}=c^T_{\alpha\beta}$.

There are also cases of perfectly elastic wave propagation that exist in an idealized thermoelastic medium simply because the wave under consideration is not coupled to thermal effects. This is the case if the only strain components V_α that occur in the wave are those for which $\gamma_\alpha=0$, for then the last term in (25.39) vanishes independently of any assumption about the heat conduction. For these cases, there is no difference between the isentropic and isothermal elastic constants governing the wave propagation.

TRUESDELL (1966, 1968) showed that there is at least one propagation direction $\boldsymbol{n}$ in which a longitudinal wave (polarization direction $\boldsymbol{u}=\boldsymbol{n}$) may exist and propagate in an elastic material of positive longitudinal elasticity. A material is said to have positive longitudinal elasticity if $\boldsymbol{n}\cdot Q(\boldsymbol{n})\,\boldsymbol{n}>0$ for all unit vectors $\boldsymbol{n}$, where Q is the acoustical tensor [cf. (28.35)]. This theorem applies to either an initially stressed medium or an unstressed one. KOLODNER (1966) showed that if the medium is unstressed (or hydrostatically stressed, in view of the discussion of Sect. 30.5), then there are at least three directions for longitudinal waves.

Even when the medium is perfectly elastic and locally isotropic, complications arise if the medium is not homogeneous. The case of periodic inhomogeneity (i.e. a spatially periodic medium) with application to composite materials has been discussed by KOHN, KRUMHANSL, and LEE (1972).

28. Initially stressed elastic medium.

28.1. Introduction. The thermostatic relations of Chap. C are *thermodynamic* relations for an idealized perfectly elastic medium. It is a fortunate simplification that many solid materials of technical importance and theoretical interest have sufficiently low damping that their elastic properties can be determined to high precision on the basis of the perfectly elastic model. Thus, unless the damping is unusually high, elastic coefficients can be determined accurately by relating them to measured small-amplitude wave speeds through formulas for wave propagation in a perfectly elastic medium. It is the purpose of this chapter to develop and discuss these relations.

The basic equation that applies is (25.36) with Q_{in} now the real matrix (27.1). However, the expression for Q_{in} for this perfectly elastic case can be derived more directly if heat flux and internal friction are ruled out from the start. We now do this, because the derivation in Sect. 25 is unnecessarily complicated for the present purpose, and because some additional considerations are necessary to describe variations of the initial state.

In certain important experiments, the transit time of small-amplitude waves is measured as a function of variable initial stress and temperature. When varying the initial stress at constant temperature, the initial configuration is conveniently described with reference to the unstressed configuration at that same temperature. When varying the initial temperature at zero stress, the initial configuration can be described with reference to the unstressed configuration at some reference temperature. Thus, four different configurations need to be identified: (1) a "master" reference configuration of coordinates A_i, conveniently taken as the unstressed configuration at some master reference temperature T_M; (2) the "natural" unstressed configuration at the temperature of the experiment, of coordinates a_i; (3) the "initial" homogeneously deformed configuration of coordinates X_i; (4) the "present" configuration of coordinates x_i. All four positions (A_i, a_i, X_i, x_i) are referred to a common Cartesian system. The description of master, natural, initial, and present states is summarized in Table 28.1. The densities in the reference configurations A_i, a_i, and X_i are denoted by ϱ_M, ϱ_N, and $\bar{\varrho}$ respectively, the initial stress by $\bar{T}_{ij}$, and the common temperature of the initial and natural states by $\bar{T}$. Thermal expansion at zero stress accounts for the difference between $\boldsymbol{a}$ and $\boldsymbol{A}$, whereas the change from $\boldsymbol{a}$ to $\boldsymbol{X}$ is by definition an elastic deformation produced by stress at constant temperature. The deformation from $\boldsymbol{X}$ to $\boldsymbol{x}$ takes place at constant static equilibrium entropy. This follows from (8.11) and the assumptions of perfect elasticity ($t_{ij}=\tau_{ij}$) and no heat flux (or sources) in the deformation from $\boldsymbol{X}$ to $\boldsymbol{x}$.

Table 28.1. *Description of reference and present states.*

State	Coordinates	Density	Stress	Temperature	Entropy
Master	$x_i=A_i$	ϱ_M	Zero	T_M	S_M
Natural	$x_i=a_i$	ϱ_N	Zero	$T_N=\bar{T}$	S_N
Initial	$x_i=X_i$	$\bar{\varrho}$	$\bar{T}_{ij}$	$\bar{T}$	$\bar{S}$
Present	$x_i=x_i$	ϱ	T_{ij}	T	$S=\bar{S}$

In any given present state, there is of course only one present stress tensor $T_{ij}(t, \boldsymbol{x})$, but there are different strain tensors and Piola-Kirchhoff stress tensors corresponding to each different choice of the reference configuration, the relations between them being purely geometric. Quantities referred to the "master" and "natural" configurations will be distinguished by the superscripts M and N respectively. Thus,

$$\begin{aligned} T_{ij} &= \frac{\varrho}{\varrho_M}\frac{\partial x_i}{\partial A_p}\frac{\partial x_i}{\partial A_q} t_{pq}^M = \frac{\varrho}{\varrho_N}\frac{\partial x_i}{\partial a_p}\frac{\partial x_j}{\partial a_q} t_{pq}^N = \frac{\varrho}{\bar{\varrho}}\frac{\partial x_i}{\partial X_p}\frac{\partial x_j}{\partial X_q} t_{pq} \\ &= \frac{\varrho}{\varrho_M}\frac{\partial x_i}{\partial A_p} P_{jp}^M = \frac{\varrho}{\varrho_N}\frac{\partial x_i}{\partial a_p} P_{jp}^N = \frac{\varrho}{\bar{\varrho}}\frac{\partial x_i}{\partial X_p} P_{jp}. \end{aligned} \tag{28.1}$$

$$\begin{aligned} V_{ij}^M &= \frac{1}{2}\left(\frac{\partial x_k}{\partial A_i}\frac{\partial x_k}{\partial A_j} - \delta_{ij}\right), \\ V_{ij}^N &= \frac{1}{2}\left(\frac{\partial x_k}{\partial a_i}\frac{\partial x_k}{\partial a_j} - \delta_{ij}\right). \end{aligned} \tag{28.2}$$

The different first Piola-Kirchhoff tensors arise naturally when the equation of motion is referred to the different reference configurations. Thus, with the body force omitted, (7.6) leads to

$$\ddot{x}_i = \frac{1}{\varrho_M}\frac{\partial P_{ij}^M}{\partial A_j} = \frac{1}{\varrho_N}\frac{\partial P_{ij}^N}{\partial a_j} = \frac{1}{\bar{\varrho}}\frac{\partial P_{ij}}{\partial X_j} = \frac{1}{\varrho}\frac{\partial T_{ij}}{\partial x_j}. \tag{28.3}$$

In the perfectly elastic model, the stress is, by assumption, at all times equal to the static equilibrium stress. This means that the second Piola-Kirchhoff stress t_{ij} is equal to τ_{ij}, given by (8.4). There is, of course, an equation like (8.4) for each choice of the reference configuration. Having chosen to refer thermodynamic functions to unit volume in the reference configuration (cf. Table 10.1), we must now deal with the additional complication that the material occupies different volumes in the various configurations. When several different reference configurations are under consideration, the advantage of referring thermodynamic functions to unit mass is overwhelming. Therefore, in this chapter let us adopt the symbol $\mathcal{U}$ for internal energy per unit *mass*. Then the forms of (8.4) appropriate to the different configurations are

$$t_{ij}^M=\varrho_M(\partial\mathcal{U}/\partial V_{ij}^M)_S;\quad t_{ij}^N=\varrho_N(\partial\mathcal{U}/\partial V_{ij}^N)_S;\quad t_{ij}=\bar{\varrho}(\partial\mathcal{U}/\partial V_{ij})_S. \tag{28.4}$$

Any two of these equations follow from the third by virtue of the geometric connections.

We next show that the assumption (28.4) makes the first Piola-Kirchhoff stress divided by the reference density a thermodynamic tension conjugate to the deformation gradient referred to the appropriate reference configuration, i.e.,

$$\begin{gathered}\frac{1}{\varrho_M}P_{qr}^M=[\partial\mathcal{U}/\partial(\partial x_q/\partial A_r)];\quad \frac{1}{\varrho_N}P_{qr}^N=[\partial\mathcal{U}/\partial(\partial x_q/\partial a_r)]_S;\\ \frac{1}{\bar{\varrho}}P_{qr}=[\partial\mathcal{U}/\partial(\partial x_q/\partial X_r)]_S.\end{gathered} \tag{28.5}$$

It is sufficient to demonstrate $(28.5)_3$. From the definition in (3.2),

$$\frac{\partial V_{ij}}{\partial(\partial x_k/\partial X_m)}=\frac{1}{2}\left(\frac{\partial x_k}{\partial X_i}\delta_{jm}+\frac{\partial x_k}{\partial X_j}\delta_{im}\right). \tag{28.6}$$

From (28.4), (28.6), and the symmetry of t_{ij},

$$\frac{\bar{\varrho}\,\partial\mathcal{U}}{\partial(\partial x_i/\partial X_m)}=\frac{\partial x_i}{\partial X_j}t_{mj}. \tag{28.7}$$

Comparison with $(4.8)_2$ verifies (28.5).

28.2. Linearization of the equation of motion. With the body force omitted, the equation of motion (6.7) is

$$\bar{\varrho}\ddot{x}_i=\partial P_{ij}/\partial X_j. \tag{28.8}$$

A linearized equation of motion can be obtained by expanding P_{ik} as a power series in the deformation gradients $\partial x_k/\partial X_m$ and the static equilibrium entropy change $(S-\bar{S})$. However, (8.11) shows this entropy change to be zero because of the assumptions of no heat flux, no heat sources, and perfect elasticity ($t_{ij}=\tau_{ij}$). Therefore,

$$P_{ij}=\bar{T}_{ij}+\bar{B}_{ijkm}(\partial x_k/\partial X_m-\delta_{km})+\cdots \tag{28.9}$$

where

$$B_{ijkm}=\left(\frac{\partial P_{ij}}{\partial(\partial x_k/\partial X_m)}\right)_S=\left(\frac{\bar{\varrho}\,\partial^2\mathcal{U}}{\partial(\partial x_k/\partial X_m)\,\partial(\partial x_i/\partial X_j)}\right)_S. \tag{28.10}$$

In (28.9), $\bar{T}_{ij}$ is the initial stress, and the bar over B_{ijkm} indicates evaluation at the initial state where $\partial x_k/\partial X_m=\delta_{km}$ and $S=\bar{S}$. By substitution from (28.9) into (28.5), keeping only the linear terms, we obtain

$$\bar{\varrho}\ddot{x}_i=\bar{B}_{ijkm}\frac{\partial^2 x_k}{\partial X_j\,\partial X_m}. \tag{28.11}$$

From (28.10), (28.7), and (28.6),

$$B_{ijkm} = \delta_{ik} t_{jm} + \frac{\partial x_i}{\partial X_p} \frac{\partial x_k}{\partial X_q} c^S_{pjqm} \tag{28.12}$$

where

$$c^S_{pjqm} = (\partial t_{pj}/\partial V_{qm})_S. \tag{28.13}$$

Hence,

$$\bar{B}_{ijkm} = \delta_{ik} \bar{T}_{jm} + \bar{c}^S_{ijkm}. \tag{28.14}$$

In writing these equations and subsequent ones, we make use of the symmetry implied by interchange of the order of differentiation.

We have emphasized that different choices of the reference configuration generate different strain components and different Piola-Kirchhoff stresses (thermodynamic tensions). The same applies to the whole hierarchy of thermodynamic coefficients obtained by differentiating a thermodynamic potential with respect to a strain component or thermodynamic tension. The elastic coefficient in (28.12) and (28.13) is a derivative with respect to the strain measured from the initial configuration. We say the coefficient in (28.12) is *referred to* the initial configuration ($\boldsymbol{X}$) but it can be evaluated at any state. In (28.14), the coefficient is both *referred to* and *evaluated at* the initial state.

When making measurements of small-amplitude wave transit time (or of a resonance frequency) as a function of the initial stress, it is ordinarily most convenient to use the *natural* state as a reference. Then (28.9) is replaced by

$$P^N_{ij} - \bar{P}^N_{ij} = \bar{A}_{ijkm} \left(\frac{\partial x_k}{\partial a_m} - \frac{\partial X_k}{\partial a_m} \right) + \cdots \tag{28.15}$$

where

$$A_{ijkm} = \left(\frac{\partial P^N_{ij}}{\partial (\partial x_k/\partial a_m)} \right)_S. \tag{28.16}$$

Here the bar still denotes evaluation at the initial configuration ($\boldsymbol{X}$). We take the initial deformation and state to be spatially homogeneous so that $\bar{P}^N_{ij}$ and the initial deformation gradients $\partial X_k/\partial a_m$ are constants. From (28.3) and (28.15), the linearized equation of motion referred to the "natural" configuration but linearized about the *initial* state is

$$\varrho_N \ddot{x}_i = \bar{A}_{ijkm} \frac{\partial^2 x_k}{\partial a_j \, \partial a_m}. \tag{28.17}$$

From (28.16), $(28.5)_2$, $(28.4)_2$, and (28.2), we find, in the same manner as (28.12),

$$A_{ijkm} = \delta_{ik} t^N_{jm} + \frac{\partial x_i}{\partial a_p} \frac{\partial x_k}{\partial a_q} c^{NS}_{pjqm} \tag{28.18}$$

with

$$c^{NS}_{pjqm} = \left(\frac{\varrho_N \, \partial^2 U}{\partial V^N_{pj} \, \partial V^N_{qm}} \right)_S. \tag{28.19}$$

Here c^{NS}_{pjqm} is simply the ordinary isentropic elastic coefficient referred to the natural state. The coefficient $\bar{A}_{ijkm}$ in the equation of motion (28.17) is obtained by evaluating (28.18) at the homogeneously deformed initial state to obtain

$$\bar{A}_{ijkm} = \delta_{ik} \bar{t}^N_{jm} + \frac{\partial X_i}{\partial a_p} \frac{\partial X_k}{\partial a_q} \bar{c}^{NS}_{pjqm}. \tag{28.20}$$

From (28.1), the relation between the initial values of t^N_{jm} and the stress T_{rs} is

$$\bar{t}^N_{jm} = \frac{\varrho_N}{\bar{\varrho}} \frac{\partial a_j}{\partial X_r} \frac{\partial a_m}{\partial X_s} \bar{T}_{rs}. \tag{28.21}$$

28.3. Solutions for small-amplitude waves. By seeking plane-wave solutions of (28.11) in the form

$$x_i - X_i = u_i \exp[j\omega(t - n_s X_s/V)], \tag{28.22}$$

one obtains the propagation condition

$$(\bar{B}_{ijkm} n_j n_m - \bar{\varrho} V^2 \delta_{ik}) u_k = 0. \tag{28.23}$$

Similarly, by assuming solutions of (28.17) in the form

$$x_i - X_i = u_i \exp[j\omega(t - N_q a_q/W)], \tag{28.24}$$

one obtains

$$(\bar{A}_{ijkm} N_j N_m - \varrho_N W^2 \delta_{ik}) u_k = 0. \tag{28.25}$$

Here $\boldsymbol{n}$ and V denote the actual direction and velocity of propagation. $\boldsymbol{N}$ and W are the corresponding *natural* direction and velocity.

THURSTON and BRUGGER (1964) have pointed out the advantages of (28.24) over (28.22) in the analysis of experiments on small-amplitude ultrasonic wave transit time as a function of applied static stress. The wave front is a material plane which has the unit normal $\boldsymbol{N}$ in the natural state. The wave front moves from the plane $\boldsymbol{N} \cdot \boldsymbol{a} = 0$ to the plane $\boldsymbol{N} \cdot \boldsymbol{a} = L_0$ in the time L_0/W, making W the wave speed referred to the dimensions in the natural configuration ($\boldsymbol{a}$). In a typical ultrasonic experiment, plane waves are reflected between opposite parallel faces of a specimen, the wave fronts being parallel to these faces. One ordinarily measures a repetition frequency f which is the inverse of the time required for a round trip between the opposite faces, i.e.,

$$W = 2L_0 f. \tag{28.26}$$

Thus, W is proportional to the measured frequency f, whereas $V = 2Lf$ involves the actual length under stress. A second advantage of (28.24) over (28.22) is that $\boldsymbol{n}$ may change with static stress, but since the propagation direction remains normal to the same faces of the specimen, $\boldsymbol{N}$ remains constant while the static stress is varied.

It follows from the propagation condition (28.25) that the possible values of $\varrho_N W^2$ for plane wave propagation normal to a material plane of natural normal $\boldsymbol{N}$ are the eigenvalues of the second rank tensor

$$S_{ik}(\boldsymbol{N}) = \bar{A}_{ijkm} N_j N_m. \tag{28.27}$$

The possible particle displacement directions are the corresponding eigenvectors. Now if we form S_{ki} by interchanging i and k in (28.27) and (28.20), and make use of the symmetry $c_{pjqm} = c_{qmpj}$, we find that S_{ik} is symmetric. Hence, at any state of strain there are three mutually perpendicular particle displacement directions for plane waves corresponding to a given $\boldsymbol{N}$. The condition for all three waves to be real is that S_{ik} be positive definite. For additional discussion of this point, the reader may refer to the article on elasticity by M. E. GURTIN in this encyclopedia, Vol. VIa/2.

TOUPIN and BERNSTEIN (1961) showed how to transform the propagation condition (28.25) to a representation that depends only on the strain, independent of the rotation from the natural state. The key is to transform the particle displacement direction $\boldsymbol{u}$ back to the *natural* direction of the material line along it by the transformation

$$u_j = (\partial X_j/\partial a_q) U_q. \tag{28.28}$$

The propagation condition (28.25) is thereby transformed to

$$\varrho_N W^2 U_r = w_{rs} U_s, \tag{28.29}$$

$$w_{rs} = \frac{\partial a_r}{\partial X_i} \frac{\partial X_k}{\partial a_s} S_{ik} = (\delta_{rs} \bar{t}^N_{jm} + \bar{C}_{sq} \bar{c}^{NS}_{pjqm}) N_j N_m, \tag{28.30}$$

$$\bar{C}_{sq} = \frac{\partial X_k}{\partial a_s} \frac{\partial X_k}{\partial a_q} = (\delta_{sq} + 2\bar{V}^N_{sq}). \tag{28.31}$$

All quantities appearing in (28.29) thus depend only on the *strain* from the natural to the initial state, independent of the rotation. The significance of (28.29) is that the possible values of $\varrho_N W^2$ for propagation normal to a material surface of natural normal $\boldsymbol{N}$ are the eigenvalues of w_{rs}, and its eigenvectors $\boldsymbol{U}$ are the natural directions of the material lines which, in the deformation from the natural to the initial state, are rotated into the actual particle displacement directions $\boldsymbol{u}$. The three eigenvectors $\boldsymbol{U}$ corresponding to a given $\boldsymbol{N}$ are not in general orthogonal.

It may be remarked that w_{rs} and S_{ik} have the same eigenvalues but different eigenvectors.

Eqs. (28.29)–(28.31) are fundamental to the determination of third-order elastic coefficients and pressure derivatives of elastic coefficients, subjects that are discussed in Sects. 29 and 31. First, however, we digress to relate the actual propagation direction $\boldsymbol{n}$ and velocity V to the *natural* direction $\boldsymbol{N}$ and velocity W.

28.4. Propagation direction and velocity. The actual and natural propagation directions $\boldsymbol{n}$ and $\boldsymbol{N}$ are connected geometrically by the rotation of material planes in the deformation from the natural to the initial configurations (from $\boldsymbol{a}$ to $\boldsymbol{X}$). The relation of $\boldsymbol{n}$ to $\boldsymbol{N}$ is therefore (3.9) with X replaced by a and x by X:

$$n_j = N_i (\partial a_i / \partial X_j) / f_{\boldsymbol{N}}, \tag{28.32}$$

$$f_N^2 = \frac{\partial a_i}{\partial X_s} \frac{\partial a_j}{\partial X_s} N_i N_j = \bar{C}^{-1}_{ij} N_i N_j \tag{28.33}$$

where $\bar{C}_{ij}$ is defined by (28.31). The actual and natural propagation velocities V and W are connected by the thickness shrink in the direction $\boldsymbol{N}$. By analogy with (3.15), the ratio of the actual to natural separation of material planes is $L/L_0 = 1/f_{\boldsymbol{N}}$. Hence, since $V = L W / L_0$, we have

$$V = W / f_{\boldsymbol{N}}. \tag{28.34}$$

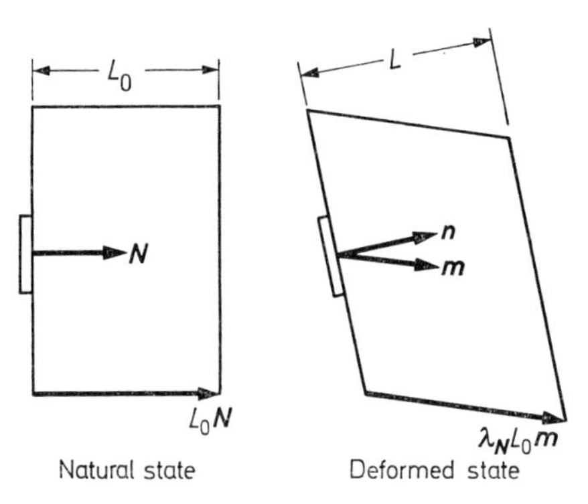

Fig. 28.1. Change of propagation direction $\boldsymbol{n}$ and path length L with change of static deformation. (The propagation direction remains perpendicular to the reflecting faces of the specimen while the material line segment $L_0 \boldsymbol{N}$ is rotated and stretched into $\lambda_{\boldsymbol{N}} L_0 \boldsymbol{m}$. The perpendicular distance between the faces changes from L_0 to $L = \lambda_{\boldsymbol{N}} L_0 \boldsymbol{m} \cdot \boldsymbol{n}$.)

The geometric relationships may be summarized with the help of Fig. 28.1: $\boldsymbol{N}$ denotes the original unit normal to a pair of parallel material planes. In the deformation from the natural to the initial configuration, the originally normal material line segment $L_0 \boldsymbol{N}$ connecting the planes is deformed into $\lambda_{\boldsymbol{N}} L_0 \boldsymbol{m}$ [cf. (3.14)] while the material planes acquire the new unit normal $\boldsymbol{n}$. The separation of the planes changes from L_0 to $L = \lambda_{\boldsymbol{N}} L_0 \boldsymbol{m} \cdot \boldsymbol{n} = L_0 / f_{\boldsymbol{N}}$.

To relate the present treatment to that of TRUESDELL (1961), we can multiply (28.25) by $1/f_{\boldsymbol{N}}^2$ and substitute for $\boldsymbol{N}$ from (28.32) to obtain the propagation condition

$$\begin{aligned} &(Q_{ik}(\boldsymbol{n}) - \varrho_N V^2 \delta_{ik}) u_k = 0, \\ &Q_{ik}(\boldsymbol{n}) = \bar{A}_{ijkm} \frac{\partial X_s}{\partial a_j} \frac{\partial X_r}{\partial a_m} n_s n_r. \end{aligned} \tag{28.35}$$

The TRUESDELL propagation condition was obtained without recourse to the assumption that a stored energy function exists.

Some additional considerations concerning waves in initially stressed media have been published by TOKUOKA and SAITO (1968) and by TOKUOKA and IWASHIMIZU (1968).

J. Ultrasonic measurements as a function of static initial stress.

29. Determination of third-order elastic coefficients.

29.1. Introduction. Third-order elastic constants are of interest in solid state physics because they enable one to evaluate the leading *anharmonic* terms of the interatomic potential or of generalized Grüneisen parameters, which enter the theories of all anharmonic phenomena, such as the interaction of acoustic waves with thermal phonons, and the equation of state.

The most precise method of measuring the third-order coefficients is through the influence of a static stress on the transit time of small-amplitude waves in a homogeneously stressed specimen [MCSKIMIN and ANDREATCH (1962)].

In a typical experiment, the repetition frequency f of Eq. (28.26) is measured as a function of the statically applied stress at constant temperature. One ordinarily considers stress states such as hydrostatic pressure or uniaxial compression—states that depend on a single scalar variable p. For example, p can represent either the hydrostatic pressure or the magnitude of a uniaxial load in some prescribed direction per unit of natural undeformed area. The value $p=0$ corresponds to the *natural* (unstressed) state.

From the *initial slope* of the curve $f(p)$ versus p, namely $f'(0)$, one can determine $W'(0)$, the initial derivative of $W(p)$: From (28.26)

$$W'(0)/W(0) = f'(0)/f(0), \tag{29.1}$$

and hence

$$\left[\frac{\partial}{\partial p} (\varrho_N W^2)\right]_{\bar{T};p=0} = 2 \varrho_N W(0)^2 f'(0)/f(0)$$

or

$$w'(0) = 2 w(0) f'(0)/f(0) \tag{29.2}$$

where we have introduced the abbreviation

$$w(p) = \varrho_N W(p)^2 \tag{29.3}$$

and the prime denotes the isothermal pressure derivative. The third-order elastic coefficients, evaluated at the natural state, can be related to the derivative in (29.2).

Eq. (29.2) relates the frequency data to the initial slope of $\varrho_N W^2$. This representation allows the zero stress value of $\varrho_N W^2$ itself to be obtained in any convenient way. By measuring ϱ_N and the path length L_0, the quantity can of course be calculated from the measured $f(0)$ through Eq. (28.26). Alternatively, one can rely on values from an independent experiment.

While only the *initial* slope of $f(p)$ versus p is needed for the evaluation of third-order elastic constants, the interpretation of experimental results on crystals is greatly simplified by the observation that $f(p)$ ordinarily appears to be linear throughout a considerable range, to within experimental error. This makes it easier to obtain a reliable experimental value of the slope, and tends to support the usefulness of the experimentally obtained numerical values. Were significant curvature easily detectable, one would be left with the feeling that any meaningful calculation of anharmonic properties should include fourth-order, and perhaps higher-order coefficients also. By making very precise measurements of frequency under initial pressures up to 10 kilobars, G. R. BARSCH and his collaborators have succeeded in obtaining reasonably accurate determinations of the curvature of $f(p)$ for some materials. This makes possible the determination of *second* pressure derivatives of the effective second-order elastic coefficients, and (equivalently) certain combinations of *fourth*-order elastic coefficients. In this work, it is necessary to determine the repetition frequency f of Eq. (28.26) to within a few parts in 10^8 [CHANG and BARSCH, 1967; BARSCH and SHULL, 1971; BARSCH and CHANG, 1971; BARSCH, SHULL, and TESTARDI, 1971].

(Even with the older data, in which $f(p)$ appears linear, one can in fact place rough bounds on the curvature of $f(p)$ at $p=0$, from which rough bounds on the numerical values of certain combinations of fourth-order coefficients can be calculated.)

29.2. Initial derivative of $\varrho_N W^2$ in terms of material properties. The next task is to relate material properties to the derivative in (29.2). In terms of the abbreviation $w(p)=\varrho_N W(p)^2$ introduced in (29.3), let us rewrite (28.29) as

$$w\,U_r = w_{rs}\,U_s. \tag{29.4}$$

In isothermal deformations under the assumed one-parameter stress system, the quantities w, $\boldsymbol{U}$, and w_{rs} are functions only of p. (The fixed temperature $\overline{T}$ common to the natural and initial states is of course present as a parameter.) By differentiation of (29.4),

$$w'\,U_r + w\,U_r' = w_{rs}'\,U_s + w_{rs}\,U_s' \tag{29.5}$$

where the prime denotes the derivative with respect to p at constant temperature. We assume without loss of generality that the eigenvector $\boldsymbol{U}$ is normalized, and hence

$$\boldsymbol{U}\cdot\boldsymbol{U}=1\,, \qquad \boldsymbol{U}\cdot\boldsymbol{U}'=0\,. \tag{29.6}$$

Multiplying (29.5) by U_r and making use of (29.6), we find

$$w' = U_r\,w_{rs}'\,U_s + U_r\,w_{rs}\,U_s'\,. \tag{29.7}$$

Now at $p=0$, w_{rs} is symmetric, being identically equal to S_{rs}. It follows that

$$(U_r\,w_{rs})_{p=0} = (w\,U_s)_{p=0} \tag{29.8}$$

and hence when (29.7) is evaluated at $p=0$, the last term vanishes, leaving only

$$w'(0) = (U_r\,w_{rs}'\,U_s)_{p=0}\,. \tag{29.9}$$

Eq. (29.9) states that the derivative of an eigenvalue of the tensor w_{rs} evaluated at $p=0$, is obtained from the corresponding eigenvector of the tensor w_{rs} and the components of w'_{rs}.

If w_{rs} has multiple eigenvalues at $p=0$, a precaution should be observed in the use of (29.9) because the direction of the eigenvector belonging to a multiple eigenvalue is not determined. It is implicit in the use of the relation $\boldsymbol{U}\cdot\boldsymbol{U}'=0$ that in this case one should use in (29.9) the limiting right eigenvector of $w_{rs}(p)$ as $p\to 0$, not just any eigenvector of $w_{rs}(0)$ belonging to the multiple eigenvalue.

Let us now relate $w'_{rs}(0)$ to second-order and third-order elastic constants by differentiation of (28.30) with respect to p at constant temperature. Quantities dependent on the strain are differentiated by following the formula

$$\left(\frac{\partial}{\partial p}\right)_T=\left(\frac{\partial t^N_{km}}{\partial p}\right)_T\frac{\partial V^N_{ij}}{\partial t^N_{km}}\left(\frac{\partial}{\partial V^N_{ij}}\right)_T=t'_{km}\,s^T_{ijkm}\left(\frac{\partial}{\partial V_{ij}}\right)_T. \tag{29.10}$$

We now understand that all quantities are referred to the *natural* state, and the reminding superscript N will be suppressed, as in the last form of (29.10). Recalling that $\boldsymbol{N}$ is independent of p, we differentiate (28.30) and evaluate the result at $p=0$ to obtain

$$w'_{rs}(0)=\left[\left(\delta_{rs}\bar{t}'_{jm}+\frac{\partial c^S_{rjsm}}{\partial p}+C'_{sq}\,c^S_{rjqm}\right)N_j\,N_m\right]_{p=0} \tag{29.11}$$

where the primes denote derivatives with respect to p. We retain the superscript S on the elastic coefficients in (29.11) as a reminder that they are *isentropic* coefficients [cf. (28.19)]. The superscript N in (28.19) and (28.30), now suppressed, was to emphasize that the coefficients are referred to the natural state.

From (29.11) and (29.9),

$$w'(0)=N_j\,N_m\left[\bar{t}'_{jm}+U_r\,U_s\left(\frac{\partial c^S_{rjsm}}{\partial p}+C'_{sq}\,c^S_{rjqm}\right)\right]_{p=0}. \tag{29.12}$$

The last term can be simplified as follows. By evaluating (28.30) and (28.29) at the natural state ($p=0$), one finds

$$\begin{aligned} w_{rs}(0)&=N_j\,N_m\,c^S_{rjsm}(0),\\ w(0)\,U^0_r&=w_{rs}(0)\,U^0_s \end{aligned} \tag{29.13}$$

and hence the last term of (29.12) reduces to

$$\begin{aligned} (N_j\,N_m\,U_r\,U_s\,C'_{sq}\,c_{rjqm})_{p=0}&=w(0)\,C'_{sq}(0)\,U^0_q\,U^0_s\\ &=w(0)\,C'_{sr}(0)\,U^0_r\,U^0_s. \end{aligned} \tag{29.14}$$

Now by (29.10),

$$C'_{sr}(0)=2s^T_{srab}\,\bar{t}'_{ab}(0), \tag{29.15}$$

$$\begin{aligned} \left(\frac{\partial c^S_{rjsm}}{\partial p}\right)_{T;p=0}&=C_{rjsmip}\,s^T_{ipab}\,\bar{t}'_{ab}(0),\\ C_{rjsmip}&=(\partial c^S_{rjsm}/\partial V_{ip})_T, \end{aligned} \tag{29.16}$$

where all quantities are understood to be both referred to and evaluated at the natural state. Substitution from (29.14)–(29.16) into (29.12) yields an equation which we choose to write in the form

$$\begin{aligned} -w'(0)&=-N_a\,N_b\,\bar{t}'_{ab}(0)+2w(0)\,F+G,\\ F&\equiv-s^T_{srab}\,U^0_s\,U^0_r\,\bar{t}'_{ab}(0),\\ G&\equiv-C_{rjsmip}\,s^T_{ipab}\,N_j\,N_m\,U^0_r\,U^0_s\,\bar{t}'_{ab}(0). \end{aligned} \tag{29.17}$$

The derivatives $\bar{t}'_{ab}(0)$ are easily evaluated for hydrostatic pressure and uniaxial compression. For hydrostatic pressure, the initial stress is $\bar{T}_{rs} = -p\,\delta_{rs}$ and hence, from (28.21),

$$\bar{t}'_{ab}(0) = -\delta_{rs}\left(\frac{\varrho_N}{\varrho}\frac{\partial a_a}{\partial X_r}\frac{\partial a_b}{\partial X_s}\right) = -\delta_{ab}. \tag{29.18}$$

For uniaxial compression normal to a material plane of natural unit normal $\boldsymbol{M}$, the initial Piola-Kirchhoff stress is

$$\bar{P}_{ik} = -p\,M_i M_k \tag{29.19}$$

where p is the compressive force per unit of original cross-sectional area. By analogy with (4.8),

$$\bar{t}_{ab} = \bar{P}_{ja}\frac{\partial a_b}{\partial X_j} = -p\,M_j M_a \frac{\partial a_b}{\partial X_j}. \tag{29.20}$$

From (29.20),

$$\bar{t}'_{ab}(0) = -M_a M_b. \tag{29.21}$$

(The same first derivative is obtained for uniaxial stress along a material line of natural direction $\boldsymbol{M}$. However, the *second* derivatives will differ unless the material line remains normal to its naturally normal material plane.) From (29.21) and (29.18),

$$\begin{aligned} -N_a N_b\,\bar{t}'_{ab}(0) &= 1 && \text{for hydrostatic pressure,} \\ &= (\boldsymbol{N}\cdot\boldsymbol{M})^2 && \text{for uniaxial compression in the direction } \boldsymbol{M}. \end{aligned} \tag{29.22}$$

The final results for

$$w'(0) \equiv \left[\frac{\partial}{\partial p}(\varrho_N W^2)\right]_{T;\,p=0}$$

will now be summarized.

For hydrostatic pressure:

$$\begin{aligned} -w'(0) &= 1 + 2F_H\,w(0) + G_H, \\ F_H &= s^T_{sraa} U^0_s U^0_r, \\ G_H &= C_{rjsmip}\, s^T_{ipaa} N_j N_m U^0_s U^0_r. \end{aligned} \tag{29.23}$$

For uniaxial compression in the direction $\boldsymbol{M}$:

$$\begin{aligned} -w'(0) &= (\boldsymbol{N}\cdot\boldsymbol{M})^2 + 2F_U\,w(0) + G_U, \\ F_U &= s^T_{srab} U^0_s U^0_r M_a M_b, \\ G_U &= C_{rjsmip}\, s^T_{ipab} N_j N_m U^0_r U^0_s M_a M_b. \end{aligned} \tag{29.24}$$

In every case, $w(0)$ is given by the following expression, obtained by multiplying $(29.13)_2$ by U^0_r:

$$w(0) = c^S_{rjsm} N_j N_m U^0_s U^0_r. \tag{29.25}$$

We emphasize that quantities entering into the final equations refer to the natural unstressed state, and the coefficients are evaluated at that state.

To obtain third-order elastic coefficients, the second-order coefficients being known, one obtains experimental determinations of the slope $f'(0)$, from which $w'(0)$ is found through (29.2). Eqs. (29.23) and (29.24) then relate such values of $w'(0)$ to certain combinations of third-order coefficients through the factors G_H and G_U. When the initial slopes have been determined for a sufficient number of independent modes of propagation, polarization, and stressing, one can solve for the values of the third-order coefficients.

The first determination of a complete set of third-order elastic constants was accomplished by BATEMAN, MASON, and McSKIMIN (1961) using formulas developed for cubic crystals by SEEGER and BUCK (1960). SEEGER and BUCK used an equation of motion with u_i measured from the *natural* state, and used expansions about the natural state in terms of the displacement gradients, all second degree terms being retained. Their result is formulas for ϱV^2 of the form $\varrho V^2 = b + mp$, in which m is related to the third-order constants. THURSTON (1964), influenced by the elegant treatments of TOUPIN and BERNSTEIN (1961) and TRUESDELL (1961), linearized the equation of motion about the *initial* state, distinguishing the deformation from the natural state to the initial state as isothermal and the deformations in the wave as isentropic, and derived general formulas relating third-order constants to the *initial slope* of the curve of $\varrho_N V^2$. versus p. THURSTON and BRUGGER (1964) accomplished an additional simplification by referring the velocity to the natural unstressed dimensions ("natural velocity" W) and published the formulas of this section relating third-order coefficients to the initial derivative of $\varrho_N W^2$ versus p. they also published tables, specializing the formulas to cubic crystals and isotropic media for several modes that are of experimental importance. BRUGGER (1964) should be credited with preventing a potential disaster of confusion in the literature concerning the definition of the third-order and higher-order coefficients in abbreviated notation by recognizing a convenient scheme and successfully advocating its use. BRUGGER's definitions (9.10) appear to be in almost universal use. In particular, they were adopted in the important compilation by BECHMANN, HEARMON, and KURTZ (1969). BRUGGER (1965) has also published a useful compendium of formulas for determining third-order elastic coefficients in crystals of various symmetry. Some additional formulas for cubic crystals, needed for the [111], $[1\bar{1}0]$, $[11\bar{2}]$ directions of propagation, polarization, and stress, have been given by ELKIN, ALTEROVITZ, and GERLICH (1970).

A summary of the experimental data to 1968 has been included in a paper by BARSCH (1968). DRABBLE (1970) has discussed elastic constants under pressure.

The third-order coefficients that appear in (29.16) and subsequent equations are of the "mixed" type. Because the static stress is varied isothermally whereas the small-amplitude wave propagates isentropically, two differentiations of the energy are at constant entropy and the third is at constant temperature. SKOVE and POWELL (1967) pointed out that these mixed coefficients do not have the full symmetry of the pure isothermal or isentropic coefficients because the pair of indices corresponding to the isothermal differentiation cannot necessarily be interchanged with the other two pairs. BRUGGER (1964) gave a formula relating the mixed coefficients to the purely isentropic ones. POWELL and SKOVE (1967) related the isothermal and mixed third-order coefficients to each other.

When the distinction pointed out by SKOVE and POWELL is observed, some of the previously published tables listing expressions for G_H and G_U of (29.23) and (29.24) need to be re-examined, as the full symmetry of the pure coefficients has been incorporated into the tables. Observation of the distinction renders the data processing somewhat more complicated, and additional data are needed on the expansion coefficient and temperature derivatives of the elastic coefficients. These data enter expressions for the difference between the pure isentropic and mixed coefficients. As an example of the difference, SKOVE and POWELL calculated $(C_{441} - C_{144})/C_{441}$ to be about 3% in NaCl. (Here the *last* subscript corresponds to the isothermal differentiation.) While the probable experimental error may often exceed this amount, the distinction should ideally be observed. GUINAN and RITCHIE (1970) have indicated the modifications needed for cubic crystals of symmetry 432, $\bar{4}3m$, and $m3m$ (Laue group CI), and have calculated estimates of the corrections for Nb, Cu, NaCl, Si, and MgO. Because of the smallness of the corrections, most workers to date have treated the mixed third-order coefficients as if they had the full symmetry of the pure coefficients.

SEKOYAN (1970) has indicated the formulas needed to account properly for the actual symmetry of the mixed third-order elastic coefficients in the case of isotropic media.

30. Effective elastic coefficients.

30.1. Introduction. The name "elastic coefficients" is a general one that is used in at least three ways that are related but different: (1) as second derivatives of the internal energy with respect to some measure of the deformation; (2) as first derivatives of the stress with respect to some measure of the deformation (or as coefficients in a linearized stress-deformation relation); (3) as coefficients in a linearized equation of motion, or, equivalently, as coefficients in formulas for the propagation velocities of small-amplitude waves. We call coefficients of type (1) *thermodynamic* elastic coefficients while (2) and (3) are *effective* coefficients for the stress-deformation relation and for wave propagation, respectively.

The subject of effective elastic coefficients has been discussed by BIRCH (1947), THURSTON (1965_1, 1965_2, 1967), BARSCH (1967), and BARSCH and CHANG (1968).

30.2. Lack of uniqueness of coefficients in equation of motion. The first point to be noted is that the linearized equations of motion (28.11) or (28.17) [or the corresponding propagation conditions (28.23) or (28.25)] fail to define a set of coefficients *uniquely*.

The coefficients A_{ijkm} and B_{ijkm} could be called effective elastic coefficients because of the way that they appear in the equations of motion and propagation conditions. However, because of the summation over the second and fourth subscripts in these equations, other sets of coefficients, say α_{ijkm} and β_{ijkm}, will yield the same equations of motion and propagation conditions, as long as the sum of two α's with second and fourth subscripts interchanged equals the corresponding sum of A's, and similarly for the β's and B's. That is,

$$\alpha_{ijkm}+\alpha_{imkj}=A_{ijkm}+A_{imkj}, \tag{30.1}$$

$$\beta_{ijkm}+\beta_{imkj}=B_{ijkm}+B_{imkj}. \tag{30.2}$$

30.3. Symmetry. A second important point concerns the symmetry of the coefficients with respect to interchange of indices. Because the strain tensor V_{ij} is symmetric and the order of differentiation is interchangeable, the thermodynamic coefficients of (9.9) have the symmetry

$$c_{ijkm}=c_{kmij}=c_{jikm}=c_{ijmk}=c_{jimk}=\cdots. \tag{30.3}$$

However, the coefficients $\bar{B}_{ijkm}$ and $\bar{A}_{ijkm}$ do not have this symmetry. For example, from (28.14),

$$\bar{B}_{ijkm}-\bar{B}_{kmij}=0, \tag{30.4}$$

but

$$\begin{aligned}
\bar{B}_{ijkm}-\bar{B}_{jikm}&=\delta_{ik}\bar{T}_{jm}-\delta_{jk}\bar{T}_{im},\\
\bar{B}_{ijkm}-\bar{B}_{ijmk}&=\delta_{ik}\bar{T}_{jm}-\delta_{im}\bar{T}_{jk},\\
\bar{B}_{ijkm}-\bar{B}_{jimk}&=\delta_{ik}\bar{T}_{jm}-\delta_{jm}\bar{T}_{ik}.
\end{aligned} \tag{30.5}$$

Now it follows from (30.2) and (30.5) that the coefficients in the linearized equation of motion (28.11) or the propagation condition (28.23) *necessarily* lack the full symmetry of (30.3) unless the initial stress is spherical. To show this, first observe that in spite of the freedom allowed by (30.2), the coefficients are uniquely defined if the second and fourth subscripts are the same. For example,

$$\begin{aligned} \beta_{1313} &= \bar{B}_{1313}, & \beta_{3131} &= \bar{B}_{3131}, \\ \beta_{1212} &= \bar{B}_{1212}, & \beta_{2121} &= \bar{B}_{2121}. \end{aligned} \tag{30.6}$$

But when $(30.5)_3$ is included,

$$\begin{aligned} \beta_{1313} - \beta_{3131} &= \bar{B}_{1313} - \bar{B}_{3131} = \bar{T}_{33} - \bar{T}_{11}, \\ \beta_{1212} - \beta_{2121} &= \bar{B}_{1212} - \bar{B}_{2121} = \bar{T}_{22} - \bar{T}_{11}. \end{aligned} \tag{30.7}$$

Since $\bar{T}_{11}$ cannot equal *both* $\bar{T}_{22}$ and $\bar{T}_{33}$ except when the initial stress is spherical (hydrostatic pressure), some of the coefficients in the equation of motion

$$\bar{\varrho}\,\ddot{x}_i = \beta_{ijkm} \frac{\partial^2 x_k}{\partial X_j \partial X_m} \tag{30.8}$$

necessarily lack the full symmetry of the cees in (30.3). A similar conclusion applies to the coefficients α_{ijkm} in

$$\varrho_N\,\ddot{x}_i = \alpha_{ijkm} \frac{\partial^2 x_k}{\partial a_j \partial a_m}. \tag{30.9}$$

30.4. Essential difference between wave propagation in unstressed and anisotropically stressed media. This departure from the familiar symmetry (under which β_{1313} would equal β_{3131}, both being β_{55} in the abbreviated Voigt notation), indicates an essential difference between wave propagation in an anisotropically stressed medium and an unstressed one. The physical interpretation is particularly simple in the case of uniaxial stress in an isotropic medium. Suppose that the uniaxial stress is in the 3 direction. Let us compare the transverse mode propagating in the direction of tension and having particle displacement in the 1 direction with the transverse mode in which these directions of propagation and particle displacement are interchanged. The equations of motion for these modes are, respectively,

$$\bar{\varrho}\,\ddot{u}_1 = \beta_{1313}(\partial^2 u_1/\partial X_3^2), \qquad \bar{\varrho}\,\ddot{u}_3 = \beta_{3131}(\partial^2 u_3/\partial X_1^2). \tag{30.10}$$

Hence, the values of $\bar{\varrho}V^2$ for these modes are, respectively, β_{1313} and β_{3131}, which, from Eq. (30.7) differ by the applied uniaxial stress.

Superficial considerations could tempt one to conclude that, as far as wave propagation is concerned, an isotropic medium under uniaxial stress should behave just like an unstressed medium with transverse isotropy. The above example proves such a conclusion to be false, because the two modes considered always propagate with the same velocity in the unstressed medium. The difference of velocities in the stressed medium enables it, in principle, to be distinguished experimentally from the unstressed medium of lower symmetry.

If this result is surprising at first, it is no longer so after a little thought. An important point omitted from the "superficial considerations" is the direct effect of initial stress on the restoring force. A uniaxial tension in an infinite medium increases the restoring force for displacements associated with the transverse wave propagating in the direction of tension, thereby increasing the propagation velocity just as in a stretched string or bar. For the same reason, a compression reduces the restoring force and lowers the velocity. This "stretched-string effect" is absent in the other mode, in which the directions of particle displacement and propagation are interchanged.

The effect is relatively small. Since the difference in $\bar{\varrho}V^2$ equals the uniaxial stress, a stress equal to about 0.02% of the shear modulus is required to produce a velocity difference of 0.01%.

TRUESDELL (1961) has discussed wave propagation in an isotropic medium subjected to a general stress. The principal axes of stress are directions of propagation and particle displacement for purely longitudinal and purely transverse modes. The two transverse modes propagating along a given principal axis will have the same propagation velocity if the corresponding transverse principal stretches are equal. From (30.10) and (30.7), two purely transverse modes with directions of propagation and particle velocity interchanged will have the same velocity if and only if the normal stresses in these directions are equal. (This last statement applies to any material symmetry, but it should be noted that the interchange of directions does not always give a possible wave.)

30.5. Effective elastic coefficients under hydrostatic pressure. When all the principal initial stresses are equal, the differences illustrated by (30.7) disappear, and one may inquire whether it is possible in this case to find coefficients in the linearized equation of motion which obey all the symmetry relations of (30.3). The answer is yes: when the initial stress is a hydrostatic pressure

$$\bar{T}_{ij} = -p\,\delta_{ij}, \tag{30.11}$$

the coefficients

$$\beta_{ijkm} = p\,(\delta_{ij}\,\delta_{km} - \delta_{im}\,\delta_{jk} - \delta_{ik}\,\delta_{jm}) + \bar{c}^S_{ijkm}, \tag{30.12}$$

which satisfy all the desired symmetry conditions, may properly be used in (30.8).

The symmetry conditions may be verified by direct inspection. To show that the coefficients may properly be used in the linearized equation of motion, we note that under hydrostatic initial stress, (28.14) shows that $\bar{B}_{ijkm}$ becomes

$$\bar{B}_{ijkm} = -p\,\delta_{ik}\delta_{jm} + \bar{c}^S_{ijkm} \tag{30.13}$$

and hence

$$\beta_{ijkm} = p\,(\delta_{ij}\,\delta_{km} - \delta_{im}\,\delta_{jk}) + \bar{B}_{ijkm}. \tag{30.14}$$

It is clear from (30.14) that the coefficients β_{ijkm} satisfy (30.2), and hence are correct coefficients for use in the linearized equation of motion (30.8).

We call the quantities β_{ijkm} the *effective elastic coefficients* under hydrostatic pressure. Because of the way they appear in (30.8), they will appear just like the quantities $\bar{B}_{ijkm}$ in the propagation condition (28.23). Thus, when the initial stress is hydrostatic pressure,

$$(\beta_{ijkm}\,n_j\,n_m - \bar{\varrho}V^2\,\delta_{ik})\,u_k = 0. \tag{30.15}$$

Because the betas retain their zero-pressure symmetry at all pressures, the effective elastic coefficients can be determined as functions of pressure by relating them to the values of $\bar{\varrho}V^2$ for various modes of propagation by the same formulas that are used at the natural state.

It should be emphasized that $\bar{c}^S_{ijkm}$ is not a constant but a function of the initial pressure, being the thermodynamic coefficient referred to and evaluated at the *initial* state.

BARSCH and CHANG (1968) have extended the concept of effective elastic coefficients under hydrostatic pressure to include higher-order coefficients.

30.6. Interpretation of the effective elastic coefficients β_{ijkm} as coefficients in a linearized stress-deformation relation. THURSTON (1965) has interpreted the effective elastic coefficients β_{ijkm} as derivatives of the *stress* T_{ij} with respect to deformation gradients from the initial state, $\partial x_k/\partial X_m$, evaluated at the initial hydrostatically stressed state $\boldsymbol{X}$. From (28.1) and (28.5), putting $J=\bar{\varrho}/\varrho$ as in (6.1),

$$T_{ij}=\frac{1}{J}\frac{\partial x_i}{\partial X_p}P_{jp}=\frac{1}{J}\frac{\partial x_i}{\partial X_p}\frac{\bar{\varrho}\,\partial U}{\partial(\partial x_j/\partial X_p)}. \tag{30.16}$$

By differentiation, employing the definition (28.10),

$$\frac{\partial T_{ij}}{\partial(\partial x_k/\partial X_m)}=\frac{1}{J}\left(\frac{\partial x_i}{\partial X_p}B_{jpkm}+P_{jp}\,\delta_{ik}\,\delta_{pm}\right)-\frac{P_{jp}}{J^2}\frac{\partial x_i}{\partial X_p}\frac{\partial J}{\partial(\partial x_k/\partial X_m)}. \tag{30.17}$$

Defining

$$E_{ijkm}\equiv\frac{\partial T_{ij}}{\partial(\partial x_k/\partial X_m)}, \tag{30.18}$$

employing (3.32), and evaluating quantities at the initial state where $\partial x_i/\partial X_j=\delta_{ij}$ and $P_{jm}=\bar{T}_{jm}$, we obtain

$$\bar{E}_{ijkm}=\bar{B}_{ijkm}+\bar{T}_{jm}\,\delta_{ik}-\bar{T}_{ji}\,\delta_{km}, \tag{30.19}$$

which, in view of $(30.5)_1$, (28.14), and the symmetry of T_{ij} may be written as

$$\begin{aligned}\bar{E}_{ijkm}&=\bar{B}_{ijkm}+\bar{T}_{im}\,\delta_{jk}-\bar{T}_{ij}\,\delta_{km}\\&=\bar{c}^{S}_{ijkm}+\bar{T}_{jm}\,\delta_{ik}+\bar{T}_{im}\,\delta_{jk}-\bar{T}_{ij}\,\delta_{km}.\end{aligned} \tag{30.20}$$

It is now easily verified that when the initial stress is the hydrostatic pressure (30.11), $\bar{E}_{ijkm}$ reduces to the β_{ijkm} of Eqs. (30.12) and (30.14).

Thurston (1965) also showed the effective elastic coefficients β_{ijkm} to be the derivatives of the stress T_{ij} with respect to the "small strains" from the initial state

$$e_{km}\equiv\tfrac{1}{2}[(\partial u_k/\partial X_m)+(\partial u_m/\partial X_k)] \tag{30.21}$$

when the initial stress is hydrostatic. The change of stress in general depends on all nine of the displacement gradients $\partial u_k/\partial X_m$, i.e., on both strain and rotation from the initial state. The stress changes due to the small rotation can be separated from those due to the small strain by introducing as independent variables e_{km} and

$$\omega_{km}=\tfrac{1}{2}[(\partial u_k/\partial X_m)-(\partial u_m/\partial X_k)]. \tag{30.22}$$

One can incorporate the symmetry of e_{km} and antisymmetry of ω_{km} into the transformation from independent variables $(\partial u_p/\partial X_q)$ to (e_{km},ω_{km}) by writing

$$\frac{\partial u_p}{\partial X_q}=e_{pq}+\omega_{pq}=\tfrac{1}{2}(e_{pq}+e_{qp})+\tfrac{1}{2}(\omega_{pq}-\omega_{qp}). \tag{30.23}$$

Then

$$\partial(\partial u_p/\partial X_q)/\partial e_{km}=\tfrac{1}{2}(\delta_{pk}\,\delta_{qm}+\delta_{qk}\,\delta_{pm}), \tag{30.24}$$

$$\partial(\partial u_p/\partial X_q)/\partial\omega_{km}=\tfrac{1}{2}(\delta_{pk}\,\delta_{qm}-\delta_{qk}\,\delta_{pm}). \tag{30.25}$$

Defining

$$R_{ijkm}=\partial T_{ij}/\partial\omega_{km}, \tag{30.26}$$

$$S_{ijkm}=\partial T_{ij}/\partial e_{km}, \tag{30.27}$$

one finds

$$R_{ijkm}=\frac{\partial T_{ij}}{\partial(\partial u_p/\partial X_q)}\frac{\partial(\partial u_p/\partial X_q)}{\partial\omega_{km}}=\frac{1}{2}(\delta_{pk}\,\delta_{qm}-\delta_{qk}\,\delta_{pm})E_{ijpq}$$

whence

$$\begin{aligned}\bar{R}_{ijkm}&=\tfrac{1}{2}(\bar{E}_{ijkm}-\bar{E}_{ijmk})\\&=\tfrac{1}{2}(\bar{T}_{jm}\,\delta_{ik}+\bar{T}_{im}\,\delta_{jk}-\bar{T}_{jk}\,\delta_{im}-\bar{T}_{ik}\,\delta_{jm})\end{aligned} \tag{30.28}$$

where the last step makes use of $(30.20)_2$. Similarly,

$$\begin{aligned}\bar{S}_{ijkm} &= \tfrac{1}{2}(\bar{E}_{ijkm}+\bar{E}_{ijmk})\\ &= \bar{c}^S_{ijkm} - \bar{T}_{ij}\,\delta_{km} + \tfrac{1}{2}(\bar{T}_{jm}\,\delta_{ik}+\bar{T}_{im}\,\delta_{jk}+\bar{T}_{jk}\,\delta_{im}+\bar{T}_{ik}\,\delta_{jm}).\end{aligned} \tag{30.29}$$

When the initial stress is hydrostatic pressure, R_{ijkm} vanishes identically and $\bar{S}_{ijkm}$ reduces to β_{ijkm}.

When evaluated at the initial state, the derivatives with respect to the true strain components from that state, V_{km}, have the same values as the derivatives with respect to e_{km}. Hence, the effective coefficients β_{ijkm} may also be interpreted as the derivatives of the stress T_{ij} with respect to strain V_{km} from the initial state, evaluated at the hydrostatically stressed initial state.

It should be emphasized that the effective elastic coefficients β_{ijkm} are the derivatives of the stresses T_{ij}, *not* of the thermodynamic tensions $t_{ij}=\tau_{ij}$. Even though they coincide at the initial state $(\bar{T}_{ij}=\bar{t}_{ij})$, the stresses and thermodynamic tensions are *different functions* and have different derivatives. Evaluated at the initial state, the derivatives of the thermodynamic tensions (second derivatives of the energy) are the coefficients $\bar{c}^S_{ijkm}$, whereas the derivatives of the stresses are the effective elastic coefficients β_{ijkm}, provided the initial stress is hydrostatic.

30.7. Relations of effective elastic coefficients to the bulk modulus and compressibility. The bulk modulus and its reciprocal the compressibility are defined by (11.14) and (11.18), and the effective elastic coefficients under hydrostatic pressure by the relations of this section, say (30.12) [or (30.18) or (30.29) with the restriction that the initial stress be hydrostatic]. THURSTON $(1965)_2$ showed that the bulk modulus *at any pressure* is the same function of effective elastic coefficients as at zero pressure. The proof is based on (3.32) and the result that β_{ijkm} reduces to $\bar{S}_{ijkm}$ when the initial stress is spherical. The latter result can be expressed as

$$\frac{\partial T_{ij}}{\partial e_{km}} = \beta_{ijkm}; \qquad \frac{\partial T_\lambda}{\partial e_\mu} = \beta_{\lambda\mu} \tag{30.30}$$

in tensor and abbreviated notation, respectively, and where the derivatives are evaluated at the initial state, understood to satisfy (30.11). The effective compliance coefficients $k_{\lambda\mu}$ and k_{ijkm} are defined such that the 6×6 matrix of effective compliances $k_{\lambda\mu}$ is the inverse of the $\beta_{\lambda\mu}$ matrix. Therefore,

$$\frac{\partial e_{ij}}{\partial T_{km}} = k_{ijkm}; \qquad \frac{\partial e_\lambda}{\partial T_\mu} = k_{\lambda\mu}. \tag{30.31}$$

One can now evaluate the compressibility at the initial state as follows:

$$\begin{aligned}\bar{\chi} &= -\left[\frac{\partial}{\partial p}(\bar{\varrho}/\varrho)\right]_{\boldsymbol{X}} = -\left(\frac{\partial T_{ij}}{\partial p}\,\frac{\partial e_{km}}{\partial T_{ij}}\,\frac{\partial(\bar{\varrho}/\varrho)}{\partial e_{km}}\right)_{\boldsymbol{X}}\\ &= -(-\delta_{ij}\,k_{kmij}\,\delta_{km}) = k_{mmjj}\\ &= k_{11}+k_{22}+k_{33}+2(k_{12}+k_{13}+k_{23}).\end{aligned} \tag{30.32}$$

This completes the proof, since the initial configuration may be chosen to correspond to any desired pressure.

Since the compressibility is always the same function of the effective compliances, so is the bulk modulus always the same function of effective stiffnesses as at zero pressure. Obviously, the same applies to the derivatives of these quantities with respect to pressure and temperature. This constitutes an advantage of the "effective" coefficients over the "thermodynamic" coefficients in dealing with the relation of ultrasonic data to the equation of state.

Thus, the isentropic compressibility and bulk modulus at any pressure p in a crystal of arbitrary symmetry can be found from the values of ϱV^2 for ultrasonic waves by the following program:

a) Use the values of ϱV^2 to find the effective elastic coefficients for wave propagation at the desired pressure.

b) Invert the 6×6 matrix of effective elastic coefficients to obtain the effective isentropic compliance coefficients.

c) Form the sum in (30.32) to obtain the isentropic compressibility, and take the reciprocal to get the isentropic bulk modulus.

30.8. Measurement of effective elastic coefficients. In principle, the effective elastic coefficients can be determined as functions of the initial pressure p from data on the transit time of small-amplitude waves as expressed by the repetition frequencies $f(p)$ in (28.26) for a sufficient number of modes (directions of propagation and polarization). For example, COOK (1957) has indicated a procedure for calculating the effective elastic coefficients at any pressure p in cubic and hexagonal crystals, making use only of the basic frequency or transit-time data and the elastic constants at zero pressure. COOK considers the measured frequencies or transit times to be arbitrary functions of the pressure. At any given pressure, the ratios of density and path length to their zero-pressure values are expressed as certain integrals involving the measured functions. The final result is a set of formulas for the effective elastic coefficients which involve only the measured functions of pressure and the zero-pressure values. The beauty of COOK's work is that it makes no assumptions of linearity of any kind. Aside from the small increment needed to convert from adiabatic to isothermal coefficients, everything is determined from the measured functions. However, this work seems to have been several years ahead of its time, since reliable ultrasonic measurements on elastic solids have until recently been made only in a pressure range where both the measured frequencies and their squares look like linear functions of pressure. This inability to distinguish experimentally between f linear and f^2 linear meant essentially that the available data were inadequate to provide a basis for calculating the effective elastic coefficients as accurate nonlinear functions of pressure. One could, however, calculate their first pressure derivatives at zero pressure, to two or three significant figures.

By restricting attention to the initial pressure derivatives from the very outset, the data processing can be simplified. It is not necessary to calculate any quantities at an elevated pressure. Rather, one can make use of direct relations between the initial slopes of the measured frequency vs. pressure curves and the initial pressure derivatives of the elastic coefficients, as indicated in detail in the next section. The procedure is analogous to that used in calculating the zero-stress values of the third-order elastic coefficients, described in Sect. 29.

31. Pressure derivatives of elastic coefficients.

31.1. Introduction. In discussing the derivatives of various coefficients with respect to pressure or temperature, it is important to define clearly the functions whose derivatives are under consideration. Considered as derivatives of the internal energy with respect to strain, elastic coefficients depend on the reference configuration. Besides this "thermodynamic" definition, one may also deal with "effective" coefficients that are defined by the formulas for wave propagation speeds. The two sets of elastic coefficients so defined are equal at the natural state, but they are different functions of pressure, and so have different pressure derivatives.

Of the many possible kinds of elastic coefficients whose pressure derivatives could be considered, only two are of importance: (1) the thermodynamic coefficients *referred to the natural state*, and (2) the effective elastic coefficients of Sect. 30.5.

For the reason explained at the end of Sect. 30, we are concerned only with the initial values of the pressure derivatives.

31.2. Pressure derivatives of thermodynamic coefficients. From (29.16) and (29.18), with p interpreted as the hydrostatic pressure at the initial state, the initial values of the pressure derivatives of the thermodynamic elastic coefficients, referred to the natural state, are given by

$$\left(\frac{\partial c^S_{ijkm}}{\partial p}\right)_{T;\,p=0} = -C_{ijkmpq}\, s^T_{pqaa} \equiv c'_{ijkm} \tag{31.1}$$

and hence G_H of Eq. (29.23) can be expressed as

$$G_H = -c'_{ijkm} N_j N_m U^0_i U^0_k \tag{31.2}$$

where (31.1) defines c'_{ijkm}. The pressure derivatives of the thermodynamic elastic coefficients can then be calculated from experimental data through (31.2), $(29.23)_1$, $(29.23)_2$, and (29.2). Alternatively, (31.1) enables the pressure derivatives to be calculated from the third-order elastic coefficients and the second-order compliances.

In abbreviated notation (31.1) reads

$$c'_{\lambda\mu} = -s^T_\nu\, C_{\lambda\mu\nu} \tag{31.3}$$

where s_ν is defined in (11.71). The difference from (11.91) is only that we are here concerned with isothermal derivatives of isentropic coefficients because of the experimental situation in which initial pressure is varied isothermally, but the wave propagates isentropically.

31.3. Pressure derivatives of effective coefficients. Turning now to the effective coefficients, we multiply (30.15) by u_i and interpret $\boldsymbol{u}$ as a unit vector to obtain

$$\bar{\varrho}\, V^2 = \beta_{ijkm} n_j n_m u_i u_k. \tag{31.4}$$

We can obtain an expression involving the pressure derivatives of the effective elastic coefficients by differentiating (31.4) with respect to the initial pressure p.

For many commonly used modes, $\boldsymbol{n}$ and $\boldsymbol{u}$ remain constant when the pressure is changed. For simplicity, we first consider only cases where $\boldsymbol{n}$ and $\boldsymbol{u}$ remain constant at their zero-pressure values, denoted by $\boldsymbol{N}$ and $\boldsymbol{U}^0$. Then

$$(\bar{\varrho} V^2)' = N_j N_m U^0_i U^0_k \beta'_{ijkm}. \tag{31.5}$$

Thus, the pressure derivatives of $\bar{\varrho} V^2$ for these modes are related to the pressure derivatives of the effective elastic coefficients by exactly the same formulas that relate $\bar{\varrho}\, V^2$ to the coefficients themselves.

Let us now relate the *initial* pressure derivative of $\bar{\varrho} V^2$ to the initial pressure derivative of the measured repetition frequency f in (28.26) or (29.2). The initial pressure derivatives of V^2/W^2 and $\bar{\varrho}/\varrho_N$ are needed. From (28.34),

$$V^2/W^2 = (1/f_N)^2 \tag{31.6}$$

where f_N is the thickness shrink from the natural to the initial state, given by (28.33). The initial pressure derivative of f_N has already been calculated in (11.84).

(In Sect. 11, the natural configuration was labeled $\boldsymbol{X}$. For the present application, we need only remember that the coefficients are referred to and evaluated at the natural state. To obtain the present notation, $\boldsymbol{X}$ would be replaced by $\boldsymbol{a}$ and $\boldsymbol{x}$ by $\boldsymbol{X}$ in the definitions of Sect. 11.) From (11.84), $(11.68)_1$, and the tensor form of $(11.69)_1$,

$$f'_{\boldsymbol{N}}(0) = -N_j N_m V'_{jm}(0) = N_j N_m s^T_{jmaa}, \tag{31.7}$$

or, in the abbreviated notation of (11.85) and (11.71), we get the result (11.88):

$$f'_{\boldsymbol{N}}(0) = \nu_\alpha s_\alpha. \tag{31.8}$$

In the tensor notation,

$$\left[\frac{\partial}{\partial p}(V^2/W^2)\right]_{T;p=0} = -2f'_{\boldsymbol{N}}(0) = -2N_j N_m s^T_{jmaa}. \tag{31.9}$$

The pressure derivative of $(\bar{\varrho}/\varrho_N)$ is simply the compressibility χ [cf. (11.14)]. From (11.23) and (11.78), we have the tensor and abbreviated expressions

$$\chi^T(0) = s^T_{iikk} = s^T_\beta \delta_\beta. \tag{31.10}$$

We now combine these expressions to obtain, at $p=0$,

$$\begin{aligned}\left[\frac{\partial}{\partial p}(\bar{\varrho}V^2)\right]_{T;p=0} &= \left[\frac{\partial}{\partial p}\left(\frac{\bar{\varrho}}{\varrho_N}\frac{V^2}{W^2}\varrho_N W^2\right)\right]_{T;p=0} \\ &= w'(0) + w(0)\left[\chi^T - 2N_j N_m s^T_{jmaa}\right]\end{aligned} \tag{31.11}$$

where $w(0)$ and $w'(0)$ are the zero-pressure values of $\varrho_N W^2$ and its pressure derivative as in Sect. 29. Finally, the pressure derivatives of $\bar{\varrho}V^2$ are related to the basic frequency data by combining (31.11) with (29.2) to obtain

$$\left[\frac{\partial}{\partial p}(\bar{\varrho}V^2)\right]_{T;p=0} = w(0)\left[2f'(0)/f(0) + \chi^T - 2N_j N_m s^T_{jmaa}\right]. \tag{31.12}$$

Eq. (31.12) is always a correct formula for the initial pressure derivative of $\bar{\varrho}V^2$, but (31.5) relates this derivative to the derivatives of the effective elastic coefficients under the assumption that the propagation direction $\boldsymbol{n}$ and particle displacement direction $\boldsymbol{u}$ are independent of pressure for the mode under consideration. The assumption concerning $\boldsymbol{u}$ can be eliminated by the same scheme that eliminated $\boldsymbol{U}'$ from (29.7): differentiation of (30.15) leads to

$$(\beta_{ijkm} n_j n_m)' u_k + (\beta_{ijkm} n_j n_m) u'_k = (\bar{\varrho}V^2)' u_i + (\bar{\varrho}V^2) u'_i. \tag{31.13}$$

Since $\beta_{ijkm} n_j n_m$ is a symmetric second rank tensor, we have

$$\beta_{ijkm} n_j n_m u_i = \bar{\varrho}V^2 u_k, \tag{31.14}$$

and hence when (31.13) is multiplied by u_i and summed over $_k$, there results only

$$u_i(\beta_{ijkm} n_j n_m)' u_k = (\bar{\varrho}V^2)' \tag{31.15}$$

where use has been made of the conditions $\boldsymbol{u}\cdot\boldsymbol{u}=1$ and $\boldsymbol{u}\cdot\boldsymbol{u}'=0$. In relating the initial derivative of $\bar{\varrho}V^2$ to derivatives of effective coefficients, it is apparently only the change with p in propagation direction $\boldsymbol{n}$ that needs to be considered, and not the change in polarization direction $\boldsymbol{u}$.

Experimentally, the propagation direction remains normal to a crystal face and hence $\boldsymbol{N}$ is constant in (28.32): $n_j = N_i(\partial a_i/\partial X_j)/f_{\boldsymbol{N}}$. To obtain an equation to replace (31.5) in the more general case, we substitute for $\boldsymbol{n}$ from (28.32) and evaluate (31.15) at the natural zero-pressure state. The result of this substitution can be expressed as

$$(\bar{\varrho} V^2 f_{\mathbf{N}}^2)_0' = (\bar{\varrho} W^2)_0' = \bar{\beta}_{irks}'(0)\, N_r N_s U_i^0 U_k^0,$$
$$\bar{\beta}_{irks} = \beta_{ijkm} \frac{\partial a_r}{\partial X_j} \frac{\partial a_s}{\partial X_m}. \tag{31.16}$$

31.4. Relation between effective and thermodynamic elastic coefficients. In the definition (30.12), the coefficients $\bar{c}_{ijkm}^S$ are referred to the initial state, i.e., they are derivatives with respect to strain from the *initial* state [cf. (28.13)]. The switch to the natural state as a reference is based on

$$\frac{\partial x_i}{\partial a_j} = \frac{\partial x_i}{\partial X_q} \frac{\partial X_q}{\partial a_j}. \tag{31.17}$$

In differentiating with respect to deformation gradients from the initial state, the deformation from $\boldsymbol{a}$ to $\boldsymbol{X}$ is held constant. For example, from (31.17),

$$\frac{\partial(\partial x_i/\partial a_j)}{\partial(\partial x_i/\partial X_q)} = \frac{\partial X_q}{\partial a_j}. \tag{31.18}$$

As applications of (31.18) it may be noted that

$$\frac{1}{\varrho} P_{iq} = \frac{1}{\varrho_N} P_{ij}^N \frac{\partial X_q}{\partial a_j}, \tag{31.19}$$

$$\frac{1}{\varrho} B_{ijkm} = \frac{1}{\varrho_N} A_{irks} \frac{\partial X_j}{\partial a_r} \frac{\partial X_m}{\partial a_s} \tag{31.20}$$

where the coefficients are defined in (28.5), (28.10). and (28.16), To relate derivatives with respect to V_{ij} and derivatives with respect to V_{ij}^N, we note from (28.2), (31.17), and (3.2) that

$$\begin{aligned} V_{pq}^N &= \frac{1}{2}\left(\frac{\partial x_s}{\partial a_p}\frac{\partial x_s}{\partial a_q} - \delta_{pq}\right) = \frac{1}{2}\left(\frac{\partial x_s}{\partial X_i}\frac{\partial x_s}{\partial X_j}\frac{\partial X_i}{\partial a_p}\frac{\partial X_j}{\partial a_q} - \delta_{pq}\right) \\ &= \frac{1}{2}\left[(\delta_{ij} + 2V_{ij})\frac{\partial X_i}{\partial a_p}\frac{\partial X_j}{\partial a_q} - \delta_{pq}\right] \end{aligned} \tag{31.21}$$

and hence

$$\frac{\partial V_{pq}^N}{\partial V_{ij}} = \frac{\partial X_i}{\partial a_p}\frac{\partial X_j}{\partial a_q}. \tag{31.22}$$

Therefore, the initial-state coefficients $\bar{c}_{ijkm}^S = \bar{\varrho}(\partial^2 U/\partial V_{ij}\,\partial V_{km})_{\boldsymbol{X}}$ are referred to the natural state by the formula

$$\bar{c}_{ijkm}^S = \bar{c}_{pqrs}^{NS} \frac{\partial X_i}{\partial a_p}\frac{\partial X_j}{\partial a_q}\frac{\partial X_k}{\partial a_r}\frac{\partial X_m}{\partial a_s}\frac{\bar{\varrho}}{\varrho_N}. \tag{31.23}$$

As before, the bar denotes evaluation at the initial state. The coefficient $\bar{c}_{pqrs}^{NS}$ is the quantity in (28.19), referred to the natural state, evaluated at the initial state while $\bar{c}_{ijkm}^S$ is both referred to and evaluated at the initial state. From (31.23) and (30.12),

$$\beta_{ijkm} = p(\delta_{ij}\delta_{km} - \delta_{im}\delta_{jk} - \delta_{ik}\delta_{jm}) + \frac{\bar{\varrho}}{\varrho_N}\frac{\partial X_i}{\partial a_p}\frac{\partial X_j}{\partial a_q}\frac{\partial X_k}{\partial a_r}\frac{\partial X_m}{\partial a_s}\bar{c}_{pqrs}^{NS}. \tag{31.24}$$

In the present context, the deformation from the natural to the initial state is taken as isothermal, while the betas govern the propagation of small-amplitude waves that are assumed to propagate isentropically. We call such betas the *isentropic* effective elastic coefficients under hydrostatic pressure. Based on the discussion of Sect. 30.6, one can also define *isothermal* effective elastic coefficients as the isothermal derivatives of the stress with respect to strain from the initial state. The only change in (31.24) is to replace the superscript S by T. Thus, a relation of the same form (31.24) relates isentropic betas to isentropic cees, and isohermal betas to isothermal cees.

THURSTON (1965) has noted the following simplification of (31.24) that applies to all crystal classes except those belonging to the monoclinic and triclinic classes. With these exceptions, the conventional rectangular axes [BOND *et al.*, 1949] are principal axes of the deformation at all pressures, since the components of $(\partial V_{ij}^N/\partial p)$ must satisfy the relations for symmetric second rank tensors indicated in Table 16.2. For all except the monoclinic and triclinic cases, the deformation resulting from hydrostatic pressure therefore has the form

$$\left\| \frac{\partial X_i}{\partial a_j} \right\| = \left\| \begin{matrix} \lambda_1 & 0 & 0 \\ 0 & \lambda_2 & 0 \\ 0 & 0 & \lambda_3 \end{matrix} \right\| . \tag{31.25}$$

Also,

$$\bar{\varrho}/\varrho_N = 1/\lambda_1 \lambda_2 \lambda_3 . \tag{31.26}$$

Subject to (31.25) and (31.26), (31.24) may be expressed by abandoning the summation convention and writing

$$\beta_{ijkm} = p(\delta_{ij}\,\delta_{km} - \delta_{im}\,\delta_{jk} - \delta_{ik}\,\delta_{jm}) + \frac{\lambda_i \lambda_j \lambda_k \lambda_m}{\lambda_1 \lambda_2 \lambda_3}\, \bar{c}_{ijkm} . \tag{31.27}$$

(Nothing summed!)

Nothing is summed in (31.27), i, j, k, and m being fixed indices. Table 31.1 displays the results (31.27), converted to the abbreviated notation.

31.5. Relation between pressure derivatives of effective and thermodynamic coefficients. The relation between the pressure derivatives of the effective and thermodynamic elastic coefficients is obtained by differentiating (31.24). Again, an essential simplification is possible in all cases except crystal classes belonging to the monoclinic and triclinic systems. From (31.25) and (28.2), the principal strains V_i^N and principal stretches λ_i are related by

$$V_i^N = \tfrac{1}{2}(\lambda_i^2 - 1) . \tag{31.28}$$

Hence, at the natural state we have

$$(dV_i^N/dp)_{p=0} = (d\lambda_i/dp)_{p=0} . \tag{31.29}$$

From (11.75) and (11.71), the derivative in (31.29) is

$$(dV_i^N/dp)_{p=0} = -(s_{i1} + s_{i2} + s_{i3}) = -s_i . \tag{31.30}$$

Having determined $(d\lambda_i/dp)_{p=0}$, the zero-pressure values of the pressure derivatives of the effective and thermodynamic coefficients can be related to each other by differentiating the entries in Table 31.1, evaluating all quantities at the natural state. The results are given in Table 31.2, taken from THURSTON (1965).

To cover the remaining monoclinic and triclinic cases, one needs expressions for the pressure derivatives of the deformation gradients with the crystal referred to the conventional Cartesian axes. If we define

$$b_{kr} = \left[\frac{\partial}{\partial p} \left(\frac{\partial X_k}{\partial a_r} \right) \right]_{p=0} , \tag{31.31}$$

the result of differentiating (31.24) takes the following form in which it is understood that the cees are referred to the natural state and all quantities are evaluated

Table 31.1. *$\beta_{\mu\nu}$ under hydrostatic pressure (monoclinic and triclinic classes excluded).*

$-p+\frac{\lambda_1^3}{\lambda_2\lambda_3}c_{11}$	$p+\frac{\lambda_1\lambda_2}{\lambda_3}c_{12}$	$p+\frac{\lambda_3\lambda_1}{\lambda_2}c_{13}$	$\lambda_1 c_{14}$	$\frac{\lambda_1^2}{\lambda_2}c_{15}$	$\frac{\lambda_1^2}{\lambda_3}c_{16}$
	$-p+\frac{\lambda_2^3}{\lambda_1\lambda_3}c_{22}$	$p+\frac{\lambda_2\lambda_3}{\lambda_1}c_{23}$	$\frac{\lambda_2^2}{\lambda_1}c_{24}$	$\lambda_2 c_{25}$	$\frac{\lambda_2^2}{\lambda_3}c_{26}$
		$-p+\frac{\lambda_3^3}{\lambda_1\lambda_2}c_{33}$	0	0	0
			$-p+\frac{\lambda_2\lambda_3}{\lambda_1}c_{44}$	$\lambda_3 c_{45}$	$\lambda_2 c_{46}$
				$-p+\frac{\lambda_3\lambda_1}{\lambda_2}c_{55}$	$\lambda_1 c_{56}$
					$-p+\frac{\lambda_1\lambda_2}{\lambda_3}c_{66}$

at the natural state ($p=0$).

$$\begin{aligned}\beta'_{ijkm}=&(\delta_{ij}\delta_{km}-\delta_{im}\delta_{jk}-\delta_{ik}\delta_{jm})+c'_{ijkm}+\chi c_{ijkm}+b_{ip}c_{pjkm}\\&+b_{jq}c_{iqkm}+b_{kr}c_{ijrm}+b_{ms}c_{ijks}.\end{aligned}\tag{31.32}$$

The problem is now to evaluate the quantities b_{kr} defined in (31.31).

Now the deformation gradients $\partial X_k/\partial a_r$ do not have unique derivatives unless some constraints are specified. We here use the conventionally chosen Cartesian axes, which are related in a specified manner to the standard crystallographic axes a, b, and c. In the general case of a triclinic crystal, the angles between the crystallographic axes will change with the pressure, but the Cartesian system always has its 3 direction along c, its 1 direction in the (a, c) plane, and its 2 direction chosen to complete a righthanded system [MASON, 1966; BOND *et al.*, 1949]. With respect to this system, then, the orientation of the (1, 3) material planes and of the material lines lying along the 3 direction are fixed. For our purpose, the needed constraints consist of this specified relation between the Cartesian and crystallographic axes. It doesn't matter whether the Cartesian system is fixed in the laboratory or not. Our specification of the reference system serves to "constrain" the crystal with respect to this system, and we may picture the crystal (and associated reference system) as being completely unconstrained with respect to the laboratory.

The above described constraints are readily expressed in terms of conditions on the deformation gradients. Because the (1, 3) material planes have a fixed orientation, X_2 does not depend on a_1 or a_3. Because material lines lying along the 3 direction are not rotated, X_1 and X_2 do not depend on a_3. Thus, if the deformation is to be described with respect to the conventionally chosen Cartesian axes, the constraints are

$$\partial X_1/\partial a_3=\partial X_2/\partial a_3=\partial X_2/\partial a_1=0.\tag{31.33}$$

To facilitate the evaluation of b_{kr}, we write

$$\partial X_k/\partial \alpha_r=\delta_{kr}+e_{kr}+\omega_{kr},\tag{31.34}$$

where

$$e_{kr}=\frac{1}{2}\left(\frac{\partial X_k}{\partial a_r}+\frac{\partial X_r}{\partial a_k}-2\delta_{kr}\right),\tag{31.35}$$

$$\omega_{kr}=\frac{1}{2}\left(\frac{\partial X_k}{\partial a_r}-\frac{\partial X_r}{\partial a_k}\right).\tag{31.36}$$

Table 31.2. *Difference of pressure derivatives of effective and thermodynamic elastic coefficients,* $\frac{d}{dp}(\beta_{\mu\nu} - c_{\mu\nu})$ *(monoclinic and triclinic classes excluded).*

$-1+(s_2+s_3-3s_1)\,c_{11}$	$1+(s_3-s_1-s_2)\,c_{12}$	$1+(s_2-s_1-s_3)\,c_{13}$	$-s_1\,c_{14}$	$(s_2-2s_1)\,c_{15}$	$(s_3-2s_1)\,c_{16}$
	$-1+(s_1+s_3-3s_2)\,c_{22}$	$1+(s_1-s_2-s_3)\,c_{23}$	$(s_1-2s_2)\,c_{24}$	$-s_2\,c_{25}$	$(s_3-2s_2)\,c_{26}$
		$-1+(s_1+s_2-3s_3)\,c_{33}$	—	—	—
			$-1+(s_1-s_2-s_3)\,c_{44}$	$-s_3\,c_{45}$	$-s_2\,c_{46}$
				$-1+(s_2-s_1-s_3)\,c_{55}$	$-s_1\,c_{56}$
					$-1+(s_3-s_1-s_2)\,c_{66}$

$s_i = s_{i1} + s_{i2} + s_{i3}, \quad i = 1, 2, 3.$

The constraints, (31.33), imply that

$$\begin{aligned}\omega_{12}&=e_{12}=-\omega_{21},\\ \omega_{23}&=-e_{23}=-\omega_{32},\\ \omega_{31}&=e_{31}=-\omega_{13}.\end{aligned} \tag{31.37}$$

Now at zero pressure, the pressure derivatives of the quantities e_{kr} are the same as the pressure derivatives of the strains, since V_{ij}^N equals e_{ij} plus terms quadratic in e_{ki} and ω_{kj}. Hence, from $(11.68)_1$, the tensor form of $(11.69)_1$, and (11.75),

$$e'_{kr}(0)=-s_{krtt}=-s_\alpha. \tag{31.38}$$

From (31.37) and (31.38), the nine pressure derivatives of the deformation gradients may be written out as

$$\begin{aligned}&b_{11}=-s_1, && b_{22}=-s_2, && b_{33}=-s_3,\\ &b_{12}=-s_6, && b_{23}=0, && b_{31}=-s_5,\\ &b_{21}=0, && b_{32}=-s_4, && b_{13}=0.\end{aligned} \tag{31.39}$$

The general expression, (31.32), may now be written in the rather cumbersome form

$$\begin{aligned}\beta'_{ijkm}=&(\delta_{ij}\,\delta_{km}-\delta_{im}\,\delta_{jk}-\delta_{ik}\,\delta_{jm})+c'_{ijkm}+\chi c_{ijkm}\\ &-\delta_{i1}(s_1\,c_{1jkm}+s_6\,c_{2jkm})-\delta_{i2}\,s_2\,c_{2jkm}\\ &-\delta_{i3}(s_3\,c_{3jkm}+s_4\,c_{2jkm}+s_5\,c_{1jkm})\\ &-\delta_{j1}(s_1\,c_{i1km}+s_6\,c_{i2km})-\delta_{j2}\,s_2\,c_{i2km}\\ &-\delta_{j3}(s_3\,c_{i3km}+s_4\,c_{i2km}+s_5\,c_{i1km})\\ &-\delta_{k1}(s_1\,c_{ij1m}+s_6\,c_{ij2m})-\delta_{k2}\,s_2\,c_{ij2m}\\ &-\delta_{k3}(s_3\,c_{ij3m}+s_4\,c_{ij2m}+s_5\,c_{ij1m})\\ &-\delta_{m1}(s_1\,c_{ijk1}+s_6\,c_{ijk2})-\delta_{m2}\,s_2\,c_{ijk2}\\ &-\delta_{m3}(s_3\,c_{ijk3}+s_4\,c_{ijk2}+s_5\,c_{ijk1}).\end{aligned} \tag{31.40}$$

The pressure derivative of any particular coefficient may now be read off from (31.40) by fixing i, j, k, m. The results, in the abbreviated notation, are listed in Table 31.3 [THURSTON, 1967]. For crystals not belonging to the monoclinic or triclinic systems, these results reduce to those already given in Table 31.2.

To distinguish explicitly between derivatives at constant temperature or entropy and between derivatives of isothermal or adiabatic coefficients, the compressibility and compliances in (31.40) and Table 31.3 may be labeled with the quantity held constant in the differentiation and the stiffnesses with the type of coefficient being differentiated. For example, the isothermal pressure derivatives of adiabatic coefficients are obtained by using $\chi=\chi^T$, $s_\lambda=s_\lambda^T$, $c_{ijkm}=c_{ijkm}^S$, and $\beta_{ijkm}=\beta_{ijkm}^S$, while interpreting the prime as a derivative with respect to pressure at constant temperature. These are the pressure derivatives that have most commonly been cited in the ultrasonics literature.

BARSCH (1967) has discussed the pressure derivatives of the effective elastic coefficients for cubic symmetry, and has related the zero-pressure values of the isothermal pressure derivatives of the isothermal coefficients, $(\partial\beta_{\mu\nu}^T/\partial p)_T$, the isothermal pressure derivatives of the isentropic coefficients $(\partial\beta_{\mu\nu}^S/\partial p)_T$, and the isentropic pressure derivatives of the isentropic coefficients $(\partial\beta_{\mu\nu}^S/\partial p)_S$. BARSCH denotes his coefficients by the letter "cee", but they are the same as the "effective" coefficients of this section as well as of BIRCH (1947) and THURSTON (1965).

Table 31.3. *Pressure derivatives of effective and thermodynamic elastic coefficients.*

$$\beta'_{11} = -1 + c'_{11} + \chi c_{11} - 4(s_1 c_{11} + s_6 c_{16})$$
$$\beta'_{22} = -1 + c'_{22} + \chi c_{22} - 4 s_2 c_{22}$$
$$\beta'_{33} = -1 + c'_{33} + \chi c_{33} - 4(s_3 c_{33} + s_4 c_{34} + s_5 c_{35})$$
$$\beta'_{44} = -1 + c'_{44} + \chi c_{44} - 2 s_2 c_{44} - 2(s_3 c_{44} + s_4 c_{24} + s_5 c_{46})$$
$$\beta'_{55} = -1 + c'_{55} + \chi c_{55} - 2(s_1 c_{55} + s_6 c_{45}) - 2(s_3 c_{55} + s_4 c_{56} + s_5 c_{15})$$
$$\beta'_{66} = -1 + c'_{66} + \chi c_{66} - 2(s_1 c_{66} + s_6 c_{26}) - 2 s_2 c_{66}$$
$$\beta'_{12} = 1 + c'_{12} + \chi c_{12} - 2(s_1 c_{12} + s_6 c_{26}) - 2 s_2 c_{12}$$
$$\beta'_{13} = 1 + c'_{13} + \chi c_{12} - 2(s_1 c_{13} + s_6 c_{36}) - 2(s_3 c_{13} + s_4 c_{14} + s_5 c_{15})$$
$$\beta'_{14} = c'_{14} + \chi c_{14} - 2(s_1 c_{14} + s_6 c_{46}) - s_2 c_{14} - s_3 c_{14} - s_4 c_{12} - s_5 c_{16}$$
$$\beta'_{15} = c'_{15} + \chi c_{15} - 2(s_1 c_{15} + s_6 c_{56}) - s_1 c_{15} - s_6 c_{14} - s_3 c_{15} - s_4 c_{16} - s_5 c_{11}$$
$$\beta'_{16} = c'_{16} + \chi c_{16} - 3 s_1 c_{16} - 2 s_6 c_{66} - s_6 c_{12} - s_2 c_{16}$$
$$\beta'_{23} = 1 + c'_{23} + \chi c_{23} - 2 s_2 c_{23} - 2(s_3 c_{23} + s_4 c_{24} + s_5 c_{15})$$
$$\beta'_{24} = c'_{24} + \chi c_{24} - 3 s_2 c_{24} - s_3 c_{24} - s_4 c_{22} - s_5 c_{26}$$
$$\beta'_{25} = c'_{25} + \chi c_{25} - 2 s_2 c_{25} - s_1 c_{56} - s_6 c_{24} - s_3 c_{25} - s_4 c_{26} - s_5 c_{12}$$
$$\beta'_{26} = c'_{26} + \chi c_{26} - 3 s_2 c_{26} - s_1 c_{26} - s_6 c_{22}$$
$$\beta'_{34} = c'_{34} + \chi c_{34} - s_2 c_{34} - 3 s_3 c_{34} - 2 s_4 c_{44} - s_4 c_{23} - 2 s_5 c_{45} - s_5 c_{36}$$
$$\beta'_{35} = c'_{35} + \chi c_{35} - 3 s_3 c_{35} - 2 s_4 c_{45} - s_4 c_{34} - 2 s_5 c_{55} - s_5 c_{15} - s_1 c_{35} - s_6 c_{34}$$
$$\beta'_{36} = c'_{36} + \chi c_{36} - 2(s_3 c_{36} + s_4 c_{46} + s_5 c_{56}) - s_1 c_{36} - s_6 c_{23} - s_2 c_{36}$$
$$\beta'_{45} = c'_{45} + \chi c_{45} - s_2 c_{45} - 2 s_3 c_{45} - s_4 c_{25} - s_4 c_{46} - s_5 c_{56} - s_5 c_{14} - s_1 c_{45} - s_6 c_{44}$$
$$\beta'_{46} = c'_{46} + \chi c_{46} - s_3 c_{46} - s_4 c_{26} - s_5 c_{66} - 2 s_2 c_{46} - s_1 c_{46} - s_6 c_{24}$$
$$\beta'_{56} = c'_{56} + \chi c_{56} - s_3 c_{56} - s_4 c_{66} - s_5 c_{16} - 2 s_1 c_{56} - s_6 c_{46} - s_6 c_{56} - s_2 c_{56}$$

BARSCH and CHANG (1967) have tabulated numerical results of the above three derivatives for 25 materials of cubic symmetry.

By extending the range of initial pressure to about 10 kilobars and refining the accuracy of the frequency measurements, a definite curvature can be detected in the frequency versus pressure data. From such data, the *second* pressure derivatives of the effective elastic constants have been calculated for CsCl, CsBr, CsI [CHANG and BARSCH, 1967] KCl [DOBRETSOV and PERESEDA, 1969], RbCl, RbBr, RbI [CHANG and BARSCH, 1971], and NaI and KI [BARSCH and SHULL, 1971]. Measurements over the same pressure range in V_3Si and V_3Ge revealed an essentially linear pressure dependence of the effective elastic constants except for $\frac{1}{2}(\beta_{11} - \beta_{12})$, which governs propagation in the [110] direction with polarization along $[1\bar{1}0]$ [CARCIA, BARSCH, and TESTARDI, 1971].

K. Analysis of ultrasonic measurements as a function of temperature.

32. Elastic coefficients as a function of temperature.

In many respects, the most convenient elastic coefficients to deal with are those referred to an unstressed ("natural") state. They appear as second derivatives of thermodynamic potentials with respect to strain *from that state*, are most simply related to values of $\bar{\varrho}\, V^2$ *at that state*, and to the compressibility and bulk modulus. When temperature changes are considered, the natural configuration $\boldsymbol{a}(\boldsymbol{A})$ changes because of thermal expansion, and the elastic coefficients referred to and evaluated at the natural state play a role with respect to temperature that is similar to that of the previously discussed effective coefficients with respect to pressure.

In other contexts, it is desirable to have a temperature-independent reference configuration to which the deformation can be referred. This is the reason for introducing the "master" reference configuration of coordinates A_i in Table 28.1. The natural coordinates a_i are functions

$$a_i = f_i(T, A_1, A_2, A_3) \tag{32.1}$$

with

$$A_i = f_i(T_M, A_1, A_2, A_3) \tag{32.2}$$

where (32.2) involves the assumption that the master configuration has been chosen to be the unstressed configuration at the temperature T_M. The temperature-dependent strain (thermal expansion) from the master to the natural configuration, denoted by $\overline{V}_{ij}^M$, is obtained from $(28.2)_1$ by substituting $\boldsymbol{a}$ for $\boldsymbol{x}$:

$$\overline{V}_{ij}^M = \frac{1}{2}\left(\frac{\partial a_k}{\partial A_i}\frac{\partial a_k}{\partial A_j} - \delta_{ij}\right). \tag{32.3}$$

With $\mathscr{U}$ and $\mathscr{F}$ denoting the internal and free energies per unit *mass*, the master isentropic and isothermal elastic coefficients are defined by

$$c_{ijkm}^{MS} = \varrho_M \left(\frac{\partial^2 \mathscr{U}}{\partial V_{ij}^M \partial V_{km}^M}\right)_S; \quad c_{ijkm}^{MT} = \varrho_M \left(\frac{\partial^2 \mathscr{F}}{\partial V_{ij}^M \partial V_{km}^M}\right)_T. \tag{32.4}$$

We are concerned here with the relation between these coefficients and those referred to the natural configuration, defined by

$$c_{ijkm}^{NS} = \varrho_N \left(\frac{\partial^2 \mathscr{U}}{\partial V_{ij}^N \partial V_{km}^N}\right)_S; \quad c_{ijkm}^{NT} = \varrho_N \left(\frac{\partial^2 \mathscr{F}}{\partial V_{ij}^N \partial V_{km}^N}\right)_T, \tag{32.5}$$

in which V_{ij}^N has already been defined in $(28.2)_2$. Now

$$\frac{\partial V_{pq}^M}{\partial V_{ij}^N} = \frac{\partial a_i}{\partial A_p}\frac{\partial a_j}{\partial A_q}, \tag{32.6}$$

as follows from (31.21) upon replacing $\boldsymbol{a}$ by $\boldsymbol{A}$ and $\boldsymbol{X}$ by $\boldsymbol{a}$.

Therefore, the natural coefficients are connected to the master coefficients by the formula

$$c_{ijkm}^{NS} = \frac{\varrho_N}{\varrho_M}\frac{\partial a_i}{\partial A_p}\frac{\partial a_j}{\partial A_q}\frac{\partial a_k}{\partial A_r}\frac{\partial a_m}{\partial A_s} c_{pqrs}^{MS}. \tag{32.7}$$

This formula applies with the coefficients evaluated at any state, but our principal application involves evaluation at the natural state. A similar relation connects c_{ijkm}^{NT} with c_{pqrs}^{MT}. It should be recognized that (32.7) involves only the definitions (28.2), (32.4), and (32.5), not the assumption that the configurations are unstressed. Therefore, a relation like (32.7) connects coefficients referred to arbitrary configurations $\boldsymbol{A}$ and $\boldsymbol{a}$. The present application is special.

The superscript 0 will indicate evaluation at the natural state. Thus,

$$c_{ijkm}^{MS0} = \varrho_M (\partial^2 \mathscr{U} / \partial V_{ij}^M \partial V_{km}^M)_{S,\boldsymbol{a}} \tag{32.8}$$

is the isentropic coefficient, referred to the master configuration, evaluated at the natural state.

When the ultrasonics literature refers to elastic constants as a function of temperature, ordinarily it is the temperature dependence of the coefficients c_{ijkm}^{NS0} (referred to and evaluated at a *natural* state) that is meant, as these are the coefficients that are simply related to data on $\bar{\varrho} V^2$ (density times square of wave speed), taken as a function of temperature at zero stress. To obtain such data, the path length and density have to be corrected for thermal expansion. It has been common practice to do this in cases where it is felt that the precision of the data justifies the correction.

As in the case of measurements under pressure, however, a quantity that is more closely related to the experimental data is a wave transit time. The calculation of path length and density at the temperature of the measurement is a complication that is endured for the sake of reporting $c_{ijkm}^{NS\,0}$.

By analogy with the analysis of THURSTON and BRUGGER (1964) for measurements under stress, one can refer wave speeds to the dimensions in the master configuration $\boldsymbol{A}$, while linearizing the equation of motion about the natural configuration $\boldsymbol{a}$. The linearization is expressed as

$$P_{ij}^{M} - P_{ij}^{M\,0} = A_{ijkm}^{MS\,0}\left(\frac{\partial x_k}{\partial A_m} - \frac{\partial a_k}{\partial A_m}\right) + \cdots \tag{32.9}$$

where

$$A_{ijkm}^{MS} = \left(\frac{\partial P_{ij}^{M}}{\partial(\partial x_k/\partial A_m)}\right)_S \tag{32.10}$$

and the superscript 0 denotes evaluation at the natural state. The equation of motion $(28.3)_1$ then takes the linearized form

$$\varrho_M\,\ddot{x}_i = A_{ijkm}^{MS\,0}\,\frac{\partial^2 x_k}{\partial A_j\,\partial A_m}. \tag{32.11}$$

By analogy with (28.18) [or from (28.1), (28.2), and (28.4)],

$$A_{ijkm}^{MS} = \delta_{ik}\,t_{jm}^{M} + \frac{\partial x_i}{\partial A_p}\,\frac{\partial x_k}{\partial A_q}\,c_{pjqm}^{MS}. \tag{32.12}$$

At a natural state, the stress and thermodynamic tensions vanish, and the deformation gradients are of course $\partial a_i/\partial A_p$. Hence, the evaluation of (32.12) at the natural state gives

$$A_{ijkm}^{MS\,0} = \frac{\partial a_i}{\partial A_p}\,\frac{\partial a_k}{\partial A_q}\,c_{pjqm}^{MS\,0}. \tag{32.13}$$

With the small-amplitude wave represented as

$$x_i - a_i = u_i\,\exp[j\,\omega\,(t - M_q A_q/M)],$$

the propagation conditions are

$$(A_{ijkm}^{MS\,0} M_j M_m - \varrho_M M^2 \delta_{ik})\,u_k = 0. \tag{32.14}$$

The direction $\boldsymbol{M}$ and wave speed M are here referred to the *master* configuration $\boldsymbol{A}$. The actual direction and speed in the natural configuration are easily obtained by analogy with (28.32)–(28.34), upon replacing $\boldsymbol{a}$ by $\boldsymbol{A}$ and $\boldsymbol{X}$ by $\boldsymbol{a}$.

The transformation

$$u_k = (\partial a_k/\partial A_q)\,\mu_q \tag{32.15}$$

in (32.14) gives

$$m_{rs}\,\mu_s = \varrho_M\,M^2\,\mu_r, \tag{32.16}$$

$$m_{rs} = c_{rjqm}^{MS\,0}\,C_{qs}^{M\,0}\,M_j M_m, \tag{32.17}$$

$$C_{qs}^{M\,0} = \frac{\partial a_k}{\partial A_q}\,\frac{\partial a_k}{\partial A_s} = \delta_{qs} + 2\,V_{qs}^{M\,0}. \tag{32.18}$$

The quantities $\varrho_M M^2$ can be viewed as the eigenvalues of either the *symmetric* tensor $A_{ijkm}^{MS0} M_j M_m$ or of m_{rs}, which is not in general symmetric. In terms of *unit* vectors $\boldsymbol{M}$ and $\boldsymbol{\mu}$,

$$\begin{aligned} \varrho_M M^2 &= c_{rjqm}^{MS0} C_{qs}^{M0} M_j M_m \mu_r \mu_s \\ &= (c_{rjsm}^{MS0} + 2\, c_{rjqm}^{MS0} V_{qs}^{M0}) M_j M_m \mu_r \mu_s . \end{aligned} \tag{32.19}$$

When quantities are referred to a *natural* configuration, the wave is represented by (28.24) and the propagation condition is simply

$$(c_{ijkm}^{NS0} N_j N_m - \varrho_N W^2 \delta_{ik}) u_k = 0 . \tag{32.20}$$

Since there is no stress in the natural state, W and $\boldsymbol{N}$ are in this case the same as the actual wave speed V and direction $\boldsymbol{n}$. If $\boldsymbol{U}$ denotes a unit vector along the polarization direction, (32.20) implies

$$\varrho_N W^2 = c_{ijkm}^{NS0} N_j N_m U_i U_k . \tag{32.21}$$

The relations of W^2 to M^2, $\boldsymbol{N}$ to $\boldsymbol{M}$, and $\boldsymbol{U}$ to $\boldsymbol{\mu}$ may be listed as

$$\begin{aligned} & M^2/W^2 = L_0^2/L^2 = f_{\boldsymbol{M}}^2 ; \\ & f_{\boldsymbol{M}}^2 = \frac{\partial A_i}{\partial a_k} \frac{\partial A_j}{\partial a_k} M_i M_j ; \qquad f_{\boldsymbol{M}}^{-2} = \frac{\partial a_i}{\partial A_k} \frac{\partial a_j}{\partial A_k} N_i N_j ; \\ & N_j = M_i (\partial A_i / \partial a_j) / f_{\boldsymbol{M}} ; \qquad M_j = N_i (\partial a_i / \partial A_j) f_{\boldsymbol{M}} ; \\ & U_k = \lambda_{\boldsymbol{\mu}} \frac{\partial a_k}{\partial A_q} \mu_q ; \qquad \lambda_{\boldsymbol{\mu}} \mu_k = \frac{\partial A_k}{\partial a_q} U_q ; \\ & 1/\lambda_{\boldsymbol{\mu}}^2 = C_{qr}^{M0} \mu_q \mu_r ; \qquad \lambda_{\boldsymbol{\mu}}^2 = \frac{\partial A_k}{\partial a_q} \frac{\partial A_k}{\partial a_r} U_q U_r . \end{aligned} \tag{32.22}$$

From the above, $\varrho_N W^2$ can be expressed as

$$\varrho_N W^2 = \frac{\varrho_N}{\varrho_M} f_{\boldsymbol{M}}^{-2} \varrho_M M^2 \tag{32.23}$$

and hence (32.21) becomes

$$\frac{\varrho_N}{\varrho_M} \frac{\partial a_i}{\partial A_k} \frac{\partial a_j}{\partial A_k} N_i N_j \varrho_M M^2 = c_{ijkm}^{NS0} N_j N_m U_i U_k . \tag{32.24}$$

The basic repetition frequency data is of course simply related to the velocity M referred to master dimensions. Thus,

$$\varrho_M M^2 = \varrho_M M_M^2 \left(\frac{f(T)}{f(T_M)} \right)^2 \tag{32.25}$$

where M_M denotes the wave speed at the reference temperature T_M, and $f(T)$ is the measured frequency that corresponds to the reciprocal of a transit time between opposite faces of the specimen.

The person who wishes to extract material coefficients from the basic data $f(T)/f(T_M)$ now has the choice of extracting master coefficients from (32.19) or natural coefficients from (32.24). Either calculation involves knowledge of the deformation (thermal expansion) from $\boldsymbol{A}$ to $\boldsymbol{a}$.

Because the deformation resulting from thermal expansion is typically small, it is ordinarily sensible to approximate the factor $\varrho_N f_{\boldsymbol{M}}^{-2}/\varrho_M$ in (32.23) only to first order in the displacement gradients from the master to the natural state. With $\bar{u} = \boldsymbol{a} - \boldsymbol{A}$, the approximate first-order expressions are

$$\begin{aligned} \varrho_N/\varrho_M &\doteq 1 - \Delta , \\ \Delta &= \partial \bar{u}_i / \partial A_i , \\ f_{\boldsymbol{M}}^{-2} &\doteq 1 + 2 N_i N_k \partial \bar{u}_i / \partial A_k . \end{aligned} \tag{32.26}$$

Hence,

$$\frac{\varrho_N}{\varrho_M} f_M^{-2} \doteq 1 - \Delta + 2 N_i N_k \, \partial \bar{u}_i / \partial A_k . \tag{32.27}$$

33. Derivatives with respect to temperature.

The derivatives with respect to temperature that enter thermodynamic formulas such as (10.29) are derivatives of coefficients that are referred to a temperature-independent reference configuration. Hence, the temperature derivatives of c_{ijkm}^{MS0} are of interest. Simple exact formulas relating the derivatives to experimental data are obtained only when the derivatives are evaluated at the reference state. By a procedure analogous to that used in the derivation of (29.12), we can operate on (32.16)–(32.18) and evaluate derivatives at the master state $(\bar{T} = T_M)$ to obtain

$$\begin{aligned}(\varrho_M M^2)'_M &= (\mu_r \, m'_{rs} \, \mu_s)_M \\ &= \left[M_j \, M_m \, \mu_r \, \mu_s \left(\frac{\partial c_{rjsm}^{MS0}}{\partial T} + \frac{\partial C_{qs}^{M0}}{\partial T} \, c_{rjqm}^{MS0} \right) \right]_{\bar{T} = T_M}\end{aligned} \tag{33.1}$$

where the prime now refers to the derivative with respect to temperature. By analogy with (29.14),

$$(M_j \, M_m \, \mu_r \, \mu_s \, c_{rjqm}^{MS0})_{\bar{T} = T_M} = (\varrho_M \, M^2 \, \mu_q \, \mu_s)_{\bar{T} = T_M} . \tag{33.2}$$

Hence, (33.1) reduces to

$$(\varrho_M M^2)_{\bar{T} = T_M} = \left[\mu_r \, \mu_s \left(2 \alpha_{rs}^{M0} \, \varrho_M \, M^2 + M_j \, M_m \, \frac{\partial c_{rjsm}^{MS0}}{\partial T} \right) \right]_{\bar{T} = T_M} , \tag{33.3}$$

since at the master state

$$\frac{\partial}{\partial T} (C_{qs}^{M0}) = \frac{\partial}{\partial T} (\delta_{qs} + 2 V_{qs}^{M0}) = 2 \alpha_{qs}^{M0} , \tag{33.4}$$

where α_{qs}^{M0} is the expansion coefficient at zero stress, referred to the master configuration. It should be remarked that since the evaluation of derivatives in (33.1)–(33.3) is at a coincident master and natural state, the directions of propagation and polarization in the final formulas can equally correctly be denoted by the $\boldsymbol{N}$ and $\boldsymbol{U}$ of (32.21). Similarly, at this state, $M = W = V$.

When (32.25) is used to obtain an experimental value of the temperature derivative $(\varrho_M M^2)'_M$, Eq. (33.3) for the determination of the temperature derivatives becomes

$$2 \varrho_M \, M_M^2 \left(\frac{f'(T_M)}{f(T_M)} - \alpha_{rs}^{M0} \, \mu_r \, \mu_s \right)_{\bar{T} = T_M} = \mu_r \, \mu_s \, M_j \, M_m \left(\frac{\partial c_{rjsm}^{MS0}}{\partial T} \right)_{\bar{T} = T_M} . \tag{33.5}$$

Eq. (33.5) is suitable for determining the temperature derivatives of the master elastic constants directly from the experimental data. However, its convenience is reduced by the necessity of choosing the master configuration to coincide with the natural one at the temperature of interest.

A review of the derivation shows that the only reason for restricting (33.1) to $\bar{T} = T_M$ is to take care of the possibility that the eigenvector $\boldsymbol{\mu}$ depends on the temperature. In many important practical cases, $\boldsymbol{\mu}$ remains constant, independent of temperature, and in such cases, (33.5) can be replaced by

$$2 \varrho_M \, M_M^2 \left(\frac{f(T) \, f'(T)}{[f(T_M)]^2} - \alpha_{rs}^{M0} \, \mu_r \, \mu_s \right) = \mu_r \, \mu_s \, M_j \, M_m \, \frac{\partial c_{rjsm}^{MS0}}{\partial T} \tag{33.6}$$

in which the expansion coefficient and temperature derivative that appear can be evaluated at a natural state of arbitrary temperature.

When $\boldsymbol{N}$ and $\boldsymbol{U}$ are both constant in (32.21), as in many cases studied experimentally, the temperature derivatives of c_{ijkm}^{NS0} are related to the temperature derivatives of $\varrho_N W^2$ by the same formulas that relate values of $\varrho_N W^2$ to the elastic coefficients:

$$(\varrho_N W^2)' = N_j N_m U_i U_k \frac{\partial c_{ijkm}^{NS0}}{\partial T}. \tag{33.7}$$

But since it is $(\varrho_M M^2)'$ that is most simply related to the frequency data, the correction for the effect of thermal expansion on path length and density may be of interest. Now

$$\frac{\partial}{\partial T}\left(\frac{\varrho_N}{\varrho_M}\right) = \frac{1}{\varrho_M}\frac{\partial \varrho_N}{\partial T} = -\beta\,\varrho_N/\varrho_M \tag{33.8}$$

where β is the expansion coefficient defined in (11.15). Since

$$f_{\boldsymbol{M}}^2 = (\bar{C}_{ij}^{M0})^{-1} M_i M_j, \tag{33.9}$$

the temperature derivative of $f_{\boldsymbol{M}}$ at the master state is

$$f'_{\boldsymbol{M}}(T_M) = -\alpha_{ij}^{M0} M_i M_j. \tag{33.10}$$

Hence, from (32.23) and the above, the connection between temperature derivatives of $\varrho_N W^2$ and $\varrho_M M^2$ at the master state is

$$(\varrho_N W^2)'_M = [(\varrho_M M^2)' + (2\,\alpha_{ij}^{M0} M_i M_j - \beta)\,\varrho_M M^2]_M. \tag{33.11}$$

By differentiating (32.21) and using (32.25) and (33.11), treating $\boldsymbol{N}$ and $\boldsymbol{U}$ as constant. we find

$$\varrho_M M_M^2 \left(2\frac{f'(T_M)}{f(T_M)} + 2\alpha_{ij}^{M0} M_i M_j - \beta\right)_{T=T_M} = N_j N_m U_i U_k \left(\frac{\partial c_{ijkm}^{NS0}}{\partial T}\right)_{T=T_M}. \tag{33.12}$$

The general relation between temperature derivatives of the master and natural coefficients can be obtained from (32.7) if the deformation gradients and their temperature derivatives are known. In terms of the volumetric expansion coefficient β of (11.15) and thermal expansion parameters

$$f_{rk} \equiv \frac{\partial}{\partial T}\left(\frac{\partial a_r}{\partial A_k}\right), \tag{33.13}$$

the result of evaluating the temperature derivative of (32.7) at the master state is

$$\begin{aligned}&\left[\frac{\partial}{\partial T}(c_{ijkm}^{NS0} - c_{ijkm}^{MS0})\right]_{T=T_M}\\ &\quad = (-\beta c_{ijkm}^S + f_{ip} c_{pjkm}^S + f_{jq} c_{iqkm}^S + f_{kr} c_{ijrm}^S + f_{ms} c_{ijks}^S)_{T=T_M}.\end{aligned} \tag{33.14}$$

On the right hand side of (33.14), the superscript M or N to indicate the reference configuration has been dropped, since at the coincident master and natural states, there is no distinction between the values of the coefficients themselves, but only between their temperature derivatives.

The relation of f_{ij} to α_{ij}^M at the master state is analogous to the relation of b_{kr} to $-s_{krtt}$ [cf. (31.31) and (31.39)]. Hence, b can be replaced by f and $-s$ by α in (31.39) to obtain

$$\begin{aligned} &f_{11} = \alpha_1, \quad f_{22} = \alpha_2, \quad f_{33} = \alpha_3,\\ &f_{12} = \alpha_6, \quad f_{23} = 0, \quad f_{31} = \alpha_5,\\ &f_{21} = 0, \quad f_{32} = \alpha_4, \quad f_{13} = 0,\end{aligned} \tag{33.15}$$

where the coefficients are understood to be referred to and evaluated at the master state. The difference between the temperature derivatives at the master state is then obtained by substitution from (33.15) into (33.14). The result is just like (31.40) without the first three terms in the Kronecker deltas and with appropriate changes of symbols ($-\beta$ for χ, α_λ for $-s_\lambda$, etc.). For all except monoclinic and triclinic cases, the results can be read from Table 31.2 by omitting the ones and substituting α_λ for $-s_\lambda$.

An identical formula connects the derivatives of the isothermal coefficients. It is only necessary to replace the superscript S by T.

THURSTON (1967) has discussed the relation of ultrasonic data to thermodynamic coefficients, including the relation between adiabatic and isothermal coefficients, while observing the distinction between specific heat and expansion coefficients at constant pressure and at constant thermodynamic tensions. Reviews of the relation of elastic coefficients to the thermal equilibrium properties of solids have been given by HOLDER and GRANATO (1971) and D. C. WALLACE (1970, 1972). Data may be found in LANDOLT-BÖRNSTEIN (1966, 1969).

L. Examples.

34. Elastic waves in crystals.

34.1. Introduction. Waves have been studied in crystals of all the eleven Laue groups but only recently have all 21 elastic coefficients of a triclinic crystal been measured. [The coefficients of the triclinic crystal $CuSO_4 \cdot 5\,H_2O$ have been determined by HAUSSÜHL and SIEGERT (1969) using Schaefer-Bergmann techniques and by KRISHNAN, RADHA, and GOPAL (1970) using the pulse echo method.] A veritable gold mine of information on wave propagation in crystals is contained in the book by F. I. FEDEROV (1968). Valuable practical accounts detailing the applications of ultrasonic measurements to solid state physics have been given by TRUELL and ELBAUM (1962) and TRUELL, ELBAUM, and CHICK (1969). The latter book includes material on experimental techniques and equipment in addition to a wealth of information on wave propagation and on loss mechanisms. ROBERT E. GREEN (1973) has provided a valuable review while TUCKER and RAMPTON (1972) treat gigahertz frequency ultrasonics as a tool in solid state physics.

A "mode" of plane wave propagation is specified by a propagation direction, which is normal to the planes of constant phase, and a polarization direction, which is the direction of the particle displacement for linearly polarized plane waves. For a given propagation direction, there are always three mutually perpendicular polarization directions, but in general, none of these directions is parallel or perpendicular to the propagation direction. For certain propagation directions, the acoustical tensor may have a double eigenvalue, in which case only one of the possible polarization directions (eigenvectors) is determined uniquely. This is the one corresponding to the remaining eigenvalue. Any direction perpendicular to the uniquely determined eigenvector is then a possible polarization direction for waves that propagate with $\bar{\varrho} V^2$ equal to the double eigenvalue.

Although waves in anisotropic media are generally neither longitudinal (polarized along the propagation direction) nor transverse (polarized perpendicularly to the propagation direction), there are special directions of propagation that allow propagation of pure longitudinal modes. It then follows from the orthogonality of the polarization directions that the other two modes are transverse. TRUESDELL (1966, 1968) has proved that there is always at least one such propagation direction in an anisotropically stressed crystal, and KOLODNER (1966) has shown that there are three such propagation directions in an unstressed crystal. Propagation directions for these pure modes have been investigated by BORGNIS (1955) and BRUGGER (1965)$_1$. BRUGGER's paper presents a complete discussion of such directions for all except the monoclinic and triclinic Laue groups. Some of the directions are determined from symmetry alone, and others depend on the values of the elastic coefficients.

In addition, there are special propagation directions that allow the propagation of one transverse mode, but the other two modes are neither longitudinal nor transverse. It can be shown [FEDOROV, 1968, Sect. 16] that if the propagation direction lies in a mirror plane or in a plane perpendicular to a rotation axis of even order, then at least one of the three modes for that direction will be transverse with its polarization direction normal to that plane. In other words, directions normal to a mirror plane or along an even-order rotation axis are *polarization* directions for transverse waves whose propagation direction can lie anywhere in the perpendicular plane. BRUGGER and THURSTON (1971) have pointed out that this fact can be put to use in the design of single-crystal folded-path ultrasonic delay lines.

When a mode is purely longitudinal, the direction of energy flux is along the propagation direction. For transverse modes, this is not necessarily the case. The direction of energy flux will be along the propagation direction if the directions of propagation and polarization can be interchanged.

In the remainder of this section, we shall consider relations that apply in cubic crystals, as these are important, and by far the easiest to discuss.

34.2. Special forms of the coefficients for cubic crystals. In the five classes of crystals called "cubic", the 6×6 matrix of elastic coefficients, referred to the crystallographic axes, has the form (cf. Table 16.4)

$$\|c_{\alpha\beta}\| = \begin{Vmatrix} c_{11} & c_{12} & c_{12} & 0 & 0 & 0 \\ c_{12} & c_{11} & c_{12} & 0 & 0 & 0 \\ c_{12} & c_{12} & c_{11} & 0 & 0 & 0 \\ 0 & 0 & 0 & c_{44} & 0 & 0 \\ 0 & 0 & 0 & 0 & c_{44} & 0 \\ 0 & 0 & 0 & 0 & 0 & c_{44} \end{Vmatrix}. \tag{34.1}$$

T. Y. THOMAS (1966) has expressed the tensor coefficient in terms of three material parameters λ, μ, α by the expression

$$\begin{gathered} c_{ijkm} = \lambda\delta_{ij}\,\delta_{km} + \mu(\delta_{ik}\,\delta_{jm} + \delta_{im}\,\delta_{jk}) + \alpha(\nu_{ai}\,\nu_{aj}\,\nu_{ak}\,\nu_{am}), \\ \lambda = c_{12}, \quad \mu = c_{44}, \quad \alpha = (c_{11} - c_{12} - 2c_{44}). \end{gathered} \tag{34.2}$$

In $(34.2)_2$, the two-subscript coefficients $c_{\alpha\beta}$ are referred to crystallographic axes, but in $(34.2)_1$ c_{ijkm} is referred to an arbitrarily oriented rectangular Cartesian system of axes, relative to which the direction of the three crystallographic axes are specified by unit orientation vectors $\boldsymbol{\nu}_a$, $a=1, 2, 3$, of components ν_{ai}. To refer $(34.2)_1$ to crystallographic axes, we may set $\nu_{ai} = \delta_{ai}$. Eq. (34.2) may be specialized to isotropic media by setting $\alpha = 0$.

In cubic crystals all second rank tensors, such as the expansion coefficients α_{ij} of Eq. (9.7), λ_{ij} of (9.8), or γ_{ij} of (10.40), are spherical (cf. Table 16.2). That is,

$$\alpha_{ij} = \tfrac{1}{3}\alpha_{kk}\,\delta_{ij}, \qquad \text{etc.} \tag{34.3}$$

Now the tensor form of (10.16) is

$$c^S_{ijkm} - c^T_{ijkm} = T\lambda_{ij}\lambda_{km}/C_V, \tag{34.4}$$

which, for cubic crystals reduces to

$$c^S_{ijkm} - c^T_{ijkm} = \frac{T}{9C_V}(\lambda_{ss})^2\,\delta_{ij}\,\delta_{km}. \tag{34.5}$$

By comparison with (34.2), we see that there is no distinction between isentropic and isothermal values of the material parameters μ and α, but

$$\lambda^S - \lambda^T = \frac{T}{9C_V}(\lambda_{ss})^2. \tag{34.6}$$

The tensor form of (10.13) can then be reduced to

$$\begin{aligned} C_r\,\lambda_{km} &= \alpha_{ij}\,c_{ijkm} = \tfrac{1}{3}\alpha_{nn}\,\delta_{ij}[\lambda^S\,\delta_{ij}\,\delta_{km} + \mu(\delta_{ik}\,\delta_{jm} + \delta_{im}\,\delta_{jk}) + \alpha\,\delta_{ai}\,\delta_{aj}\,\delta_{ak}\,\delta_{am}] \\ &= \tfrac{1}{3}\alpha_{nn}(3\lambda^S + 2\mu + \alpha)\,\delta_{km} = \beta B^S\,\delta_{km}, \end{aligned} \tag{34.7}$$

where

$$\begin{aligned} B^S &= \tfrac{1}{3}(3\lambda^S + 2\mu + \alpha), \\ \beta &= \alpha_{nn}. \end{aligned} \tag{34.8}$$

It is easily verified that B^S is the isentropic bulk modulus defined by (11.18), and that the sum α_{nn}, the trace of the expansion coefficient matrix, is the volumetric expansion coefficient, β. From (34.7),

$$\lambda_{ss} = 3\beta B^S/C_r \tag{34.9}$$

and finally

$$\lambda^S - \lambda^T = \frac{T}{9C_V}(\lambda_{ss})^2 = T\beta^2(B^S)^2/C_r^2\,C_V = T\beta^2(B^S)/C_r\,C_p. \tag{34.10}$$

Consistent with (34.8) and the analogous expression for B^T, the difference in (34.10) is precisely the difference $B^S - B^T$, as is easily verified from (11.43) and (11.44). It follows easily from (34.2) that the differences $c^S_{11} - c^T_{11}$ and $c^S_{12} - c^T_{12}$ also have this same value:

$$\begin{aligned} c^S_{11} - c^T_{11} &= c^S_{12} - c^T_{12} = \lambda^S - \lambda^T = B^S - B^T = B^S(C_r - 1)/C_r, \\ C_r &= 1 + T\beta\gamma, \\ \gamma &\equiv \beta B^S\,J/C_p. \end{aligned} \tag{34.11}$$

Here C_r is the specific heat ratio C_p/C_V, and γ is the dimensionless Grüneisen constant as defined also by (11.43). The specific heat C_p is here referred to unit reference volume.

If the initial stress is zero, the Taylor series expansion of $U(S_0, V_{ij}) - U(S_0, 0)$ begins

$$U(S_0, V_{ij}) - U(S_0, 0) = c^S_{ijkm}\,V_{ij}\,V_{km}, \tag{34.12}$$

where the coefficients c^S_{ijkm} are the *constant* initial values.

34.3. Conditions for a positive definite strain energy in cubic crystals. Let us now find the conditions on (λ^S, μ, α) in (34.2) that make (34.12) positive for arbitrary strains V_{ij}. From (34.2), with $\nu_{ai} = \delta_{ai}$,

$$c^S_{ijkm}\,V_{ij}\,V_{km} = \lambda^S(V_{ss})^2 + 2\mu V_{jk}\,V_{jk} + \alpha(V_{11}^2 + V_{22}^2 + V_{33}^2). \tag{34.13}$$

Separating V_{ij} into its spherical and deviator parts by writing

$$V_{ij}=\tfrac{1}{3}V_{ss}\,\delta_{ij}+v_{ij}, \tag{34.14}$$

where v_{ss}, the trace of v_{ij}, is zero, we put (34.13) in the form

$$c^S_{ijkm}\,V_{ij}\,V_{km}=B^S V_{ss}^2+2\mu v_{jk}\,v_{jk}+\alpha(v_{11}^2+v_{22}^2+v_{33}^2). \tag{34.15}$$

Now $v_{jk}\,v_{jk}$ is the sum of the squares of the elements of the deviator v_{ij}. Thus,

$$v_{jk}\,v_{jk}=(v_{11}^2+v_{22}^2+v_{33}^2)+2(v_{12}^2+v_{23}^2+v_{31}^2). \tag{34.16}$$

Finally (34.15) becomes

$$c^S_{ijkm}\,V_{ij}\,V_{km}=B^S V_{ss}^2+4\mu(v_{12}^2+v_{23}^2+v_{31}^2)+(\alpha+2\mu)(v_{11}^2+v_{22}^2+v_{33}^2). \tag{34.17}$$

Now the functions of strain in (34.17) are never negative, and they can be varied independently. Hence, we can read off the following condition for a positive definite quadratic contribution to the strain energy:

$$\begin{aligned} B^S=\tfrac{1}{3}(\alpha+3\lambda^S+2\mu)&>0,\\ \mu&>0,\\ \alpha+2\mu&>0. \end{aligned} \tag{34.18}$$

If $\alpha=0$, these conditions reduce to those given by GURTIN (1972)$_2$ [this encyclopedia, Vol. VIa/2, p. 87] for the isotropic case.

34.4. Acoustical tensor for arbitrary propagation directions in a cubic crystal. Let the initial stress be zero, and consider wave propagation in the direction $\boldsymbol{N}$ of direction cosines $N_1=l$, $N_2=m$, $N_3=n$. For the purely elastic case, (25.39) becomes

$$\begin{aligned} Q_{jm}=c_{ijkm}\,N_i\,N_k&=l^2\,c_{1j1m}+m^2\,c_{2j2m}+n^2\,c_{3j3m}\\ &+lm(c_{1i2n}+c_{2i1n})+mn(c_{2i3n}+c_{3i2n})+nl(c_{3i1n}+c_{1i3n}). \end{aligned} \tag{34.19}$$

Eq. (34.19) is for any symmetry. When restricted to cubic symmetry (34.1), the matrix Q_{jm} takes the form

$$Q=\begin{Vmatrix} l^2\,c_{11}+(m^2+n^2)\,c_{44} & lm(c_{12}+c_{44}) & nl(c_{12}+c_{44})\\ lm(c_{12}+c_{44}) & m^2\,c_{11}+(l^2+n^2)\,c_{44} & mn(c_{12}+c_{44})\\ nl(c_{12}+c_{44}) & mn(c_{12}+c_{44}) & n^2\,c_{11}+(l^2+m^2)\,c_{44} \end{Vmatrix} \tag{34.20}$$

or, in terms of (λ,μ,α),

$$Q=\begin{Vmatrix} \mu+l^2(\alpha+\lambda+\mu) & lm(\lambda+\mu) & nl(\lambda+\mu)\\ lm(\lambda+\mu) & \mu+m^2(\alpha+\lambda+\mu) & mn(\lambda+\mu)\\ nl(\lambda+\mu) & mn(\lambda+\mu) & \mu+n^2(\alpha+\lambda+\mu) \end{Vmatrix}. \tag{34.21}$$

34.5. Directions for purely longitudinal and transverse waves in cubic crystals. Longitudinal waves can exist when the propagation direction (l,m,n) is an eigenvector of the matrix Q. It is apparent from (34.21) that Q can be decomposed into

$$Q=\mu I+(\lambda+\mu)Q^{(1)}+\alpha Q^{(2)} \tag{34.22}$$

where I is the unit matrix, and

$$Q^{(1)}=\begin{bmatrix} l^2 & lm & nl\\ lm & m^2 & mn\\ nl & mn & n^2 \end{bmatrix},\quad Q^{(2)}=\begin{bmatrix} l^2 & 0 & 0\\ 0 & m^2 & \\ 0 & 0 & n^2 \end{bmatrix}. \tag{34.23}$$

Recall that the eigenvectors of $Q(\boldsymbol{N})$ are the possible particle displacement directions for a wave propagating in the direction $\boldsymbol{N}$, and the eigenvalues are the values of $\bar{\varrho}\, c^2$ for the corresponding modes. Since any vector is an eigenvector of I, vectors that are eigenvectors of both $Q^{(1)}$ and $Q^{(2)}$ are eigenvectors of Q. Moreover, if $\lambda^{(1)}$ and $\lambda^{(2)}$ are respectively the eigenvalues of $Q^{(1)}$ and $Q^{(2)}$ belonging to some eigenvector, it follows from (34.22) that

$$\bar{\varrho}\, c^2 = \mu + (\lambda+\mu)\,\lambda^{(1)} + \alpha\,\lambda^{(2)} \tag{34.24}$$

is the corresponding eigenvalue of Q.

Now $Q^{(1)}$ is the outer product $N\tilde{N}$ where N is the single-column matrix of the components (l, m, n) of $\boldsymbol{N}$, and $\tilde{N}$ denotes its transpose. Since $\tilde{N}N=1$, it follows that $N\tilde{N}N=N$, and hence N is an eigenvector of $Q^{(1)}$ with eigenvalue 1. Moreover any vector b that is orthogonal to N is an eigenvector of $Q^{(1)}$ with eigenvalue zero because $\tilde{N}b=0$. If $\alpha=0$ (isotropic medium), it then follows from (34.22) that N and b are eigenvectors of Q with eigenvalues $(\lambda+2\mu)$ and μ respectively. This result corresponds to the well-known fact that longitudinal waves are propagated in an isotropic medium with speed c given by $\bar{\varrho}\, c^2=\lambda+2\mu$, and transverse waves with $\bar{\varrho}\, c^2=\mu$.

When $\alpha \neq 0$ (cubic crystal), longitudinal waves can be propagated in directions (l, m, n) for which (l, m, n) is an eigenvector of $Q^{(2)}$. There are three such cases that are distinguishable:

(1) Propagation in a $\langle 100\rangle$-direction,* i.e., along a crystal axis. The six possible values of $(l. m, n)$ are $(1, 0, 0)$, $(0, 1, 0)$, $(0, 0, 1)$ and their negatives. In this case, the eigenvalue of $Q^{(2)}$ belonging to N is 1, and the corresponding eigenvalue of Q is therefore $(\lambda+2\mu+\alpha)$. Thus, longitudinal waves are propagated along a crystal axis with speed c given by $\bar{\varrho}\, c^2=(\lambda+2\mu+\alpha)$. Any vector orthogonal to N is an eigenvector (of both $Q^{(1)}$ and $Q^{(2)}$) with eigenvalue zero, and hence linearly polarized transverse waves with $\bar{\varrho}\, c^2=\mu$ may be propagated along a crystal axis with direction of polarization lying anywhere in the plane orthogonal to N.

(2) Propagation along a $\langle 111\rangle$-direction, i.e., along the cube body-diagonal. In this case, $Q^{(2)}=I/3$, and any vector is an eigenvector with eigenvalue $\frac{1}{3}$. Thus in view of (34.22), the eigenvalue of Q for longitudinal waves along a $\langle 111\rangle$-direction is $\bar{\varrho}\, c^2=\frac{1}{3}\alpha+\lambda+2\mu$. Similarly, transverse waves linearly polarized along any direction perpendicular to $\langle 111\rangle$ have $\bar{\varrho}\, c^2=\frac{1}{3}\alpha+\mu$.

(3) Propagation along a $\langle 110\rangle$-direction, i.e., along a cube face-diagonal. In this case, the eigenvalue of $Q^{(2)}$ belonging to N is $\frac{1}{2}$. Therefore, longitudinal waves are propagated along a $\langle 110\rangle$-direction with $\bar{\varrho}\, c=\frac{1}{2}\alpha+\lambda+2\mu$. $Q^{(2)}$ has the distinct eigenvalue zero belonging to the eigenvector along the crystal axis perpendicular to the $\langle 110\rangle$-direction under consideration, and the double eigenvalue $\frac{1}{2}$ applies to any vector orthogonal to this axis. Hence, transverse waves linearly polarized along $[001]$ can be propagated along $[110]$ with $\bar{\varrho}\, c^2=\mu$, while transverse waves linearly polarized along $[1\bar{1}0]$ can be propagated along $[110]$ with $\bar{\varrho}\, c^2=\frac{1}{2}\alpha+\mu$.

The above three cases exhaust the number of propagation directions for which there are *three* pure modes, one longitudinal and two transverse. There are other propagation directions that allow a single purely transverse mode, but the other two modes are neither purely longitudinal nor purely transverse. It is easy to see that if $l=0$, then $(1, 0, 0)$ is an eigenvector of Q with eigenvalue μ. It follows that a

* The Miller indices are enclosed in parentheses ( ), braces { }, brackets [ ], or angular parentheses ⟨ ⟩ for a plane, set of planes equivalent by symmetry, a direction, or set of equivalent directions, respectively [KITTEL, 1956, p. 34].

purely transverse wave linearly polarized along a crystal axis can exist with propagation direction anywhere in the plane perpendicular to that axis. For such a wave, $\bar{\varrho} c^2 = \mu$. Further, it can be seen that the [110]-direction is an eigenvector for any propagation direction normal to it. The corresponding eigenvalue of $Q^{(2)}$ is $\frac{1}{2}(1-n^2)$. Hence, for any propagation direction normal to a $\langle 110 \rangle$-direction, there is a pure transverse wave, linearly polarized along that $\langle 110 \rangle$-direction with $\bar{\varrho} c^2 = \frac{1}{2}(1-n^2)\alpha + \mu$. Here n is the cosine of the angle between the propagation direction and the crystal axis that is normal to the polarization direction. As this angle can range from 0° to 360°, the value of $\bar{\varrho}\, c^2$ for this family of modes ranges from μ (for propagation along $\langle 001 \rangle$) to $\mu + \frac{1}{2}\alpha$ (for propagation along $\langle 110 \rangle$).

Because the $\langle 111 \rangle$ and $\langle 110 \rangle$ directions have special properties, samples in the form of rectangular solids are sometimes prepared with facets having the three mutually perpendicular directions $[111]$, $[1\bar{1}0]$, $[11\bar{2}]$. For propagation along $[11\bar{2}]$, we have a special case of the above pure transverse wave polarized along $[1\bar{1}0]$ with $n^2 = \frac{2}{3}$. Hence, $\bar{\varrho} c^2 = \mu + \alpha/6$. (Unless $\alpha = 0$, these directions of propagation and polarization cannot be interchanged, because for propagation along $[1\bar{1}0]$, the possible polarization directions are uniquely determined as $[1\bar{1}0]$, $[110]$, and $[001]$. The $[11\bar{2}]$ direction is *not* a possible polarization direction for propagation along $[1\bar{1}0]$ unless $\alpha = 0$.) For propagation along $[11\bar{2}]$, the other two possible polarization directions must lie in the plane normal to $[1\bar{1}0]$ because any symmetric 3×3 matrix Q has three mutually perpendicular eigenvectors. However, neither $[11\bar{2}]$ nor $[111]$ is an eigenvector of Q in general, and hence the other two modes are neither purely longitudinal nor transverse. The one whose polarization direction lies closer to the propagation direction is called quasi-longitudinal, and the other quasi-transverse, or quasi-shear.

Pure modes are ordinarily "cleaner" to work with experimentally because it is difficult to excite and detect a quasi-mode without also exciting and detecting its partner.

The preceding results on pure modes in cubic crystals are summarized in Table 34.1.

Table 34.1. *Pure modes in cubic crystals.*

Propagation direction	Displacement direction	$\lambda^{(1)}$	$\lambda^{(2)}$	$w(\lambda, \mu, \alpha)$	$w \equiv \bar{\varrho} c^2$ $w(c_{11}, c_{12}, c_{44})$
[100]	[100]	1	1	$\lambda + 2\mu + \alpha$	c_{11}
[100]	Any direction ⊥ to [100]	0	0	μ	c_{44}
[111]	[111]	1	$\frac{1}{3}$	$\lambda + 2\mu + \frac{1}{3}\alpha$	$\frac{1}{3}(c_{11} + 2c_{12} + 4c_{44})$
[111]	Any direction ⊥ to [111]	0	$\frac{1}{3}$	$\mu + \frac{1}{3}\alpha$	$\frac{1}{3}(c_{11} - c_{12} + c_{44})$
$[1\bar{1}0]$	$[1\bar{1}0]$	1	$\frac{1}{2}$	$\lambda + 2\mu + \frac{1}{2}\alpha$	$\frac{1}{2}(c_{11} + c_{12} + 2c_{44})$
Any direction ⊥ to [001]	[001]	0	0	μ	c_{44}
$[1\bar{1}0]$	[110]	0	$\frac{1}{2}$	$\mu + \frac{1}{2}\alpha$	$\frac{1}{2}(c_{11} - c_{12})$
Any direction ⊥ to $[1\bar{1}0]$	$[1\bar{1}0]$	0	$\frac{1}{2}(1 - n^2)$	$\mu + \frac{1}{2}(1 - n^2)\alpha$	$n^2 c_{44} + \frac{1}{2}(1 - n^2)(c_{11} - c_{12})$
[001]	$[1\bar{1}0]$	0	0	μ	c_{44}
$[11\bar{2}]$	$[1\bar{1}0]$	0	$\frac{1}{6}$	$\mu + \alpha/6$	$\frac{1}{6}(c_{11} - c_{12} + 4c_{44})$
[111]	$[1\bar{1}0]$	0	$\frac{1}{3}$	$\mu + \alpha/3$	$\frac{1}{3}(c_{11} - c_{12} + c_{44})$
[110]	$[1\bar{1}0]$	0	$\frac{1}{2}$	$\mu + \alpha/2$	$\frac{1}{2}(c_{11} - c_{12})$

34.6. Determination of elastic constants. One convenient sample orientation for the determination of elastic constants of cubic crystals is with (110), $(1\bar{1}0)$, and (001) facets. From measurements on the three modes that can be propagated along [110] or $[1\bar{1}0]$, and the two independent ones along [001], the three independent coefficients can be calculated with two "cross-checks".

Another possibility makes use of the (111), $(1\bar{1}0)$, $(11\bar{2})$ triad. In this case, three modes can be propagated along $[1\bar{1}0]$, two independent ones along [111], and one pure mode along $[11\bar{2}]$. Here the coefficients can be determined from measurements on transverse waves alone, whereas in the other case, both transverse and longitudinal waves have to be used.

An important practical consideration is the effect of crystal misorientation on measured wave speeds. This problem has been admirably covered by TRUELL, ELBAUM, and CHICK (1969), who provide tables for practical use.

There are often experimental difficulties associated with measurements on the transverse modes propagating along $\langle 111\rangle$. This has to do with sensitivity to slight misorientation. Even with a perfectly oriented crystal facet, the wave front itself cannot be a perfect plane because of the finite size of the sample and transducer. Hence, it is inevitable that portions of the wave front will deviate from the desired direction. In the case of transverse (or quasi-transverse) waves propagating along a direction near $\langle 111\rangle$, the wave speed does not have a maximum or minimum with respect to changes of propagation direction [WATERMAN, 1959; TRUELL, ELBAUM, and CHICK, 1969]. This causes the received signal to be more spread out in time than is the case when the velocity has an extremum in the propagation direction.

The transverse waves along a $\langle 111\rangle$ direction also constitute examples of modes in which the direction of energy flux does not coincide with the propagation direction. As the polarization direction is varied in the plane perpendicular to [111], the energy flux direction describes a cone of circular cross section. This phenomenon, termed internal conical refraction, has been a subject of both experimental and theoretical interest [WATERMAN, 1959; MCSKIMIN and BOND, 1966; FEDOROV, 1968, Sects. 22 and 23; TRUELL, ELBAUM, and CHICK, 1969, p. 36].

35. Thermoviscoelastic waves in cubic crystals.

Let us consider small-amplitude time-sinusoidal waves in cubic crystals subject to the special thermoviscoelastic model of Sect. 23. Under the additional assumption that Y_{ijkm} of (25.39) can be expressed in the form (34.2), it can be shown that the directions of propagation and polarization for linearly polarized pure modes in the initially unstressed elastic case, indicated in Table 34.1, are also directions for linearly polarized pure modes in the initially unstressed thermoviscoelastic case. This result follows from the form of Q in (25.39), which, subject to the stated assumption, can still be represented in the form (34.22), but now the scalar coefficients are complex functions of the frequency ω.

The thermal term in (25.39) has the same eigenvectors as $Q^{(1)}$ in (34.22) because it is a scalar function of ω times $\gamma_{ij}\gamma_{km}N_i N_k$. For cubic crystals, from (34.7) and Table (10.1), in view of $\gamma_{km}=\lambda_{km}/C_V=C_r\,\lambda_{km}/C_p$,

$$\gamma_{ij}\gamma_{km}N_i N_k=N_j N_m(\beta B^S/C_p)^2. \tag{35.1}$$

But $N_j N_m$ is simply the matrix $Q^{(1)}$. Since all the polarization directions previously considered are eigenvectors of $Q^{(1)}$, they are also polarization directions for the thermoelastic case.

In the thermoviscoelastic case, let us generalize the decomposition (34.22) to

$$Q=(\mu+\mu^V)I+(\lambda+\mu+\lambda^V+\mu^V+h)Q^{(1)}+(\alpha+\alpha^V)Q^{(2)} \tag{35.2}$$

where h is the thermal contribution to the coefficient of $Q^{(1)}$, and μ^V, λ^V, and α^V are derived from Y_{ijkm} in the same way that μ, λ, and α are related to c_{ijkm} in (34.2). To find the viscoelastic and thermal contributions to the eigenvalues of Q, we need only identify h, μ^V, λ^V, α^V, and then apply the rule (34.24) to the generalized decomposition (35.2). That is, instead of (34.24),

$$\bar{\varrho} c^{*2} = \mu + \mu^V + (\lambda + \mu + \lambda^V + \mu^V + h)\,\lambda^{(1)} + (\alpha + \alpha^V)\,\lambda^{(2)}. \tag{35.3}$$

The eigenvalues of $Q^{(1)}$ and $Q^{(2)}$, namely $\lambda^{(1)}$ and $\lambda^{(2)}$, can be read off from Table (34.1) as the coefficients of λ and α respectively.

The thermal contribution to the coefficient of $Q^{(1)}$ in (35.2) is, from (35.1) and (25.39),

$$h = \frac{(\beta B^S/C_p)^2 \, T_0 \, j\omega \bar{\varrho} L_{\boldsymbol{N}}}{\bar{\varrho} c^{*2} - j\omega \bar{\varrho} L_{\boldsymbol{N}}/C_V}. \tag{35.4}$$

$L_{\boldsymbol{N}}$ is defined by (25.15) in which $\bar{L}(s)$ is given by (24.17) for the special model under consideration. For cubic crystals, we assume the 3×3 matrices q and $\varkappa$ to have the spherical symmetry listed for cubic crystals in Table 16.2. In this case, (24.7) reduces to a scalar coefficient times the unit matrix. Allowing the symbols q and $\varkappa$ to play the dual role of matrices in (24.17) and now scalar coefficients in

$$q_{ij} = q\,\delta_{ij}, \qquad \varkappa_{ij} = \varkappa\,\delta_{ij}, \tag{35.5}$$

(24.17) becomes

$$\bar{L}(s) = \frac{q\varkappa I}{(s+q)}, \quad \text{or} \quad \bar{L}_{ki}(s) = \frac{q\varkappa}{s+q}\,\delta_{ki}. \tag{35.6}$$

Since $\delta_{ki} N_k N_i = 1$, independent of the direction $\boldsymbol{N}$, $L_{\boldsymbol{N}}(\omega)$ in (25.15) is simply

$$L_{\boldsymbol{N}}(\omega) = \frac{q\varkappa}{q + j\omega}. \tag{35.7}$$

The assumed symmetry has eliminated the dependence of $L_{\boldsymbol{N}}$ on $\boldsymbol{N}$. From (35.4) and (35.7), after some rearrangement and making use of (11.43) and (11.44), putting $J=1$ and noting that $C_V = C_J$ for cubic crystals, we find

$$h = B\,\frac{j\beta T_0 \gamma \dfrac{\bar{\varrho}\omega\varkappa}{\bar{\varrho} c^{*2} C_V}}{1 + j\left(\dfrac{\omega}{q} - \dfrac{\bar{\varrho}\omega\varkappa}{\bar{\varrho} c^{*2} C_V}\right)}. \tag{35.8}$$

Let us next identify μ^V, λ^V, and α^V. Y_{ijkm} is the tensor representation of $Y(j\omega)$ where $Y(s)$ is the 6×6 matrix in $(24.2)_2$. Assuming the 6×6 matrices p and A to have the symmetry indicated for cubic crystals, we first work out the three distinct nonzero elements of $(sI+p)^{-1}_{\alpha\beta}$. They are

$$\begin{aligned}
(sI+p)^{-1}_{11} &= \frac{(s+p_{11}+p_{12})}{(s+p_{11}+2p_{12})\,(s+p_{11}-p_{12})},\\
(sI+p)^{-1}_{12} &= \frac{-p_{12}}{(s+p_{11}+2p_{12})\,(s+p_{11}-p_{12})},\\
(sI+p)^{-1}_{44} &= \frac{1}{s+p_{44}}.
\end{aligned} \tag{35.9}$$

Then the three distinct nonzero elements of $Y_{\alpha\beta}(s)$ are found from $Y_{\alpha\beta}(s) = s\,(sI+p)^{-1}_{\alpha\beta}\,A_{\gamma\beta}$ to be

$$\begin{aligned}
Y_{11}(s) &= \frac{s\,[(s+p_{11}+p_{12})\,A_{11} - 2p_{12}\,A_{12}]}{(s+p_{11}+2p_{12})\,(s+p_{11}-p_{12})},\\
Y_{12}(s) &= \frac{s\,[(s+p_{11})\,A_{12} - p_{12}\,A_{11}]}{(s+p_{11}+2p_{12})\,(s+p_{11}-p_{12})},\\
Y_{44}(s) &= \frac{s\,A_{44}}{s+p_{44}}.
\end{aligned} \tag{35.10}$$

By analogy with $(34.2)_2$, we now put

$$\begin{aligned}\lambda^V &= Y_{12}(j\,\omega), \qquad \mu^V = Y_{44}(j\omega),\\ \alpha^V &= Y_{11}(j\omega) - Y_{12}(j\omega) - 2Y_{44}(j\omega).\end{aligned} \tag{35.11}$$

Since λ^V, μ^V, α^V, and h (for $c^* \neq 0$) all approach zero as $\omega \to 0$, it can be seen immediately from (35.11), (35.10), (35.8), and (35.3) that as $\omega \to 0$, $\bar{\varrho} c^{*2}$ approaches the real elastic value given by (34.24), or the entries in Table 34.1. In the present case, this limiting value will be denoted by v in order to distinguish it from the real phase velocity c. (It will be seen that the thermal contribution introduces a second value of c^{*2} that goes to zero as $\omega \to 0$.)

We next consider in more detail the modes having the directions of propagation and polarization indicated in Table 34.1.

As the simplest example, consider one of the transverse modes for which $\bar{\varrho} c^2 = \mu$ in the ideal elastic case. Since $\lambda^{(1)}$ and $\lambda^{(2)}$ are both zero, (35.3) yields simply

$$\bar{\varrho} c^{*2} = \mu + \mu^V = \mu + \frac{j\omega A_{44}}{j\omega + p_{44}}. \tag{35.12}$$

By comparison with Table 20.1, it can be seen that these modes conform to the behavior predicted for the "standard linear solid". The easiest way to make the comparison is through (20.5): $\bar{\varrho} c^{*2} = j\omega \bar{G}(j\omega)$. Then, recalling that in this case v in Table 20.1 is the limiting value of c^* as $\omega \to 0$, we have, from (35.12),

$$\frac{\bar{G}(s)}{\mu} = \frac{1}{s} + \frac{A_{44}/\mu}{s + p_{44}}, \tag{35.13}$$

which agrees with the expression for the "standard linear solid" in Table 20.1 with

$$\bar{\varrho} v^2 = \mu, \qquad K^2 = A_{44}/\mu, \qquad 1/\tau_0 = p_{44}. \tag{35.14}$$

The exact formulas for phase velocity and attenuation are included in Table 20.1.

The next-simplest cases are the transverse modes for which $\lambda^{(1)} = 0$ but $\lambda^{(2)} \neq 0$. Both μ^V and α^V now enter, but the thermal contribution h is still absent. From (35.10),

$$Y_{11} - Y_{12} = \frac{s(A_{11} - A_{12})}{s + p_{11} - p_{12}},$$

and hence (with $s = j\omega$),

$$\alpha^V = \frac{s(A_{11} - A_{12})}{s + p_{11} - p_{12}} - 2\mu^V = \frac{s(A_{11} - A_{12})}{s + p_{11} - p_{12}} - \frac{2sA_{44}}{s + p_{44}}. \tag{35.15}$$

Then $\bar{\varrho} c^{*2}$ follows from (35.3). Explicit formulas for phase velocity and attenuation in terms of the material parameters can be worked out in a straightforward manner. In the context of "one-dimensional" representations like those of Table 20.1, the appropriate $\bar{G}(s)$ can be described as the sum of an elastic part (proportional to $1/s$) and *two* "relaxation terms". Explicitly

$$\begin{aligned}\bar{\varrho} c^{*2} &= \mu + \mu^V + (\alpha + \alpha^V)\,\lambda^{(2)}\\ &= \mu + \mu^V (1 - 2\lambda^{(2)}) + (\alpha + \alpha^V + 2\mu^V)\,\lambda^{(2)}\\ &= \mu + \alpha\lambda^{(2)} + \frac{j\omega A_{44}}{j\omega + p_{44}}(1 - 2\lambda^{(2)}) + \frac{j\omega(A_{11} - A_{12})}{j\omega + p_{11} - p_{12}}\,\lambda^{(2)}.\end{aligned} \tag{35.16}$$

Then

$$\frac{\bar{G}(s)}{\bar{\varrho} v^2} = \frac{1}{s} + \frac{b_1}{s + p_{44}} + \frac{b_2}{s + p_{11} - p_{12}} \tag{35.17}$$

in which

$$b_1=\frac{A_{44}(1-2\lambda^{(2)})}{\bar{\varrho}v^2}, \qquad b_2=\frac{(A_{11}-A_{12})\lambda^{(2)}}{\bar{\varrho}v^2},$$
$$\bar{\varrho}v^2=\mu+\alpha\lambda^{(2)}. \tag{35.18}$$

In the case of the longitudinal waves, $\lambda^{(1)}=1$, and λ^V and h both enter. To identify the relaxation terms arising from λ^V, we rewrite $Y_{12}(s)$ from $(35.10)_2$ as

$$\frac{1}{s}Y_{12}(s)=\frac{\frac{1}{3}(A_{11}+2A_{12})}{s+p_{11}+2p_{12}}-\frac{\frac{1}{3}(A_{11}-A_{12})}{s+p_{11}-p_{12}}. \tag{35.19}$$

Let us also introduce s into (35.8) by the substitutions

$$s=j\omega, \qquad s\bar{G}(s)=\bar{\varrho}c^{*2}. \tag{35.20}$$

Then (35.8) can be expressed as

$$h(s)=\frac{B\beta T_0\gamma s\tau_\Theta\bar{\varrho}v^2/s\bar{G}(s)}{1+\frac{s}{q}-s\tau_\Theta\bar{\varrho}v^2/s\bar{G}(s)}, \tag{35.21}$$

where τ_Θ, to be interpreted as a thermal relaxation time, is defined by

$$\tau_\Theta=\frac{\bar{\varrho}\varkappa}{\bar{\varrho}v^2C_V} \tag{35.22}$$

and $\bar{\varrho}v^2$ is the limiting (real) value of $\bar{\varrho}c^{*2}$ as $\omega\to 0$, which can be obtained from (34.24) or Table 34.1 with c replaced by v. Clearly, τ_Θ and $s\bar{G}(s)$ depend on the mode under consideration as well as on the material properties. Eq. (35.3) can now be expressed as

$$s\bar{G}(s)=\bar{\varrho}v^2+\frac{sA_{44}}{s+p_{44}}(1+\lambda^{(1)}-2\lambda^{(2)})+\frac{s(A_{11}-A_{12})}{s+p_{11}-p_{12}}\left(\lambda^{(2)}-\frac{1}{3}\lambda^{(1)}\right)$$
$$+\left(\frac{\frac{1}{3}s(A_{11}+2A_{12})}{s+p_{11}+2p_{12}}+\frac{B\beta T_0\gamma s\tau_\Theta\bar{\varrho}v^2/s\bar{G}(s)}{1+s/q-s\tau_\Theta\bar{\varrho}v^2/s\bar{G}(s)}\right)\lambda^{(1)} \tag{35.23}$$

with

$$\bar{\varrho}v^2=\mu+(\lambda+\mu)\lambda^{(1)}+\alpha\lambda^{(2)}. \tag{35.24}$$

With appropriate choice of $\lambda^{(1)}$ and $\lambda^{(2)}$, this result applies to all modes for which the polarization direction is an eigenvector of both $Q^{(1)}$ and $Q^{(2)}$ in (34.23). This includes all the modes discussed in Sect. 34 and included in Table 34.1. To obtain a more concise representation, let us introduce the abbreviations

$$\begin{aligned}
&K_1=A_{44}(1+\lambda^{(1)}-2\lambda^{(2)})/\bar{\varrho}v^2,\\
&K_2=(A_{11}-A_{12})\,(\lambda^{(2)}-\tfrac{1}{3}\lambda^{(1)})/\bar{\varrho}v^2,\\
&K_3=\tfrac{1}{3}(A_{11}+2A_{12})\,\lambda^{(1)}/\bar{\varrho}v^2,\\
&K_\theta=B\beta T_0\gamma\lambda^{(1)}/\bar{\varrho}v^2,\\
&\qquad\tau_1=p_{44}, \quad \tau_2=p_{11}-p_{12}, \quad \tau_3=p_{11}+2p_{12}.
\end{aligned} \tag{35.25}$$

Then (35.23) becomes

$$\frac{s\bar{G}(s)}{\bar{\varrho}v^2}=1+\frac{sK_1}{s+\tau_1}+\frac{sK_2}{s+\tau_2}+\frac{sK_3}{s+\tau_3}+\frac{s\tau_\theta K_\Theta\bar{\varrho}v^2/s\bar{G}(s)}{1+\frac{s}{q}-s\tau_\Theta\bar{\varrho}v^2/s\bar{G}(s)}. \tag{35.26}$$

This is effectively a quadratic equation for $s\bar{G}(s)/\bar{\varrho}v^2$ whose solutions can be obtained by straightforward algebraic methods. The problem is reminiscent of the analysis of TRUESDELL (1953) concerning waves in a fluid with Newtonian viscosity and heat conduction according to FOURIER's law.

If the thermoelastic coupling is put to zero, for example by letting the thermal expansion go to zero so that $\gamma=0$ and therefore $K_\Theta=0$, the second root is lost from (35.26) and one can interpret the remaining $s\bar{G}(s)$ as an elastic part plus three viscoelastic relaxation terms. In the uncoupled case, the energy relation (25.17) must be investigated separately. Because of the assumed vanishing expansion coefficient, γ_{mn} vanishes also [cf. (10.41)] and (25.18) yields

$$(\Delta S)^* = \frac{C_V}{T}(\Delta T)^*. \tag{35.27}$$

Considered simultaneously, (35.27) and (25.17) imply

$$\frac{\Gamma^2 L_{\boldsymbol{N}}(\omega)}{j\omega C_V} = 1. \tag{35.28}$$

From (35.28) and (35.7),

$$\Gamma^2 = \frac{j\,\omega\, C_V}{L_{\boldsymbol{N}}(\omega)} = \frac{j\,\omega\, C_V(j\,\omega+q)}{q\,\varkappa}. \tag{35.29}$$

Recalling $\Gamma=j\omega/c^*$ from (25.35), we find that (35.29) can be expressed as

$$\bar{\varrho}c^{*2} = -\frac{\omega^2\bar{\varrho}}{\Gamma^2} = \frac{j\,\omega\,\bar{\varrho}\,q\,\varkappa/C_V}{j\,\omega+q}, \tag{35.30}$$

or

$$s\bar{G}(s) = \frac{s\,\bar{\varrho}\,q\,\varkappa/C_V}{s+q} = \bar{\varrho}v^2\frac{s\,q\,\tau_\Theta}{s+q} \tag{35.31}$$

where $\bar{\varrho}\,v^2$ still denotes the *elastic* value, given by (35.24). With $s=j\,\omega$, (35.31) describes the behavior of thermal waves when the expansion coefficient vanishes, and it probably characterizes in some sense their limiting behavior when the expansion coefficient is small but finite, and they are governed by one of the roots of (35.26). It appears also that as $s\to 0$, one of the roots of (35.26) goes to the elastic case $s\bar{G}(s)=\bar{\varrho}\,v^2$ while the other behaves like (35.31). At low frequencies ($\omega\ll q$), the behavior described by (35.30) is essentially like that described by FOURIER's law, or the diffusion equation. FOURIER's law is obtained by letting $q\to\infty$. Then

$$\begin{aligned} c^{*2} &= j\omega\varkappa/C_V;\\ \frac{1}{c} &= \frac{\alpha}{\omega} = (C_V/2\varkappa\omega)^{\frac{1}{2}};\\ Q &= \frac{\omega}{2\alpha c} = \frac{1}{2}. \end{aligned} \tag{35.32}$$

The "waves" are so highly damped that they hardly deserve the name. The amplitude drops by a factor $1/e$ every *penetration depth*

$$\delta = \frac{1}{\alpha} = (2\varkappa/\omega C_V)^{\frac{1}{2}} = \frac{c}{\omega} \doteq \frac{\lambda}{2\pi}. \tag{35.33}$$

This means that the amplitude falls off by the factor $e^{2\pi}=535$ in only one wave length. This rapid attenuation applies to all "waves" that are governed by a diffusion equation. On the other hand, with q finite, the departure from FOURIER's law at frequencies $\omega\gg q$ is drastic. From (35.30) it can be seen that in the limit as $\omega\to\infty$, c^{*2} tends to the real value $q\varkappa/C_V$, and hence the uncoupled thermal waves tend to be undamped in the high frequency limit. This behavior is the result of generalizing the constitutive relation for the heat flux to (23.5).

With $K_\Theta \neq 0$, there are two meaningful values of $\bar{\varrho} c^{*2}$ for each longitudinal wave direction in Table 34.1. One of them corresponds to the expected viscoelastic wave modified by thermal effects, and the other to a thermal wave modified by coupling to viscoelastic effects. The ratio of the temperature amplitude to the strain amplitude, given by (25.20), will be different for the two waves because of the Γ^2 in (25.20).

The viscoelastic terms in (35.26) can be specialized to the limiting case of Newtonian viscosity as described in Sect. 26. The matrix $p_{\xi\alpha}^{-1}$ is obtained by putting $s=0$ in (35.9). The resulting *viscosity coefficients* μ_{11}, μ_{12}, and μ_{44} then appear as the coefficients of s on the right hand side of (35.10), but without the esses that were put to zero in (35.9). Thus,

$$\begin{aligned} \mu_{11} &= \frac{(p_{11}+p_{12})\,A_{11}-2p_{12}\,A_{12}}{(p_{11}+2p_{12})\,(p_{11}-p_{12})}, \\ \mu_{12} &= \frac{p_{11}\,A_{12}-p_{12}\,A_{11}}{(p_{11}+2p_{12})\,(p_{11}-p_{12})}, \\ \mu_{44} &= A_{44}/p_{44}. \end{aligned} \tag{35.34}$$

Then $Y_{\alpha\beta}(s) = s\mu_{\alpha\beta}$, consistent with (26.4). The final result in (35.26) is simply to convert a term like $sK_1/(s+\tau_1)$ into sK_1/τ_1. That is, s drops out of the *denominator* in the viscoelastic relaxation terms.

In this Section we have dealt with directions of propagation for which the polarization direction could be deduced from the assumed symmetry of the material coefficients, and for which the values of $\bar{\varrho} c^{*2}$ could be calculated from the acoustical tensor Q and these polarization directions. In these cases, the polarization directions are seen to be eigenvectors of the entire complex acoustical tensor, and linearly polarized waves result. For a general propagation direction, the corresponding eigenvectors are not eigenvectors of $Q^{(1)}$ and $Q^{(2)}$, and it seems that the material coefficients would have to satisfy special relations in order for the eigenvectors of Q to be real. [If the real part of Q commutes with the imaginary part, then they have the same eigenvectors (which can be taken as real) and linearly polarized waves result. Cf. the discussion of Eq. (25.3).] Therefore, to allow linearly polarized waves for an arbitrary propagation direction places a special constraint on the constitutive relation, as noted by LEITMAN and FISHER (1972) [this encyclopedia, Vol. VIa/3, p. 109].

In Sect. 23, the special model of viscoelasticity on which the present results are based, was presented *ad hoc*. However the model is not without a physical basis. That FOURIER's law needs generalization on physical grounds has long been recognized. The viscoelastic assumption (23.1) is one way of representing the results of a theory concerning the contribution of thermal vibrations to internal friction. This theory, initiated by AKHIESER (1939), has been further developed by MASON (1960), WOODRUFF and EHRENREICH (1961), MASON and BATEMAN (1964, 1966), and discussed by PROHOFSKY (1967) and MARIS (1971). THURSTON (1967, 1968) put the theory in a general setting and related it to the relaxation integral law of macroscopic viscoelasticity, expressing the relaxation function in terms of microscopic parameters, namely (1) the strain derivatives of the normal mode frequencies in the harmonic approximation of lattice dynamics, (2) the isentropic strain derivatives of the equilibrium values of the ensemble-averaged occupation numbers for the thermal modes, and (3) characteristic relaxation times in an equation of evolution that describes how the ensemble-averaged occupation numbers return to equilibrium after they are perturbed. The work of MASON (1960), MASON and BATEMAN (1964, 1966), M. F. LEWIS (1968), and others, relating the ultrasonic attenuation in a large number of crystals to their nonlinear elastic properties through similar relations can leave little doubt that the theory contains a grain of truth, although it certainly does not always describe the dominant mechanism.

This model, in common with almost all others, gives an attenuation α that increases in proportion to ω^2 in the low-frequency limit. This makes the "quality" $Q=\omega/2\alpha c$ increase as $1/\omega$ as $\omega\to 0$. One would normally expect this limiting behavior to be achieved in the ultrasonic or acoustic range of frequencies. However, experimental observations reveal that many materials appear to have nearly constant Q over a disturbingly large range of low frequencies. MASON (1971) has recently proposed a mechanism to account for this.

36. Piezoelectrically excited vibrations.

Piezoelectric excitation is one of the most widely used methods of exciting waves for physical investigations. Conducting electrodes may be placed on the material and connected in a circuit to apply a voltage, or the piezoelectric material may be placed in a microwave cavity [BOMMEL and DRANSFELD, 1960; JACOBSEN, 1960, ILUKOR and JACOBSEN, 1968]. The needed fundamental equations have been developed in Sects. 13 and 14. TIERSTEN (1969) has provided an excellent discussion of much of the theory needed for the application to plate vibrations. Surface wave excitation by interdigital electrodes has been described by COQUIN and TIERSTEN (1967), JOSHI and WHITE (1968), SMITH, GERARD, COLLINS, REEDER, and SHAW [$(1969)_1$, $(1969)_2$], DRANSFELD and SALZMANN (1970), and AULD and KINO (1971).

36.1. Thickness-shear vibrations of an infinite plate. Following TIERSTEN (1963, 1969) we consider a plate bounded by two parallel planes at $X_3=\pm h$, and extending to infinity in the X_2 and X_1 directions. The equation of motion is (13.70) in which the last term can be dropped and the other part reduced (for small displacement gradients) to

$$\bar{\varrho}\,\ddot{u}_i=\frac{\partial}{\partial X_q}\left(\frac{\partial U}{\partial V_{iq}}\right)=\partial\tau^*_{iq}/\partial X_q. \tag{36.1}$$

We work in an approximation that is *quasistatic* from the point of view of electromagnetic theory, so that

$$\boldsymbol{E}=-\nabla\Phi. \tag{36.2}$$

With plate surfaces at $X_3=\pm h$ coated with conducting electrodes and a voltage applied across them, the most convenient starting equations are (14.25).

By *thickness* vibrations, one means solutions in which $\boldsymbol{u}$ depends on X_3 only, independent of X_1 and X_2. The strain components, linearized in the displacement gradients then reduce to

$$\begin{aligned} &V_1=0; \quad V_2=0; \quad V_3=u_{3,3};\\ &V_4=u_{2,3}; \quad V_5=u_{1,3}; \quad V_6=0 \end{aligned} \tag{36.3}$$

where the subscript comma denotes differentiation with respect to the X-coordinate whose number or numbers follow the comma. From (36.1)–(36.3) and (14.25), the linearized equations of motion become

$$\begin{bmatrix}\bar{\varrho}\,\ddot{u}_1\\ \bar{\varrho}\,\ddot{u}_2\\ \bar{\varrho}\,\ddot{u}_3\end{bmatrix}=\begin{bmatrix}c_{35} & c_{45} & c_{55} & e_{35}\\ c_{34} & c_{44} & c_{45} & e_{34}\\ c_{33} & c_{34} & c_{35} & e_{33}\end{bmatrix}\begin{bmatrix}u_{3,33}\\ u_{2,33}\\ u_{1,33}\\ \Phi_{,33}\end{bmatrix}. \tag{36.4}$$

The elastic coefficients are in this section understood to be at constant E unless otherwise specified, i.e., $c_{44}=c_{44}^E$. Similarly, the dielectric constants that appear below are understood to be at constant strain V unless otherwise specified, i.e., $\varepsilon_{33}=\varepsilon_{33}^V$. An additional equation that applies on $-h<X_3<h$ is Div $\boldsymbol{D}\equiv D_{i,i}=0$, which reduces to $D_{3,3}=0$. From (14.25) and (36.3), this condition is

$$e_{33}\,u_{3,33}+e_{34}\,u_{2,33}+e_{35}\,u_{1,33}-\varepsilon_{33}\,\Phi_{,33}=0. \tag{36.5}$$

For an electrically driven plate, the boundary conditions are (1) that the potential difference across the electrodes at $X_3=\pm h$ is prescribed, and of course each electrode (presumed infinitely thin) is an equipotential surface: thus $\Phi(X_3)$ satisfies

$$\Phi(\pm h)=\pm\,\Phi_0\cos\omega t, \tag{36.6}$$

and (2), that the surface tractions vanish at $X_3=\pm h$. The surface tractions can be satisfactorily approximated by τ_{3i}^*, $i=1, 2, 3$; or in abbreviated notation τ_5^*, τ_4^*, τ_3^*. From (14.25) and (36.3), these conditions are expressed as follows.

$$\begin{bmatrix} c_{33} & c_{34} & c_{35} & e_{33}\\ c_{34} & c_{44} & c_{45} & e_{34}\\ c_{35} & c_{45} & c_{55} & e_{35}\end{bmatrix}\begin{bmatrix} u_{3,3}\\ u_{2,3}\\ u_{1,3}\\ \Phi_{,3}\end{bmatrix}=\begin{bmatrix}0\\0\\0\\0\end{bmatrix}. \tag{36.7}$$

Now consider the elastic, piezoelectric, and dielectric coefficients, referred to the chosen axes, to have the symmetry indicated for point group 2 in the monoclinic system. This does not restrict the crystal to the monoclinic system, since the present X_3 direction need not be along the conventional 3-direction for the crystal. For example, TIERSTEN (1969, p. 53) states that the coefficients for "rotated Y-cut quartz" referred to axes in and normal to the plate conform to this symmetry. From Tables 16.3 and 16.4, $e_{33}=e_{35}=0$ and $c_{34}=c_{45}=0$. Then (36.4) shows that there is no direct electrical coupling to u_1 and u_3. Nor do u_1 and u_3 interact with u_2 or Φ through (36.5) or (36.7). The equations therefore permit a solution in which

$$u_1=u_3=0. \tag{36.8}$$

Since $u_3=0$, the mode of vibration is called *thickness shear*. In view of (36.8), the equations for u_2 on $-h<X_3<h$ then simplify to

$$\begin{aligned}\bar{\varrho}\,\ddot{u}_2-c_{44}\,u_{2,33}-e_{34}\,\Phi_{,33}&=0,\\ D_{3,3}=e_{34}\,u_{2,33}-\varepsilon_{33}\,\Phi_{,33}&=0,\end{aligned} \tag{36.9}$$

while the boundary condition (36.7) at $X_3=\pm h$ becomes

$$c_{44}\,u_{2,3}+e_{34}\,\Phi_{,3}=0,\qquad X_3=\pm h. \tag{36.10}$$

Now let

$$\begin{aligned}u_2(X_3,t)&=u(X_3)\exp(j\omega t),\\ \Phi(X_3,t)&=\varphi(X_3)\exp(j\omega t).\end{aligned} \tag{36.11}$$

Integration of $(36.9)_2$ with respect to X_3 gives

$$D_3=e_{34}\,u_{2,3}-\varepsilon_{33}\,\Phi_{,3}=-\sigma\exp(j\omega t) \tag{36.12}$$

in which the constant of integration σ can be interpreted as the amplitude of the charge density needed to terminate the field D_3. (The assumptions force D_3 to be independent of X_3.) Another integration yields an expression for $\varphi(X_3)$, while $(36.9)_2$ can be used to eliminate the potential from $(36.9)_1$. Thus, from (36.9),

(36.11), and (36.12),

$$\varphi = \frac{e_{34}}{\varepsilon_{33}} u + \frac{\sigma}{\varepsilon_{33}} X_3 + L,$$
$$\bar{\varrho}\omega^2 u + \bar{c}_{44} u_{,33} = 0, \tag{36.13}$$
$$\bar{c}_{44} \equiv c_{44} + e_{34}^2/\varepsilon_{33}.$$

L is an additional constant of integration. The boundary conditions (36.10) and (36.6) may now be written as

$$c_{44} u_{,3}(\pm h) + e_{34} \varphi_{,3}(\pm h) = 0,$$
$$\varphi(\pm h) = \pm \varphi_0. \tag{36.14}$$

From $(36.13)_1$ and (36.14),

$$\bar{c}_{44} u_{,3}(\pm h) + e_{34} \sigma/\varepsilon_{33} = 0,$$
$$\frac{e_{34}}{\varepsilon_{33}} u(\pm h) \pm \frac{h\sigma}{\varepsilon_{33}} + L = \pm \varphi_0. \tag{36.15}$$

From $(36.15)_1$, $u_{,3}$ must have the same value at $X_3 = +h$ as at $X_3 = -h$. Therefore, an appropriate solution of $(36.13)_2$ is

$$u = A \sin \eta X_3,$$
$$\eta^2 = \bar{\varrho}\omega^2/\bar{c}_{44}. \tag{36.16}$$

It can now be seen from $(36.15)_2$ that $L=0$. By substituting from $(36.16)_1$ into $(36.15)_2$, σ is found to be

$$\sigma = \varepsilon_{33} \varphi_0/h - \frac{e_{34}}{h} A \sin \eta h. \tag{36.17}$$

From (36.17) and $(36.15)_1$,

$$A \left[\bar{c}_{44} \eta h \cos \eta h - \frac{e_{34}^2}{\varepsilon_{33}} \sin \eta h\right] = -e_{34} \varphi_0. \tag{36.18}$$

Eq. (36.18) determines the amplitude A in terms of the applied voltage. The coefficient of A depends on the excitation frequency through η of $(36.16)_2$. Resonances occur at frequencies for which the coefficient of A vanishes in (36.18). This condition is

$$\frac{\tan \eta_R h}{\eta_R h} = \frac{\bar{c}_{44} \varepsilon_{33}}{e_{34}^2} \tag{36.19}$$

where η_R is the value of η at a resonance frequency.

36.2. Electromechanical coupling coefficient. Let us now consider the *static* behavior. In the static case, (36.9) reduces to $\bar{c}_{44} u_{2,33} = 0$, so $u_{2,3} = V_4$ is independent of X_3, and so also, from $(36.9)_2$ is $\Phi_{,3} = E_3$. Since $u_{2,3}$ is the only nonvanishing displacement gradient, V_4 is the only nonvanishing strain component. From (14.25) and assumptions already stated,

$$\tau_4^* = c_{44} V_4 - e_{34} E_3,$$
$$D_3 = e_{34} V_4 + \varepsilon_{33} E_3. \tag{36.20}$$

Solving $(36.20)_2$ for E_3 and substituting the result into $(36.20)_1$, we obtain

$$E_3 = -\frac{e_{34}}{\varepsilon_{33}} V_4 + \frac{1}{\varepsilon_{33}} D_3,$$
$$\tau_4^* = \bar{c}_{44} V_4 - \frac{e_{34}}{\varepsilon_{33}} D_3, \tag{36.21}$$

where $\bar{c}_{44}$ is still given by (36.13).

For the deformation and electric field under consideration, a static electromechanical coupling coefficient k can be defined by

$$k^2 = \frac{\left(\frac{\partial^2 \hat{U}}{\partial V_4\, \partial D_3}\right)^2}{\left(\frac{\partial^2 \hat{U}}{\partial V_4^2}\right)\left(\frac{\partial^2 \hat{U}}{\partial D_3^2}\right)}, \tag{36.22}$$

where $\hat{U} = \hat{U}(V_4, D_3)$ is the internal energy in the special situation under discussion here. Since $E_3 = \partial\hat{U}/\partial D_3$ and $\tau_4^* = \partial\hat{U}/\partial V_4$, (36.21) enables us to evaluate k^2 as

$$k^2 = \frac{e_{34}^2}{\bar{c}_{44}\, \varepsilon_{33}^V}. \tag{36.23}$$

With $\bar{c}_{44}$ from (36.13), it follows also that

$$\frac{k^2}{1-k^2} = \frac{e_{34}^2}{c_{44}\, \varepsilon_{33}^V}, \tag{36.24}$$

where

$$c_{44}/\bar{c}_{44} = 1 - k^2. \tag{36.25}$$

To elucidate the significance of k^2, note that two limiting mechanical conditions are "free" ($\tau_4^* = 0$) and "clamped" ($V_4 = \text{const.}$). Two limiting electrical conditions are "short-circuit" ($E_3 = 0$) and "open-circuit" ($D_3 = \text{const.}$). Let us introduce the following notation:

$$\begin{aligned} W_{ME} &= \text{mechanical work done with } E_3 = 0,\\ W_{MD} &= \text{mechanical work done with } D_3 = 0,\\ W_{E\tau} &= \text{electrical work done with } \tau_4^* = 0,\\ W_{EV} &= \text{electrical work done with } V_4 = 0. \end{aligned} \tag{36,26}$$

Each of these quantities can be considered as a function of one variable, which may be any one of the four E_3, D_3, τ_4^*, V_4 that is not held to zero. For example, $W_{E\tau}$ can be considered as a function of E_3, D_3, or V_4. $W_{E\tau}(V_4)$ would mean the electrical work needed to bring about a state characterized by $\tau_4^* = 0$ and an arbitrary value of V_4. With this notation, it can be verified that

$$k^2 = \frac{W_{ME}(V_4)}{W_{E\tau}(V_4)} = \frac{W_{MD}(\tau_4^*)}{W_{EV}(\tau_4^*)} = \frac{W_{E\tau}(D_3)}{W_{ME}(D_3)} = \frac{W_{EV}(E_3)}{W_{MD}(E_3)}, \tag{36.27}$$

$$\frac{k^2}{1-k^2} = \frac{W_{ME}(\tau_4^*)}{W_{EV}(\tau_4^*)} = \frac{W_{MD}(V_4)}{W_{E\tau}(V_4)} = \frac{W_{E\tau}(E_3)}{W_{MD}(E_3)} = \frac{W_{EV}(D_3)}{W_{ME}(D_3)}. \tag{36.28}$$

The denominator in each of the above expressions represents either the electrical work to bring about a state specified by a mechanical variable (with no mechanical work done) or the mechanical work to bring about a state specified by an electrical variable (with no electrical work done). The numerators represent either the mechanical work to bring about a state specified by a mechanical variable with no electrical work done, or the electrical work to bring about a state specified by an electrical variable with no mechanical work done. One can also consider the ratios of mechanical work on open and short circuit, and of the electrical work when clamped and free. The results are

$$(1-k^2) = \frac{W_{MD}(\tau_4^*)}{W_{ME}(\tau_4^*)} = \frac{W_{ME}(V_4)}{W_{MD}(V_4)} = \frac{W_{EV}(E_3)}{W_{E\tau}(E_3)} = \frac{W_{E\tau}(D_3)}{W_{EV}(D_3)} = \frac{c_{44}}{\bar{c}_{44}} = \frac{\varepsilon_{33}^V}{\bar{\varepsilon}_{33}}, \tag{36.29}$$

where $\bar{\varepsilon}_{33}$ is the "free" dielectric constant,* obtained by putting $\tau_4^*=0$ in $(36.20)_1$, solving for V_4, and substituting that result in $(36.21)_1$:

$$\begin{aligned} D_3 &= \varepsilon_{33}\left(1+\frac{e_{34}^2}{c_{44}}\right)E_3=\bar{\varepsilon}_{33}E_3, \\ \bar{\varepsilon}_{33} &= \varepsilon_{33}\left(1+\frac{e_{34}^2}{\varepsilon_{33}c_{44}}\right)=\frac{\varepsilon_{33}}{1-k^2}. \end{aligned} \tag{36.30}$$

We can also relate k^2 to energy ratios in a piezoelectric generating cycle: (1) Deform the material on open circuit by doing the mechanical work $W_{MD}(V_4)$. (2) Deliver electrical energy (reversibly) while clamped, until the state (V_4, $E_3=0$) is reached. (3) On short-circuit, return to the original state, recovering the mechanical work $W_{ME}(V_4)$. By conservation of energy, the electrical energy delivered is $W_{MD}(V_4)-W_{ME}(V_4)$. The ratio of electrical energy delivered to the initial mechanical energy input in this cycle is

$$\frac{W_{MD}(V_4)-W_{ME}(V_4)}{W_{MD}(V_4)}=1-(1-k^2)=k^2. \tag{36.31}$$

Clearly, $k^2<1$.

A caution is necessary in connection with (36.22) because the equation contains a trap for the unwary. It is presented here in a context in which the stored energy is regarded as a function of only two independent variables: D_3 and V_4. The strain components other than V_4 are zero, but since $\varepsilon_{13}\neq 0$, E_3 invokes a component D_1 through $(14.25)_2$. From $(14.25)_2$ and $(36.21)_1$,

$$D_1=e_{14}V_4+\varepsilon_{13}E_3=\left(e_{14}-\frac{\varepsilon_{13}}{\varepsilon_{33}}e_{34}\right)V_4+\frac{\varepsilon_{13}}{\varepsilon_{33}}D_3. \tag{36.32}$$

Therefore, the partial derivatives in (36.22) are not the same as the derivatives at constant D_1 in Table 14.1. In (36.22), $\partial^2\hat{U}/\partial V_4\,\partial D_3$ is $-e_{34}/\varepsilon_{33}$, not $-h_{34}$, $\partial^2\hat{U}/\partial V_4^2$ is $\bar{c}_{44}$, not c_{44}^D, and $\partial^2\hat{U}/\partial D_3^2$ is $1/\varepsilon_{33}^S$, not β_{33}^V. The coupling coefficient in (36.22) can be related to the quantities in Table 14.1 through (36.32) and the equality

$$\hat{U}(V_4, D_3)=U[V_4, D_3, D_1(V_4, D_3)]. \tag{36.33}$$

The relations between β_{ij} and ε_{ij} for this symmetry are

$$\beta_{33}=\frac{\varepsilon_{11}}{\Delta},\quad \beta_{11}=\frac{\varepsilon_{33}}{\Delta},\quad \beta_{13}=\frac{-\varepsilon_{13}}{\Delta},\quad \beta_{22}=\frac{1}{\varepsilon_{22}}. \tag{36.34}$$

With the help of these relations and the connections between $h_{i\beta}$ and $e_{j\beta}$ from Table 14.3, we find

$$e_{14}-\frac{\varepsilon_{13}}{\varepsilon_{33}}e_{34}=e_{14}+\frac{\beta_{13}}{\beta_{11}}e_{34}=h_{14}/\beta_{11}$$

and hence (36.32) can be expressed as

$$D_1=\frac{1}{\beta_{11}}(h_{14}V_4-\beta_{13}D_3). \tag{36.35}$$

[The relation in this form also follows easily from $(14.22)_1$ by setting $E_1=0$.] It follows that

$$\begin{aligned} \partial^2\hat{U}/\partial D_3^2 &= 1/\varepsilon_{33}=\beta_{33}-\beta_{13}^2/\beta_{11}, \\ \partial^2\hat{U}/\partial D_3\,\partial V_4 &= -e_{34}/\varepsilon_{33}=h_{34}-\beta_{13}h_{14}/\beta_{11}, \\ \partial^2\hat{U}/\partial V_4^2 &= \bar{c}_{44}=c_{44}^D-h_{14}^2/\beta_{11}=c_{44}^E+e_{34}^2/\varepsilon_{33}, \end{aligned} \tag{36.36}$$

* Here "free" refers only to the vanishing of τ_4^*. $\bar{\varepsilon}_{33}$ is not the true ε_{33}^T.

from which k^2 can be expressed in terms of β_{ij}, h_{14}, h_{34}, and c_{44}^D. TIERSTEN (1963) has emphasized the distinction between $\bar{c}_{44}$ and c_{44}^D, in general.

The definition (36.22) is often presented as a ratio of energies rather than a ratio of partial derivatives. Including quadratic terms, $\hat{U}$, expanded about the state of zero V_4 and D_3, may be represented as

$$\hat{U} = U_M + U_E + 2U_m \tag{36.37}$$

where

$$U_M = \frac{1}{2}\left(\frac{\partial^2 \hat{U}}{\partial V_4^2}\right) V_4^2, \quad U_E = \frac{1}{2}\left(\frac{\partial^2 \hat{U}}{\partial D_3^2}\right) D_3^2, \quad U_m = \frac{1}{2}\left(\frac{\partial^2 \hat{U}}{\partial V_4\, \partial D_3}\right) V_4 D_3. \tag{36.38}$$

The three terms in (36.37) may be interpreted as mechanical energy with $D_3 = 0$, electrical energy with $V_4 = 0$, and "mutual energy", Clearly, the definition (36.22) is equivalent to

$$k^2 = \frac{U_m^2}{U_M U_E}, \tag{36.39}$$

which is an often-cited form [BERLINCOURT *et al.*, 1964; MASON, 1966].

We have dealt here with a very specific system in which the energy ratios in (36.27) are simply expressed in terms of material coefficients. Coupling is a general unifying concept that applies to a great variety of systems. The *measure* of coupling based on (36.22) can be generalized in at least two different directions: (1) Preserve the coefficient as a fundamental property of piezoelectric materials, but apply it to other components of strain and electric displacement, or (2) allow V_4 and D_3 to be generalized to other independent variables that are appropriate for the system under consideration. In the case of the first kind of generalization, appropriate coefficients are obtained for various other modes of vibration of crystal and piezoelectric ceramic plates [BERLINCOURT, CURRAN, and JAFFE, 1964]. In the case of the second kind of generalization, the coefficient usually ceases to be a property of the material alone. It becomes a *system* property which may depend on the boundary conditions and geometry of the system. As such, it is widely used as an index of performance potential in evaluating new or proposed transducers, and is regarded as a powerful tool for revealing possible changes in performance resulting from changes in design [WOOLLETT, 1957, 1966]. This usefulness stems from the fact that an appropriate generalization of the ratio in (36.22), which can be calculated from static quantities, ordinarily bears an important influence on the dynamical performance of a transducer.

36.3. Electrical impedance of a vibrating piezoelectric plate. We now return to the dynamic problem. The charge per unit area on an electrode is given by (36.12) with the charge amplitude σ given by (36.17) in which A can be obtained from (36.18). With the help of (36.23), the result can be expressed as

$$\begin{aligned} \sigma &= \frac{\varepsilon_{33}\,\varphi_0}{h}\left[1 + \frac{k^2}{\eta\, h \cot \eta\, h - k^2}\right] \\ &= \frac{\varepsilon_{33}\,\varphi_0}{h}\left[\frac{\eta\, h \cot \eta\, h}{\eta\, h \cot \eta\, h - k^2}\right]. \end{aligned} \tag{36.40}$$

The alternating current per unit area of plate is $j\omega\sigma \exp(j\omega t)$ and therefore the electrical impedance of an area S of the infinite plate is

$$Z = \frac{2\varphi_0}{j\omega\sigma S} = \frac{1}{j\omega C}\left[1 - \frac{k^2 \tan \eta\, h}{\eta\, h}\right] \tag{36.41}$$

where

$$C = S\, \varepsilon_{33}/2\,h \tag{36.42}$$

is the "clamped" capacitance of the area S. The resonance condition (36.19), which, in view of (36.23) can be written as $k^2 \tan \eta h = \eta h$, corresponds to zero impedance. The frequencies of infinite impedance, given by $\eta h = (n+\frac{1}{2})\pi$. $n=0, 1, 2, \ldots$, are called frequencies of antiresonance. There is a resonance frequency below each antiresonance frequency.

Writing $\eta = \omega/v$, we express the frequencies of antiresonance, ω_{An}, through

$$\frac{\omega_{\mathrm{An}} h}{v} = \left(n + \frac{1}{2}\right)\pi, \qquad n = 0, 1, 2, 3, \ldots. \tag{36.43}$$

When the coupling is small ($k^2 \ll 1$), an approximate expression for the resonance frequencies can be obtained by expanding $\tan \eta h$ in the neighborhood of antiresonance. Write

$$\begin{aligned} \eta h &= \frac{\omega h}{v} = \frac{\omega_{\mathrm{An}} h}{v}\left(1 + \frac{\omega - \omega_{\mathrm{An}}}{\omega_{\mathrm{An}}}\right) = \left(n + \frac{1}{2}\right)\pi\left(1 + \frac{\omega - \omega_{\mathrm{An}}}{\omega_{\mathrm{An}}}\right) \\ &= \pi\left(n + \frac{1}{2}\right) + \Theta_n \end{aligned} \tag{36.44}$$

where

$$\Theta_n = \pi\left(n + \frac{1}{2}\right)\frac{\omega_{\mathrm{An}} - \omega}{\omega_{\mathrm{An}}}. \tag{36.45}$$

Then, since $\tan[\pi(n+\frac{1}{2})]$ is infinite,

$$\tan \eta h = \frac{-1}{\tan \Theta_n} = \frac{-1}{\Theta_n + \frac{1}{3}\Theta_n^3 + \cdots}. \tag{36.46}$$

The resonance condition $k^2 = \eta h / \tan \eta h$ can now be expressed as

$$-k^2 = [\pi(n+\tfrac{1}{2}) + \Theta_n]\,(\Theta_n + \tfrac{1}{3}\Theta_n^3 + \cdots). \tag{36.47}$$

In the lowest approximation, $\Theta_n \doteq -k^2/[\pi(n+\frac{1}{2})]$ or

$$\frac{\omega_{\mathrm{An}} - \omega_{\mathrm{Rn}}}{\omega_{\mathrm{An}}} \doteq \frac{k^2}{\pi^2 (n+\frac{1}{2})^2} \tag{36.48}$$

where ω_{Rn} denotes the resonance frequency just below the antiresonance ω_{An}. The smaller the coupling, the closer the frequencies of resonance are to the frequencies of antiresonance.

Without any algebraic approximation, of course, one can calculate k^2 from the value of ηh at resonance. In general, $h\eta = \pi(n+\frac{1}{2})\,\omega/\omega_{\mathrm{An}}$. Hence,

$$\begin{aligned} h\eta_{\mathrm{Rn}} &= \pi\left(n + \frac{1}{2}\right)\frac{\omega_{\mathrm{Rn}}}{\omega_{\mathrm{An}}}, \\ k^2 &= h\eta_{\mathrm{Rn}}/\tan h\eta_{\mathrm{Rn}}. \end{aligned} \tag{36.49}$$

Under the assumption that experimental results for a plate of finite size can be approximated by these analytical results for an infinite plate, it is interesting to see what material coefficients can be determined from electrical measurements.

(1) In the low-frequency limit, $Z(\omega)$ from (36.41) behaves like

$$\underset{\omega \to 0}{Z(\omega)} = \frac{1 - k^2}{j\omega C}, \tag{36.50}$$

the reactance of a capacitor of capacitance $C/(1-k^2)$. From (36.42) and $(36.30)_2$, a measurement of low-frequency capacitance thus enables $\bar{\varepsilon}_{33}$ to be determined:

$$\bar{\varepsilon}_{33} = 2\frac{h}{S} C_{\text{measured}}. \tag{36.51}$$

(2) Identification of the lowest antiresonance frequency ($\eta h = \pi/2$) determines

$$v = (\bar{c}_{44}/\bar{\varrho})^{\frac{1}{2}} = 2\omega_{A0} h/\pi = 4 f_{A0} h \tag{36.52}$$

from (36.43) and (36.16). The quantity v can be identified as a wave speed for transverse waves polarized along X_2, propagating in the 3-direction. Their speed is governed, however, by a "piezoelectrically stiffened" elastic constant $\bar{c}_{44}$, that, as we have seen, is in general neither equal to c_{44}^E nor to c_{44}^D.

(3) Identification of the lowest resonance frequency ω_{R0} allows k^2 to be found from (36.49).

Assuming the density $\bar{\varrho}$ to be known, one can of course get $\bar{c}_{44}$ from (36.52) and c_{44}^E from (36.25). Knowing ε_{33}, $\bar{c}_{44}$, and k^2, one can obtain the piezoelectric coefficient e_{34} from (36.23). These formulas are suggestive of the "dynamic" (resonator) method of determining material coefficients. For more details, the reader may refer to the works of CADY (1946), MASON $(1950)_2$, BECHMANN (1950, 1966, 1969), BECHMANN and AYERS (1953), and to the IEEE Standards [No. 176 (1949), No. 178 (1958), No. 179 (1961), No. 177 (1966)].

In practice, dissipative effects, entirely neglected in the preceding analysis, cause the impedance to depart from the ideal reactance of (36.41) by introducing a real part. As a result, there are not just two kinds of critical frequencies, resonance and antiresonance. Instead, one can identify frequencies of zero reactance near resonance and antiresonance, frequencies of minimum impedance and resistance near resonance, a frequency of maximum impedance below antiresonance, etc. These effects are traditionally accounted for by introducing resistances into an equivalent circuit of the resonator. Impedance measurements then yield the values of these resistances as well as of the reactive elements that are related to the elastic and piezoelectric coefficients of the material. Relatively little has been done on the logically next step of relating the measurements to material coefficients that describe dissipation in the constitutive equations.

36.4. Equivalent circuit of a piezoelectric transducer. The equivalent circuit of a resonator [K. S. VAN DYKE (1925, 1928)], considered as a one-port network, can be derived from (36.41). However, more insight is gained by considering the case of a piezoelectric transducer that may receive power electrically and deliver it mechanically, or vice versa.

To adapt the analysis of Sect. 36.1 to this case, it is necessary to leave the surface traction $\tau_{32}^* (= \tau_4^*)$ unspecified instead of putting it to zero. The result is that (36.10) is revised to

$$\left.\begin{aligned} c_{44}\, u_{2,3}(h) + e_{34}\, \Phi_{,3}(h) &= \tau_4^*(h), \\ c_{44}\, u_{2,3}(-h) + e_{34}\, \Phi_{,3}(-h) &= \tau_4^*(-h) \end{aligned}\right\} \tag{36.53}$$

with the consequent changes that zero is replaced by the amplitude of $\tau_4^*(\pm h)$ on the right hand side of $(36.14)_1$ and $(36.15)_1$. Unless $\tau_4^*(h) = \tau_4^*(-h)$, the solution (36.16) must now be generalized to

$$u = A \sin \eta X_3 + B \cos \eta X_3, \tag{36.54}$$

with $\eta^2 = \bar{\varrho}\omega^2/\bar{c}_{44}$ as before. Instead of $L = 0$, we now get

$$L = -\frac{e_{34}}{\varepsilon_{33}} B \cos \eta h, \tag{36.55}$$

and the effect of this is to leave (36.17) unchanged. The reason for no change in (36.17) is that the added term $B \cos \eta X_3$ introduces a symmetrically distributed strain field $u_{2,3}$ in the plate, which does not affect the charge density on the electrodes. Letting

$$\tau_4^* = \tau_4 \, e^{j\omega t} \tag{36.56}$$

so that τ_4 denotes the amplitude of τ_4^*, we generalize $(36.15)_1$ to

$$\bar{c}_{44} \, u_{,3}(\pm h) + e_{34} \, \sigma/\varepsilon_{33} = \tau_4(\pm h). \tag{36.57}$$

By substituting from (36.54) and (36.17) into (36.57), we find

$$\begin{aligned} B/h &= \frac{\tau_4(-h) - \tau_4(h)}{2\bar{c}_{44}\eta h \sin \eta h}, \qquad \sin \eta h \neq 0 \\ A \bar{c}_{44}(\eta h \cos \eta h - k^2 \sin \eta h) &= \frac{h}{2} [\tau_4(h) + \tau_4(-h)] - e_{34} \varphi_0. \end{aligned} \tag{36.58}$$

With A from $(36.58)_2$ and σ from (36.17), the current to an area S of the infinite plate becomes

$$I = j\omega\sigma S = j\omega C \, 2\varphi_0 \left[\frac{\eta h \cot \eta h}{\eta h \cot \eta h - k^2}\right] - \frac{e_{34}}{2\bar{c}_{44}} \left[\frac{\tau_4(h) + \tau_4(-h)}{\eta h \cot \eta h - k^2}\right], \tag{36.59}$$

in which C is still given by (36.42). From (36.54), with A and B from (36.58), the middle plane of the plate $(X_3 = 0)$ has the velocity

$$j\omega u(0) = j\omega B = \frac{j\omega [\tau_4(-h) - \tau_4(h)]}{2\bar{c}_{44}\eta \sin \eta h}. \tag{36.60}$$

The velocity amplitudes at the upper and lower surfaces are

$$\begin{aligned} v_2 &= j\omega u(+h) = j\omega (B \cos \eta h + A \sin \eta h), \\ v_1 &= j\omega u(-h) = j\omega (B \cos \eta h - A \sin \eta h), \end{aligned} \tag{36.61}$$

where these equations serve to define v_1 and v_2. It may be noted that

$$\begin{aligned} v_1 + v_2 &= 2j\omega B \cos \eta h, \\ v_2 - v_1 &= 2j\omega A \sin \eta h, \end{aligned} \tag{36.62}$$

while from (36.17) and (36.42), the current I may be expressed as

$$I = j\omega\sigma S = 2j\omega C \varphi_0 - \frac{j\omega S e_{34}}{h} A \sin \eta h. \tag{36.63}$$

From (36.63) and (36.62),

$$I = 2\varphi_0 j\omega C + \frac{S e_{34}(v_1 - v_2)}{2h}. \tag{36.64}$$

Eq. (36.64) shows that the current I driven by the voltage $2\varphi_0$ differs from the current in a capacitor of "clamped" capacitance C by an amount proportional to the velocity difference $v_1 - v_2$.

Let us denote by F_2 and F_1 the amplitudes of the traction on an area S of the upper and lower surfaces:

$$F_2 = S\tau_4(h), \qquad F_1 = S\tau_4(-h). \tag{36.65}$$

"Circuit equations" connecting v_1, v_2, I, F_1, F_2, φ_0 follow from the above. From (36.58) and (36.65),

$$\begin{aligned} j\omega B \cos \eta h &= \frac{j\omega h (F_1 - F_2) \cos \eta h}{2S\bar{c}_{44}\eta h \sin \eta h} = \frac{(F_2 - F_1)}{2Z_1}, \\ j\omega A \sin \eta h &= \frac{F_2 + F_1 - 2F}{2Z_2}, \end{aligned} \tag{36.66}$$

in which

$$Z_1 = jS\,\bar{c}_{44}(\eta/\omega)\tan\eta h, \qquad F = S e_{34}\,\varphi_0/h,$$
$$Z_2 = \frac{S\bar{c}_{44}(\eta h\cot\eta h - k^2)}{j\omega h}. \tag{36.67}$$

From (36.66) and (36.62),

$$F_2 - F_1 = (v_2+v_1)\,Z_1, \qquad F_2+F_1-2F = (v_2-v_1)\,Z_2, \tag{36.68}$$

whence

$$(F_1-F) = -\tfrac{1}{2}(Z_1+Z_2)\,v_1 + \tfrac{1}{2}(Z_2-Z_1)\,v_2,$$
$$F_2 - F = -\tfrac{1}{2}(Z_2-Z_1)\,v_1 + \tfrac{1}{2}(Z_1+Z_2)\,v_2. \tag{36.69}$$

With $(V_1, V_2) \sim (F_1-F, F_2-F)$ and $(I_1, I_2) \sim (-v_1, v_2)$, Eqs. (36.69) have the form

$$V_1 = z_{11}\,I_1 + z_{12}\,I_2,$$
$$V_2 = z_{12}\,I_1 + z_{22}\,I_2, \tag{36.70}$$

for which the T form of the equivalent circuit is as shown in Fig. 36.1a. With the indicated analogy, we have

$$z_{11} = z_{22} - \tfrac{1}{2}(Z_1+Z_2),$$
$$z_{12} = \frac{1}{2}(Z_2-Z_1) = -\frac{1}{2}\,j\,S\bar{\varrho}\,v\left(\tan\frac{\omega h}{v} + \cot\frac{\omega h}{v} - \frac{k^2 v}{\omega h}\right), \tag{36.71}$$
$$z_{11}-z_{12} = z_{22}-z_{12} = Z_1 = jS\,\bar{\varrho}\,v\,\tan(\omega h/v),$$

where we have substituted $\eta = \omega/v$ and $\bar{c}_{44} = \bar{\varrho}\,v^2$ [cf. (36.16) and (36.52)]. Now

$$\tan\Theta + \cot\Theta = \frac{2}{\sin 2\Theta}.$$

With k^2 from (36.23) and C from (36.42), the term in (36.71) involving k^2 becomes the impedance of a negative compliance:

$$Z_k = -\frac{S\,e_{34}^2}{2j\omega h\,\varepsilon_{33}^V} = -\left(\frac{S\,e_{34}}{2h}\right)^2 \frac{1}{j\omega C} = -\frac{\Phi^2}{j\omega C}, \qquad \Phi^2 \equiv (S e_{34}/2h)^2. \tag{36.72}$$

Hence, z_{12} can be expressed as

$$z_{12} = -\frac{jS\bar{\varrho}\,v}{\sin\dfrac{2\omega h}{v}} - \frac{\Phi^2}{j\omega C}. \tag{36.73}$$

Thus, Eqs. (36.69) have the equivalent circuit representation shown in Fig. 36.1b.

The quantity F, defined in (36.67), is dimensionally a force that is always proportional to the voltage $2\varphi_0$. With Φ from (36.72),

$$F = \frac{S e_{34}}{2h}\,2\varphi_0 = 2\varphi_0\,\Phi. \tag{36.74}$$

The force-to-voltage ratio $\Phi = S e_{34}/2h$ has already appeared in (36.72) and (36.64). Φ and Φ^2 have the dimensions

$$[\Phi] = [\text{force/voltage}] = [\text{current/velocity}],$$
$$[\Phi^2] = \frac{[\text{force/velocity}]}{[\text{voltage/current}]} = \frac{[\text{mechanical impedance}]}{[\text{electrical impedance}]}. \tag{36.75}$$

In an equivalent circuit, the conversion between electrical and mechanical quantities is conveniently represented by an ideal electromechanical transformer, defined

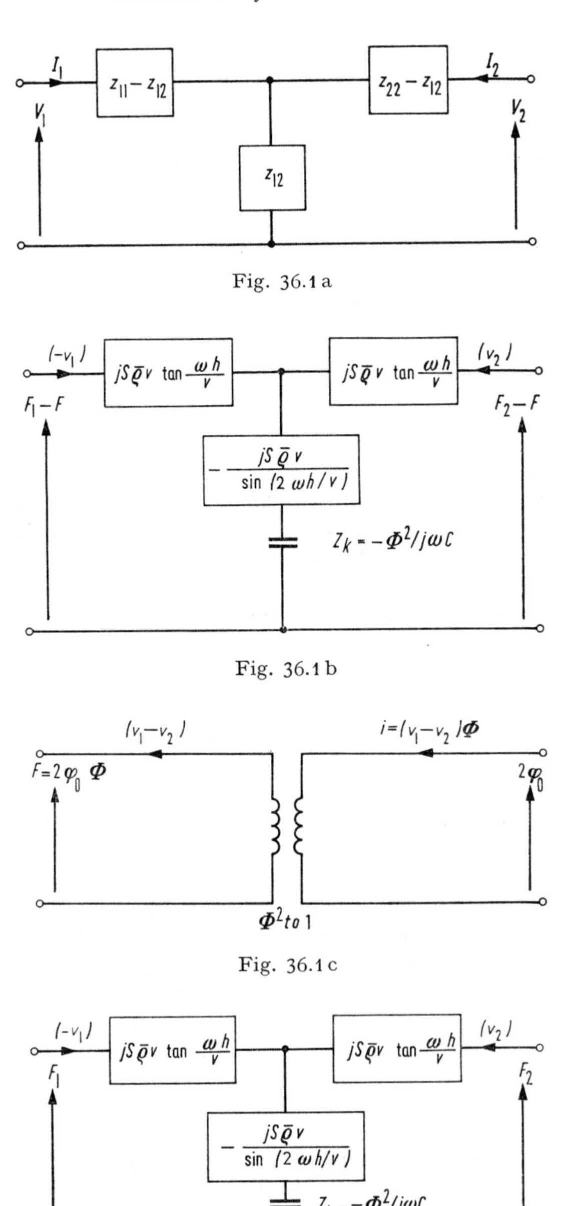

Fig. 36.1 a

Fig. 36.1 b

Fig. 36.1 c

Fig. 36.1 d

Fig. 36.1 a–d. Equivalent circuits. a) Equivalent T representation of Eqs. (36.70). The arrows represent directions of currents and voltage *rise*. b) Equivalent T representation of Eqs. (36.69). c) Ideal electromechanical transformer of impedance transformation ratio Φ^2. The voltage $2\varphi_0$ is transformed to a force $F = 2\varphi_0 \Phi$ and the current i to a velocity i/Φ. d) Complete equivalent circuit of thickness-shear transducer. [Eqs. (36.64), (36.69), (36.74).]

such that the ratio of force on the mechanical side to voltage on the electrical side is always Φ, the same as the ratio of current on the electrical side to velocity on the mechanical side. Since the product force $\times$ velocity is equal to the product voltage $\times$ current, these relations do not permit any power loss in the transformer itself. Fig. 36.1c shows the present application of the ideal electromechanical transformer to represent (36.74) and the associated current-velocity relation. The circuit element is characterized by its *impedance* transformation ratio, Φ^2 to 1. With the help of the ideal electromechanical transformer, the complete electromechanical equivalent circuit representing (36.59) or (36.64), (36.69), and (36.74), is shown in Fig. 36.1d. In this form, the circuit is very general, as the boundary conditions on the major faces have not yet been specified.

To obtain the equivalent circuit of a transducer free on one face and driving a load at the other, F_1 can be put to zero to obtain the form shown in Fig. 36.2a. Following MASON (1948), we simplify this form by using the general network equivalence shown in Fig. 36.2b, due to E. L. NORTON (1928). In the present application, we have $Z_A = Z_B = j S \bar{\varrho} v \tan(\omega h/v)$ so $\varphi = 2$ $(\varphi^2 = 4)$ and $Z_X = Z_Y = 2j S \bar{\varrho} v \tan(\omega h/v)$. This puts Fig. 36.2a into the form of Figure 36.2c. The final form of Fig. 36.2d is obtained by combining the two transformere into a single one of impedance transformation ratio $4\Phi^2$. The impedances Z_k and $-j S \bar{\varrho} v/\sin(2\omega h/v)$ have to be multiplied by four when they are moved to the other side of the 1 to 4 transformer. One then makes use of the identity.

$$-\frac{4}{\sin 2\Theta} + 2\tan\Theta \equiv -2\cot\Theta$$

to obtain the final representation in Fig. 36.2d.

The form applicable to a resonator is obtained by putting $F_2 = 0$, i.e., "shorting" the right hand side in Fig. 36.2d. The result is merely a circuit representation of the impedance Z in (36.41), as may be verified upon noting that $2\Phi^2/\bar{\varrho} v \omega C S = k^2/\eta h$.

Equivalent circuits have proved useful in the design of transducer driving systems for various applications [MASON, 1948], and in the application of piezoelectric crystals as resonators in filters and oscillators. The very sharply rising inductive reactance characteristic of a freely vibrating high Q crystal, represented by (36.41) between a resonance and the next higher antiresonance, cannot be achieved practically with coils, and hence freely vibrating piezoelectric crystals, or resonators, have found application as electrical impedance elements, even when there is no use for the vibrations other than their influence on the electrical impedance of the crystal.

36.5. Thickness-longitudinal vibrations of an infinite piezoelectric plate. An electric field can couple with the normal component of strain in the same direction through piezoelectric coefficients such as e_{11}, e_{22}, or e_{33}. Important cases in which such coupling occurs are X-cut quartz and polarized piezoelectric ceramics.

X-cut means the plate has its thickness direction along the conventional X-axis, or X_1. This is then the direction of the applied field, and the thickness dilatational mode is driven through the coefficient e_{11}. Quartz belongs to the point group 32, and it can be seen from Table 16.3 that e_{12} and e_{14} are also nonzero. The presence of e_{14} indicates a piezoelectric coupling to *face shear*, which is elastically coupled also, through the elastic coefficient c_{14}. The presence of e_{12} indicates a piezoelectric coupling to longitudinal motion in the 2-direction, which would also be elastically coupled in any case through c_{12}. These cross-coupling do not prevent the usefulness of X-cut quartz plates in exciting longitudinal vibrations for ultra-

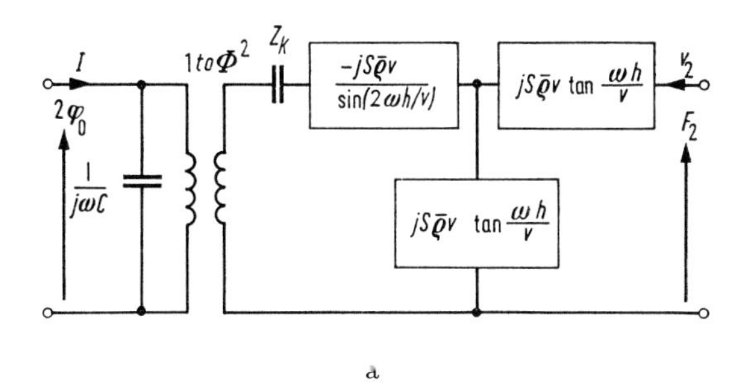

a

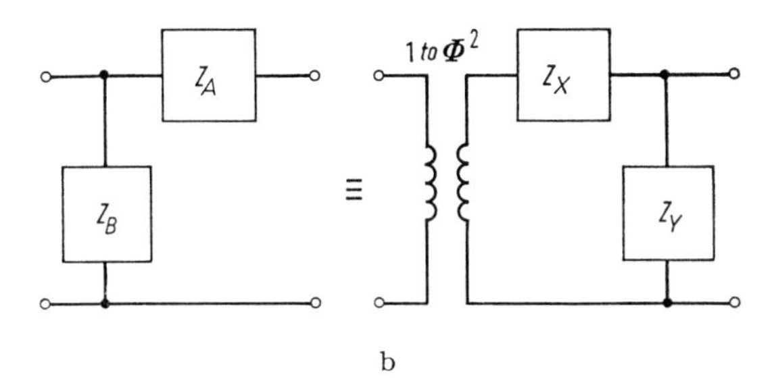

b

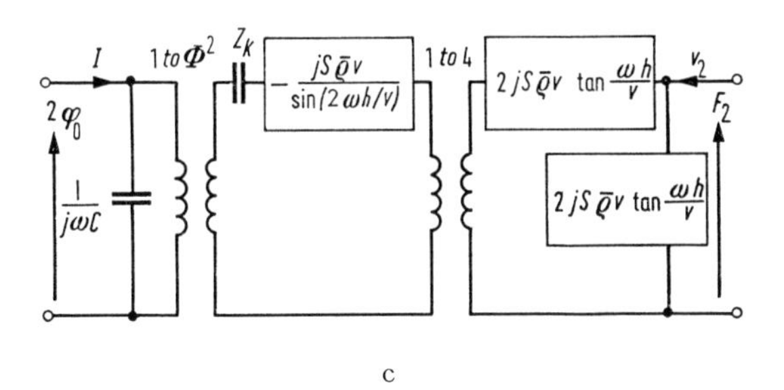

c

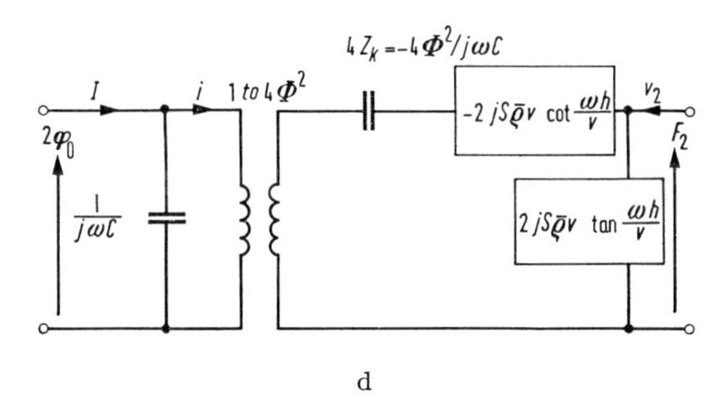

d

Fig. 36.2a–d. Equivalent circuit of a transducer free on one end. a) Form obtained directly from Fig. 36.1d) by putting F_1 to zero. b) Network equivalence. $\varphi=(Z_A+Z_B)/Z_B$, $Z_X=\varphi Z_A$, $Z_Y=\varphi Z_B$. c) Form obtained from a) by the use of b). d) Final form of the equivalent circuit of a thickness-shear transducer free on one end and driving a load at the other.

sonic studies, although the actual motion may be more complicated than a simple alternating extension in the 1-direction. [SHAW and SUJIRA, 1960]. X-cut quartz plates have been used for frequency control in oscillators, but other cuts exhibiting smaller changes of frequency with temperature are now preferred [W. P. MASON, 1950_2, p. 96].

The ceramics are made from crystalline powders by pressing and sintering. Within each sufficiently small grain, the material is like a single crystal, but since the grains are randomly oriented, the unpolarized bulk material appears macroscopically isotropic. However, the application of a high electric field at an elevated temperature can establish a preferred direction by causing the polarization in each crystallite to lie predominantly along that possible direction which is closest to that in which the field was applied. Viewed macroscopically, the bulk material then retains a remanent polarization in the direction the field had, even after the field is removed. Call this preferred direction the 3-direction. The material has the macroscopic symmetry called transverse isotropy (∞m) with "3" the preferred direction.

An infinite plate allows thickness vibrations in which the only nonvanishing displacement component is u_3. The analysis is essentially like that of Sect. 36.1 with u_3 and τ_3^* replacing u_2 and τ_4^*. The components τ_1^* and τ_2^* do not vanish, but since they are independent of X_1 and X_2, they do not affect the solution for an infinite plate. With electrodes on the faces at $X_3 = \pm h$, E_3 is the only nonvanishing component of $\boldsymbol{E}$. The pertinent equations of (14.25) then reduce to

$$\begin{aligned} \tau_3^* &= c_{33}^E V_3 - e_{33} E_3, \\ D_3 &= e_{33} V_3 + \varepsilon_{33}^V E_3. \end{aligned} \tag{36.76}$$

From $(36.76)_2$,

$$E_3 = \frac{1}{\varepsilon_{33}^V} (D_3 - e_{33} V_3), \tag{36.77}$$

and substitution of this into $(36.76)_1$ gives

$$\begin{aligned} \tau_3^* &= \bar{c}_{33} V_3 - e_{33} D_3/\varepsilon_{33}^V, \\ \bar{c}_{33} &= c_{33}^E + e_{33}^2/\varepsilon_{33}^V = c_{33}^D. \end{aligned} \tag{36.78}$$

The piezoelectrically stiffened coefficient $\bar{c}_{33}$ is the same as $\bar{c}_{33}^D$ in the present case because e_{13} and e_{23} are both zero, the nondiagonal components of the dielectric tensor vanish, and $\beta_{33}^V = 1/\varepsilon_{33}^V$. Since all variables are assumed to be independent of X_1 and X_2, and since $D_{3,3} = 0$, the linearized equation of motion is

$$\bar{\varrho}\, \ddot{u}_3 = \bar{c}_{33}\, u_{3,33}. \tag{36.79}$$

For the assumed motion, the internal energy can be reduced to a function of the two variables, say $\hat{U}(V_3, D_3)$, such that

$$E_3 = \partial \hat{U}/\partial D_3, \qquad \tau_3^* = \partial \hat{U}/\partial V_3. \tag{36.80}$$

From (36.77) and (36.78), the second derivatives of $\hat{U}$ are

$$\partial \tau_3^*/\partial V_3 = \bar{c}_{33}, \qquad \partial E_{33}/\partial D_3 = 1/\varepsilon_{33}^V, \qquad \partial^2 \hat{U}/\partial V_3\, \partial D_3 = -e_{33}/\varepsilon_{33}^V. \tag{36.81}$$

By analogy with (36.22), the static electromechanical coupling factor for the assumed mode of motion is then

$$k^2 = \frac{e_{33}^2}{\bar{c}_{33}\, \varepsilon_{33}^V}. \tag{36.82}$$

The remainder of the analysis for an infinite plate goes through just as in the preceding sections.

The solution for an infinite plate of X-cut quartz is essentially the same with the 3-direction replaced by 1.

The actual "thickness" vibrations of a plate or disk of finite size are considerably more complicated than this analysis suggests because the boundary conditions at the periphery force quantities to vary with X_1 and X_2, contrary to the above assumption. In the *infinite* plate, we can take τ_1^* and τ_2^* to be whatever $(14.25)_1$ says, independent of X_1 and X_2; but the normal stress must vanish at the edge of an actual plate, and this results in a coupling of thickness and transverse motions.

The actual vibratory motion of circular disks has been investigated experimentally by E. A. G. SHAW (1956) in a comprehensive survey covering 25 different diameter/thickness ratios ranging from 1.14 to 6.63. Of twelve modes that were studied, there is no single one that can be identified as corresponding to the fundamental dilatational thickness resonance. An effective "dynamical" electromechanical coupling coefficient, based on an appropriate modification of (36.48), was measured for each mode, and it was found that as the diameter/thickness ratio is varied, each mode in turn exhibits a maximum of minimum of effective coupling in the thickness resonance region. For any given diameter/thickness ratio, two or three different modes have high coupling in this region. The vibration patterns have well-defined forms, but none approaches uniform piston-like motion. In a later study of *loaded* disks [SHAW and SUJIR, 1960] it was found that the discrete modes observed with an unloaded disk begin to merge so that the vibration patterns show a continuous transition with changing frequency in passing from one mode to the next in the region of thickness resonance.

37. Radial motion of thin circular piezoelectric ceramic disks.

The solution for the radial motion of a thin circular piezoelectric ceramic disk polarized along the thickness and provided with electrodes on the major surfaces is of considerable interest because such a disk is a convenient specimen to use in the evaluation of material properties of the ceramic. Moreover, it is primarily the radial resonances that couple with the thickness motion to prevent the occurrence of a" pure" piston-like thickness motion.

To gain some insight into the radial motion, let it be assumed that E_3 is the only nonvanishing component of $\boldsymbol{E}$ and that τ_1^* and τ_2^* are the only nonvanishing components of stress. Then, in view of the symmetry of the material coefficients for transverse isotropy in Tables 16.2, 16.3, and 16.4, Eqs. (14.23) give V_γ and D_i as

$$\begin{bmatrix} V_1 \\ V_2 \\ V_3 \\ D_3 \end{bmatrix} = \begin{bmatrix} s_{11}^E & s_{12}^E & d_{31} \\ s_{12}^E & s_{11}^E & d_{31} \\ s_{13}^E & s_{13}^E & d_{33} \\ d_{31} & d_{31} & \varepsilon_{33}^\tau \end{bmatrix} \begin{bmatrix} \tau_1^* \\ \tau_2^* \\ E_3 \end{bmatrix}. \tag{37.1}$$

From the equations for V_1 and V_2,

$$\begin{aligned} V_1 - V_2 &= (s_{11}^E - s_{12}^E)(\tau_1^* - \tau_2^*), \\ V_1 + V_2 &= (s_{11}^E + s_{12}^E)(\tau_1^* + \tau_2^*) + 2d_{31} E_3 \end{aligned} \tag{37.2}$$

whence

$$\begin{aligned} \tau_1^* &= \bar{c}_{11} V_1 + \bar{c}_{12} V_2 - e E_3, \\ \tau_2^* &= \bar{c}_{12} V_1 + \bar{c}_{11} V_2 - e E_3, \end{aligned} \tag{37.3}$$

where

$$\bar{c}_{11} = \frac{s_{11}}{s_{11}^2 - s_{12}^2}, \qquad \bar{c}_{12} = \frac{-s_{12}}{s_{11}^2 - s_{12}^2}, \qquad e = \frac{d_{31}}{s_{11} + s_{12}} \tag{37.4}$$

and the compliances are understood to be at constant E.

To apply (37.2) locally at any point, we can choose the 1-direction as radial and the 2-direction as circumferential in a cylindrical polar coordinate system. With u denoting the radial component of displacement, the radial and circumferential strain components are

$$V_r = \partial u / \partial r, \qquad V_\Theta = u/r. \tag{37.5}$$

In terms of radial and circumferential stresses T_r and T_Θ, the equation of radial motion is

$$\bar{\varrho}\ddot{u} = \frac{1}{r}\left[\frac{\partial}{\partial r}(r T_r) - T_\Theta\right] = \frac{\partial T_r}{\partial r} + (T_r - T_\Theta)/r. \tag{37.6}$$

In view of (37.3) and (37.5), noting that E_3 is independent of r for a fully electroded disk, we reduce the equation of motion to

$$\bar{\varrho}\ddot{u} = \bar{c}_{11}\left[\frac{\partial^2 u}{\partial r^2} + \frac{1}{r}\left(\frac{\partial u}{\partial r} - \frac{u}{r}\right)\right]. \tag{37.7}$$

The terms multiplying $\bar{c}_{12}$ drop out because $\partial V_\Theta / \partial r = (V_r - V_\Theta)/r$.

From $(37.2)_2$ and the equation for D_3 in (37.1),

$$D_3 = e(V_1 + V_2) + \varepsilon E_3 = e(V_r + V_\theta) + \varepsilon E_3 \tag{37.8}$$

where

$$\varepsilon \equiv \varepsilon_{33}^T - 2e d_{31} = \varepsilon_{33}^T - \frac{2d_{31}^2}{s_{11} + s_{12}}, \tag{37.9}$$

and e has already been defined in (37.4). From (37.8),

$$E_3 = \frac{1}{\varepsilon}[D_3 - e(V_1 + V_2)] = \frac{1}{\varepsilon}[D_3 - e(V_r + V_\Theta)]. \tag{37.10}$$

There is a *static* solution of (37.7) in which u is a constant times r so that $\partial^2 u/\partial r^2 = 0$ and $V_r = V_\Theta$. (Also, $T_r = T_\Theta$.) For this kind of deformation, the internal energy can be expressed as $\hat{U}(V_r, D_3)$, a function of the *two* variables V_r and D_3. With this constraint, $(37.3)_1$ and (37.10) yield

$$T_r = \left(\bar{c}_{11} + \bar{c}_{12} + \frac{2e^2}{\varepsilon}\right) V_r - \frac{e}{\varepsilon} D_3. \tag{37.11}$$

Now the thermodynamic tension conjugate to V_r is not T_r but $2T_r$ under the present constraints, since an increment dV_r implies an equal increment dV_Θ and the work $T_r dV_r + T_\Theta dV_\Theta = 2T_r dV_r$. Thus,

$$\frac{\partial \hat{U}}{\partial V_r} = 2T_r, \qquad \frac{\partial \hat{U}}{\partial D_3} = E_3, \tag{37.12}$$

and the appropriate static electromechanical coupling coefficient, defined by analogy with (36.22), can be evaluated from (37.10)–(37.12) and (37.4) and (37.9) as

$$k_p^2 = \frac{\left(\dfrac{\partial^2 \hat{U}}{\partial V_r \partial D_3}\right)^2}{\left(\dfrac{\partial^2 \hat{U}}{\partial V_r^2}\right)\left(\dfrac{\partial^2 \hat{U}}{\partial D_3^2}\right)} = \frac{2e^2/\varepsilon}{(\bar{c}_{11} + \bar{c}_{12} + 2e^2/\varepsilon)} = \frac{2d_{31}^2}{(s_{11}^E + s_{12}^E)\,\varepsilon_{33}^T}. \tag{37.13}$$

The subscript p in k_p is for "planar", as k_p is commonly called the planar coupling coefficient. The formula for k_p^2 is sometimes expressed in the form [BERLINCOURT, CURRAN, and JAFFE, 1964]

$$k_p^2 = \frac{2Y d_{31}^2}{(1-\nu)\,\varepsilon_{33}^T} \tag{37.14}$$

in which

$$Y = 1/s_{11}^E, \qquad \nu = -s_{12}^E/s_{11}^E. \tag{37.15}$$

Y and ν are YOUNG's modulus and POISSON's ratio for directions in the plane of the disk.

Let us now consider vibratory solutions of (37.7) The assumption

$$u = U(\omega r/v)\exp j\omega t \tag{37.16}$$

reduces (37.6) to

$$U'' + \frac{1}{\alpha_r}U' + \left(1 - \frac{1}{\alpha_r^2}\right)U = 0, \qquad \alpha \neq 0, \tag{37.17}$$

with

$$\begin{aligned} \alpha_r &= \omega r/v, \\ v^2 &= \bar{c}_{11}/\varrho. \end{aligned} \tag{37.18}$$

Eq. (37.17) is BESSEL's equation of first order, of which an appropriate solution is

$$U = A\,J_1(\alpha_r). \tag{37.19}$$

Now the charge density σ on an electrode is $\sigma = -D_3$ on the upper electrode at $X_3 = +h$ and of opposite sign on the other electrode. In the present approximation, both D_3 and E_3 are independent of X_3, and E_3 is independent of r. From (37.8) and (37.5), the total charge on the upper electrode of radius a is

$$Q = 2\pi\int_0^a e\,\sigma\,r\,dr = -\pi a^2 \varepsilon E_3 - 2\pi e\int_0^a \frac{\partial}{\partial r}(r u)\,dr = CV - 2\pi e\,a\,u(a) \tag{37.20}$$

where V is the voltage, in terms of which

$$E_3 = -V/2h, \quad \text{and} \quad C = \pi a^2 \varepsilon/2h. \tag{37.21}$$

The disk thickness has been taken as $2h$. From above, the amplitude Q_0 of the oscillating charge in the expression $Q = Q_0 \exp j\omega t$ is

$$Q_0 = C V_0 - 2\pi a e A J_1(\alpha) \tag{37.22}$$

where V_0 is the voltage amplitude and

$$\alpha \equiv \omega a/v. \tag{37.23}$$

The radial stress T_r is obtained from $(37.3)_1$, (37.5), (37.19), and (37.21). At $r = a$, the result for the stress amplitude is

$$T_r^0(a) = A\left[\bar{c}_{11}\frac{\omega}{v}J_1'(\alpha) + \bar{c}_{12}\frac{1}{a}J_1(\alpha)\right] + eV_0/2h,$$

or

$$a\,T_r^0(a) = A\bar{c}_{11}[\alpha J_1'(\alpha) + \nu J_1(\alpha)] + a e V_0/2h, \tag{37.24}$$

where $\nu = -s_{12}^E/s_{11}^E = \bar{c}_{12}/\bar{c}_{11}$. Unless the disk is constrained radially, the boundary condition at $r = a$ is that the radial stress should vanish. Now among the recursion relations for Bessel functions is

$$J_1'(\alpha) = J_0(\alpha) - \frac{1}{\alpha}J_1(\alpha). \tag{37.25}$$

Hence, the boundary condition can be expressed as

$$A\bar{c}_{11}[\alpha J_0(\alpha)-(1-\nu)J_1(\alpha)]=-a e V_0/2h. \tag{37.26}$$

With A from (37.26), (37.22) becomes

$$Q_0=V_0 C\left[1+\frac{2e^2}{\varepsilon\bar{c}_{11}}\frac{J_1(\alpha)}{[\alpha J_0(\alpha)-(1-\nu)J_1(\alpha)]}\right]. \tag{37.27}$$

From (37.13),

$$\frac{2e^2}{\varepsilon(\bar{c}_{11}+\bar{c}_{12})}=\frac{2e^2}{\varepsilon\bar{c}_{11}(1+\nu)}=\frac{k_p^2}{1-k_p^2}, \tag{37.28}$$

and hence the admittance $j\omega Q_0/V_0$ can be expressed as

$$\frac{j\omega Q_0}{V_0}=j\omega C\left[1+\frac{k_p^2}{(1-k_p^2)}\frac{(1+\nu)J_1(\alpha)}{[\alpha J_0(\alpha)-(1-\nu)J_1(\alpha)]}\right]. \tag{37.29}$$

The frequencies of resonance on short-circuit, for which the coefficient of A vanishes in (37.26), are frequencies of infinite admittance. The frequencies of antiresonance, or zero admittance, can be identified as natural frequencies of unforced vibration under open circuit conditions.

Eq. (37.29) is basic to the determination of material properties of piezoelectric ceramics from measurements on thin disks. A low-frequency capacitance measurement determines C and hence ε, since the dimensions can be measured. POISSON's ratio ν can be determined from the ratio of successive resonance frequencies [MEITZLER, OBRYAN, and TIERSTEN, 1973], while the absolute value of the lowest resonance then enables $v=(\bar{c}_{11}/\varrho)^{\frac{1}{2}}$ to be determined in $\alpha_R=\omega_R a/v$. Hence, $\bar{c}_{11}$ can be determined as $\bar{c}_{11}=\varrho v^2$. The planar coupling factor k_p^2 can be determined from the lowest frequency of resonance and the next antiresonance. From (37.13) and (37.9),

$$\varepsilon/\varepsilon_{33}^{T}=1-k_p^2, \tag{37.30}$$

and hence ε_{33}^{T} can be determined. Now ν and $\bar{c}_{11}$ determine

$$\bar{c}_{11}+\bar{c}_{12}=\bar{c}_{11}(1+\nu)=\frac{1}{s_{11}^E+s_{12}^E}=\frac{1}{s_{11}^E(1-\nu)}=\frac{Y}{1-\nu}. \tag{37.31}$$

Finally, d_{31} can be determined from (37.13) or (37.14). In practice, the situation is complicated by dielectric and mechanical losses, which cause the impedance (37.29) to have a real part, and by coupling to thickness dilatational modes. These expected effects are more or less amenable to analysis and calculation [IEEE Standards No. 177 (1966), No. 178 (1958), No. 179 (1961)]. An unexpected compilcation is that the low-frequency capacitance of certain ceramics has been reported not to have any well-defined limit as $\omega\to 0$ [GERSON, 1960]. This is presumed to be a polarization effect due to the voltage applied during the capacitance measurement. As the frequency is lowered, there is more time for re-orientation of the polarization, an effect which suggests a dependence of the polarization on the *history* of the field in these materials. Because of experimental difficulties, it is not known whether lowering the measurement voltage to arbitrarily small values would eliminate the effect. In practice, it is usual to report the dielectric constant measured at 1 kHz.

For small coupling, one sometimes works with an approximate version of (37.29) in the neighborhood of a resonance, obtained by expanding about a zero of the denominator. Define

$$G(\alpha)=\alpha J_0(\alpha)-(1-\nu)J_1(\alpha). \tag{37.32}$$

Then

$$G'(\alpha)=J_0(\alpha)+\alpha J_0'(\alpha)-(1-\nu)\, J_1'(\alpha). \tag{37.33}$$

In view of (37.25) and the relation

$$J_0'(\alpha)=-J_1(\alpha), \tag{37.34}$$

(37.33) becomes

$$\begin{aligned} G'(\alpha)&=\nu J_0(\alpha)+\left(\frac{1-\nu}{\alpha}-\alpha\right) J_1(\alpha)\\ &=\frac{\nu}{\alpha} G(\alpha)+\left(\frac{1-\nu^2}{\alpha}-\alpha\right) J_1(\alpha). \end{aligned} \tag{37.35}$$

From (37.29) and (37.32), the antiresonance frequencies satisfy

$$\frac{G(\alpha)}{J_1(\alpha)}=-(1+\nu)\frac{k_p^2}{1-k_p^2}. \tag{37.36}$$

Expanding about $\alpha_R=\omega_R\, a/v$, a zero of $G(\alpha)$, writing

$$\frac{G(\alpha_A)}{J_1(\alpha_A)}=(\alpha_A-\alpha_R)\frac{G'(\alpha_R)}{J_1(\alpha_R)}+\cdots, \tag{37.37}$$

one obtains

$$\frac{\omega_A-\omega_R}{\omega_R}=\frac{1+\nu}{\alpha_R^2-(1-\nu^2)}\,\frac{k_p^2}{1-k_p^2}. \tag{37.38}$$

The planar coupling coefficient involved in the radial mode is also the basis for generating axially symmetric flexural vibrations in layered disks. If two piezoelectric ceramic disks, oppositely polarized through the thickness, are bonded back-to-back, or with a thin metal layer sandwiched between them, an electric field through the thickness will make one of them tend to expand radially and the other contract, resulting in axially symmetric flexure. Flexure can also be generated by a single piezoelectric ceramic disk bonded to a passive metal one. In this case, "attempted" radial vibrations of the ceramic, constrained on one face, are converted to axially symmetric flexure. Such arrangements have been considered for voice-frequency microphones.

The equivalent circuit of a radially vibrating disk has been discussed by LAZUTKIN and MIKHAILOV (1972), who also indicate formulas for the electromechanical coupling factor.

38. Waves of finite amplitude in elastic media.

The aim of the next two sections is to relate higher-order elastic coefficients to measurements on distortion of finite-amplitude waves (Sect. 38) and measurements of wave velocity versus particle velocity in shock waves (Sect. 39). It is regretted that a discussion of nonlinear dissipative and dispersive phenomena and nonlinea propagation in lattices is not included. An introduction to some of the modern work on these subjects may be obtained from the papers of N. J. ZABUSKY (1967, 1973).

Elastic nonlinearity, as expressed through nonlinearity of the equation of motion, manifests itself in (1) the interaction of waves with each other, (2) distortion of a wave of finite amplitude, and (3) a dependence of the transit time of small-amplitude waves on the initial static stress or strain. Observations on each of these phenomena can in principle be used to deduce information concerning the nature of the nonlinearities, and, in particular, to obtain experimental values of some or all of the third-order elastic coefficients.

Still a fourth method of determining higher-order elastic coefficients is through observations on shock wave velocity and the associated particle velocity [FOWLES, 1967; GRAHAM, 1972]. The interaction of acoustic waves in solids has been studied experimentally [ROLLINS, TAYLOR, and TODD, 1964] in various materials and for various angles of the intersecting beams, and the subject has been reviewed by ROLLINS (1965). The most reliable information is obtained from measurements of the stress dependence of the transit time of small-amplitude waves using the analysis of Sect. 29. Another interesting technique is the observation of finite-amplitude distortion of an initially sinusoidal wave. The distortion can be detected in transparent materials by light diffraction techniques [MELNGAILIS, MARADUDIN, and SEEGER, 1963], and in any case by a capacitance probe [GAUSTER and BREAZEALE, 1966] at the face of the specimen opposite the transducer. This technique has been used succesfully to obtain certain combinations of third-order elastic constants. It is the purpose of this section to relate higher-order elastic constants to the growth of harmonics in an initially sinusoidal wave.

38.1. Equation of motion. The exact equation of motion (6.7) with the body force omitted, in view of (4.8), (8.4), and the assumption of ideal thermoelasticity that the Piola-Kirchhoff stress t_{ij} is always equal to its static equilibrium value, can be written as

$$\bar{\varrho}\ddot{x}_i = \frac{\partial}{\partial X_k}\left(\frac{\partial x_i}{\partial X_s}\frac{\partial U}{\partial V_{ks}}\right). \tag{38.1}$$

This equation is to be expanded in powers of the displacement gradients from the natural state under the assumption that the deformation takes place at constant entropy. First, the internal energy is expanded in powers of the strain from the natural state:

$$U(V_{ij}, S) - U(0, S) = \tfrac{1}{2} c_{ijkm} V_{ij} V_{km} + \tfrac{1}{6} c_{ijkmpq} V_{ij} V_{km} V_{pq} + \cdots. \tag{38.2}$$

The coefficients are the elastic constants, referred to and evaluated at the natural state. Carrying out the indicated differentiations in (38.1), expressing the strain components and their derivatives in terms of the displacement gradients through (28.6), (3.2), and (3.4), we develop the expansion

$$\bar{\varrho}\ddot{u}_i = \frac{\partial^2 u_k}{\partial X_i \partial X_m}\left[c_{ijkm} + \frac{\partial u_p}{\partial X_q}\left(M_{ijkmpq} + \frac{\partial u_r}{\partial X_s} M_{ijkmpqrs} + \cdots\right)\right], \tag{38.3}$$

where

$$M_{ijkmpq} = c_{ijkmpq} + \delta_{kp} c_{ijmq} + \delta_{ik} c_{jmpq} + \delta_{ip} c_{jkmq}, \tag{38.4}$$

$$\begin{aligned} M_{ijkmpqrs} = {} & \tfrac{1}{2} c_{ijkmpqrs} + \delta_{ir} c_{sjkmpq} + \delta_{kp} c_{ijmqrs} \\ & + \tfrac{1}{2}\delta_{pr} c_{ijkmqs} + \tfrac{1}{2}\delta_{ik} c_{jmpqrs} + \delta_{kp}\delta_{ir} c_{jqms} + \tfrac{1}{2}\delta_{pr}\delta_{ik} c_{jmqs}. \end{aligned} \tag{38.5}$$

Our application of (38.3) is to purely longitudinal motion. In anisotropic media, such motion can occur only along certain directions. If X (witout a subscript) is the coordinate in one of these directions, and u the displacement, then the one-dimensional form of Eq. (38.3) is

$$\bar{\varrho}\ddot{u} = \frac{\partial^2 u}{\partial X^2}\left[M_2 + M_3 \frac{\partial u}{\partial X} + M_4\left(\frac{\partial u}{\partial X}\right)^2 + \cdots\right] \tag{38.6}$$

where M_2 is a linear combination of second-order coefficients, M_3 a linear combination of second-order and third-order coefficients, and M_4 a linear combination of second-order, third-order, and fourth-order coefficients. Explicitly, if $\boldsymbol{N}$ denotes a unit vector in the direction of the one-dimensional motion under consideration, then

$$
\begin{aligned}
M_2 &= c_{ijkm} N_i N_j N_k N_m, \\
M_3 &= M_{ijkmpq} N_i N_j N_k N_m N_p N_q, \\
M_4 &= M_{ijkmpqrs} N_i N_j N_k N_m N_p N_q N_r N_s.
\end{aligned}
\tag{38.7}
$$

Because of the way in which both second-order and third-order elastic coefficients enter the constant M_3, it is frequently broken down into [BREAZEALE and FORD, 1965].

$$M_3 = K_3 + 3K_2 \tag{38.8}$$

where

$$K_2 \equiv M_2 = c_{ijkm} N_i N_j N_k N_m, \tag{38.9}$$

$$K_3 = c_{ijkmpq} N_i N_j N_k N_m N_p N_q. \tag{38.10}$$

The equation of longitudinal motion thus has the form

$$\ddot{u} = \frac{\partial^2 u}{\partial X^2} g(\partial u/\partial X) \tag{38.11}$$

in which the function $g(\partial u/\partial X)$ can be expanded as

$$g(\xi) = \frac{1}{\bar{\varrho}} (M_2 + M_3\xi + M_4\xi^2 + M_5\xi^3 + \cdots + M_{n+2}\xi^n + \cdots). \tag{38.12}$$

Now the equation of one-dimensional motion of a fluid can be expressed in this same form under the assumption that the pressure is a function of the density. Since much work has been done in fluids, it is worth indicating the correspondence of the commonly used nonlinearity parameters for fluids with the coefficients M_i in (38.6) and (38.12). The exact equation of motion of a fluid in which $T_{ij} = -p\,\delta_{ij}$ is, from (6.5) with the body force omitted,

$$\varrho\,\ddot{u} = -\frac{\partial p}{\partial x_i} = -\frac{\partial X_j}{\partial x_i}\frac{\partial p}{\partial X_j}. \tag{38.13}$$

For one-dimensional motion in the direction indicated by the coordinate X (without a subscript),

$$\bar{\varrho}/\varrho = \partial x/\partial X = 1 + (\partial u/\partial X), \tag{38.14}$$

so that the exact one-dimensional equation becomes

$$\bar{\varrho}\,\ddot{u} = -\frac{\partial p}{\partial X}. \tag{38.15}$$

With the additional assumption that the pressure is a function of the density, we have

$$\frac{\partial p}{\partial X} = \frac{dp}{d\varrho}\frac{\partial \varrho}{\partial X} = -\frac{\varrho^2}{\bar{\varrho}}\frac{dp}{d\varrho}\frac{\partial}{\partial X}\left(\frac{\bar{\varrho}}{\varrho}\right) = -\frac{\varrho^2}{\bar{\varrho}}\frac{dp}{d\varrho}\frac{\partial^2 u}{\partial X^2}, \tag{38.16}$$

where (38.14) has been used to evaluate $\partial(\bar{\varrho}/\varrho)/\partial X$. From (38.14)–(38.16)

$$\ddot{u} = \frac{\varrho^2}{\bar{\varrho}^2}\frac{dp}{d\varrho}\frac{\partial^2 u}{\partial X^2} = \left(1 + \frac{\partial u}{\partial X}\right)^{-2}\frac{dp}{d\varrho}\frac{\partial^2 u}{\partial X^2}. \tag{38.17}$$

This equation appears in the classical literature [H. LAMB, 1932, p. 481; J. W. STRUTT, Lord RAYLEIGH, 1929]. It has been customary to express the pressure p in the form

$$p = p_0 + A s + \frac{B s^2}{2} + \frac{C s^3}{3!} + \frac{D s^4}{4!} + \cdots \tag{38.18}$$

where

$$s \equiv (\varrho - \bar{\varrho})/\bar{\varrho}. \tag{38.19}$$

A special case that can be considered is that of a fluid with constant pressure derivative of the bulk modulus:

$$\frac{d}{dp}\left(\varrho \frac{dp}{d\varrho}\right) = \gamma = \text{constant}. \tag{38.20}$$

An example of this is the behavior of an ideal gas at constant entropy. By comparing the expansions of the equation of motion for these cases, one finds [THURSTON and SHAPIRO, 1967],

$$\begin{aligned} -\frac{M_3}{M_2} &= 2 + \frac{B}{A} = \gamma + 1, \\ \frac{M_4}{M_2} &= 3 + \frac{3B}{A} + \frac{C}{2A} = \frac{1}{2}(\gamma+1)(\gamma+2), \\ -\frac{M_5}{M_2} &= 4 + \frac{6B}{A} + \frac{2C}{A} + \frac{D}{6A} = \frac{1}{6}(\gamma+1)(\gamma+2)(\gamma+3). \end{aligned} \tag{38.21}$$

These associations of nonlinearity parameters are helpful in correlating the treatments of one-dimensional waves of finite amplitude in solids, perfect gases, and other fluids.

38.2. Characteristics of the equation of one-dimensional longitudinal motion. The partial differential equation satisfied by purely longitudinal motion at constant entropy is (38.11) with the function g determined by the equation of state of the medium, here expressed by (38.12). We assume $g > 0$. The characteristic curves of Eq. (38.11) are readily shown to have slope $dX/dt = \pm\sqrt{g}$ in the (t, X)-plane, while the changes of $\partial u/\partial t$ and $\partial u/\partial X$ along a characteristic must obey

$$d\left(\frac{\partial u}{\partial t}\right) = \pm\sqrt{g}\, d\left(\frac{\partial u}{\partial X}\right), \tag{38.22}$$

where the $+$ sign pertains to the integration along characteristics with slope $+\sqrt{g}$, and the $-$ sign with slope $-\sqrt{g}$. Defining

$$\lambda \equiv -\int_{\xi_0}^{\partial u/\partial X} [g(\xi)]^{\frac{1}{2}}\, d\xi, \tag{38.23}$$

and noting that the particle velocity is

$$v \equiv \partial u/\partial t, \tag{38.24}$$

we see that (38.22) becomes

$$d(v \pm \lambda) = 0. \tag{38.25}$$

A pair of "Riemann invariants" (r, s) can be defined by [COURANT and FRIEDRICHS, 1948]

$$2r \equiv v + \lambda, \qquad -2s \equiv v - \lambda. \tag{38.26}$$

Then, by integration of (38.22) along the characteristics, we see that r is constant along the characteristics of positive slope, and s is constant along the characteristics of negative slope. If either r or s is the same constant *throughout a region* of the (t, X)-plane, the solution in that region is called a *simple wave.*

38.3. Simple-wave solution. Let us consider the half-space $X \geqq 0$ to be initially unstrained and at rest until $t = 0$, after which the motion of the face $X = 0$ is prescribed.

A solution describing a continuous transition of velocity from a constant initial state necessarily contains a simple-wave region [COURANT and FRIEDRICHS, 1948]. In the present case, this is the region of the (t, X)-plane covered by those negative-sloping characteristics that lead back to the zone of constant initial conditions (constant v and λ) without crossing any discontinuities in v or λ. We shall be concerned with the solution only in this region. Clearly, s is constant throughout this region, and each characteristic of positive slope carries constant values of v and λ, and therefore also a constant value of $\partial u/\partial X$. But $\partial u/\partial X$ determines the slope of the characteristics through the function $g(\partial u/\partial X)$. Since a line of constant slope is straight, the positive-sloping characteristics are straight lines. Because $(v-\lambda)$ is constant throughout the simple wave, we may write

$$v=\frac{\partial u}{\partial t}=k-\int_0^{\partial u/\partial X}\sqrt{g(\xi)}\,d\xi. \tag{38.27}$$

The constant k is the value of v when $\partial u/\partial X=0$. Since the medium is initially unstrained and at rest, $k=0$. The important properties of the simple wave are (38.27) and the fact that both v and $\partial u/\partial X$ are constant along straight lines of slope $\partial X/\partial t=[g(\partial u/\partial X)]^{\frac{1}{2}}$.

This last property can be described by saying that surfaces of constant $\partial u/\partial t$ or constant $\partial u/\partial X$ are *propagated* with the speed $|\delta X/\delta t|=g^{\frac{1}{2}}$. We denote this quantity by W:

$$W=g^{\frac{1}{2}}. \tag{38.28}$$

In the present notation X is a "Lagrangian" or material coordinate. Hence, W is the speed of propagation in the medium relative to the *reference positions* of the particles, i.e., relative to their positions in a configuration of constant density $\bar{\varrho}$. The corresponding speed *in space* may be obtained as $V=v+c$ where $c=W\partial x/\partial X=W(1+\partial u/\partial X)$. This quantity c may be interpreted as a local propagation speed relative to the moving medium. W and c correspond respectively to the natural and actual speeds of propagation of small-amplitude waves in a stressed medium [cf. (28.34). V in (28.34) plays the role of the present c].

Because the particle velocity is constant along a straight line of slope $W=g^{\frac{1}{2}}$, the velocity at any point (t, X) is equal to the velocity of the face $X=0$ at the earlier time $\tau=t-X/W$. Thus, in the simple-wave region,

$$\begin{aligned} v(t, X)&=v(\tau, 0),\\ \tau&=t-X/W=t-X g^{-\frac{1}{2}}. \end{aligned} \tag{38.29}$$

In (38.29), it has to be understood that $W=[g(\partial u/\partial X)]^{\frac{1}{2}}$ must be evaluated at the value of $\partial u/\partial X$ that corresponds, through (38.27), to the v under consideration. With $g(\partial u/\partial X)$ and $v(\tau, 0)$ given, (38.27) and (38.29) provide an implicit relation for $v(t, X)$ in the simple-wave region. Special considerations are necessary when $v(\tau, 0)$ undergoes a jump.

The nature of simple-wave solutions has been known for a long time. BLACKSTOCK (1962) and COURANT and FRIEDRICHS (1948) have written clear treatments with references to the early work of POISSON, STOKES, EARNSHAW, RIEMANN, and others.

38.4. The discontinuity distance. Many workers have calculated the distance at which the simple-wave solution breaks down. Unless g is constant, the waveform becomes distorted as it progresses through the medium. In a hypothetical lossless nonlinear medium such as that described by the present equations, the waveform becomes distorted to such an extent that the velocity-distance curve, plotted for some constant time, acquires a vertical tangent, i.e., $\partial v/\partial X \to \infty$. The solution cannot be continued analytically beyond the corresponding value of X. In terms of characteristics, the simple wave cannot persist indefinitely because positive-sloping characteristics corresponding to different values of v would intersect. The vertical tangent $\partial v/\partial X \to \infty$ signals the onset of a nonphysical multivaluedness that can be associated with the intersection of characteristics carrying different values of v and $\partial u/\partial X$. Hence, we are interested in the shortest distance X at which positive-sloping characteristics intersect.

Consider the positive-sloping characteristics ($dX/dt = +W$) through $(t_1, 0)$ and $(t_2, 0)$. When $t_2 \to t_1$, the value of X at which these characteristics intersect tends to

$$X_1 = \frac{2g^{\frac{3}{2}}}{g' \, \partial v/\partial X} = -\frac{2g^2}{g' \, \ddot{u}(t_1, 0)} \tag{38.30}$$

where g and g' are to be evaluated on the characteristic through $(t_1, 0)$. When $g' < 0$, which is the usual case, the positive solutions of (38.30) correspond to positive acceleration $\ddot{u}$. From (38.27), (38.28), and the equation of motion (38.11), $\partial v/\partial X$ is a negative quantity $(-1/W)$ times the acceleration. Hence, the growth of $\partial v/\partial X$ (and of the acceleration) corresponds in this case to the familiar situation in which the velocity crests gain on the troughs so that the portions of the wave profile for which $\partial v/\partial X < 0$ become progressively steeper. In a medium with $g' > 0$, the troughs gain on the crests, resulting in a steepening of the portion of the profile with $\partial v/\partial X > 0$. Stable rarefaction shocks ($g' > 0$) have been observed in vitreous silica [BARKER and HOLLENBACH (1970)]. We assume $g' < 0$ in the following discussion.

If the motion starts with a velocity jump V_0, positive-sloping characteristics intersect already at the origin. If $V_0 > 0$, the jump propagates as a shock with a speed $> W_0$, and there is strictly no simple-wave region. If $V_0 < 0$, the corresponding propagation speed $< W_0$, and the velocity jump gets immediately smoothed out into a continuous rarefaction in the region between the characteristics $X = t\,W_0$ and $X = t\,W^+$ where W^+ is the value of $g^{\frac{1}{2}}$ that corresponds, through (38.27), to the particle velocity V_0. The positive-sloping characteristics ($dX/dt = +W$) that cover this region all intersect at the origin of the (t, X)-plane and the initial singularity is limited to that single point. The disturbance in this region is called a centered simple wave [COURANT and FRIEDRICHS, 1948].

An approximation to the minimum positive value of X_1 satisfying (38.30) can be obtained by evaluating g and g' at $\partial u/\partial X = 0$, and using the maximum surface acceleration ωv_0. The value of X_1 so obtained is called the "discontinuity length", L:

$$L = -\frac{2[g(0)]^2}{\omega v_0 \, g'(0)}. \tag{38.31}$$

Eq. (38.31) may be written as

$$L = -\frac{2W_0^4 \bar{\varrho}}{\omega v_0 M_3} = -\frac{2W_0^2 M_2}{\omega v_0 M_3} \tag{38.32}$$

because, from Eq. (38.12),

$$g(0) = W_0^2 = M_2/\bar{\varrho}, \quad g'(0) = M_3/\bar{\varrho}. \tag{38.33}$$

Other forms of (38.32) may be obtained by replacing M_3/M_2 as indicated in (38.21). In the cases defined by (38.18) and (38.20), the well-known results are

$$L = \frac{2W_0^2}{\omega v_0(\gamma+1)}, \quad L = \frac{2W_0^2}{\omega v_0(2+B/A)}. \tag{38.34}$$

Eq. (38.31) is indicated here as an *approximation* to the exact minimum of (38.30). However, if the motion starts with zero initial velocity and positive initial acceleration ωv_0, the point at which the solution first breaks down is given exactly by $X=L$, $t=L/c_0$. BLACKSTOCK (1962) has calculated, for perfect gases, the distance that corresponds to the exact minimum of (38.30) for nonzero initial velocity at $X=0$. PETER D. LAX (1964) provides an alternative treatment.

When the motion starts with zero velocity but nonzero initial acceleration, a surface of discontinuity in the acceleration (and in $\partial v/\partial X$) is initiated. It is interesting that (38.30) can be connected with a corresponding result in the theory of propagation of singular surfaces across which there is a jump in the acceleration of initial magnitude ωv_0. The magnitude of a plane jump in the acceleration propagating into a homogeneous region at rest has been shown to decay if the jump has the same sign as $g'(0)$, but to become infinite in the finite time L/W_0 if the jump and $g'(0)$ are of opposite sign [THOMAS, 1957; COLEMAN and GURTIN, 1965; CHEN, 1972].

After a certain time, the simple-wave solution breaks down even near $X=0$, for it may not be continued into the range of influence of the shock that presumably originates where $\partial v/\partial X\to\infty$. Thus, with a shock forming at $X\doteq L$, $t\doteq L/W_0$, the solution near $X=0$ would be expected to break down for t greater than about $2L/W_0$, the approximate time required for reflections off the shock to arrive back at $X=0$.

BREAZEALE (1966) has presented the results of calculations of the discontinuity distance in various solids.

38.5. Particle velocity in the oscillating simple wave. To investigate the growth of harmonics in an initially sinusoidal wave, the displacement at $X=0$ can be prescribed as

$$\begin{aligned} &u(t,0)=0, && t<0,\\ &u(t,0)=tV_0+u_0(1-\cos\omega t), && t>0. \end{aligned} \tag{38.35}$$

With $V_0=0$, this covers the case treated by FUBINI-GHIRON (1935) referred to in the tutorial reviews of BLACKSTOCK (1962) and BEYER (1965). THURSTON and SHAPIRO (1967) included the constant velocity V_0 (which can always be put to zero) in order to reveal additional features of the solution. The initial conditions are that the half space is unstrained and at rest at $t=0$. It is well known [COURANT and FRIEDRICHS, 1948] that this problem has no solution with a continuous velocity field on the entire region ($0\leqq X<\infty$, $0<t<\infty$), except in the case of a hypothetical linear medium having $g(\partial u/\partial X)=$constant. However, there is a simple-wave solution that is valid for short times, and the predicted harmonic growth near $X=0$ can be presumed relevant to experiments on finite-amplitude distortion.

By differentiation of (38.35),

$$\begin{aligned} &v(\tau,0)=0, && \tau<0,\\ &v(\tau,0)=(V_0+\omega u_0\sin\omega\tau), && \tau>0. \end{aligned} \tag{38.36}$$

Substitution into (38.29) yields the following relation valid in the oscillating part of the simple-wave region.

$$v(t,X)=V_0+\omega u_0\sin[\omega(t-Xg^{-\frac{1}{2}})],\qquad t>X/W^+. \tag{38.37}$$

If $V_0=0$, $W^+=W_0$. If $V_0<0$, there is a centered simple wave on $X/W_0<t<X/W^+$ as noted previously. Thus, the characteristic $X=tW^+$ marks the beginning of (38.37). The other boundary is determined by the formation of a discontinuity,

and probably lies somewhere near $t=(2L-X)/W_0$, the approximate position of the first negative-sloping characteristic ($dX/dt=-W$) that intersects the shock.

Eq. (38.37) contains $g^{-\frac{1}{2}}=1/W$, which can be expressed as a power series in v through (38.12) and (38.27). The calculation is facilitated by noting that

$$\frac{dW}{dv}=\frac{d(\partial u/\partial X)}{dv}\,\frac{dW}{d(\partial u/\partial X)} \tag{38.38}$$

and, from (38.28) and (38.27)

$$\frac{dW}{d(\partial u/\partial X)}=\frac{1}{2}g'\,g^{-\frac{1}{2}}=g'/2W, \tag{38.39}$$

$$\frac{d(\partial u/\partial X)}{dv}=-g^{-\frac{1}{2}}=-1/W. \tag{38.40}$$

By straightforward calculation,

$$\begin{aligned} \frac{W_0}{W}&=1-\frac{1}{2}c_3 w+\frac{1}{2}c_4 w^2-\frac{1}{6}c_5 w^3+\cdots,\\ w&=v/W_0,\\ c_3&=-M_3/M_2,\\ c_4&=c_3^2-M_4/M_2,\\ c_5&=\tfrac{1}{2}(-6c_3^3+13\,c_4 c_3)-3M_5/M_2. \end{aligned} \tag{38.41}$$

THURSTON and SHAPIRO (1967) assumed $c_3\neq 0$ and defined

$$\begin{aligned} m_3&\equiv c_3, \quad m_4=c_4/c_3^2, \quad m_5=c_5/c_3^3,\\ \alpha&=c_3 M, \quad M=\omega u_0/W_0, \end{aligned} \tag{38.42}$$

to put (38.41) in the form

$$\begin{aligned} \frac{W_0}{W}&=1-\frac{1}{2}\alpha y+\frac{1}{2}m_4\alpha^2 y^2-\frac{1}{6}m_5\alpha^3 y^3+\cdots,\\ y&\equiv v/\omega u_0. \end{aligned} \tag{38.43}$$

Making the definitions

$$z\equiv(v-V_0)/\omega u_0, \quad r\equiv -V_0/\omega u_0, \tag{38.44}$$

they found

$$W_0/W=\beta-a_1 z-a_2 z^2-a_3 z^3-\cdots \tag{38.45}$$

in which

$$\begin{aligned} \beta&=1+\tfrac{1}{2}\alpha r+\tfrac{1}{2}m_4\alpha^2 r^2+\tfrac{1}{6}m_5\alpha^3 r^3+\cdots,\\ a_1&=\tfrac{1}{2}\alpha+m_4\,\alpha^2 r+\tfrac{1}{2}m_5\,\alpha^3 r^2+\cdots,\\ a_2&=-\tfrac{1}{2}\alpha^2(m_4+m_5\,\alpha r+\cdots),\\ a_3&=\tfrac{1}{6}\alpha^3 m_5+\cdots. \end{aligned} \tag{38.46}$$

Since the characteristic $X=tW^+$ carries the value $v=V^0$, it can be seen from (38.45) that

$$\beta=W_0/W^+. \tag{38.47}$$

Eq. (38.37) can now be written as

$$\begin{aligned} z&=\sin\left[T-Y(\beta-a_1 z-a_2 z^2-a_3 z^3+\cdots)\right],\\ &\omega X/W^+<T<\,? \end{aligned} \tag{38.48}$$

where

$$T\equiv\omega t, \quad Y\equiv\omega X/W_0, \quad \beta Y=\omega X/W^+. \tag{38.49}$$

T and Y are the nondimensionalized "time" and "distance" appropriate for the problem. In (38.48), the end of the simple-wave region, presumed near $T=\omega(2L-X)/W_0$, is indicated by the question mark.

To investigate the growth of harmonics in the oscillating part of the simple-wave region, z can be written as a Fourier series in the variable

$$\Theta \equiv T-\beta Y=\omega(t-X/W^+). \tag{38.50}$$

Thus,

$$z\equiv\frac{v-V_0}{v_0}=\sum_{n=1}^{\infty}(A_n\cos n\Theta+B_n\sin n\Theta), \tag{38.51}$$

in which the coefficients A_n and B_n are given by the standard formulas

$$A_n=\frac{1}{\pi}\int_0^{2\pi} z\cos n\Theta\,d\Theta,\qquad B_n=\frac{1}{\pi}\int_0^{2\pi} z\sin n\Theta\,d\Theta. \tag{38.52}$$

With the change of variable

$$z=\sin\xi, \tag{38.53}$$

(38.48) can be expressed as

$$\Theta=\xi-Y(a_1\sin\xi+a_2\sin^2\xi+a_3\sin^3\xi+\cdots). \tag{38.54}$$

From (38.52) and (38.53), the Fourier coefficients become [THURSTON and SHAPIRO, 1967]

$$\begin{aligned}
&A_0=0,\\
&A_1=-\tfrac{1}{8}\alpha^2 Y(m_4+\alpha r m_5)+\cdots,\\
&A_2=-\tfrac{1}{8}\alpha^3 m_4 Y^2+\cdots,\\
&A_3=\tfrac{1}{8}\alpha^2 Y(m_4+\alpha r m_5)+\cdots,\\
&A_4=\tfrac{1}{8}\alpha^3 m_4 Y^2+\cdots,\\
&B_1=1-\tfrac{1}{32}\alpha^2(1+4\alpha m_4 r)Y^2+\cdots,\\
&B_2=\tfrac{1}{4}\alpha[1+2\alpha m_4 r+m_5\alpha^2(r^2+\tfrac{1}{6})]Y-\tfrac{1}{48}\alpha^3Y^3+\cdots,\\
&B^2=\tfrac{3}{32}\alpha^2(1+4\alpha m_4 r)Y^2+\cdots,\\
&B_4=-\tfrac{1}{48}\alpha^3 m_5 Y+\tfrac{1}{24}\alpha^3Y^3+\cdots.
\end{aligned} \tag{38.55}$$

38.6. Solution for the displacement. An integration of (38.51) with respect to t [recall $\Theta=\omega(t-X/W^+)$] gives the particle displacement as

$$\begin{aligned}
&u=u_0\sum_{n=1}^{\infty}\left(\frac{A_n}{n}\sin n\Theta-\frac{B_n}{n}\cos n\Theta\right)+tV_0+F(X),\\
&\Theta=\omega(t-X/W^+),\qquad X/W^+<t<\,?
\end{aligned} \tag{38.56}$$

The function $F(X)$ can be determined by matching (38.56) to the value along the bounding characteristic $t=X/W^+$. Since the boundary $t=X/W^+$ corresponds to $\Theta=0$, the condition for determining $F(X)$ is

$$(u)_{t=X/W^+}=-u_0\sum_{n=1}^{\infty}\frac{B_n}{n}+XV_0/W^++F(X). \tag{38.57}$$

In the special case $V_0=0$, we have $u=0$ along the bounding characteristic and

$$F(X)=u_0\sum_{n=1}^{\infty}(B_n/n). \tag{38.58}$$

In this case (38.56) can be written

$$u = u_0 \sum_{n=1}^{\infty} \left[\frac{A_n}{n} \sin n\Theta + \frac{B_n}{n} (1 - \cos n\Theta) \right]. \tag{38.59}$$

To handle the case $V_0 \neq 0$, one can expand $\partial u/\partial X$ in powers of $w \equiv v/W_0$ or $y \equiv v/\omega u_0$, obtaining [THURSTON and SHAPIRO, 1967]

$$\begin{aligned} \frac{\partial u}{\partial X} &= -w + \frac{1}{4} m_3 w^2 - \frac{1}{6} m_3^2 m_4 w^3 + \frac{1}{24} m_3^3 m_5 w^4 + \cdots \\ &= -M y + \frac{1}{4} m_3 M^2 y^2 - \frac{1}{6} m_4 m_3^2 M^3 y^3 + \frac{1}{24} m_5 m_3^3 M^4 y^4 + \cdots. \end{aligned} \tag{38.60}$$

Along the bounding characteristic, $v = V_0$ so that $y = -r$ where r is defined in (38.44). Thus, (38.60) becomes

$$\frac{1}{M} \left(\frac{\partial u}{\partial X} \right)_{t=X/W^+} \equiv e_0 = r \left(1 + \frac{1}{4} \alpha r + \frac{1}{6} m_4 \alpha^2 r^2 + \frac{1}{24} m_5 \alpha^3 r^3 + \cdots \right), \tag{38.61}$$

where e_0 is introduced simply as an abbreviation. Now the derivative of (38.56) with respect to X, evaluated along the bounding characteristic $X = tW^+$, is

$$\begin{aligned} \left(\frac{\partial u}{\partial X} \right)_{t=X/W^+} &= -\frac{\omega u_0}{W^+} \sum A_n - u_0 \sum \frac{1}{n} \frac{\partial B_n}{\partial X} + F'(X) \\ &= -u_0 \sum \frac{1}{n} \frac{\partial B_n}{\partial X} + F'(X) \end{aligned} \tag{38.62}$$

since $\sum A_n = 0$. That $\sum A_n$ is zero can be seen from (38.50) and the fact that $\Theta = 0$ corresponds to the bounding characteristic $X = tW^+$ along which $v = V_0$. Then $\sum A_n$ is zero because it has to equal the value of z along this characteristic. By combining (38.62) and (38.61), taking B_n from (38.55), one finds, since $u_0 Y = MX$

$$\begin{aligned} F'(X) &= u_0 \sum \frac{1}{n} \frac{\partial B_n}{\partial X} + M e_0 = M \left[\sum \frac{1}{n} \frac{\partial B_n}{\partial Y} + e_0 \right] \\ &= M \left[r + \frac{1}{8} \alpha (1 + 2r^2) + \frac{1}{4} m_4 \alpha^2 r \left(1 + \frac{2}{3} r^2 \right) \right. \\ &\quad \left. + \frac{1}{64} m_5 \alpha^3 \left(1 + 8r^2 + \frac{8}{3} r^4 \right) + \cdots \right]. \end{aligned} \tag{38.63}$$

To the present approximation,

$$F(X) = X F'(X) + C \tag{38.64}$$

where $F'(X)$ is the constant in (38.63) and C is a constant that can be determined to make (38.56) and (38.55) yield $u(0^+, 0) = 0$. This gives

$$C = u_0 \sum (B_n/n)_{X=0} = u_0. \tag{38.65}$$

The complete expression for the displacement in the oscillating part of the simple wave can then be expressed as

$$\begin{aligned} &u = u_0 \left[1 + \sum_{n=1}^{\infty} (D_n \sin n\Theta + E_n \cos n\Theta) \right] + tV_0 + XF'(X), \\ &D_n = A_n/n, \quad E_n = -B_n/n, \quad \Theta = \omega (t - X/W^+), \quad X/W^+ < t < ? \end{aligned} \tag{38.66}$$

38.7. Relation of harmonic growth to higher-order elastic coefficients [THURSTON and SHAPIRO, 1967]. Let it be desired to determine m_3, and ultimately the quantity M_3, from measurements on the second harmonic. We shall consider the driving amplitude u_0 as known, and suppose that the experimenter measures the amplitude of the second harmonic component of the displacement

$$H_2 \equiv u_0 (D_2^2 + E_2^2)^{\frac{1}{2}}. \tag{38.67}$$

From (38.55)

$$\begin{aligned} D_2 &= \tfrac{1}{2} A_2 = -\tfrac{1}{16} m_3 m_4 Y^2 M^3 + \cdots, \\ E_2 &= -\tfrac{1}{2} B_2 = -\tfrac{1}{8} m_3 Y M (1 + 2 r m_3 m_4 M + \cdots) + \cdots. \end{aligned} \tag{38.68}$$

The terms not written explicitly in E_2 are of order M^3 and higher. For fixed Y, m_3 may be determined independently of m_4 and r by finding the initial slope of the curve of (H_2/u_0) versus M, or alternatively, by finding the intercept of the curve of $(H_2/u_0 M)$ versus M, for it follows from (38.67) and (38.68) that

$$\left(\frac{d\,(H_2/u_0)}{dM}\right)_{M=0} = \left(\frac{H_2}{u_0 M}\right)_{M=0} = \frac{1}{8} m_3 Y. \tag{38.69}$$

Thus, although for any finite driving amplitude, H_2 depends in principle on elastic constants of order higher than the third, the limiting values indicated in Eq. (38.69) do not. Hence, with second-order elastic constants already known, certain combinations of third-order constants can be determined by making measurements on the second harmonic as a function of driving amplitude and then making the indicated extrapolation to $M=0$.

There is a similarity in the relation of data to nonlinear elastic properties in this case and in the case of measurements on the transit time of small-amplitude waves as a function of static stress. As is clear from Sects. 28 and 29, the small-amplitude wave transit time corresponding to any nonzero static stress depends in principle on elastic coefficients of all orders, but the initial derivative with respect to static stress is independent of coefficients of order higher than the third.

Let us now consider the third harmonic. We have

$$\begin{aligned} D_3 &= \tfrac{1}{24} m_3^2 m_4 Y M^2 + \cdots, \\ E_3 &= -\tfrac{1}{32} m_3^2 Y^2 M^2 + \cdots. \end{aligned} \tag{38.70}$$

The terms not written explicitly in (38.70) are of order M^3 and higher. Defining the amplitude of the third harmonic as

$$H_3 = u_0 (D_3^2 + E_3^2)^{\frac{1}{2}}, \tag{38.71}$$

it can be seen that

$$\left(\frac{H_3}{u_0 M^2}\right)_{M=0} = \frac{1}{32} m_3^2 Y^2 \left(1 + \frac{16 m_4^2}{9 Y^2}\right)^{\frac{1}{2}}. \tag{38.72}$$

If Y is sufficiently large that $(16 m_4^2/9 Y^2) \ll 1$, then m_4 has negligible influence on the quantity in (38.72). In this case, measurement of the third harmonic amplitude as a function of M would enable m_3 to be determined from (38.72), thus providing a check on the determination from the second harmonic in (38.69). It is conceivable that sufficiently accurate measurements on both the second and third harmonics would enable some bounds to be placed on m_4.

Much of the literature on finite-amplitude distortion deals either with the case of an ideal gas or with the case of a constitutive relation (38.12) that is truncated after the term involving M_3. In the latter case, the results are presumed relevant to the determination of third-order elastic constants, but in the absence of precise theorems, there can be some uncertainty as to how, if at all, fourth-order and higher-order coefficients affect the results. By including all the higher-order coefficients and expanding in powers of the source Mach number $M=\omega u_0/W_0$, THURSTON and SHAPIRO obtained the preceding definitive results relating experimental data to third-order coefficients, independently of the fourth-order coefficients.

The relation of finite-amplitude waves to third-order elastic constants has been analyzed also by BUCK and THOMPSON (1966). A paper by FENLON (1972) illustrates an approach that has proved fruitful in acoustics.

The reflection of finite-amplitude waves has been considered by THOMPSON, TENNISON, and BUCK (1968) and by BREAZEALE (1965, 1968), and by VAN BUREN and BREAZEALE (1968).

38.8. Hypothetical linear medium for one-dimensional longitudinal motion. If $g(\partial u/\partial X)$=constant, independent of $\partial u/\partial X$, the exact equation of longitudinal motion (38.11) is linear, and finite-amplitude distortion does not occur. By rewriting the equation of motion for this hypothetical case as

$$\bar{\varrho}\ddot{u}_3=\frac{\partial P_{33}}{\partial X_3}=\frac{\partial}{\partial X_3}\left(\frac{\partial u_3}{\partial X_3}g\bar{\varrho}\right), \tag{38.73}$$

it can be seen that the first Piola-Kirchhoff component P_{33} is in this case the constant g times $\partial u_3/\partial X_3$. From (4.2),

$$T_{33}=P_{33}, \tag{38.74}$$

since $\partial x_3/\partial X_1=\partial x_3/\partial X_2=0$ and $J=\partial x_3/\partial X_3$ for the one-dimensional longitudinal geometry under consideration. The equality (38.74) is also apparent from the interpretation of P_{ik} as force per unit reference area, since material surface elements normal to the motion remain unstretched. Thus, the one-dimensional constitutive equation that gives rise to a linear equation of longitudinal motion is simply that the stress is proportional to the extension:

$$T_{33}=\frac{\partial u_3}{\partial X_3}g\bar{\varrho} \tag{38.75}$$

where g is a constant.

This condition also leads to a small-amplitude longitudinal wave transit time that is independent of the extension, i.e., $\bar{\varrho}W^2=\bar{\varrho}g$=constant. These and other consequences of the linearity of the equation of motion for this special constitutive behavior have been discussed by THURSTON (1969). An interesting side-issue is that the exact equation of motion expressed in *spatial* coordinates remains nonlinear, yet necessarily has distortionless traveling-wave solutions.

38.9. One-dimensional longitudinal stress-extension relation. In real media, the connection of $g(\xi)$ to the relation between the longitudinal stress T and extension $\xi=\partial u/\partial X$ is readily obtained by comparing (38.11) with the equation of one-dimensional longitudinal motion in the form

$$\ddot{u}=\frac{1}{\varrho}\frac{\partial T}{\partial x}=\frac{1}{\varrho}\frac{\partial X}{\partial x}\frac{\partial T}{\partial X}=\frac{1}{\bar{\varrho}}\frac{\partial T}{\partial X}=\frac{1}{\bar{\varrho}}\frac{\partial T}{\partial \xi}\frac{\partial \xi}{\partial X}=\frac{1}{\bar{\varrho}}\frac{\partial T}{\partial \xi}\frac{\partial^2 u}{\partial X^2}. \tag{38.76}$$

By comparison with (38.11),

$$\frac{\partial T}{\partial \xi}=\bar{\varrho}g(\xi). \tag{38.77}$$

Thus, the derivatives of $T(\xi)$ are simply related to those of $g(\xi)$. The relation to higher-order elastic constants and other nonlinearity parameters can be unraveled through (38.4)–(38.12) and (38.21).

39. Longitudinal shock waves in elastic solids.

A discussion of shock waves in solids, with primary emphasis on their growth or decay has been included along with many results on acceleration waves in the article by CHEN (1972) in this encyclopedia. CHEN's Sect. 17, pp. 354–356, summarizes the relation of experimental data to material parameters in the constitutive equation of a material with memory. Cf. also the article by NUNZIATO, WALSH, SCHULER and BARKER, preceding in this volume. However, it is appropriate to indicate here a procedure that can be used to relate shock wave data to higher-order elastic constants in elastic crystals. There are data on shock waves that appear to be consistent with the elastic model, and this has provided an interesting basis for determining certain fourth-order elastic coefficients [FOWLES, 1967; GRAHAM, 1972a, 1972b].

A shock wave is treated as a propagating singular surface across which there is a discontinuity in particle velocity, strain, stress, density, and associated thermodynamic variables. The basic equations of balance that apply across such a surface have been discussed by TRUESDELL and TOUPIN (1960) and reiterated by CHEN (1972). As the surface propagates, material passes through it from the front (+) side to the back (—). Field values just in front of the surface will be denoted by the superscript +, and those in back by the superscript —. The magnitude of the discontinuity or jump in any quantity ψ will be denoted by enclosing the symbol in bold-face brackets:

$$[\psi] \equiv \psi^- - \psi^+ . \tag{39.1}$$

We shall discuss only the one-dimensional "normal" shock case. Consider a plane shock wave normal to the X-direction, propagating in the X-direction with natural speed V, referred to the material in its reference configuration. This means that if the material has undergone the deformation $(\partial x/\partial X)$, then the propagation speed relative to this material is $V(\partial x/\partial X)$. With the particle velocity of the material denoted by $\dot{x}$, the actual speed of the shock in the laboratory is then $\dot{x}+V(\partial x/\partial X)$. Since this relation is applicable both in front of and behind the shock, one can equate the two expressions so obtained:

$$\dot{x}^- + V(\partial x/\partial X)^- = \dot{x}^+ + V(\partial x/\partial X)^+ , \tag{39.2}$$

or

$$V[\partial x/\partial X] = -[\dot{x}] . \tag{39.3}$$

39.1. Relations that apply across a shock propagating into a medium at rest in its reference configuration. For the present purpose, it is sufficient to consider a shock wave propagating into an initially unstrained medium that is at rest, in which case the right-hand side of (39.2) reduces to V. In this case V is the same as the velocity of the shock in the laboratory, and (39.3) reduces to

$$V(\partial u/\partial X) = -v , \tag{39.4}$$

where $\partial u/\partial X$ and v denote the extension and particle velocity *behind* the shock $(\partial u/\partial X)^-$ and v^-. Per unit time, an area A of the shock passes through an amount of material of reference volume AV, mass $\bar{\varrho} AV$, imparting to it a velocity v. As a result of the velocity jump, the material in a cylinder with axis normal to the shock and cross-sectional area A has its momentum changing at the rate $\bar{\varrho} AVv$, which implies the resultant force $A(T^+ - T^-)$ equal to

$$A(T^+ - T^-) = \bar{\varrho} AVv ,$$

or, with $T^+ = 0$ and $T^- = T$, the normal stress behind the shock is

$$T = -\bar{\varrho} Vv . \tag{39.5}$$

This normal stress does work at the rate $-ATv$, which, in the absence of heat conduction and body forces, accounts for the rate of increase of internal energy and kinetic energy. With internal energy per unit mass denoted by $\mathcal{U}$, the total internal and kinetic energy of the above cylinder increases at the rate $\bar{\varrho} A V(\frac{1}{2}v^2 + \mathcal{U}^- - \mathcal{U}^+)$ and hence the equation of balance of energy is

$$-Tv = \bar{\varrho} V(\tfrac{1}{2}v^2 + [\mathcal{U}]) . \tag{39.6}$$

Taken together, (39.4) and (39.5) show that the ratio of stress to extension behind the shock is $\bar{\varrho} V^2$ while their product is $\bar{\varrho} v^2$:

$$\bar{\varrho} V^2 = \frac{T}{\partial u/\partial X}, \qquad \bar{\varrho} v^2 = T \partial u/\partial X . \tag{39.7}$$

39.2. The connection of the stress-extension relation to the curve of shock velocity versus particle velocity. Simultaneous measurements of shock velocity V and particle velocity v determine points on the curve of normal stress T versus extension $\partial u/\partial X$ through (39.4) and (39.5). This is clear since each pair (V, v) corresponds to definite values of

$$T = -\bar{\varrho} V v, \qquad \partial u/\partial X = -v/V . \tag{39.8}$$

It is of some interest to note that a strictly linear shock velocity versus particle velocity relation of the form [RUOFF (1967)].

$$V = W_0 + kv \tag{39.9}$$

would imply

$$\begin{aligned} v &= -W_0 \frac{\partial u}{\partial X}\left(1 + k\frac{\partial u}{\partial X}\right)^{-1}, \qquad V = W_0\left[1 - k\frac{\partial u}{\partial X}\left(1 + k\frac{\partial u}{\partial X}\right)^{-1}\right], \\ T &= \bar{\varrho} W_0^2 \frac{\partial u}{\partial X}\left(1 + k\frac{\partial u}{\partial X}\right)^{-2}. \end{aligned} \tag{39.10}$$

In any case, first and higher derivatives of V with respect to v can be related to first and higher derivatives of $T/(\partial u/\partial X)$ with respect to $(\partial u/\partial X)$. As an abbreviation, let

$$\xi = \partial u/\partial X . \tag{39.11}$$

Then the assumption that V and v are differentiable functions of ξ leads to

$$\frac{dV}{d\xi} = \frac{dv}{d\xi}\,\frac{dV}{dv} . \tag{39.12}$$

Differentiation of (39.4) and substitution from (39.12) gives

$$\frac{dv}{d\xi} = -V - \xi\frac{dV}{d\xi} = -V - \xi\frac{dv}{d\xi}\,\frac{dV}{dv} \tag{39.13}$$

which can be solved for $dv/d\xi$ to obtain

$$\frac{dv}{d\xi} = \frac{-V}{1 + \xi dV/dv} . \tag{39.14}$$

Differentiation of $(39.7)_1$ with respect to ξ and use of (39.12) and (39.14) now gives

$$\frac{d}{d\xi}\left(\frac{T}{\xi}\right) = 2\bar{\varrho} V\frac{dV}{d\xi} = -2\bar{\varrho} V^2\frac{dV}{dv}\left(1 + \xi\frac{dV}{dv}\right)^{-1}. \tag{39.15}$$

The result for the second derivative is

$$\frac{d^2}{d\xi^2}(\bar{\varrho} V^2) = \bar{\varrho} V^2\left[6\left(\frac{V'}{1 + \xi V'}\right)^2 + \frac{2VV''}{(1 + \xi V')^3}\right] \tag{39.16}$$

where $V' \equiv dV/dv$. Finally, with

$$w_0 = (\bar{\varrho} V^2)_{\xi=0} = \bar{\varrho} W_0^2, \tag{39.17}$$

the results for the first three derivatives, evaluated at $\xi = 0$, are listed below.

$$\begin{gathered}\frac{d}{d\xi}(\bar{\varrho} V^2)_0 = -2w_0 V_0', \quad \left[\frac{d^2}{d\xi^2}(\bar{\varrho} V^2)\right]_0 = w_0[6(V_0')^2 + 2W_0 V_0''],\\ \left[\frac{d^3}{d\xi^3}(\bar{\varrho} V^2)\right]_0 = -w_0[24(V_0')^3 + 24 W_0 V_0' V_0'' + 2W_0^2 V_0'''].\end{gathered} \tag{39.18}$$

From $(39.18)_1$ and $(39.7)_1$, the initial slope of the shock velocity versus particle velocity curve $V_0' \equiv (dV/dv)_0$ is related to the initial slope of the curve of (stress/extension) versus extension. That is, since $(39.7)_1$ gives $T/\xi = \bar{\varrho} V^2$, we have

$$\left[\frac{d}{d\xi}(T/\xi)\right]_0 = -2w_0 V_0'. \tag{39.19}$$

These initial slopes involve third-order elastic coefficients. Eqs. $(39.18)_2$ and $(39.18)_3$ provide corresponding relations for the second and third derivatives. The second derivatives involve fourth-order coefficients, the third involve fifth, etc.

39.3. Thermodynamic considerations. Thermodynamic relations across a shock are neither isothermal nor isentropic. They are often called Hugoniot relations. For weak shocks, the assumption of constant entropy can be a good approximation because, as we shall see, the entropy jump is of third order in the extension ξ. The corresponding result for fluids is well known [COURANT and FRIEDRICHS, 1948], and it holds in solids for essentially the same thermodynamic reasons.

The fact that conditions across a shock are neither isothermal nor isentropic produces a complication in the relation of shock wave data to material coefficients at constant temperature or entropy. The jump in the thermodynamic variables has to be compatible with the balance of energy (39.6) which, in view of (39.5), can be expressed as

$$\mathcal{U}^- - \mathcal{U}^+ = [\mathcal{U}] = \tfrac{1}{2} v^2. \tag{39.20}$$

With internal energy referred to unit reference volume as in Sect. 10, the corresponding equation is

$$(U^- - U^+)/\bar{\varrho} = [U/\bar{\varrho}] = \tfrac{1}{2} v^2. \tag{39.21}$$

Note that these equations do not express constancy of the sum of kinetic and internal energy. Rather, the kinetic energy and internal energy per unit mass both increase by equal amounts when the shock passes. In the present context, since only internal energy *changes* enter the equations, there is nothing to prevent the arbitrary assignment $\mathcal{U}^+ = 0$, in which case (39.21) becomes

$$U^- = \tfrac{1}{2}\bar{\varrho} v^2 = \tfrac{1}{2}\bar{\varrho}\xi^2 V^2 \tag{39.22}$$

where $(39.8)_2$ has been used to obtain the final form. Thus, measurements of (V, v) determine the internal energy increment across the shock. Through (39.22) and $(39.7)_2$, the stress T behind the shock can be related explicitly to the energy increment U^- and the extension:

$$2U^- = T\frac{\partial u}{\partial X}. \tag{39.23}$$

In terms of the thermodynamic tension of Sect. 10, with X_1 assumed to be the direction of propagation,

$$t_{11} = \tau_{11} = \tau = J T_{ij}\frac{\partial X_1}{\partial x_i}\frac{\partial X_1}{\partial x_j} = T/(1+\xi) \tag{39.24}$$

and hence

$$2U^{-}=\tau\xi(1+\xi) \tag{39.25}$$

with $\xi=\partial u/\partial X$.

The relation (39.25) provides a basis for determining the entropy jump across the shock and, together with $(39.7)_1$, for relating the derivatives in (39.18) to material coefficients.

To find the needed relations, we make use of the expansions of U and τ in powers of the strain and entropy jump. Comparison with (39.25) enables the entropy jump to be determined as a power series in the strain, after which the derivatives of $\bar{\varrho}V^2$ with respect to ξ can be found from $(39.7)_1$, rewritten as

$$\bar{\varrho}V^2=T/\xi=(1+\xi)\,\tau/\xi. \tag{39.26}$$

A minor nuisance is the fact that material coefficients are conventionally defined in terms of derivatives with respect to the strain components (3.5), whereas the extension $\xi\equiv\partial u/\partial X$ appears as a most convenient variable in the present one-dimensional situation. Hence, we shall have occasion to go back and forth between expansions in powers of the strain and expansions in powers of the extension.

From (3.5), the longitudinal strain component in the X-direction is

$$\eta=\xi(1+\tfrac{1}{2}\xi). \tag{39.27}$$

It may be noted that

$$(1+2\eta)=(1+\xi)^2. \tag{39.28}$$

The meaningful root for ξ is

$$\xi=-1+(1+2\eta)^{\frac{1}{2}}=\eta-\tfrac{1}{2}\eta^2+\tfrac{1}{2}\eta^3-\tfrac{5}{8}\eta^4+\tfrac{7}{8}\eta^5-\cdots. \tag{39.29}$$

Consequently, (39.25) can be expressed as

$$2U^{-}=\tau(\eta+\tfrac{1}{2}\eta^2-\tfrac{1}{2}\eta^3+\tfrac{5}{8}\eta^4-\tfrac{7}{8}\eta^5+\cdots). \tag{39.30}$$

The next task is to write the expansions for U and τ in powers of η and the entropy jump. Let us introduce the following additional notation:

$$\begin{aligned}
S&\equiv[S]=S^{-}-S^{+},\\
\Theta&=(\partial U/\partial S)^{+},\\
a&\equiv(\partial^2 U/\partial S\,\partial\eta)^{+},\\
b&\equiv(\partial^3 U/\partial\eta^2\,\partial S)^{+},\\
c_n&\equiv(\partial^n U/\partial\eta^n)^{+}.
\end{aligned} \tag{39.31}$$

S is the entropy jump, Θ the temperature ahead of the shock, the c_n are elastic coefficients, and from Table 10.1, a can be expressed as $-\Theta\lambda_1/C_V=-\Theta\gamma_1$. We now write the expansions of U and τ in powers of the strain and entropy change from the initial unstressed and unstrained state as follows:

$$U=\Theta S+a\eta S+\frac{1}{2}b\eta^2 S+\frac{1}{2}c_2\eta^2+\frac{1}{6}c_3\eta^3+\cdots+\frac{1}{n!}c_n\eta^n+\cdots, \tag{39.32}$$

$$\tau=\frac{\partial U}{\partial\eta}=aS+b\eta S+c_2\eta+\frac{1}{2}c_3\eta^2+\frac{1}{6}c_4\eta^3+\frac{1}{24}c_5\eta^4+\cdots. \tag{39.33}$$

Substitution from (39.33) into (39.30) and comparison of the results with (39.32) [since U^{-} must be equal to $U(\eta, S)$] determines the entropy jump S as

$$S=a_3\eta^3+a_4\eta^4+a_5\eta^5+\cdots \tag{39.34}$$

with

$$\begin{aligned} a_3 &= \frac{\Theta^{-1}}{12}(c_3+3c_2), \\ a_4 &= \frac{\Theta^{-1}}{24}\left[c_4+3c_3-6c_2-a\Theta^{-1}(c_3+3c_2)\right] \\ a_5 &= \frac{\Theta^{-1}}{240}\left[3c_5+10c_4-30c_3+75c_2\right] \\ &\quad -\frac{a\Theta^{-2}}{48}\left[c_4+2c_3-9c_2-a\Theta^{-1}(c_3+3c_2)\right]. \end{aligned} \tag{39.35}$$

With S determined from (39.34), the expansion (39.33) now becomes an expansion in η alone:

$$\tau = c_2\eta+\tfrac{1}{2}c_3\eta^2+\eta^3\left(\tfrac{1}{6}c_4+a a_3\right)+\eta^4\left(\tfrac{1}{24}c_5+a a_4+b a_3\right)+\cdots. \tag{39.36}$$

For the remainder of the calculation, it appears most convenient to return to the extension $\xi=\partial u/\partial X$ as a variable through (39.27). The result for τ then becomes

$$\begin{aligned} \tau &= c_2\xi+\tfrac{1}{2}\xi^2(c_2+c_3)+\xi^3\left(\tfrac{1}{6}c_4+\tfrac{1}{2}c_3+a a_3\right) \\ &\quad +\xi^4\left[\tfrac{1}{24}c_5+\tfrac{1}{4}c_4+\tfrac{1}{8}c_3+a\left(a_4+\tfrac{3}{2}a_3\right)+b a_3\right]+\cdots. \end{aligned} \tag{39.37}$$

One now has all the relations needed for expressing the shock wave parameters T, $\bar{\varrho}V^2$, $\bar{\varrho}v^2$, etc., in terms of material coefficients [defined as derivatives of $U(\eta, S)$] and powers of the extension ξ. In particular, substitution into (39.26) results in

$$\begin{aligned} \bar{\varrho}V^2 &= c_2+\tfrac{1}{2}\xi(3c_2+c_3)+\tfrac{1}{6}\xi^2(c_4+6c_3+3c_2+6a a_3) \\ &\quad +\xi^3\left[\tfrac{1}{24}(c_5+10c_4+15c_3)+a\left(a_4+\tfrac{5}{2}a_3\right)+b a_3\right]+\cdots. \end{aligned} \tag{39.38}$$

It may be noted that the difference between isentropic conditions and the actual "Hugoniot" conditions required by the balance of energy does not appear in (39.38) until the coefficient of ξ^2, which involves fourth-order elastic coefficients. This is a consequence of the fact that $a_3\eta^3$ (or $a_3\xi^3$) is the first term in the expansion (39.34) for the entropy change.

It is now a straightforward matter to relate shock wave data to third-order and higher-order elastic coefficients. The shock wave data are expressed through the derivatives V_0', V_0'', V_0''' in (39.18). Thus, from (39.17), (39.18), and (39.38),

$$c_2=\bar{\varrho}W_0^2=w_0, \tag{39.39}$$

$$\tfrac{1}{4}(3+c_3/c_2)=-V_0', \tag{39.40}$$

$$c_4/c_2+6c_3/c_2+3+6a a_3/c_2=18(V_0')^2+6W_0V_0'', \tag{39.41}$$

$$\begin{aligned} &\left[\tfrac{1}{4}(c_5+10c_4+15c_3)+6a\left(a_4+\tfrac{5}{2}a_3\right)+b a_3\right]/c_2 \\ &\quad =-\left[24(V_0')^3+24W_0V_0'V_0''+2W_0^2V_0'''\right]. \end{aligned} \tag{39.42}$$

40. Reflection of longitudinal waves at normal incidence.

To illustrate reflection and transmission in a simple situation, uncomplicated by any nonlinear effects, let us consider an interface separating two hypothetical linear media for longitudinal motion. According to the definition in Sect. 38.8, each such medium is characterized by a constant value of the natural propagation speed W. The longitudinal stress is given exactly by

$$T=\bar{\varrho}W^2\frac{\partial u}{\partial X} \tag{40.1}$$

and the exact equation of longitudinal motion is the linear wave equation

$$\ddot{u} = W^2 \frac{\partial^2 u}{\partial X^2}. \tag{40.2}$$

Let the medium in the left half-space ($X<0$) be characterized by reference density $\bar{\varrho}_1$ and natural propagation speed W_1, while $\bar{\varrho}$ and W characterize the right half-space ($X>0$).

40.1. Reflection of a continuous disturbance. Let a longitudinal wave prescribed by

$$\begin{aligned} u_i &= f(W_1 t - X), \qquad X<0, \\ f(\Theta) &= 0, \qquad \Theta \leqq 0, \end{aligned} \tag{40.3}$$

be incident on the interface from the left. In (40.3), the subscript i is for "incident", and the condition $f(\Theta)=0$ on $\Theta \leqq 0$ is stated in order to specify that the incident disturbance is confined to $(W_1 t - X) > 0$, i.e., the incident wave does not reach the interface before $t=0$. At $t=0^-$, the entire disturbance consists of the incident wave (40.3). For later times, we expect a transmitted wave in the right half-space,

$$\begin{aligned} u_t &= g(Wt - X), \qquad X>0 \\ g(\Theta) &= 0, \qquad \Theta \leqq 0, \end{aligned} \tag{40.4}$$

and a reflected wave on the left,

$$\begin{aligned} u_r &= f_r(W_1 t + X), \qquad X<0 \\ f_r(\Theta) &= 0, \qquad \Theta \leqq 0. \end{aligned} \tag{40.5}$$

The resultant disturbance in medium 1 is the sum $u_i + u_r$. Thus, the displacement, velocity, and normal stress may be represented as

$$\begin{aligned} u &= f(W_1 t - X) + f_r(W_1 t + X), \qquad X<0, \\ u &= g(Wt - X), \qquad X>0, \end{aligned} \tag{40.6}$$

$$\begin{aligned} v &= W_1 [f'(W_1 t - X) + f'_r(W_1 t + X)], \qquad X<0, \\ v &= W g'(Wt - X), \qquad X>0, \end{aligned} \tag{40.7}$$

$$\begin{aligned} T &= \bar{\varrho}_1 W_1^2 [-f'(W_1 t - X) + f'_r(W_1 t + X)], \qquad X<0 \\ T &= -\bar{\varrho} W^2 g'(Wt - X), \qquad X>0. \end{aligned} \tag{40.8}$$

One of the boundary conditions at $X=0$ is that the two media should not separate or interpenetrate. This is expressed by requiring that u be continuous across the boundary. Continuity of u for all time implies continuity of the velocity and acceleration across the boundary also, except possibly at isolated instants corresponding to the passage of shock or acceleration waves, treated as propagating singular surfaces. The other boundary condition at $X=0$ is that the stress be continuous there, except possibly at isolated instants corresponding to the passage of shock waves. A jump in the stress from one finite value to another, and the resulting velocity jump (infinite acceleration) implied by the balance of momentum can persist only across a *propagating* surface, not a material surface [cf. TRUESDELL and TOUPIN, 1960, CFT, Sect. 185].

The conditions on continuity of u and T at $X=0$ yield

$$f(W_1 t) + f_r(W_1 t) = g(Wt), \tag{40.9}$$

$$\bar{\varrho}_1 W_1^2 [f'(W_1 t) - f'_r(W_1 t)] = \bar{\varrho} W^2 g'(Wt). \tag{40.10}$$

Differentiation of (40.6) results in

$$W_1[f_r'(W_1 t)+f'(W_1 t)]=W g'(W t). \tag{40.11}$$

This same equation (40.11) can be seen from (40.7) to express continuity of the velocity at $X=0$. Eqs. (40.10) and (40.11) can be solved simultaneously for $W_1 f_r'(W_1 t)$ and $W g'(W t)$. In terms of the characteristic impedances

$$Z_1 \equiv \bar{\varrho}_1 W_1, \quad Z \equiv \bar{\varrho} W, \tag{40.12}$$

and the quantities

$$v_i = W_1 f'(W_1 t), \quad v_r = W_1 f_r'(W_1 t), \quad v_t = W g'(W t), \tag{40.13}$$

the equations take the form

$$\begin{gathered} Z_1(v_i - v_r) = Z v_t, \\ v_i + v_r = v_t. \end{gathered} \tag{40.14}$$

The functions v_i, v_r, and v_t are the incident, transmitted, and reflected velocities at $X=0$. The solution of (40.14) may be expressed as

$$\begin{aligned} &v_r = r v_i, \qquad v_t = t v_i, \\ &r = \frac{Z_1 - Z}{Z_1 + Z}, \quad t = \frac{2Z_1}{Z + Z_1} = 1 + r. \end{aligned} \tag{40.15}$$

The quantities r and t are reflection and transmission coefficients for velocity.

The stress at $X=0, T_t = \bar{\varrho} W^2 g'(W t)$, can be decomposed into the incident and reflected parts

$$T_i = -\bar{\varrho}_1 W_1^2 f'(W_1 t), \quad T_r = \bar{\varrho}_1 W_1^2 f_r'(W_1 t). \tag{40.16}$$

Clearly,

$$T_i = -Z_1 v_i, \quad T_r = Z_1 v_r, \quad T_t = -Z v_t. \tag{40.17}$$

Hence,

$$T_r = -r T_i, \quad T_t = (1-r) T_i. \tag{40.18}$$

The quantity $(-r)$ is the reflection coefficient for stress.

It can be seen from (40.17) that the characteristic impedances are the ratios

$$Z_1 = -\frac{T_i}{v_i} = \frac{T_r}{v_r}, \quad Z = -\frac{T_t}{v_t}. \tag{40.19}$$

The signs in these equations can be understood from the following definition of characteristic impedance. Consider a material plane normal to the propagation direction of a pure traveling wave such as u_i or u_r, and denote the sides by "rear" and "front," such that the direction of travel of the wave is from the rear to the front. Then the characteristic impedance of the medium is the force per unit area which the rear material exerts on the front material, divided by the velocity in a pure traveling wave. For example, in the case of u_t from (40.4), we have

$$\frac{\text{Force}}{\text{Velocity}} = \frac{-T}{\dot{u}_t} = \frac{\bar{\varrho} W^2 g'(W t - X)}{W g'(W t - X)} = \bar{\varrho} W,$$

and in the case of u_r, since the rearward direction is $+X$,

$$\frac{\text{Force}}{\text{Velocity}} = \frac{T}{\dot{u}_r} = \frac{\bar{\varrho}_1 W_1^2 f_r'(W_1 t + X)}{W_1 f_r'(W_1 t + X)} = \bar{\varrho}_1 W_1.$$

This definition of characteristic impedance is possible in the ideal elastic case in which the above ratios have constant values that characterize the medium. A definition based on the ratio of complex amplitudes of stress and velocity for sinusoidal waves has been given in (18.6). In the present ideal elastic case, $Z^*(\omega)$, defined in (18.6), becomes independent of frequency and reduces to the present $\bar{\varrho} W$.

Three special cases of (40.15) and (40.17) are worth mentioning. The case of a free surface, corresponding to absence of the right-hand medium, is obtained with $Z=0$. This gives $r=1$. The reflected wave serves to double the incident velocity and reduce the stress to zero. The case of a rigid boundary is obtained with $Z=\infty$. In this case $r=-1$ and the velocity is reduced to zero while the stress is doubled. The case of matched impedance ($Z=Z_1$) gives $r=0$. This condition is achieved with $\bar{\varrho}_1 W_1=\bar{\varrho} W$, even though the velocities and densities be different.

40.2. Reflection of a shock wave. In the hypothetical linear medium shock waves are propagated with a natural speed W that is constant, independent of the strength of the shock, and their behavior upon normal incidence at an interface is described in essence by (40.15) and (40.18), the same equations as for continuous disturbances. These results will now be demonstrated.

When the wave propagates into a medium that may have initial stress, (39.5) is generalized to $[T]=-\bar{\varrho} V[v]$. Eq. (39.3) and this generalization of (39.5) then constitute two fundamental relations on which the other results are based. According to sign conventions already laid down at the beginning of Sect. 39, these equations apply with V and $\dot{x}$ positive in the same sense. When V is regarded as an unsigned wave speed, the equations are

$$\begin{aligned} \pm V[\partial x/\partial X] &= -[v], \\ \pm[T] &= -\bar{\varrho} V[v], \end{aligned} \tag{40.20}$$

in which the $+$ sign of the $\pm$ applies to waves propagating in the $+X$-direction, and the other sign to the other direction. For a wave in either direction, (40.20) yields

$$\frac{[T]}{[\xi]}=\bar{\varrho} V^2 \tag{40.21}$$

where $\xi \equiv \partial u/\partial X$, since the jump in $\partial x/\partial X$ is the same as the jump in $\partial u/\partial X$. Because of (40.21), the hypothetical linear constitutive relation (40.1) results in the natural shock wave speed V being a constant W, independent of the shock strength.

Now consider the same two linear media as in Sect. 40.1, and let a shock wave be incident on the interface $X=0$ at the time $t=0$. At this instant, the right half-space is unstrained, unstressed, and at rest while the left half-space has the uniform stress T_i, extension ξ_i, and velocity $\dot{x}=v_i$ satisfying

$$T_i=\bar{\varrho}_1 W_1^2 \xi_1=-\bar{\varrho}_1 W_1 v_i. \tag{40.22}$$

It is found that the boundary conditions at $X=0$ can be satisfied by a transmitted shock and a reflected shock. The transmitted shock propagates into the right half-space with speed W, bringing the stress, extension, and velocity to values T, ξ, and v satisfying

$$T=\bar{\varrho} W^2 \xi=-\bar{\varrho} W v. \tag{40.23}$$

The reflected shock propagates back into the left half-space bringing the stress and velocity to these same values. Between the transmitted and reflected shocks, the material is left with the constant velocity v and stress T. The stress jump and velocity jump across the reflected shock satisfy $[T]=\bar{\varrho}_1 W_1[v]$, or

$$T_i-T=\bar{\varrho}_1 W_1(v_i-v). \tag{40.24}$$

These equations parallel (40.17), as can be seen by defining

$$T_r = T - T_i, \quad v_r = v - v_i \tag{40.25}$$

so that the relations between stress and velocity in (40.22)–(40.24) become

$$T_i = -Z_1 v_i, \quad T_r = Z_1 v_r, \quad T = -Z v. \tag{40.26}$$

The simultaneous solution yields

$$v_r = r v_i, \quad v = \frac{2Z_1}{Z + Z_1} v_i, \tag{40.27}$$

which are the same as (40.15), and

$$T_r = -r T_i, \quad T = \frac{2Z}{Z + Z_1} T_i = (1 - r) T_i, \tag{40.28}$$

which are the same as (40.18).

From (40.23), the final extension in the right half-space is

$$\xi = -v/W. \tag{40.29}$$

The final value of extension ξ_f following the reflected wave in medium 1 is obtained from $T = \bar{\varrho}_1 W_1^2 \xi_f$. One finds

$$\xi_f = (1 - r)\xi_i = \frac{2Z}{Z + Z_1} \xi_i, \tag{40.30}$$

and

$$\xi_i - \xi_f = r \xi_i = \frac{Z_1 - Z}{Z_1 + Z} \xi_i. \tag{40.31}$$

For an incident compression shock, $\xi_i < 0$. The condition for the reflected wave to be a compression shock also is $\xi_f < \xi_i$, which requires $r < 0$. Hence, an incident compression shock is reflected as a compression shock if $Z > Z_1$ (in which case $v < v_i$) and as an expansion if $Z < Z_1$ (in which case $v > v_i$).

Both compression shocks and expansion shocks can propagate in the hypothetical linear medium, in which the characteristic curves of Eq. (38.11) are the families of parallel lines of slope $dX/dt = \pm W$, independent of the state of the medium. On the other hand, real media typically have $\bar{\varrho}\, g'(\xi) = d^2T/d\xi^2 < 0$, permitting the propagation of compression shocks but not expansion shocks [cf. CHEN, 1972, p. 359, footnote 75]. In this case, the relatively compressed state on the front side of an expansion shock corresponds to a greater propagation speed W than the expanded state on the rear. The result is that an expansion shock, if initiated in an otherwise undisturbed homogeneous medium, does not persist as a shock, but immediately becomes a centered simple wave as noted in Sect. 38.4. This is the expected result in real media if $Z < Z_1$ in the case of reflection considered here. [Fused silica is an exception. See BARKER and HOLLENBACH (1972)].

While nonlinearity of $T(\xi)$ can have a marked effect on the details of the reflection phenomenon, the present simple results for a hypothetical linear medium serve as a valuable guide concerning the gross effects. In many situations a centered simple wave and a shock, which are so different mathematically, can give rise to essentially the same experimental effects.

The reflection and transmission of large-amplitude pulses at an interface has been discussed by CEKRGE and VARLEY (1973).

41. Longitudinal motion of a piezoelectric material.

41.1. Series expansions of the stress and electric field. In a piezoelectric material, the nonlinear dielectric and piezoelectric properties can be investigated by experiments either on shock wave propagation or on the effect of superimposed electrostatic fields on the speed of small-amplitude waves. Let us consider the relations that apply in the purely longitudinal motion of a piezoelectric medium.

Suppose that the only nonzero component of electric field and electric displacement is along the direction of propagation X, and that the motion is strictly one-dimensional and longitudinal. By this we mean that u, the X-component of displacement, is the only nonvanishing one, and that it is independent of Y and Z, the coordinates perpendicular to X. For this case, we rewrite (14.2)–(14.4) as

$$\begin{aligned} E = &-qS + D\tilde{\beta}(D) - \eta\tilde{h}(\eta) + \beta_0 m D\eta \\ &+ \tfrac{1}{2}h_3 D^2\eta + \tfrac{1}{2}w D\eta^2 + \cdots, \end{aligned} \tag{41.1}$$

$$\begin{aligned} \tau^* = &-\Theta_0 S\gamma - D h_1(D) + \eta c(\eta) + \beta_0 b_2 D\eta \\ &+ \tfrac{1}{2}w D^2\eta + \tfrac{1}{2}\beta_0 b_3 D\eta^2 + \cdots \end{aligned} \tag{41.2}$$

in which

$$\begin{aligned} \tilde{\beta}(D) &= \beta + \tfrac{1}{2}\beta_0 r D + \tfrac{1}{6}\beta_0 f D^2 + \cdots, \\ \tilde{h}(\eta) &= h - \tfrac{1}{2}\beta_0 b_2\eta - \tfrac{1}{6}\beta_0 b_3\eta^2 + \cdots, \\ h_1(D) &= h - \tfrac{1}{2}\beta_0 m D - \tfrac{1}{6}\beta_0 h_3 D^2 + \cdots, \\ c(\eta) &= c_2 + \tfrac{1}{2}c_3\eta + \tfrac{1}{6}c_4\eta^2 + \cdots. \end{aligned} \tag{41.3}$$

S denotes the entropy change from the undeformed state. The constant coefficients in these power series expansions are understood to be defined from the context, and may be expressed in terms of tensor coefficients by referring to Eqs. (14.2)–(14.4) and Table 14.1. We shall again have occasion to use (39.27)–(39.29) to go back and forth between the longitudinal strain component η and the extension $\xi \equiv \partial u/\partial X$.

The longitudinal stress T^* that corresponds to the thermodynamic tension τ^* is

$$T^* = (1+\xi)\tau^* \tag{41.4}$$

in conformity with (39.24). From (13.70) the equation of motion is

$$\bar{\varrho}\ddot{u} = \frac{\partial T^*}{\partial X} \tag{41.5}$$

in which the term $J\frac{\partial}{\partial t}(\varepsilon_0 \mathscr{E} \times \mathscr{B})$ has been dropped because it is expected to be negligible.

The expressions for T^* and E that follow from (41.1)–(41.4) are

$$\begin{aligned} T^* = &(c_2 + D n_1 + D^2 n_2 + \cdots)\xi + \xi^2(\tfrac{1}{2}M_3 + D n_3 + \cdots) \\ &+ \tfrac{1}{3}M_4\xi^3 - D h_1(D) - \Theta_0\gamma S + \cdots, \end{aligned} \tag{41.6}$$

$$\begin{aligned} E = &D\tilde{\beta}(D) - \xi\frac{d}{dD}[D h_1(D)] + \frac{1}{2}\xi^2(n_1 + 2n_2 D + \cdots) \\ &+ \frac{1}{3}n_3\xi^3 - qS + \cdots \end{aligned} \tag{41.7}$$

in which

$$\begin{aligned} n_1 &= \beta_0 b_2 - h, \\ n_2 &= \tfrac{1}{2}(\beta_0 m + w), \\ M_3 &= 3c_2 + c_3, \\ n_3 &= \tfrac{1}{2}\beta_0(3b_2 + b_3), \\ M_4 &= \tfrac{1}{2}(c_4 + 6c_3 + 3c_2). \end{aligned} \tag{41.8}$$

It is worth noting that in the special one-dimensional situation under consideration, the stored energy per unit reference volume U can be expressed as

$$\begin{aligned} U=&\int_0^D s\tilde{\beta}(s)\,ds+\tfrac{1}{2}\xi^2(c_2+n_1D+n_2D^2+\cdots)\\ &+\tfrac{1}{3}\xi^3(\tfrac{1}{2}M_3+n_3D+\cdots)+\tfrac{1}{12}M_4\xi^4-\xi D h_1(D)\\ &+\Theta_0 S(1-\gamma\xi+\cdots)-qDS+\cdots \end{aligned} \tag{41.9}$$

in terms of which

$$T^*=\partial U/\partial\xi, \qquad E=\partial U/\partial D, \qquad \Theta=\partial U/\partial S. \tag{41.10}$$

Here Θ denotes the temperature.

Because there is only one component of D, its solenoidal character requires it to be independent of X in the piezoelectric medium:

$$\partial D/\partial X=0. \tag{41.11}$$

With these assumptions and the assumption of isentropic propagation, (41.5) becomes

$$\bar{\varrho}\ddot{u}=\bar{\varrho}g(\xi;D)\frac{\partial^2 u}{\partial X^2}, \qquad \xi=\partial u/\partial X, \tag{41.12}$$

$$\bar{\varrho}g(\xi;D)=\frac{\partial T^*}{\partial\xi}=c_2+n_1D+n_2D^2+\cdots+\xi(M_3+2n_3D+\cdots)+M_4\xi^2+\cdots \tag{41.13}$$

while the natural speed V of shock waves propagating into an unstrained medium is found from (41.6) and (40.21) with T replaced by T^* to be

$$\begin{aligned} \bar{\varrho}V^2=&c_2+n_1D+n_2D^2+\cdots+\xi(\tfrac{1}{2}M_3+n_3D+\cdots)\\ &+\tfrac{1}{3}M_4\xi^2-(\gamma\Theta_0/\xi)(S^--S^+)+\cdots. \end{aligned} \tag{41.14}$$

The natural speed of small-amplitude waves is $W=g^{\frac{1}{2}}$. Both this speed and the natural speed of shock waves depend on D, as is apparent from (41.13) and (41.14). Under strictly open-circuit conditions, D is constant in both space and time, and its effect is easily grasped. Otherwise, the behavior of a circuit provides D with a time dependence that influences the waves as they propagate.

41.2. Entropy jump across a shock in a piezoelectric material. There is no jump in D across the shock, but there is an entropy jump. The entropy jump across a shock in a piezoelectric material remains of order ξ^3, and can be obtained from the balance of energy, which we take as [cf. CHEN, 1972, Eq. $(5.27)_1$]

$$-V[U+\tfrac{1}{2}\bar{\varrho}v^2]=[vT^*]. \tag{41.15}$$

Elimination of T^{*-} through $[T^*]=-\bar{\varrho}V[v]$ puts (41.15) in the form

$$-V[U+\tfrac{1}{2}\bar{\varrho}v^2]=(T^{*+}-\bar{\varrho}Vv^-)[v]. \tag{41.16}$$

When the wave propagates into a stationary medium ($v^+=0$) so that $[v]=v^-=v$, we have, recalling $(40.20)_1$,

$$[U]=\frac{1}{2}\bar{\varrho}v^2-\frac{v}{V}T^{*+}=\frac{1}{2}\bar{\varrho}V^2\xi^2+\xi T^{*+}. \tag{41.17}$$

From (41.17), (41.14), (41,9), and (41.6), we have

$$\begin{aligned} 0=&\xi T^{*+}+\tfrac{1}{2}\bar{\varrho}V^2\xi^2-[U]=\tfrac{1}{12}\xi^3(M_3+2n_3D+\cdots)\\ &+\tfrac{1}{12}M_4\xi^4+(qD-\Theta_0+\gamma\Theta_0\xi)(S^--S^+)+\cdots. \end{aligned} \tag{41.18}$$

Taking $S=0$ when D and ξ are both zero, we expand the entropy change S across the shock as

$$S^- - S^+ = a_3 \xi^3 (1 + d_1 D + \cdots) + a_4 \xi^4 + \cdots. \tag{41.19}$$

Substitution into (41.18) puts that equation into the form of an expansion in (ξ, D). The condition that all the coefficients vanish determines a_3, d_1, and a_4 to be

$$\begin{aligned} a_3 &= M_3/12\Theta_0, \\ d_1 &= q/\Theta_0 + 2n_3/M_3, \\ a_4 &= (M_4 + \gamma M_3/2)/12\Theta_0. \end{aligned} \tag{41.20}$$

41.3. Approximate solution for the passage of a shock wave through a short-circuited piezoelectric slab. This problem is of interest because it has been investigated experimentally [GRAHAM, 1972b] and provides a way of determining both linear and nonlinear elastic, piezoelectric, and dielectric properties.

Let a piezoelectric slab be situated between the planes $X=0$ and $X=L$, and let a shock wave propagating in the $+X$-direction enter the material at the time $t=0$.

It is important to understand that short-circuiting does not necessarily reduce the electric field to zero, but only the voltage across the short-circuited electrodes. This is obtained from (41.7) as

$$\begin{aligned} \Phi(L) - \Phi(0) = &- \int_0^L E\,dX = -L D \tilde{\beta}(D) + (l - L) \frac{d}{dD} [D h_1(D)] \\ &- \frac{1}{2} (n_1 + 2n_2 D + \cdots) \int_0^L \xi^2\,dX \\ &- \frac{1}{3} M_4 \int_0^L \xi^3\,dX + q \int_0^L S\,dX + \cdots \end{aligned} \tag{41.21}$$

where l is the present thickness $x(0) - x(L)$.

"Uncoupled" approximation. As a first approximation, one can ignore the effects of the electrical variables on the strain and wave speed, taking the velocity and extension behind the shock to be v and $\xi = -v/V$, independent of X. In this approximation, the position of the wave at any time t during its first traversal is $X = Vt$, and taking $S^+ = 0$ we have

$$\begin{aligned} l - L &= -vt = -\xi V t, \\ \int_0^L \xi^2\,dX &= \xi^2 V t, \\ \int_0^L \xi^3\,dX &= \xi^3 V t, \\ \int_0^L S\,dX &= V t S^- = a_3 \xi^3 V t + \cdots. \end{aligned} \tag{41.22}$$

Substitution into (41.21) yields

$$\begin{aligned} \Phi(L) - \Phi(0) + L D \tilde{\beta}(D) = \xi V t \Big\{ &- \frac{d}{dD} [D h_1(D)] - \frac{1}{2} \xi (n_1 + 2n_2 D + \cdots) \\ &- \frac{1}{3} M_4 \xi^2 + q a_3 \xi^2 + \cdots \Big\}. \end{aligned} \tag{41.23}$$

We take the initial value of D to be zero. Under short-circuit conditions, $\Phi(L)=\Phi(0)$, and D can in principle be obtained as a function of time from (41.23). If powers of D higher than the second are dropped, (41.23) can be solved for D by the quadratic formula. The gross effect, obtained by neglecting squares and higher powers of D, is

$$D \doteq -\xi V t \frac{h+\frac{1}{2}\xi n_1+(\frac{1}{3}M_4-q a_3)\xi^2+\cdots}{L\beta-\xi V t(\beta_0 m-\xi n_2+\cdots)}. \tag{41.24}$$

Because the strain ξ is small, it can be presumed that the denominator in (41.24) is nearly equal to βL, and D increases roughly linearly with time, corresponding to a roughly square pulse of current during the passage of the wave.

One way to obtain an expression for the current (proportional to dD/dt) is to differentiate (41.23) with respect to D and then invert dt/dD to obtain dD/dt. The resulting expression contains D but is dominated by the D-independent part.

In this way, one finds

$$\begin{aligned}\frac{dD}{dt} &= \frac{\xi V\left\{-\frac{d}{dD}\left[D h_1(D)-\frac{1}{2}\xi(n_1+2n_2 D+\cdots)-\left(\frac{1}{3}M_4-q a_3\right)\xi^2+\cdots\right\}}{L[\tilde{\beta}(D)+D\,d\tilde{\beta}/dD]+\xi V t\left\{\frac{d^2}{dD^2}[D h_1(D)]+\xi n_2+\cdots\right\}}\\ &= \frac{-\xi V[h+\frac{1}{2}\xi n_1+(\frac{1}{3}M_4-q a_3)\xi^2+\cdots-D(\beta_0 m-n_2\xi+\cdots)-\frac{1}{2}\beta_0 h_3 D^2+\cdots]}{L(\beta+\beta_0 r D+\frac{1}{2}\beta_0 f D^2+\cdots)-\xi V t(\beta_0 m-n_2\xi+\cdots+\beta_0 h_3 D+\cdots)}.\end{aligned} \tag{41.25}$$

The initial height of the short-circuit current jump (per unit area of piezoelectric slab) is obtained by putting $D=0$ and $t=0$ in (41.25). This is the short-circuit current at $t=0^+$. The "final" value at $t=L/V$ when the shock wave has completed its transit can be obtained approximately by substituting the appropriate value of D from (41.24). It can be seen that the "final" value may differ from the initial value because of the nonlinear terms included in (41.25). An approximation to the explicit time dependence of dD/dt can be obtained by substituting from (41.24) into (41.25).

"Coupled" approximation. The preceding approximation is called "uncoupled" because the effect of D on the strain field was neglected. What actually happens as D starts to change is strongly dependent on the mechanical boundary conditions at the major surfaces. If the crystal could be constrained to constant thickness, a spatially homogeneous field would produce a spatially homogeneous stress with no other mechanical effect. In an actual case, the conditions of continuity of stress and displacement cause the surfaces to act as localized time-dependent sources for waves that are produced in order to keep the boundary conditions satisfied.

The effects of "linear piezoelectricity" can be artificially separated from thermal and higher order effects by considering the problem in a hypothetical medium that is linear with respect to longitudinal dielectric, piezoelectric, and elastic properties. For longitudinal motion, such a medium has the stored energy

$$U=\tfrac{1}{2}\bar{\varrho}W^2\xi^2+\tfrac{1}{2}\beta D^2-h D\xi, \tag{41.26}$$

in which W, β, and h are constants and $\xi\equiv\partial u/\partial X$. This is a special case of (41.9) in which certain of the coefficients are put to zero and thermal effects are ignored. From (41.26) and (41.10),

$$T^*=\bar{\varrho}W^2\xi-hD, \qquad E=\beta D-h\xi. \tag{41.27}$$

In a region of constant D, the equation of motion (41.5) becomes the linear wave equation

$$\ddot{u}=W^2\partial^2 u/\partial X^2. \tag{41.28}$$

Both shock waves and acceleration waves have the constant speed W. From (41.27), the voltage across a crystal of natural thickness L is

$$V=\Phi(L)-\Phi(0)=-\int_0^L E\,dx=-\beta D L+(l-L)h. \tag{41.29}$$

The charge on an area A of the electrode at L is $Q=-DA$. From (41.29), Q satisfies

$$Q=CV-\frac{(l-L)Ah}{\beta L}, \qquad C\equiv A/\beta L. \tag{41.30}$$

As soon as the shock wave enters the piezoelectric material, the thickness of the slab begins to change, and in the short-circuited case, charge begins to flow and D begins to change. This causes additional strain waves to originate at the boundaries in order to satisfy the boundary conditions.

Let us suppose the front and back surfaces of the piezoelectric slab to be in contact with linear elastic media like those considered in Sect. 40. Let the medium on $X<0$ be characterized by $\bar{\varrho}_1$, W_1, and that on $X>L$ by $\bar{\varrho}_2$, W_2. Suppose that at $t=0$, a shock wave is reflected from the interface $X=0$ as described in Sect. 40.2. At $t=0^+$ the shock reflection-transmission phenomenon leaves this interface and its neighborhood with velocity v_0 and stress T_0 satisfying (40.23):

$$T_0=-\bar{\varrho}\,W v_0. \tag{41.31}$$

Behind the reflected shock propagating back into medium 1. We expect the additional disturbance

$$u_r=f_r(W_1t+X), \qquad X<0, \qquad W_1t+X>0, \tag{41.32}$$

which brings the velocity from v_0 to $v_0+W_1 f_r'(W_1t+X)$ and the stress from T_0 to $T_0+\bar{\varrho}_1 W_1^2 f_r'(W_1t+X)$. Behind the transmitted shock propagating into the piezoelectric medium, the additional disturbance

$$u_t=g(Wt-X), \qquad X>0, \qquad Wt+X>0, \tag{41.33}$$

brings the velocity from v_0 to $v_0+Wg'(Wt-X)$ and the stress T^* from T_0 to $T_0-\bar{\varrho}\,W^2 g'(Wt-X)-hD$, in accordance with $(41.27)_1$. Continuity of velocity and stress at $X=0$ implies

$$f_r'(W_1t)=Wg'(Wt)/W_1, \qquad g'(Wt)=hD/(Z_1+Z)W, \tag{41.34}$$

where $Z_1=\bar{\varrho}_1 W_1$ and $Z=\bar{\varrho}\,W$. The velocity at $X=0$ is then

$$v(t,0)=v_0+Wg'(Wt)=v_0-hD/(Z_1+Z). \tag{41.35}$$

The disturbances that originate at $X=L$ when D begins to change can be represented as

$$\begin{aligned} u&=f_L(Wt+X-L), & X&\lesseqgtr L, & Wt+X-L&>0,\\ &=g_L(W_2t-X+L), & X&\geqq L, & Wt-X+L&>0. \end{aligned} \tag{41.36}$$

The associated stress in the piezoelectric medium is $\bar{\varrho}\,W^2 f_L'(Wt+X-L)-hD$ and the velocity is $Wf_L'(Wt+X-L)$. The corresponding quantities in medium 2 are $-\bar{\varrho}_2 W_2^2 g_L'(W_2t-X+L)$ and $W_2 g_L'(W_2t-X+L)$. Continuity of velocity and stress at $X=L$ implies

$$\begin{aligned} g_L'(W_2t)&=Wf_L'(Wt)/W_2,\\ f_L'(Wt)&=hD/(Z_2+Z)W, \end{aligned} \tag{41.37}$$

where $Z_2 = \bar{\varrho}_2 W_2$. The velocity at $X = L$ is then

$$v(t, L) = W f'_L(Wt) = hD/(Z_2 + Z). \tag{41.38}$$

Eqs. (41.35) and (41.38) hold during the time interval that the shock wave is making its first traversal of the piezoelectric slab. In the hypothetical linear case in which the wave speed is independent of D, this is simply $0 \leqq t \leqq L/W$. The rate of change of thickness of the slab is obtained from the velocities at the faces:

$$dl/dt = v(t, L) - v(t, 0) = hD\left(\frac{1}{Z_2 + Z} + \frac{1}{Z_1 + Z}\right) - v_0. \tag{41.39}$$

By differentiation of (41.30) and substitution from (41.39), recalling $Q = -DA$, we find the differential equation for the charge Q:

$$\frac{dQ}{dt} = C\frac{dV}{dt} + a + bQ, \tag{41.40}$$

where

$$a \equiv v_0\, hA/\beta L, \qquad b \equiv \frac{h^2 B}{\beta L Z}, \qquad B = \frac{1}{1 + Z_2/Z} + \frac{1}{1 + Z_1/Z}. \tag{41.41}$$

Based on (41.26), the static electromechanical coupling coefficient for longitudinal deformations can be defined as

$$k^2 = \frac{\left(\dfrac{\partial^2 U}{\partial \xi\, \partial D}\right)^2}{\left(\dfrac{\partial^2 U}{\partial \xi^2}\right)\left(\dfrac{\partial^2 U}{\partial D^2}\right)} = \frac{h^2}{\bar{\varrho} W^2 \beta} = \frac{h^2}{\beta Z W}, \tag{41.42}$$

and hence b can be expressed as

$$b = k^2 W B/L. \tag{41.43}$$

On short-circuit, the solution of (41.40) that has $Q(0) = 0$ is

$$Q(t) = a(e^{bt} - 1)/b, \qquad 0 \leqq t \leqq L/W. \tag{41.44}$$

A similar result has been reported by STUETZER (1967) for the special cases of mechanically matched terminations ($Z_1 = Z_2 = Z$) and a free end bonded to a matched line ($Z_1 = Z$ but $Z_2 = 0$).

Eq. (41.44) is an exact solution for the case of the hypothetical linear piezoelectric medium. Of somewhat more interest is the approximate correction that piezoelectric coupling introduces into the "uncoupled" but nonlinear approximate solution exemplified by (41.25). If $Q_u(t)$ represents the "uncoupled" solution, we infer that an approximation to the corresponding coupled solution can be obtained from (41.40) with $a = dQ_u/dt$. Again considering the short-circuit case, we find that the correction

$$q(t) \equiv Q(t) - Q_u(t) \tag{41.45}$$

satisfies

$$\frac{dq}{dt} - bq = bQ_u(t). \tag{41.46}$$

Eq. (41.46) has the integrating factor $\exp(-bt)$ and the pertinent solution valid during the first transit of the shock wave is

$$q(t) = b e^{bt} \int_0 e^{-bp} Q_u(p)\, dp. \tag{41.47}$$

The additional current contributed by piezoelectric coupling is therefore

$$\frac{dq}{dt} = b\left[Q_u(t) + q(t)\right]. \tag{41.48}$$

There is no correction to the "initial" current at $t=0^+$, but the correction at later times can be significant.

41.4. Determination of coefficients. In the experiments described by GRAHAM (1972b), planar elastic shock waves are introduced by precisely controlled planar impacts. Various shock strengths, characterized by the strain behind the shock, are achieved by experiments at pre-selected projectile velocities. The velocity of the projectile immediately prior to impact is measured, and the "short-circuit" current pulse through a low-impedance resistive circuit is recorded. The projectile is faced with material identical to the sample and having the same orientation so that the particle velocity behind the shock in the stationary sample can be taken as one-half the projectile velocity at impact. The wave transit time is apparent from the duration of the current pulse. Higher-order elastic constants are determined essentially as described in Sect. 39, the modification arising from piezoelectric effects being relatively small. Analysis of the current pulses corresponding to various shock strengths provide information concerning the piezoelectric coefficient and various electromechanical nonlinearities. GRAHAM (1972b) analyzed his data on X-cut quartz in terms of a model somewhat simpler than that indicated here, and obtained determinations of the second-order, third-order, and fourth-order elastic constants c_{11}, c_{111}, and c_{1111}, the piezoelectric stress coefficient e_{11}, its strain derivative $\partial e_{11}/\partial\eta$, and the logarithmic strain derivative of the permittivity $\varepsilon_{11}^{-1}\partial\varepsilon_{11}/\partial\eta$.

References.

AKHIESER, A.: On the absorption of sound in solids. J. Phys. USSR. (1), 277–287 (1939).

AULD, B. A., and G. S. KINO: Normal mode theory for acoustic waves and its application to the interdigital transducer. IEEE Trans. Electron Devices, Vol. ED-18, 898–908 (1971).

BARKER, L. M., and R. E. HOLLENBACH: Shock-wave studies of PMMA, fused silica, and sapphire. J. Appl. Phys. **41**, 4208–4226 (1970).

BARSCH, G. R.: Adiabatic, isothermal, and intermediate pressure derivatives of the elastic constants for cubic symmetry I: Basic formulae. Phys. Stat. Sol. **19**, 129–138 (1967).

BARSCH, G. R.: Relation between third-order elastic constants of single crystals and polycrystals. J. Appl. Phys. **39**, 3780–3793 (1968).

BARSCH, G. R., and Z. P. CHANG: Adiabatic, isothermal, and intermediate pressure derivatives of the elastic constants for cubic symmetry II: Numerical results for 25 materials. Phys. Stat. Sol. **19**, 139–151 (1967).

BARSCH, G. R., and H. E. SHULL: Pressure dependence of elastic constants and crystal stability of alkali halides: NaI and KI. Phys. Stat. Sol. (b) **43**, 637–649 (1971).

BATEMAN, T. B., W. P. MASON, and H. J. MCSKIMIN: Third-order elastic moduli of germanium. J. Appl. Phys. **32**, No. 5, 928–936 (1961).

BECHMANN, R.: Determination of the elastic and piezoelectric coefficients of monoclinic crystals with particular reference to ethylene diamine tartrate. Proc. Phys. Soc. (London) B **63**, 577–589 (1950).

BECHMANN, R., and S. AYERS: Theory of Dynamical Determination of Elastic and Piezoelectric constants of Anisotropic Crystals. Post Office Engineering Department, Research Report No. 13773. Post Office Research Station, Dollis Hill, London N. W. 2.

BECHMANN, R., R. F. S. HEARMON, and S. K. KURTZ: Elastic, piezoelectric, piezooptic, electropic constants and nonlinear dielectric susceptibilities of crystals, in: LANDOLT-BÖRNSTEIN, III/2, Numerical Data and Functional Relationships in Science and Technology. New Series (K.-H. HELLWEGE and A. M. HELLWEGE, eds.), Group III: Crystal and Solid State Physics, Vol. 2, Supplement and Extension to Vol. 1. Berlin-Heidelberg-New York: Springer 1969.

BERLINCOURT, D. A., D. R. CURRAN, and H. JAFFE: Piezoelectric and Piezomagnetic Materials and their Function in Transducers. In: Physical Acoustics-Principles and Methods (W. P. MASON, ed.), Vol. 1A, pp. 169–270. New York: Academic Press 1964.

BEYER, R. T.: Nonlinear acoustics. In: Physical Acoustics—Principles and Methods (W. P. MASON, ed.), Vol. 2: Properties of Polymers and Nonlinear Acoustics, Chap. 10, pp. 231–264. New York: Academic Press, Inc. 1965.

BIRCH, F.: Finite elastic strain of cubic crystals. Phys. Rev. **71**, 809–824 (1947).

BIRSS, R. R.: Symmetry and Magnetism. Amsterdam: North-Holland Publishing Co. 1964.

BLACKSTOCK, DAVID: Propagation of plane sound waves of finite amplitude in nondissipative fluids. J. Acoust. Soc. Am. **34**, 9–30 (1962).

BOMMEL, H. E., and K. DRANSFELD: Excitation and attenuation of hypersonic waves in quartz. Phys. Rev. **117**, 1245–1252 (1960).

BOND, W. L., *et al.*: Standards on piezoelectric crystals. Proc. I.R.E. **37**, 1378–1395 (1949).

BORGNIS, F. E.: Specific directions of longitudinal wave propagation in anisotropic media. Phys. Rev. **98**, 1000–1005 (1955).

BREAZEALE, M. A.: Finite-Amplitude Waves in Liquids and Solids. Paper D-18, Proc. Fifth Int. Congress on Acoustics, Liège (D. E. COMMINS, ed.), Vol. 1a. Liège: Imprimerie Georges Thone 1965

BREAZEALE, M. A.: Corrected values for the ultrasonic discontinuity distance in nonlinear solids. J. Appl. Phys. **37**, 3332 (1966).

BREAZEALE, M. A.: Distorted Wave Interaction at Boundaries. Paper H5–12, Reports of the Sixth Int. Congress on Acoustics, Tokyo, 1968.

BREAZEALE, M. A., and J. FORD: Ultrasonic studies of the nonlinear behavior of solids. J. Appl. Phys. **36**, 3486–3490 (1965).

BRUGGER, K.: Thermodynamic definition of higher order elastic coefficients. Phys. Rev. **133**, A1611–1612 (1964).

BRUGGER, K.: Pure modes for elastic waves in crystals. J. Appl. Phys. **36**, 759–768 (1965).

BRUGGER, K.: Determination of third-order elastic coefficients in crystals. J. Appl. Phys. **36**, 768–773 $(1965)_2$.

BRUGGER, K., and R. N. THURSTON: Single crystal folded-path delay lines utilizing ultrasonic pure shear modes. J. Appl. Phys. **42**, 2159–2161 (1971).

BUCHEN, PETER W.: Plane waves in linear viscoelastic media. Geophys. J. R. astr. Soc. **23**, 531–542 (1971).

BUCK, O., and D. O. THOMPSON: Relation of finite amplitude waves to third-order elastic constants. Mater. Sci. Eng. **1**. 177–180 (1966).

CADY, W. G.: Piezoelectricty. New York: McGraw-Hill 1946.

CARCIA, P. F., G. R. BARSCH, and L. R. TESTARDI: Pressure dependence of elastic constants and of shear-mode instability in V_3Si and V_3Ge. Phys. Rev. Letters **27**, 944–947 (1971).

CEKRGE, H. M., and E. VARLEY: Large-amplitude waves in bounded media. 1. Reflexion and transmission of large amplitude shockless pulses at an interface. Phil. Trans. Roy. Soc. (London), Ser. A **273**, No. 1234, 261–313 (1973).

CHANG, Z. P., and G. R. BARSCH: Nonlinear pressure dependence of elastic constants and fourth-order elastic constants of cesium halides. Phys. Rev. Letters **19**, 1381–1382 (1967).

CHANG, Z. P., and G. R. BARSCH: Pressure dependence of the elastic constants of RbCl, RbBr and RbI. J. Phys. Chem. Solids. **32**, 27–40 (1970).

CHEN, P. J.: Growth and decay of waves in solids. Handbuch der Physik (S. FLÜGGE, chief ed.) Vol. VIa/3 (C. TRUESDELL, ed.), pp. 303–402. Berlin-Heidelberg-New York: Springer 1973.

COLEMAN, B. D.: The thermodynamics of materials with memory. Arch. Rational. Mech. Anal. **17**, 1–46 (1964).

COLEMAN, B. D., and M. E. GURTIN: Waves in materials with memory II. On the growth and decay of one-dimensional acceleration waves. Arch. Rational Mech. Anal **19**, 239–265 (1965), pp. 251–252.

COLEMAN, B. D., and W. NOLL: The thermodynamics of elastic materials with heat conduction and viscosity. Arch. Rational Mech. Anal. **13**, 167–178 (1963).

COOK, R. K.: Variations of elastic constants and static strains with hydrostatic pressure: A method for calculation from ultrasonic measurements. J. Acoust. Soc. Am. **29**, 445–449 (1957).

COOPER, H. F., Jr.: Reflection and transmission of oblique plane waves at a plane interface between viscoelastic media. J. Acoust. Soc. Am. **42**, 1064–1069 (1967).

COQUIN, G. A., and H. F. TIERSTEN: Analysis of the excitation and detection of piezoelectric surface waves in quartz by means of surface electrodes. J. Acoust. Soc. Am. **41**, 921–939 (1967).

COURANT, R., and K. O. FRIEDRICHS: Supersonic Flow and Shock Waves. New York: Interscience Publishers, Inc. 1948.

DOBRETSOV, A. I., and G. I. PERESEDA: Dependence of the elastic constants of KCl on pressure. Fiz. Tverd. Tela 11, 1728 (1969); Soviet Phys.-Solid State 11, 1401–1402 (1969).
DRABBLE, J. R.: Elastic constants under pressure. In: Mechanical Behaviour of Materials Under Pressure (H. Ll. D. PUGH, ed.). Amsterdam-London-New York: Elsevier Publishing Co. 1970.
DRANSFELD, K., and E. SALZMANN: Excitation, detection, and attenuation of high-frequency elastic surface waves. In: Physical Acoustics—Principles and Methods (W. P. MASON and R. N. THURSTON, eds.), Vol. 7, Chap. 4, pp. 219–272. New York-London: Academic Press Inc. 1971.
ELKIN, S., S. ALTEROVITZ, and D. GERLICH: Third-order elastic moduli of cubic crystals. J. Acoust. Soc. Am. 47, 937–938 (1970).
ERICKSEN, J. L.: Tensor Fields. In: Handbuch der Physik (S. FLÜGGE, ed.), Vol. III/1, pp. 794–858. Berlin-Göttingen-Heidelberg: Springer 1960.
FEDOROV, F. I.: Theory of Elastic Waves in Crystals. New York: Plenum Press 1968 [English Translation of Teoriya Uprugikh Voln v Kristallakh, Moskow, 1965 (revised and updated by the author)].
FENLON, F. H.: An extension of the Bessel-Fubini series for a multiple-frequency CW acoustic source of finite amplitude. J. Acoust. Soc. Am. 51, 284–289 (1972).
FOWLES, R.: Dynamic compression of quartz. J. Geophys. Res. 72, 5729–5742 (1967).
FUBINI-GHIRON, E.: Anomalie nella propagazione di onde acustiche di grande ampiezza. Alta Frequenza 4, 530–581 (1935).
GAUSTER, W. B., and M. A. BREAZEALE: Detector for measurement of ultrasonic strain amplitudes in solids. Rev. Sci. Instr. 37, 1544–1548 (1966).
GERSON, R: Dielectric properties of lead titanate zirconate ceramics at very low frequencies. J. Appl. Phys. 31, 1615–1617 (1960).
GRAHAM, R. A.: Determination of third- and fourth-order longitudinal elastic constants by shock compression techniques—application to sapphire and fused quartz. J. Acoust. Soc. Am. 51, 1576–1581 (1972).
GREEN, R. E.: Treatise on Materials Science and Technology, Vol. 3: Ultrasonic Investigation of Mechanical Properties. New York: Academic Press 1973.
GROTH, P.: Physikalische Kristallographie und Einleitung in die Kristallographische Kenntnis der wichtigsten Substanzen, 4th ed. Leipzig: Engelmann 1905. See pp. 335–337.
GUINAN, M. W., and A. D. RITCHIE: Evaluation of third-order elastic constants for cubic crystals. J. Appl. Phys. 41, 2256–2258 (1970).
GURTIN, M. E.: On the thermodynamics of materials with memory. Arch. Rational Mech. Anal. 28, 40–50 (1968).
GURTIN, M. E.: Time-reversal and symmetry in the thermodynamics of materials with memory. Arch. Rational Mech. Anal. 44, 387–399 $(1972)_1$.
GURTIN, M. E.: Linear theory of elasticity. In: Handbuch der Physik (S. FLÜGGE, chief ed.), Vol. VIa/2 (C. TRUESDELL, ed.). Berlin-Heidelberg-New York: Springer 1972_2.
GURTIN, M. E., and A. C. PIPKIN: A general theory of heat conduction with finite wave speeds. Arch. Rational Mech. Anal. 31, 112–126 (1968).
GURTIN, M. E., and E. STERNBERG: The linear theory of viscoelasticity. Arch. Rational Mech. Anal. 11, 291–356 (1962).
HAUSSUHL, S., and H. SIEGERT: Bestimmung des Elastizitätstensors Trikliner Kristalle: Beispiel $CuSO_4 \cdot 5H_2O$. Z. Krist. 129, 142–146 (1969).
HERMANN, C.: Z. Krist. 68, 257 (1928);—69, 226, 250, 533 (1929);—75, 159 (1930).
HOLDER, J., and A. V. GRANATO: Third-order elastic constants and thermal equilibrium properties of solids. In: Physical—Acoustics-Principles and Methods (W. P. MASON and R. N. THURSTON, ed.), Vol. VIII, Chap. 5, pp. 237–277. New York: Academic Press Inc. 1971.
IEEE Standard No. 176 (1949): Piezoelectric crystals. Archival reference: Standards on piezoelectric crystals. Proc. I.R.E. 37, 1378–1395 (1949).
IEEE Standard No. 178 (1958): Piezoelectric crystals, determination of the elastic, piezoelectric, and dielectric constants, and also the electromechanical coupling factor. Archival reference: Proc. I.R.E. 46, 764–778 (1958).
IEEE Standard No. 179 (1961): Piezoelectric ceramics, measurements of. Archival reference: IRE standards on piezoelectric crystals: Measurements of piezoelectric ceramics. Proc. I.R.E. 1161–1169, (1961).
IEEE Standard No. 177 (1966). Piezoelectric vibrators, definitions and methods of measurement of. Revision of original archival reference by the same title published in: Proc. I.R.E. 45, 353–358, (1957).
IRE (1961): IRE standards on piezoelectric crystals: Measurements of piezoelectric ceramics. Proc. IRE 49, 1162–1169 (1961). IEEE Standard No. 179: Piezoelectric ceramics, measurements of.

ILUKOR, J. O., and E. H. JACOBSEN: Coherent elastic wave propagation in quartz at ultramicrowave frequencies. In: Physical Acoustics — Principles and Methods (W. P. MASON, ed.), Vol. V, pp. 221–231. New York: Academic Press 1968.

JACOBSEN, E. H.: Sources of sound in piezoelectric crystals. J. Acoust. Soc. Am. **32**, 949–953 (1960).

JAGODZINSKI, H.: Kristallographie. In: Handbuch der Physik (S. FLÜGGE, ed.), Vol. VII/1: Crystal Physics I, pp. 1–103. Berlin-Göttingen-Heidelberg: Springer 1955.

JOSHI, S. G., and R. M. WHITE: Excitation and detection of surface elastic waves in piezoelectric crystals. J. Acoust. Soc. Am. **46**, 17–27 (1969).

KITTEL, C.: Introduction to Solid State Physics. 2nd ed. New York: Wiley and Sons. 1956.

KOHN, W., J. A. KRUMHANSL, and E. H. LEE: Variational methods for dispersion relations and elastic properties of composite materials. J. Appl. Mech. **39**, Series E, No. 2, 327–336 (1972).

KOLODNER, I. I.: Existence of longitudinal waves in anisotropic media. J. Acoust. Soc. Am. **40**, 730–731 (L) (1966).

KRISHNAN, R. S., V. RADHA, and S. R. GOPAL: The twenty-one elastic constants of triclinic copper sulphate pentahydrate. J. Indian Inst. Sci. **52**, 115–132 (1970).

LAMB, H.: Hydrodynamics. New York: Dover reprint, 1945, p. 481. Republication of Sixth Ed., Cambridge University Press 1932.

LANDAU, L.D., and E. M. LIFSHITZ: Electrodynamics of Continuous Media. [English translation by J. B. SYKES and J. S. BELL]. Oxford-London-New York-Paris: Pergamon Press 1960. Distributed in the U. S. by Addison-Wesley, Reading, Mass.

LANDOLT-BÖRNSTEIN, III/1: Numerical Data and Functional Relationships in Science and Technology. New Series (K. H. HELLWEGE and A. M. HELLWEGE, eds.). Group III: Crystal and Solid State Physics, Vol. I: Elastic, Piezoelectric, Piezooptic and Electroptic Constants of Crystals (complied by R. BECHMANN, R. F. S. HEARMON, and S. K. KURTZ). Berlin-Heidelberg-New York: Springer 1966.

LANDOLT-BÖRNSTEIN, III/2: Numerical Data and Functional Relationships in Science and Technology. New Series (K. H. HELLWEGE and A. M. HELLWEGE, eds.). Group III: Crystal and Solid State Physics, Vol. 2, Supplement and Extension to Vol. 1: Elastic, Piezoelectric, Piezooptic, Electrooptic Constants and Nonlinear Susceptibilities of Crystals (complied by R. BECHMANN, R. F. S. HEARMON, and S. K. KURTZ). Berlin-Heidelberg-New York: Springer 1969.

LAX, M., and D. F. NELSON: Linear and nonlinear electrodynamics in elastic anisotropic dielectrics. Phys. Rev. B **4**, No. 10, 3694–3731 (1971).

LAX, PETER D.: Development of singularities of solutions of nonlinear hyperbolic partial differential equations. J. Math. Phys. **5**, 611–613 (1964).

LAZUTKIN, V. N., and A. I. MIKHAILOV: Equivalent circuit of a radially vibrating disk. Akust. Zh. **18**, 58–62 (1972). [English translation in: Soviet Phys.-Acoust. **18**, 45–48].

LEITMAN, M. J., and G. M. C. FISHER: The linear theory of viscoelasticity. In: Handbuch der Physik (S. FLÜGGE, chief ed.), Vol. VI a/3 (C. TRUESDELL, ed.), pp. 1–123. Berlin-Heidelberg New York: Springer 1973.

LEWIS, M. F.: Attenuation of high-frequency elastic waves in quartz and fused silica. J. Acoust. Soc. Am. **44**, 713–716 (1968).

LJAMOV, V. E.: Nonlinear acoustical parameters of piezoelectric crystals. J. Acoust. Soc. Am. **52**, 199–202 (1972).

LOVE, A. E. H.: A Treatise on the Mathematical Theory of Elasticity, 4th ed. London-New-York: Cambridge Univ. Press 1927. [Reprinted by Dover, New York.]

MARIS, H. J.: Interaction of sound waves with thermal phonons in dielectric crystals. In: Phys. Acoustics — Principles and Methods (W. P. MASON, ed.), Vol. 8, Chap. 6, pp. 280–345. New York: Academic Press, Inc. 1972.

MASON, W. P.: Piezoelectric crystals and their application to ultrasonics. New York: D. Van Nostrand 1950_2.

MASON, W. P.: Optical properties and the electro-optic and photoelastic effects in crystals expressed in tensor form. Bell System Tech. J. **29**, 161–188 (1950).

MASON, W. P.: Physical Acoustics and the Properties of Solids. Princeton, N. J.: D. Van Nostrand Co., Inc. 1958.

MASON, W. P.: Phonon viscosity and its effect on acoustic wave attenuation and dislocation motion. J. Acoust. Soc. Am. **32**, 458–472 (1960).

MASON, W. P.: Crystal physics of interaction processes. New York-London: Academic Press 1966.

MASON, W. P.: Internal friction at low frequencies due to dislocations: Applications to metals and rock mechanics. In: Phys. Acoustics — Principles and Methods (W. P. MASON, ed.), Vol. 8, Chap. 7, pp. 347–371. New York: Academic Press, Inc. 1971.

MASON, W. P., and T. B. BATEMAN: Ultrasonic-wave propagation in pure silicon and germanium. J. Acoust. Soc. Am. **36**, 644–652 (1964).

MASON, W. P., and T. B. BATEMAN: Relation between third-order elastic moduli and the thermal attenuation of ultrasonic waves in nonconducting and metallic crystals. J. Acoust. Soc. Am. **40**, 852–862 (1966).

MAUGUIN, CH.: Z. Kristallog. **76**, 542 (1931).

McMAHON, D. H.: Harmonic generation in piezoelectric crystals. J. Acoust. Soc. Am. **44**, 1007–1013 (1968).

McSKIMIN, H. J., and P. ANDREATCH: Analyisis of the pulse superposition method for measuring ultrasonic wave velocities as a function of temperature and pressure. J. Acoust. Soc. Am. **34**, 609–615 (1962).

McSKIMIN, H. J., and W. L. BOND: Conical refraction of transverse ultrasonic waves in quartz. J. Acoust. Soc. Am. **39**, 499–505 (1966).

MEITZLER, A. H., H. M. O'BRYAN, JR., and H. F. TIERSTEN: Definition and measurement of coupling factors in piezoelectric ceramic materials with large variations in Poisson's ratio. IEEE Transactions on Sonics and Ultrasonics SU 20 (1973). To be published.

MELNGAILIS, J., A. A. MARADUDIN, and A. SEEGER: Diffraction of light by ultrasound in anharmonic crystals. Phys. Rev. **131**, 1972–1975 (1963).

NELSON, D. F., and M. LAX: Theory of acoustically induced optical harmonic generation. Phys. Rev. B**3**, 2795–2812 (1971).

NIGGLI, P.: Geometrische Kristallographie des Diskontinuums. Leipzig 1919.

NORTON, E. L.: Transformer Band Filters. U.S. Patent 1,681,554, August 21, 1928.

POWELL, B. E., and M. J. SKOVE: Relation between isothermal and mixed third-order elastic constants. J. Appl. Phys. **38**, No. 1, 404 (1967).

PROHOFSKY, E. W.: A simple approach to high-temperature ultrasonic attenuation in complex insulators. IEEE Trans. Sonics and Ultrasonics SU-14, No. 3, 109–111 (1967).

RAYLEIGH. See STRUTT, J. W., LORD RAYLEIGH.

ROLLINS, F. R., JR., L. H. TAYLOR, and P. H. TODD, JR.: Ultrasonic study of three-phonon interactions II: Experimental results. Phys. Rev. **136A**, 597–601 (1964).

ROLLINS, F. R., JR.: Phonon interactions at ultrasonic frequencies. Proc. IEEE **53**, 1534–1539 (1965).

RUOFF, A. L.: Linear shock-velocity-particle-velocity relationship. J. Appl. Phys. **38**, 4976–4980 (1967).

SCHOENFLIES, A.: Theorie der Kristallstrukturen. Berlin 1923.

SEEGER, A., and O. BUCK: Die experimentelle Ermittlung der elastischen Konstanten höherer Ordnung. Z. Naturforsch. **15**a, 1056–1067 (1960).

SEKOYAN, S. S.: Calculation of the third-order elastic constants from the results of ultrasonic measurements. Akust. Zh. **16**, 453–457 (1970); — Soviet Phys. Acoust. **16**, 377–380 (1971).

SHAW, E. A. G.: On the resonant vibrations of thick barium titanate disks. J. Acoust. Soc. Am. **28**, 38–50 (1956).

SHAW, E. A. G., and R. J. SUJIR: Vibration patterns of loaded barium titanate and quartz disks. J. Acoust. Soc. Am. **32**, 1463–1467 (1960).

SKOVE, M. J., and B. E. POWELL: Symmetry of mixed third-order elastic constants. J. Appl. Phys. **38**, No. 1, 404 (1967).

SMITH, W. R., H. M. GERARD, J. H. COLLINS, T. M. REEDER, and H. J. SHAW: Analysis of interdigital surface wave transducers by use of an equivalent circuit model. IEEE Trans. Microwave Theory and Techniques, Vol. MTT-17 $(1969)_1$.

SMITH, W. R., H. M. GERARD, J. H. COLLINS, T. M. REEDER, and H. J. SHAW: Design of surface wave delay lines with interdigital transducers. IEEE Trans. Microwave Theory and Techniques, Vol. MTT-17 $(1969)_2$.

STRATTON, J. A.: Electromagnetic Theory, pp. 78–82. New York-London: McGraw-Hill 1941.

STRUTT, J. W., LORD RAYLEIGH: Theory of Sound. New York: Dover reprint, 1945, Vol. 2, p. 32. Republication of reissue by the Macmillan Co. 1929.

STUETZER, O. M.: Secondary stresses in a stress-pulse-activated piezoelectric element. J. Appl. Phys. **38**, 3901–3904 (1967).

THOMAS, T. Y.: The growth and decay of sonic discontinuities in ideal gases. J. Math. Mech. **6**, 455–469 (1957), p. 464.

THOMAS, T. Y.: On the stress-strain relations for cubic crystals. Proc. Nat. Acad. Sci. U.S. **55**, 235–239 (1966).

THOMPSON, D. O., M. A. TENNISON, and O. BUCK: Reflections of harmonics generated by finite-amplitude waves. J. Acoust. Soc. Am. **44**, 435–436 (1968).

THURSTON, R. N., and K. BRUGGER: Third-order elastic constants and the velocity of small amplitude elastic waves in homogeneously stressed media. Phys. Rev. **133**, A1604–1610 (1964).

THURSTON, R. N.: Wave propagation in fluids and normal solids. In: Phys. Acoustics-Principles and Methods (W. P. MASON, ed.), Vol. 1: Methods and Devices, Pt. A, Chap. 1, pp. 1–110. New York: Academic Press Inc. 1964.

THURSTON, R. N.: Effective elastic coefficients for wave propagation in crystals under stress. J. Acoust. Soc. Am. **37**, pp. 348–356 $(1965)_1$.

THURSTON, R. N.: Ultrasonic data and the thermodynamics of solids. Proc. IEEE **53**, 1320–1336 $(1965)_2$.

THURSTON, R. N.: Calculation of lattice-parameter changes with hydrostatic pressure from third-order elastic constants. J. Acoust. Soc. Am. **41**, 1093–1111 $(1967)_1$.

THURSTON, R. N.: The connection between nonlinear elasticity and the damping of elastic waves by thermal vibrations. IEEE Symp. on Sonics and Ultrasonics, paper E-1, Vancouver, Canada 1967_2.

THURSTON, R. N., and M. J. SHAPIRO: Interpretation of ultrasonic experiments on finite-amplitude waves. J. Acoust. Soc. Am. **41**, 1112–1125 (1967).

THURSTON, R. N.: A connection between nonlinear elasticity and ultrasonic attenuation. Reports of the Sixth Int. Congress on Acoustics, Tokyo, 1968. Paper H-5-5.

THURSTON, R. N.: Definition of a linear medium for one-dimensional longitudinal motion. J. Acoust. Soc. Am. **45**, 1329–1341 (1969).

THURSTON, R. N.: Thermodynamic substate variables for a solid. Phys. Rev. B **2**, 5012–5015 (1970).

TIERSTEN, H. F.: Wave propagation in an infinite piezoelectric plate. J. Acoust. Soc. Am. **35**, 234–239 (1963).

TIERSTEN, H. F.: Thickness vibrations of piezoelectric plates. J. Acoust. Soc. Am. **35**, 53–58 (1963).

TIERSTEN, H. F.: Linear piezoelectric plate vibrations. New York: Plenum Press 1969.

TIERSTEN, H. F.: On the nonlinear equations of thermoelectroelasticity. Int. J. Engng. Sci. **9**, 587–604 (1971).

TIERSTEN, H. F., and C. F. TSAI: On the interaction of the electromagnetic field with heat conducting deformable insulators. J. Math. Phys. **13**, 361–378 (1972).

TOKUOKA, T., and Y. IWASHIMIZU: Acoustical birefringence of ultrasonic waves in deformed isotropic elastic materials. Int. J. Solids Structures **4**, 383–389 (1968).

TOKUOKA, T., and M. SAITO: Elastic wave propagations and acoustical birefringence in stressed crystals. J. Acoust. Soc. Am. **45**, 1241–1246 (1968).

TOUPIN, R. A., and B. BERNSTEIN: Sound waves in deformed perfectly elastic materials. Acoustoelastic effect. J. Acoust. Soc. Am. **33**, 216–225 (1961).

TOUPIN, R. A.: A dynamical theory of elastic dielectrics. Int. J. Engng. Sci. **1**, 101–126 (1963).

TRUELL R., C. ELBAUM, and B. CHICK: Ultrasonic Methods in Solid State Physics. New York: Academic Press 1969.

TRUELL, R., and C. ELBAUM: High-frequency ultrasonic stress waves in solids: In: Handbuch der Physik (S. FLÜGGE, ed.), Vol. XI/2, Berlin-Göttingen-Heidelberg: Springer 1962.

TRUESDELL, C.: Precise theory of the absorption and dispersion of forced plane infinitesimal waves according to the Navier-Stokes equations. J. Rational Mech. Anal. **2**, 643–741 (1953).

TRUESDELL, C.: General and exact theory of waves in finite elastic strain. Arch. Rational Mech. Anal. **8**, No. 4, 263–296 (1961).

TRUESDELL, C., and W. NOLL: The nonlinear field theories of mechanics. In: Handbuch der Physik (S. FLÜGGE, ed.), Vol. III/3, pp. 1–602. Berlin-Heidelberg-New York: Springer 1965.

TRUESDELL, C., and R. A. TOUPIN: The classical field theories. In: Handbuch der Physik (S. FLÜGGE, ed.), Vol. III/1, pp. 226–793. Berlin-Göttingen-Heidelberg: Springer 1960.

TRUESDELL, C.: Existence of longitudinal waves. J. Acoust. Soc. Am. **40**, No. 3, 729–730 (1966).

TRUESDELL, C.: Comment on longitudinal waves. J. Acoust. Soc. Am. **43**, 170 (L) (1968).

TUCKER, J. W., and V. W. RAMPTON: Microwave Ultrasonics in Solid State Physics. Amsterdam: North-Holland 1972. New York: American Elsevier 1972.

VAN BUREN, A. L., and M. A. BREAZEALE: Reflection of finite-amplitude ultrasonic waves. I. Phase shift. II. Propagation. J. Acoust. Soc. Am. **44**, 1014–1027 (1968).

VAN DYKE, K. S.: The electric network equivalent of a piezoelectric resonator. Phys. Rev. **25**, 895 (1925). — The piezoelectric resonator and its equivalent network. Proc. I.R.E. **16**, 742–764 (1928).

VIKTOROV, I. A.: Rayleigh and Lamb Waves: Physical Theory and Applications. [Translation of 1966 Russian edition.] New York: Plenum Press 1967.

VOIGT, W.: Lehrbuch der Kristallphysik. Leipzig: Teubner 1928.

WALLACE, D. C.: Thermoelastic theory of stressed crystals and higher-order elastic constants. In: Solid State Physics-Advances in Research and Applications (F. SEITZ, D. TURNBULL, and H. EHRENREICH, eds.), Vol. 25, pp. 302–404. New York: Academic Press, Inc. 1970.

WALLACE, D. C.: Thermodynamics of Crystals. New York-London-Sydney-Toronto: John Wiley and Sons, Inc. 1972.
WATERMAN, P. C.: Orientation dependence of elastic waves in single crystals. Phys. Rev. **113**, 1240–1253 (1959).
WOODRUFF, T. O., and H. EHRENREICH: Absorption of sound in insulators. Phys. Rev. **123**, 1553–1559 (1961).
WOOLLETT, R. S.: Effective coupling factor of single-degree-of-freedom transducers. J. Acoust. Soc. Am. **40**, 1112–1123 (1966).
WOOLLETT, R. S.: Transducer comparison methods based on the electromechanical coupling-coefficient concept. I.R.E. National Convention Record, Vol. 5, Part 9, pp. 23–27.
ZABUSKY, N. J.: A synergetic approach to problems of nonlinear wave propagation and interaction. In: Nonlinear Partial Differential Equations (W. AMES, ed.), pp. 223–258. New York: Academic Press, Inc. 1967.
ZABUSKY, N. J.: Solitons and energy transport in nonlinear lattices. Computer Physics Communications **5**, 1–10 (1973).

Namenverzeichnis. — Author Index.

Sachverzeichnis.

(Deutsch-Englisch.)

Bei gleicher Schreibweise in beiden Sprachen sind die Stichwörter nur einmal aufgeführt.

Subject Index.

(English-German.)

Where English and German spellings of a word are identical the *German* version is omitted.

Springer-Verlag
Berlin Heidelberg New York
München Johannesburg London
Madrid New Delhi Paris
Rio de Janeiro Sydney Tokyo
Utrecht Wien

Handbuch der Physik
Encyclopedia of Physics

Herausgegeben von /
Chief Editor:
S. Flügge, Freiburg i.Br.

Band VII / Teil 1
Kristallphysik I
Crystal Physics I
321 Figuren. VIII, 687 Seiten
(69 Seiten in Englisch). 1955
Geb. DM 154,–; US $62.90
Bei Subskription auf das
Gesamtwerk
Geb. DM 123,20; US $50.30
ISBN 3-540-01916-2

Band VII / Teil 2
Kristallphysik II
Crystal Physics II
190 Figuren. IV, 273 Seiten. 1958
Geb. DM 97,–; US $39.60
Bei Subskription auf das
Gesamtwerk
Geb. DM 77,60; US $31.70
ISBN 3-540-02291-0

Band VIII / Teil 1
Strömungsmechanik I
Fluid Dynamics I
Co-Editor: C. Truesdell
186 figures. VI, 471 pages
(235 pages in German). 1959
Cloth DM 167,–; US $68.20
Subscription price
Cloth DM 133,60; US $54.60
ISBN 3-540-02411-5

Band VIII / Teil 2
Strömungsmechanik II
Fluid Dynamics II
Co-Editor: C. Truesdell
177 figures. VI, 696 pages
(389 pages in French, 68 pages
in German). 1963
Cloth DM 250,–; US $102.00
Subscription price
Cloth DM 200,–; US $81.60
ISBN 3-540-02997-4

Band IX
Strömungsmechanik III
Fluid Dynamics III
Co-Editor: C. Truesdell
248 figures. VI, 815 pages
(64 pages in French, 18 pages
in German). 1960
Cloth DM 250,–; US $102.00
Subscription price
Cloth DM 200,–; US $81.60
ISBN 3-540-02548-0

Preisänderungen vorbehalten
Prices are subject to change
without notice